Minkowski geometry is a non-Euclidean geometry in a finite number of dimensions that is different from elliptic and hyperbolic geometry (and from the Minkowskian geometry of space-time). Here the linear structure is the same as the Euclidean one but distance is not "uniform" in all directions. Instead of the usual sphere in Euclidean space, the unit ball is a general symmetric convex set. Therefore, although the parallel axiom is valid, Pythagoras' theorem is not.

This book begins by presenting the topological properties of Minkowski spaces, including the existence and essential uniqueness of Haar measure, followed by the fundamental metric properties – the group of isometries, the existence of certain bases and the existence of the Löwner ellipsoid. This is followed by characterizations of Euclidean space among normed spaces and a full treatment of two-dimensional spaces. The three central chapters present the theory of area and volume in normed spaces. The author describes the fascinating geometric interplay among the isoperimetrix (the convex body which solves the isoperimetric problem), the unit ball and their duals, and the ways in which various roles of the ball in Euclidean space are divided among them. The next chapter deals with trigonometry in Minkowski spaces and the last one takes a brief look at a number of numerical parameters associated with a normed space, including J. J. Schäffer's ideas on the intrinsic geometry of the unit sphere. Each chapter ends with a section of historical notes, and the book ends with a list of 50 unsolved problems.

Minkowski Geometry will appeal to students and researchers interested in geometry, convexity theory and functional analysis.

ENCYCLOPEDIA OF MATHEMATICS AND ITS APPLICATIONS

EDITED BY G.-C. ROTA

Editorial Board

R. Doran, M. Ismail, T.-Y. Lam, E. Lutwak, R. Spigler

Volume 63

Minkowski Geometry

ENCYCLOPEDIA OF MATHEMATICS AND ITS APPLICATIONS

4   W. Miller, Jr.   *Symmetry and separation of variables*
6   H. Minc   *Permanents*
11 W. B. Jones and W. J. Thron   *Continued fractions*
12 N. F. G. Martin and J. W. England   *Mathematical theory of entropy*
18 H. O. Fattorini   *The Cauchy problem*
19 G. G. Lorentz, K. Jetter, and S. D. Riemenschneider   *Birkhoff interpolation*
21 W. T. Tutte   *Graph theory*
22 J. R. Bastida   *Field extensions and Galois theory*
23 J. R. Cannon   *The one-dimensional heat equation*
25 A. Salomaa   *Computation and automata*
26 N. White (ed.)   *Theory of matroids*
27 N. H. Bingham, C. M. Goldie, and J. L. Teugels   *Regular variation*
28 P. P. Petrushev and V. A. Popov   *Rational approximation of real functions*
29 N. White (ed.)   *Combinatorial geometries*
30 M. Pohst and H. Zassenhaus   *Algorithmic algebraic number theory*
31 J. Aczel and J. Dhombres   *Functional equations containing several variables*
32 M. Kuczma, B. Chozewski, and R. Ger   *Iterative functional equations*
33 R. V. Ambartzumian   *Factorization calculus and geometric probability*
34 G. Gripenberg, S.-O. Londen, and O. Staffans   *Volterra integral and functional equations*
35 G. Gasper and M. Rahman   *Basic hypergeometric series*
36 E. Torgersen   *Comparison of statistical experiments*
37 A. Neumaier   *Interval methods for systems of equations*
38 N. Korneichuk   *Exact constants in approximation theory*
39 R. A. Brualdi and H. J. Ryser   *Combinatorial matrix theory*
40 N. White (ed.)   *Matroid applications*
41 S. Sakai   *Operator algebras in dynamical systems*
42 W. Hodges   *Model theory*
43 H. Stahl and V. Totik   *General orthogonal polynomials*
44 R. Schneider   *Convex bodies*
45 G. Da Prato and J. Zabczyk   *Stochastic equations in infinite dimensions*
46 A. Bjorner, M. Las Vergnas, B. Sturmfels, N. White, and G. Ziegler   *Oriented matroids*
47 E. A. Edgar and L. Sucheston   *Stopping times and directed processes*
48 C. Sims   *Computation with finitely presented groups*
49 T. Palmer   *Banach algebras and the general theory of *-algebras*
50 F. Borceux   *Handbook of categorical algebra I*
51 F. Borceux   *Handbook of categorical algebra II*
52 F. Borceux   *Handbook of categorical algebra III*
54 A. Katok and B. Hassleblatt   *Introduction to the modern theory of dynamical systems*
55 V. N. Sachkov   *Combinatorial methods in discrete mathematics*
56 V. N. Sachkov   *Probabilistic methods in discrete mathematics*
57 P. M. Cohn   *Skew fields*
58 Richard J. Gardner   *Geometric tomography*
59 George A. Baker, Jr., and Peter Graves-Morris   *Padé approximants*
60 Jan Krajicek   *Bounded arithmetic, propositional logic, and complexity theory*
61 H. Groemer   *Geometric applications of Fourier series and spherical harmonics*
62 H. O. Fattorini   *Infinite dimensional optimization and control theory*

ENCYCLOPEDIA OF MATHEMATICS AND ITS APPLICATIONS

Minkowski Geometry

A. C. THOMPSON
Dalhousie University

Published by the Press Syndicate of the University of Cambridge
The Pitt Building, Trumpington Street, Cambridge CB2 1RP
40 West 20th Street, New York, NY 10011-4211, USA
10 Stamford Road, Oakleigh, Melbourne 3166, Australia

© Cambridge University Press 1996

First published 1996

Printed in the United States of America

Library of Congress Cataloging-in-Publication Data
Thompson, Anthony C., 1937–
Minkowski geometry / A.C. Thompson.
p. cm. – (Encyclopedia of mathematics and its applications:
v. 63)
Includes bibliographical references.
ISBN 0-521-40472-X (hc)
1. Minkowski geometry. I. Title. II. Series.
QA685.T48   1996
516.3'74 – dc20   95-46491
                                                                         CIP

A catalog record for this book is available from the British Library.

ISBN 0-521-40472-X hardback

CONTENTS

Preface		page ix
Acknowledgements		xv
0	**The algebraic properties of linear spaces and convex sets**	**1**
	0.1 Linear spaces	1
	0.2 Convex sets	6
	0.3 Notes	9
1	**Norms and norm topologies**	**13**
	1.1 Norm topologies	14
	1.2 The unique linear topology on $\mathbb{R}^d$	27
	1.3 The Hahn–Banach theorem	32
	1.4 The existence and uniqueness of Haar measure	36
	1.5 Notes	42
2	**Convex bodies**	**45**
	2.1 Separation and support theorems	46
	2.2 Support functions and polar reciprocals	48
	2.3 Volumes and mixed volumes	53
	2.4 Various derived metrics	60
	2.5 Approximation of convex sets and the Blaschke selection theorem	64
	2.6 Notes	71
3	**Comparisons and contrasts with Euclidean space**	**75**
	3.1 The Mazur–Ulam theorem	76
	3.2 Normality in Minkowski space	77
	3.3 The Löwner ellipsoid	80
	3.4 Characterizations of Euclidean space	85
	3.5 Notes	94

v

4	**Two-dimensional Minkowski spaces**	**99**
	4.1 Inscribed regular hexagons and other constructions	100
	4.2 Sets of constant width and equichordal sets	106
	4.3 Lengths of curves, perimeter of the unit ball	111
	4.4 The isoperimetric problem in a Minkowski plane	118
	4.5 Isoperimetric inequalities	123
	4.6 Transversality	125
	4.7 Radon curves	127
	4.8 Notes	129
5	**The concept of area and content**	**135**
	5.1 Requirements and examples	137
	5.2 The role of the function σ_B	141
	5.3 The properties and the normalization of **I**	145
	5.4 The isoperimetrices that arise from Examples 5.1.4	150
	5.5 Further properties of **I**	171
	5.6 Notes	182
6	**Special properties of the Holmes–Thompson definition**	**187**
	6.1 The convexity of the area function σ	187
	6.2 Properties of the mapping **I**	195
	6.3 Cauchy's formula for surface areas	201
	6.4 Integral geometry in Minkowski spaces	205
	6.5 Bounds for the surface area of B	212
	6.6 Miscellaneous properties	215
	6.7 Notes	222
7	**Special properties of the Busemann definition**	**229**
	7.1 The convexity of the area function σ	229
	7.2 Properties of the mapping **I**	233
	7.3 Area and Hausdorff measures	237
	7.4 Bounds for the surface area of B	242
	7.5 Notes	245
8	**Trigonometry**	**251**
	8.1 The functions cm and sm	251
	8.2 The function α	258
	8.3 Trigonometric formulas	260
	8.4 Differentiation of the trigonometric functions	264
	8.5 Notes	271
9	**Various numerical parameters**	**275**
	9.1 Projection constants	276
	9.2 Macphail's constant	283

	9.3	The inner metric	286
	9.4	The girth, perimeter, inner radius and inner diameter of X	288
	9.5	Five examples in $\mathbb{R}^d$	293
	9.6	Relationships with the Banach–Mazur distance and extreme values	300
	9.7	Notes	304
10	**Fifty problems**		307

References   313

Notation index   331

Author index   335

Subject index   339

PREFACE

In choosing a title for this volume I faced two problems. Firstly, should it be "Minkowski" or "Minkowskian Geometry"? Secondly, how could I avoid accusations of false advertising from students of relativity who expect Minkowski(an) geometry to deal with one time-like and several space-like dimensions?

In an attempt to resolve the first problem I made two very long lists, including the following items:

Abelian group	Banach space
Boolean algebra	Blaschke sum, product
Euclidean geometry	Fourier series
Gaussian integer	Galois theory
Hamiltonian circuit	Hausdorff measure, topology
Hermitian matrix	Hilbert space
Newtonian mechanics	Lebesgue integral
Riemannian metric	von Neumann algebra
...	...

I then convinced myself that I detected two slight trends: the second list tended to predominate in more recent times, and this tendency was less pronounced in the applied or physical areas. My proposed solution to both problems is to suggest the use of *Minkowski geometry* for the present topic, *i.e.* the theory of finite dimensional normed linear spaces, which (as Dunford and Schwartz [130], p. 372, say) "is primarily due to Minkowski", and to use *Minkowskian geometry* for that other creation of Minkowski's, the theory of linear spaces with an indefinite inner product. To the linguistic purists, I apologize for the juxtaposition of the terms *Euclidean geometry* and *Minkowski geometry* in several places.

Space to Euclid and Newton was uniform and "isotropic" – the same in all directions. Such a notion flies in the face of daily experience, where the connotation of "up" and "down" is different from that of "east" and "west". There are preferred directions. Another good example is the preferred directions that cause crystals to grow as polyhedra and not spherically like soap-bubbles. This book is about that kind of geometry – a geometry in which, to someone peering in from outside, it appears that the unit for measuring length is different in different directions and hence unit "circles" and "spheres" are not the familiar round objects from Euclidean geometry but are some other convex shape (called the *unit ball*). Beings living inside the geometry would have a harder time detecting the preferred directions because yard-sticks, compasses and other measuring devices would, as they were rotated, adjust (in the view of someone outside) to the change. Nevertheless, internal measurements can be made to verify that this geometry is not Euclidean. One is the circumference of a unit "circle" (or the ratio of the circumference to the radius of any "circle"). This measurement yields a number somewhere between 6 and 8 but not usually the familiar 2π. Moreover, this number varies as the plane of the circle varies in space, indicating clearly that space is not uniform.

The most fundamental change from Euclidean geometry is the lack of a satisfactory idea of "perpendicular". There are shortest distances from a point to a line or plane. Unfortunately, this minimum may be attained by many points and the perpendicular relationship it yields is not symmetric. If "perpendicular" and "right angle" have gone then that staple of high school geometry, Pythagoras' theorem, has also gone. All these topics are examined in this book. However, the core of the book is Chapters 5, 6 and 7, which deal with the notion of surface area and the solution of the "isoperimetric problem". This problem asks which surface, enclosing a fixed volume, has the smallest surface area. The solution, the isoperimetrix, is a fundamental object in the geometry, and it is the interplay between this object and the unit ball which gives the geometry its special interest and flavour.

The geometry is "affine" in the sense that the properties examined are independent of the choice of basis vectors. Thus, although we are dealing with finite dimensional vector spaces over $\mathbb{R}$ which can always be regarded as $\mathbb{R}^d$, we try to avoid coordinates whenever possible. Another view of this aspect is to say that the properties examined are invariant under invertible linear transformations. Therefore, as far as Minkowski geometry is concerned, all d-dimensional ellipsoids are equivalent as unit balls. They correspond to Euclidean geometry because, with a suitable choice of coordinates, an ellipsoid is the familiar unit sphere. Likewise all d-dimensional parallelotopes are equivalent to the unit cube as unit ball.

The subject is a 20th century one. It originated in the late 19th century with the work of Brunn in convexity theory and the work of the Italian school (see Monna's book [391]) on abstract linear spaces. These strands were brought together by Minkowski. In part, Minkowski geometry shares its development with the theory of infinite dimensional normed spaces (Banach spaces) and, in part, it has much in common with the theory of convexity. The subject received large impetus from the work of Herbert Busemann in the middle of the century. He showed how significant the solution to the isoperimetric problem is for the geometry of the space and how the roles of the unit ball in Euclidean space are shared between the unit ball and the convex body that is the (normalized) solution to the isoperimetric problem. The connections with convexity theory have become stronger in recent years with the development of the "duality" between projection bodies and intersection bodies which Erwin Lutwak has vigorously pursued. Since Dvoretzky's theorem [133] appeared, the connections with Banach space theory have largely revolved around the "local theory" of Banach spaces.

The emphasis throughout the book is on geometry in a classical sense. There is, for example, frequent use of figures to illustrate an example or part of a proof. One of Busemann's aims was to develop enough of the theory of Minkowski spaces in a synthetic way to enable the theory of Finsler spaces (which correspond to Minkowski spaces in the same way that Riemannian spaces correspond to Euclidean ones) to proceed with less emphasis on coordinates and analytic expressions. I hope that the present volume helps to further Busemann's goal.

The book is aimed at readers who wish to be introduced to the geometry of Minkowski spaces. The general level of background knowledge required is not more than that of a senior undergraduate or beginning graduate student. For most of the book what is needed are: (a) a familiarity with the ideas of linear algebra, in particular the concepts of vector space, subspace, convex set, linear transformation, linear functional and dual space; and (b) a familiarity with the ideas of metric space topology, in particular the concepts of continuity, norm, open, closed and compact sets.

The level of mathematical sophistication and mathematical background needed is not, however, constant. There are three areas where I have used more than the minimum just listed. The first is the development of Haar measure in Chapter 1. This concept is of such fundamental significance to the subsequent development that I thought a rather full treatment should be included. In $\mathbb{R}^d$, Haar measure is usually viewed as a multiple of Lebesgue measure and because Lebesgue measure is generally developed with the tools of Euclidean geometry in hand, I felt it would be a good idea to make clear that those tools are not necessary for its development. Nevertheless, I have limited the discussion to the commutative case appropriate for $\mathbb{R}^d$. The second is the Brunn–Minkowski theory of convex bodies, which is used heavily in Chapters 5–7. The particular topics needed are Minkowski's theory of mixed volumes and a number of deep inequalities. This theory is introduced in Chapter 2, where I have tried to present clear statements of the results needed but with either no proofs or only brief outlines of proofs. The third area is the use of multilinear algebra in parts of Chapters 5 and 6. On this topic I have included a rather sketchy outline of the background needed. Readers who find some of the sections on these topics difficult are advised to skip them on a first reading. It is possible to read later sections without having absorbed all that has gone before.

A glance at the Table of Contents will give the reader a good idea of what is discussed. Unfortunately, many topics have been left out. The main criteria for deciding what to include were:

(i) the view that this is a geometry book,
(ii) the knowledge and competence of the author,
(iii) the limitations of space and
(iv) the availability of other treatments.

Some readers may be disappointed to find very little about the "local theory of Banach spaces", but there are treatments of this subject in, among other places, the books by Pisier [430] and by Tomczak-Jaegermann [513]. I would have liked very much to include some sections on differential geometry at the end of Chapter 8 but, for lack of space, have not done so. For the same reason the planned chapter on bodies of constant width in Minkowski spaces was reduced to a section of Chapter 4. The second criterion led to the exclusion of anything about Finsler geometry and anything about minimal surfaces with boundary. I have tried to compensate for some of these omissions with the brief notes at the end of each chapter and the reference list at the end of the book, but I am conscious of having an inadequate knowledge of the literature. To those readers who find some of their favourite topics either not covered or not sufficiently covered and citations missing from the reference list, I apologize.

A few comments about notation. When an equation is used to define the quantity on the left-hand side, the symbol := is used. I have tried to be systematic about usage and letter styles because I believe that such systematization is a great help to the reader. For example, most small Greek letters are either real numbers or real-valued functions. Uppercase Roman letters serve a variety of purposes: some are arbitrary sets, some (K, B) are convex sets, X and Y are linear spaces, L and M are subspaces and T is a linear operator. There is some duplication; $e.g.$ P is sometimes a polytope, sometimes a parallelotope or parallelogram and sometimes (in Chapter 9) a projection. Most lowercase Roman letters are vectors (points). Exceptions are d (see below) and $i, \ldots, n$, which are integral indices. At times accepted usage overcomes these conventions. For example, I have used λ for Lebesgue measure in various dimensions including d-dimensional measure in a d-dimensional space. But I have used V for mixed volumes, which makes Minkowski's inequality for mixed volumes appear a little unusual. As just indicated, I have used d rather than n for the dimension of a typical Minkowski space. This frees n for its traditional role as a typical integer. However, since the dimension d appears in a number of integral formulas, the reader should be on guard not to confuse it with the symbol for the differential. The numbers d and $d-1$ also appear frequently as superscripts to denote the particular measure to be considered (volume or area); the reader should be sure not to confuse this usage with an exponent. I hope the Notation Index will be helpful.

Definitions, propositions, lemmas, theorems and corollaries are all numbered consecutively in each section. Thus Corollary 4.3.5 is the fifth item in §4.3. The long §5.5 has been divided into subsections but this does not affect the numbering of items. Equations and figures are numbered consecutively in each chapter. Thus (6.24) denotes the 24th equation in Chapter 6.

A final comment on terminology. Chapters 6 and 7 discuss the consequences of two particular definitions of Minkowski surface area. One is due to Busemann, or at least he is responsible for the development of the concept and much of the theory. The other, as far as I know, first appeared in [262]. Descriptive terminology (the intersection definition and the projection definition) did not seem helpful. The first draft, by various circumlocutions, avoided giving names to either definition. I accepted Erwin Lutwak's criticism of the clumsiness of this solution. With reluctance I have adopted his suggestion.

ACKNOWLEDGEMENTS

A book of this kind is not an independent production. Many people have contributed and helped over the years. I have learned mathematics from many people in many places, too numerous to name. My colleague Michael Edelstein gave much encouragement and advice during the years that we were together at Dalhousie. A former colleague, Raymond Holmes, was instrumental in initiating the work that now forms Chapter 6. Erwin Lutwak suggested I write this volume and his encouragement at the other end of a telephone line has been indispensable.

The first draft was written mostly while I was on sabbatical leave in Freiburg im Breisgau. Thanks are due to Dalhousie University for that leave, which was essential for the book. It is a great pleasure to acknowledge the warm hospitality of the people in Freiburg who made our stay there in 1990–1991 and again in the summer of 1992 very memorable: Dr. Rolf Schneider and Dr. O. Kegel of the Mathematisches Institut, Albert-Ludwigs Universität, not only for their kindness in providing office space, access to the fine library and all the ancillary services but also for their personal hospitality; Dr. John Wieacker for many conversations on the subject matter of this book and on many topics of wider interest over good coffee; and the many friends my wife and I made in Freiburg.

A large number of people helped with the collection of information for the book by sending me copies of preprints and offprints. Particularly helpful in this respect were Richard Gardner, Helmut Groemer, Heinrich Guggenheimer and Horst Martini, who provided drafts of the manuscripts for [172], [206], [227] and [49] respectively. Erwin Lutwak and Horst Martini also assisted greatly by pointing out many of the works included in the reference list. Peter Gruber was particularly helpful in providing comments and references for §3.4.

Several people read all or parts of the book at various stages. I would like to thank all of them for their help and advice. They include Rolf Clack, who was the only one I allowed to see the very first attempts. He spent many hours trying to make sure that the book really was aimed at the beginning graduate student, that it was relatively free of non sequiturs, sadistic twists of logic and half-hidden pitfalls. Erwin Lutwak, Rolf Schneider and Dean Tsaltas patiently read later drafts and made numerous suggestions for improvements as well as indicating errors. I thank them for their time and effort. The first draft was carefully and ably typed in LaTeX by Sabine Linsenbold in Freiburg and Maria Fe Elder in Halifax. Both had to contend with poor handwriting as well as difficult mathematical symbols and terminology. Ms. Linsenbold deserves special credit for handling all this so well in a foreign language. Various figures were produced using both Mathematica and MAPLE by Brian Ingalls and Jennifer Overington (both former honours students at Dalhousie). The remaining figures were drawn by Hugh Thompson using AutoCAD. Keith Fordham helped me to produce the indices. Lauren Cowles at Cambridge University Press showed remarkable patience as various deadlines passed, and the production staff has done an excellent job of producing the volume.

The later part of the work was done with some financial support from NSERC (Canada) under research grant 4066. I am indebted to BCS Associates (Moscow, Idaho) and Springer (Berlin) for permission to reproduce the quotation on p. 10.

0
The algebraic properties of linear spaces and convex sets

This short chapter gives the background from linear algebra and the basic facts about convex sets that are needed throughout the book. The first section describes the notation and the conventions that will be used for concepts from linear algebra. The substantive items concern dual spaces, dual bases, the second dual and adjoint linear maps.

A large variety of facts about convex sets are needed in the later chapters. The definition of a convex set and some examples of such sets are already needed in Chapter 1, where norms are introduced. Therefore, the second section of this chapter introduces the idea of convexity via a series of examples and definitions. However, most of the important facts about convex sets (*e.g.* the separation and support theorems) are more easily discussed using such topological notions as *closed, bounded, interior* and *continuous linear mapping*. Consequently, after dealing with topological ideas in Chapter 1, we return to a fuller study of convex sets in Chapter 2.

0.1 Linear spaces

Throughout the book X (and occasionally Y) will be used to denote a finite dimensional vector space over the field $\mathbb{R}$ of real numbers. The dimension of X will usually be d. Small Roman letters v, w, x, y, z will be used for vectors and small Greek letters α, β, γ for scalars.

Subspaces of X will be denoted by capital Roman letters such as L, M; other capital Roman letters (*e.g.* B, C, K) will be used for convex sets, and T will indicate a linear transformation between vector spaces. The identity map is denoted by **1**. A basis for X will often be denoted by $(b_1, b_2, \ldots, b_d)$ and the usual basis for $\mathbb{R}^d$ will always be $(e_1, e_2, \ldots, e_d)$.

The vector space of linear functionals on X, *i.e.* all linear mappings from X into $\mathbb{R}$ with the usual pointwise operations, is called the *dual space of X* and denoted by X^*. Elements of X^* will be indicated by the letters f, g (often with subscripts). If $(b_1, b_2, \ldots, b_d)$ is a basis for X then a linear functional is entirely determined by its values at each b_i and, on the other hand, assigning arbitrary values to each b_i determines a linear functional on X. Those linear functionals $b_i^* (i = 1, 2, \ldots, d)$ defined by

$$b_i^*(b_j) := \delta_{ij} \qquad (i, j = 1, \ldots, d) \tag{0.1}$$

are evidently linearly independent and span X^* because $f(b_j) = \phi_j$ if and only if $f = \sum_{i=1}^{d} \phi_i b_i^*$. Thus $(b_1^*, b_2^*, \ldots, b_d^*)$ is a basis for X^* and is called the *dual basis* to $\{b_1, b_2, \ldots, b_d\}$. If a vector x has coordinates $(\xi_1, \xi_2, \ldots, \xi_d)$ relative to $\{b_1, b_2, \ldots, b_d\}$ and if f in X^* has coordinates $(\phi_1, \phi_2, \ldots, \phi_d)$ relative to $\{b_1^*, b_2^*, \ldots, b_d^*\}$ then

$$f(x) = \phi_1 \xi_1 + \phi_2 \xi_2 + \cdots + \phi_d \xi_d. \tag{0.2}$$

When we use coordinates in this way, in order to preserve the distinction between X and X^* we shall always regard elements of X as column vectors and elements of X^* as row vectors. Then an element of X^* acts on an element of X by the usual process of matrix multiplication; e.g. the element $f := (1, 2, 3)$ in $(\mathbb{R}^3)^*$ evaluated at

$$x := \begin{pmatrix} 3 \\ 2 \\ 1 \end{pmatrix}$$

in $\mathbb{R}^3$ is

$$f(x) = (1, 2, 3) \begin{pmatrix} 3 \\ 2 \\ 1 \end{pmatrix} = 10. \tag{0.3}$$

However, to save space, a column vector will usually be written as the transpose of a row vector, thus:

$$x = (3, 2, 1)^t.$$

At this point it is usual to omit the transpose, identify row and column vectors and think of Equation (0.3) as an *inner product*:

$$\langle (1, 2, 3), (3, 2, 1) \rangle = 10.$$

In general, if $(b_1, b_2, \ldots, b_d)$ is a basis for X then two vectors x and y in X have representations

$$x = (\xi_1, \xi_2, \ldots, \xi_d)^t, \qquad y = (\eta_1, \eta_2, \ldots, \eta_d)^t$$

relative to this basis. One can identify the vector y with the linear functional

$$f_y := (\eta_1, \eta_2, \ldots, \eta_d) = \sum_{i=1}^{d} \eta_i b_i^*$$

and hence define the *inner product*

$$\langle y, x \rangle := f_y(x) = \sum \eta_i \xi_i$$

(as in Equation (0.2)).

The relationship between y and f_y can be described by saying that X and X^* are identified by means of the linear map that sends each basis vector b_i in X to the corresponding dual basis vector b_i^* in X^*. An inner product defined in this way from a particular choice of basis for X is called a *Euclidean structure* or a *Euclidean geometry* on X. Since the basis $(b_1, b_2, \ldots, b_d)$ is entirely arbitrary and different bases give rise to different Euclidean structures, this process of constructing an inner product is rather artificial.

In the same way that a choice of basis gives rise to an inner product on X, it also gives rise to one on X^*. It follows that if x and y are vectors in X which correspond (as above) to linear functionals f_x and f_y in X^* then $\langle x, y \rangle = \langle f_x, f_y \rangle$. Thus the virtue of a Euclidean structure is that the mapping from X to X^* defined above not only is a linear isomorphism but also preserves inner products and, hence, is an isometry relative to the metrics in both X and X^* generated by the two inner products. By contrast, in a more general Minkowski geometry the map bears little or no relation to the metric structure and is, therefore, of much less use.

The *null space* or *kernel* of a linear functional f in X^* is denoted by $f^\perp$, i.e.

$$f^\perp := \{x \in X : f(x) = 0\}.$$

If x is in $f^\perp$ then we can say that f is *perpendicular* to x.

With a Euclidean structure the identification of X and X^* allows the relationship of perpendicularity to be construed as a relationship between elements of X. We say that vectors y and x are *orthogonal* if

$$\langle y, x \rangle = 0.$$

The notion of orthogonality is central to much of elementary Euclidean geometry. A great deal of the rest of this book can be viewed as finding partial substitutes for this relationship. In summary, then, there is always a notion of perpendicularity between certain elements of X and X^*. The notion of orthogonality between elements of X is special to the Euclidean case. This is the prime reason why it is important in Minkowski geometry to preserve the distinction between elements of X and elements of X^*.

Despite the foregoing discussion, for the purposes of calculation it is frequently convenient to adopt an auxiliary Euclidean structure. By this we mean an arbitrary choice of basis and the inner product derived from that basis by the corresponding identification of X and X^*.

If L_1 is a subspace of X then, since a basis for L_1 can be extended to a basis for X, we can write X as the direct sum of L_1 and a complementary subspace L_2:

$$X = L_1 \oplus L_2.$$

In X^*, two subspaces are defined by the equations

$$L_i^\perp := \{f \in X^* : f(y) = 0, \ \forall y \in L_i\}, \qquad i = 1, 2.$$

Proposition 0.1.1 *With the preceding notation, we have $X^* = L_2^\perp \oplus L_1^\perp$ and L_1^*, L_2^* are isomorphic to $L_2^\perp, L_1^\perp$ respectively.*

Proof. First, it is clear that $L_1^\perp \cap L_2^\perp = \{0\}$. If $f \in X^*$ and $x \in X$ then $x = x_1 + x_2$ for unique elements x_i in L_i. Define f_i on X by $f_i(x) := f(x_i)$. Then $f_1 \in L_2^\perp$, $f_2 \in L_1^\perp$ and $f = f_2 + f_1$ so that $X^* = L_2^\perp \oplus L_1^\perp$. Each functional $g \in L_2^\perp$ can be regarded as an element g' of L_1^* by setting $g'(x_1) := g(x_1)$, i.e. $g' = g_{|L_1}$. Conversely, each functional $g' \in L_1^*$ can be extended to an element g of $L_2^\perp$ by letting $g(x_1 + x_2) = g'(x_1)$. It is easy to show that this defines an isomorphism between L_1^* and $L_2^\perp$ and, similarly, for L_2^* and $L_1^\perp$. ∎

We repeat and extend the definition of $f^\perp$.

Definition 0.1.2 *If $f \in X^*$ then the subspace of X that is annihilated by f is denoted by $f^\perp$, i.e. $f^\perp := \{x \in X : f(x) = 0\}$. Similarly, if $x \in X$ then $x^\perp := \{f \in X^* : f(x) = 0\}$. Furthermore, each translate of the kernel of a non-zero linear functional, i.e. $\{x \in X : f(x) = \alpha\}$, is called a **hyperplane** in X.*

Having defined X^* and seen that it is another vector space, it is logical to consider $X^{**} := (X^*)^*$. Each vector x in X, however, defines an element Jx in X^{**} by

$$(Jx)(f) := f(x). \tag{0.4}$$

If $Jx = 0$ then $f(x) = 0$ for every linear functional f in X^* and hence $x = 0$. Thus $J : X \mapsto X^{**}$ is an injection. Since, for finite dimensional spaces, X and X^* have the same dimension, it follows that so do X and X^{**} and hence that J maps X onto X^{**}. Therefore, X and X^{**} are isomorphic. Moreover, the isomorphism J is natural in the sense that it is independent of basis. In fact, J can be defined for arbitrary vector spaces and it is a surjection if and only if the space is finite dimensional. We shall see later that J is also natural with respect to the norms on X and X^{**} in that it is always an isometry. There is no problem, therefore, in always identifying X and X^{**}. For this reason we shall sometimes use a pairing notation $\langle f, x \rangle$ to denote the value of f at x in order to emphasize the symmetry between f and x in this relationship. Using this notation we shall write $\langle x, f \rangle$ in place of $\langle Jx, f \rangle$.

If T is a linear transformation between vector spaces X and Y then there exists a dual transformation T^* between Y^* and X^* defined by

$$\langle T^* f, x \rangle := \langle f, Tx \rangle \qquad (0.5)$$

for all f in Y^* and x in X.

If, in this situation, T is an isomorphism and $(b_1, b_2, \ldots, b_d)$ is a basis for X then $(Tb_1, Tb_2, \ldots, Tb_d)$ is a basis for Y. The dual basis $(b_1^*, b_2^*, \ldots, b_d^*)$ for X^* is defined by Equations (0.1). Since $T^* T^{*-1} b_i^* = b_i^*$ we have

$$\langle T^* T^{*-1} b_i^*, b_j \rangle = \delta_{ij}$$

and hence, using (0.5),

$$\langle T^{*-1} b_i^*, T b_j \rangle = \delta_{ij}.$$

Thus $(T^{*-1} b_i^* : i = 1, 2, \ldots, d)$ is the basis in Y^* dual to $(Tb_1, Tb_2, \ldots, Tb_d)$ in Y.

Equation (0.5) provides another reason for writing elements of X as column vectors and elements of dual spaces as row vectors. If the spaces X and Y have dimensions d and n respectively and if bases for X and Y have been chosen then, relative to these bases, a vector x in X has a representation

$$x = (\xi_1, \xi_2, \ldots, \xi_d)^t$$

and T has a representation as an $n \times d$ matrix M_T. Then Tx is represented by

$$Tx = M_T (\xi_1, \xi_2, \ldots, \xi_d)^t.$$

If Y^* is given the basis dual to that for Y then f in Y^* is represented as

$$f = (\phi_1, \phi_2, \ldots, \phi_n)$$

and hence, from (0.2), the right-hand side of (0.5) is

$$\langle f, Tx \rangle = (\phi_1, \phi_2, \ldots \phi_n)(M_T (\xi_1, \xi_2, \ldots, \xi_d)^t). \qquad (0.6)$$

It follows that this same matrix product, with the multiplication associated differently, must also represent the left-hand side of (0.5). Thus

$$T^* f = (\phi_1, \phi_2, \ldots, \phi_n) M_T \qquad (0.7)$$

and the matrix for T^* (relative to the two dual bases) is the same matrix M_T but written on the right of the row vectors.

0.2 Convex sets

Schneider remarks in [475] that it is surprising how rich a theory is generated by the simple definition of convexity. In this section we begin with the definition and then give a list of examples of convex sets. This is followed by various related definitions. The part of the theory of convexity that we shall need is postponed until Chapter 2 after we have discussed norms and norm topologies. Nevertheless, a few topological words (*closed* and *interior*) creep in at the end of this section. It may be supposed that the finite dimensional space under consideration is equipped with an inner product which generates a Euclidean metric and that these terms refer to the topology derived from that metric. This will be clarified in the next chapter.

Definition 0.2.1 *A non-empty subset K of a linear space X is said to be **convex** if $\alpha x + (1 - \alpha)y$ is in K whenever x and y are in K and $0 \leq \alpha \leq 1$.*

The set $\{\alpha x + (1 - \alpha)y : \alpha \in [0, 1]\}$ is called the *line segment* joining x and y and is denoted by $[x, y]$. Thus K is convex if it contains the line segment joining any two of its points.

The vector operations on X can be extended to the collection of convex sets as follows:

$$K_1 + K_2 := \{x : x = x_1 + x_2, \ x_i \in K_i\},$$
$$\alpha K_1 := \{x : x = \alpha y, \ y \in K_1\}.$$

It is easy to see that if K_1 and K_2 are convex, so are all linear combinations of K_1 and K_2.

It is also clear that the associativity and commutativity of the vector operations imply the same for the extended operations. Furthermore, we have

$$\alpha(K_1 + K_2) = \alpha K_1 + \alpha K_2$$

and, if $\alpha \geq 0$ and $\beta \geq 0$,

$$(\alpha + \beta)K = \alpha K + \beta K.$$

The first of these is easy. For the second, if $y = (\alpha + \beta)x \in (\alpha + \beta)K$ then $y = \alpha x + \beta x$ is also in $\alpha K + \beta K$. On the other hand, if $y = \alpha x_1 + \beta x_2$ with $\alpha, \beta \geq 0$ then $y = (\alpha + \beta)[\alpha/(\alpha + \beta)x_1 + \beta/(\alpha + \beta)x_2]$ is in $(\alpha + \beta)K$ because $\alpha(\alpha + \beta)^{-1}x_1 + \beta(\alpha + \beta)^{-1}x_2 \in [x_1, x_2] \subseteq K$.

Because this distributive law applies only to non-negative scalars, the linear combinations of convex sets we consider will usually have non-negative coefficients. Also, for many of the convex sets that we shall consider, multiplication by negative scalars is irrelevant because they are symmetric in the following sense.

Definition 0.2.2 *A set K (which need not be convex) is said to be **symmetric** (with respect to 0) if* $(-1)K = K$.

The unqualified adjective "symmetric" will *always* mean "symmetric with respect to 0".

Example 0.2.3 *(convex sets)*

(i) A single point $\{x\}$ is convex.

(ii) Any subspace M of X is convex. Hence every translate of a subspace (*i.e.* a set of the form $M + \{x\}$) is convex. Such a set is called a *flat*. The dimension of a flat is the dimension of the subspace of which it is a translate. Each hyperplane in X is a flat of dimension $d - 1$.

(iii) A line segment $[x, y]$ is convex.

(iv) A sum of line segments is convex.

In particular, if $\{x_1, x_2, \ldots, x_k\}$ is a linearly independent set in X then $\sum_{i=1}^{k}[0, x_i]$ (or just $\sum[0, x_i]$) is convex and is called the *parallelotope* spanned by the vectors x_i. This idea is especially important when the vectors form a basis.

More generally, if $\{v_i : i = 1, 2, \ldots, n\}$ is an arbitrary set of vectors we let $[v_i] := [-2^{-1}v_i, 2^{-1}v_i]$ (so that $[v_i]$ is symmetric and of the same length as $[0, v_i]$) and then let

$$Z[v_i]_{i=1}^n = Z[v_1, v_2, \ldots, v_n] := \sum_{i=1}^{n}[v_i]. \tag{0.8}$$

This convex set is called the *zonotope* spanned by the vectors v_i. Each zonotope is symmetric. When $\{x_1, x_2, \ldots, x_k\}$ is linearly independent, the zonotope $\sum_{i=1}^{k}[x_i]$ is a translate of the parallelotope $\sum_{i=1}^{k}[0, x_i]$ by the vector $-2^{-1}\sum_{i=1}^{k}x_i$.

The class of zonotopes will play an important role in what follows.

(v) If $\{x_1, x_2, \ldots, x_k\}$ does not lie in any $(k - 2)$-dimensional flat in X then we call the set $\{y \in X : y = \sum_{i=1}^{k}\alpha_i x_i \text{ with } \alpha_i \geq 0 \text{ and } \sum_{i=1}^{k}\alpha_i = 1\}$ the *simplex* spanned by the vectors x_i and use the notation $S[x_1, x_2, \ldots, x_k]$ or $S[x_i]_{i=1}^k$ for this set. In $\mathbb{R}^d$, the special simplex $S[0, e_1, e_2, \ldots, e_d]$ with vertices at 0 and the standard basis vectors will be denoted by S_d. It is straightforward algebra to show that a simplex is convex.

The example of a simplex is a special case of a general phenomenon. An arbitrary intersection of convex sets is again convex, for if both x and y are in every member of the intersection so is the line segment joining them. Hence if A is a subset of X then we may consider first the family of all convex sets containing A (this is non-empty since it contains X itself) and then the intersection of this family. This is the smallest convex set containing A and is called the *convex hull* of A and denoted by conv A. One can show that

$$\text{conv } A = \left\{ y \in X : y = \sum_{i=1}^{n} \alpha_i x_i; \alpha_i \geq 0, \sum_{i=1}^{n} \alpha_i = 1, x_i \in A, n \text{ arbitrary} \right\}. \quad (0.9)$$

It follows that $S[x_1, x_2, \ldots, x_n] = \text{conv}\{x_1, x_2, \ldots, x_n\}$.

It is an important theorem of Carathéodory [90] that if $\dim X = d$ then the sums in (0.9) can be restricted by considering only $n = d + 1$. In other words, conv A is the union of all those simplices with vertices in A.

We now return to the list of examples of convex sets.

Example 0.2.4

(i) If $\{v_1, v_2, \ldots, v_n\}$ is an arbitrary finite set of vectors in X then the convex set conv$\{v_1, v_2, \ldots, v_n\}$ is called a *polytope*. In particular, a simplex and a zonotope are both polytopes. Throughout the book a polytope is a convex set.

(ii) If $f \in X^*$ and if α is a real number then the closed half-space $\{x \in X : f(x) \geq \alpha\}$ and the open half-space $\{x \in X : f(x) > \alpha\}$ are both convex. We shall use the notation $H[f, \alpha]^+$ to denote the closed half-space and $H(f, \alpha)^+$ for the open half-space. If the functional f is obvious we may suppress reference to it, and when $\alpha = 0$ we may sometimes write H^+ for the set $\{x \in X : f(x) \geq 0\}$.

It follows that every intersection of half-spaces is also convex. In Chapter 2 we shall show that the following converse of this is true: each closed convex set is an intersection of closed half-spaces.

Definition 0.2.5 *The **dimension** of a convex set K is the dimension of the smallest flat containing K.*

An alternative approach is first to define the dimension of a simplex $S[x_1, x_2, \ldots, x_k]$ to be $(k - 1)$; (recall that in Example 0.2.3(v) $\{x_1, x_2, \ldots, x_k\}$ does not lie in any $(k - 2)$-dimensional flat so this definition is in agreement with Definition 0.2.5). Then the dimension of a convex set K is the dimension of a maximal simplex contained in K.

Since a simplex has non-empty relative interior (*i.e.* interior points relative to the smallest flat containing it) the same is true of an arbitrary convex set. It follows that there is no real loss of generality in supposing that a single convex set has non-empty interior since we can usually restrict discussion to the flat containing the set.

Definition 0.2.6

(i) *A convex subset E of a convex set K is said to be an **extreme subset** of K if, whenever $x \in E$ and $x = \alpha x_1 + (1 - \alpha)x_2$ with $x_1, x_2 \in K$ and $\alpha \in (0, 1)$, then $x_1, x_2 \in E$.*

*(ii) If P is a polytope then the extreme subsets of P are called the **faces** of P. If P is of dimension d then the (d − 1)-dimensional faces are called the **facets** and the 0-dimensional faces are called the **vertices** of P.*

Example 0.2.7 If K is a cube in $\mathbb{R}^3$, then the extreme subsets (faces) of K are (i) K itself (of dimension 3); (ii) the facets of K (of dimension 2); (iii) the edges of K (of dimension 1); and (iv) the vertices of K (of dimension 0).

Definition 0.2.8 *A point x of K such that {x} is an extreme subset of K is called an **extreme point** of K.*

If P is a polytope, the extreme points of P coincide with the vertices of P. Every closed and bounded convex set K has a relative abundance of extreme points, in the sense that such a set is the convex hull of its extreme points. In other words, each point of K can be represented as a point of a simplex whose vertices are extreme points of K.

Definition 0.2.9 *A closed convex set is said to be **strictly convex** if its only extreme sets (other than itself) are extreme points (i.e. if there are no line segments – or larger extreme sets – in the boundary).*

0.3 Notes

Although the concepts from linear algebra are well known, their origins and history are not. A very readable account of the development of the idea of a linear space is contained in Chapter 2 of Monna's book, *Functional Analysis in Historical Perspective* [391]. The chief names and dates he gives are Bolzano (1804), Laguerre (1867), Grassmann (1844 and 1847) but especially the book *Calcolo geometrico...* [404], written by Peano in 1888, which, in Chapter 9, gives the first definition of a linear space.

An inner product on a real vector space can also be introduced axiomatically as a bilinear functional which is symmetric and positive definite. It is not always immediately obvious that, in the finite dimensional case, every such functional arises from a choice of basis in the way described in §0.1. As an example, consider the usual inner product

$$\langle p, q \rangle := \int_0^1 p(x) q(x) \, dx$$

defined on the d-dimensional space of real polynomials of degree not exceeding $(d-1)$. However, given an inner product, the familiar Gram–Schmidt orthogonalization process produces a basis with respect to which the inner product conforms to the situation described in §0.1.

For an infinite dimensional, complete, inner product space X the Riesz representation theorem asserts that the dual space (now consisting of all the *continuous* linear functionals on X) is isomorphic to the space itself. This theorem is due, independently, to both Riesz [440] and Fréchet [163, 164] in 1907. Their proofs were for the space $L_2[0, 1]$; for the proof for an abstract inner product space see Stone [495], p. 62, or Riesz [443].

For a more detailed account of dual bases and the isomorphism between X and X^{**} see, *e.g.*, the book by Halmos [240]. In particular, Proposition 0.1.1 can be found in §20. A full discussion of the place finite dimensional spaces occupy among all vector spaces, including the theorem that X is finite dimensional if and only if it is isomorphic to its second dual, is contained in Köthe [301], §9.5.

The origins of the study of convexity are clearer. In 13 pages Gruber [213] gives a masterly account of the history of the subject, and there is another good survey of a similar length by Fenchel [150]. However, for a succinct summary one cannot do better than quote the initial paragraph of the foreword to Bonnesen and Fenchel [51]:

Convex figures have always played an important role in geometry. Brunn however was the first to make a comprehensive investigation of those figures that are characterized by the convexity property alone. In two works: "Über Ovale und Eiflächen" [60] and "Über Kurven ohne Wendepunkte" [61] that appeared in 1887 and 1889 he proved, along with a variety of results about convex regions and bodies, a theorem on the areas of the intersections of a convex body with parallel planes. This theorem subsequently turned out to be fundamental. The significance of this theorem was emphasized by Minkowski. In several works, in particular in "Volumen und Oberfläche" [388] (1903) and "Zur Theorie der konvexen Körper" [390] which was laid out on a large scale but remained unfinished, Minkowski introduced basic concepts such as support function, mixed volume, and so on. Here he created the formal tools appropriate for problems about convex regions and bodies. Minkowski above all opened the way to various applications, especially to the isoperimetric (isepiphane) and other extremal problems. Furthermore, he discovered the close connection of these concepts and theorems with the question of determining convex surfaces by means of their Gaussian curvature and proved profound theorems in this regard.

Following Minkowski were, among others, Carathéodory [90], whose basic theorem on convex hulls we mentioned, Radon [433, 434] and Helly [255]. The latter two both proved theorems of a combinatorial type which are closely related to that of Carathéodory. Radon proved that every set containing $(d + 2)$ points in $\mathbb{R}^d$ can be expressed as the union of two disjoint sets whose convex hulls have a common point. Helly's theorem states that if $A_1, A_2, \ldots, A_k$ are convex sets in $\mathbb{R}^d$ and if every $(d + 1)$ of these sets have a common point then the intersection of all of them is non-empty. These three theorems and the connections between them can be found in Schneider [479]. The article by Danzer, Grünbaum and Klee [120] is a good survey of these theorems and subsequent generalizations. This article begins with an interesting short biography of Helly which explains the circumstances during and after the First World War that led to Helly's theorem, proved in 1913, being first published by Radon [433]. A very recent survey with up-to-date material on these theorems is the article by Eckhoff [135].

The result that a closed, bounded convex set is the convex hull of its extreme points is, for finite dimensional spaces, due to Minkowski [390]. The generalization of this result to infinite dimensional topological vector spaces states that a compact convex set is the closed convex hull of its extreme points and is due (for normed spaces) to Krein and Milman [305]. This result can be regarded as the starting point for the Choquet theory of convex sets, the aim of which is to represent points of a convex set by means of integrals with respect to a measure located on the "boundary" of the set (see, *e.g.*, Phelps [427]).

1
Norms and norm topologies

Minkowski geometry is the study of the interplay between two structures. The first is a finite dimensional, real linear structure and the second is a metric structure derived from a norm. The preceding chapter was concerned with the linear structure. However, before we look at the detailed metric properties derived from a specific norm, we turn to the more general topological properties induced by a norm. The essential fact is that there is only one Hausdorff linear topology with which a finite dimensional space can be endowed. This implies that between any two d-dimensional Minkowski spaces there is a linear homeomorphism or, more briefly, an isomorphism. Any property of the one space which depends only on the topology and the linear structure carries over to the other via the isomorphism. Stated more succinctly, the isomorphic theory of Minkowski spaces is rather uninteresting; the dimension is the only isomorphic invariant and all Minkowski spaces are completely classified up to isomorphism by this invariant. On the other hand, as we shall see, the isometric classification is too complicated. There is a vast array of non-isometric Minkowski spaces, even in two dimensions. Perhaps this is why Minkowski spaces have received less attention than they deserve – on the one hand, they are too simple and, on the other, too complicated.

The chapter has four sections and, for the most part, the ideas discussed are the familiar ones presented in a course on metric space topology or at the beginning of a course on functional analysis. The exception is the material in the final section.

The first section gathers together the elementary facts about norms and the metric space topologies they generate. Important relationships discussed are, first, that between a norm and its unit ball; second, that between continuity and boundedness for linear maps; and third, that between a norm and the dual norm on the dual space. The section ends with a variety of examples of norms on $\mathbb{R}^d$.

The second section is devoted to showing that the topology on any d-dimensional Minkowski space is the familiar Euclidean topology. This is the first of several uniqueness results that make finite dimensional spaces special. All these results can be thought of as consequences of the local compactness of Minkowski spaces.

Since the topology is the familiar Euclidean one, it follows that closed and (Euclidean) bounded sets are compact (in the Minkowski norm topology). In fact, more is true: any two norms on a d-dimensional space are *equivalent*. This implies that not only the topologies but also the uniformities are the same. Thus, the collection of bounded sets is the same for all norms and we can say that a set is compact if and only if it is closed and bounded. Moreover, the collection of Cauchy sequences is independent of the norm and all Minkowski spaces are complete. Another consequence is that all linear maps between finite dimensional spaces are continuous and, in particular, the algebraic dual space and the topological dual space coincide.

The centrepiece of the third section is the Hahn–Banach theorem on extending a continuous linear functional from a subspace to the whole space. The important point here is not the existence of a continuous extension – that is straightforward because all linear maps are continuous – but that there is one which preserves the norm. This theorem has a variety of consequences dealing with supporting and separating hyperplanes, which will come in Chapter 2. The most important consequence for the third section is that every finite dimensional normed space is isometrically isomorphic to its second dual space.

The final section contains less standard material. It deals with another uniqueness theorem stemming from the local compactness. On $\mathbb{R}^d$ there is only one Hausdorff linear topology, only one linear uniformity and, up to a scalar normalizing factor, only one translation-invariant measure – Haar measure. The essential ideas for this result are from group theory. The space $(\mathbb{R}^d, +)$ is a commutative, locally compact, topological group. The central chapters of the book (Chapters 5–7) will depend fundamentally on this result. It is often convenient (especially when making detailed calculations using Haar measure) to suppose that a Minkowski space has an auxiliary Euclidean structure. We would like to emphasize, however, that this is only a convenience and a Euclidean structure is not inherent in the development. For this reason, at this point we avoid talking about Lebesgue measure and give an account of the construction of Haar measure.

1.1 Norm topologies

Definition 1.1.1 *A **norm** on a real linear space X is a mapping $\|.\|$ from X into $\mathbb{R}^+$ that satisfies the following axioms:*

(i) $\|x\| \geq 0$ *with equality if and only if $x = 0$,*
(ii) $\|\alpha x\| = |\alpha| \|x\| (\alpha \in \mathbb{R})$,
(iii) $\|x + y\| \leq \|x\| + \|y\|$.

The last condition is called the *triangle inequality*. From the norm we can define a metric $\delta(x, y) := \|x - y\|$ and then the triangle inequality for δ,

$$\delta(x, z) \leq \delta(x, y) + \delta(y, z),$$

follows directly from (iii) because

$$\|x - z\| = \|(x - y) + (y - z)\| \leq \|x - y\| + \|y - z\|.$$

Condition (i) implies that $\delta(x, y) \geq 0$ with equality if and only if $x = y$ and the symmetry of δ follows from (ii) with $\alpha = -1$. Thus δ is a metric which generates a topology on X called the *norm topology*. We shall denote this metric space by $(X, \|.\|)$. A *Minkowski space* is a pair $(X, \|.\|)$ in which X is finite dimensional.

Definition 1.1.2 The **unit ball** $B = B[0, 1]$ in $(X, \|.\|)$ is the set

$$B[0, 1] := \{x \in X : \|x\| \leq 1\}.$$

We shall use B_X if the dependence on X needs to be explicit. More generally, $B[x, \rho] := \{y \in X : \|x - y\| \leq \rho\}$. The **unit sphere** in $(X, \|.\|)$ is the boundary of the unit ball, ∂B. Thus

$$\partial B := \{x \in X : \|x\| = 1\}.$$

Definition 1.1.3 An *isometry* from the normed space $(X, \|.\|_X)$ to $(Y, \|.\|_Y)$ is a mapping T such that for all x_1, x_2 in X we have

$$\|x_1 - x_2\|_X = \delta_X(x_1, x_2) = \delta_Y(Tx_1, Tx_2) = \|Tx_1 - Tx_2\|_Y.$$

Proposition 1.1.4 If $(X, \|.\|)$ and $(Y, \|.\|)$ are two Minkowski spaces of the same dimension and if T is a linear isometry of X onto Y then

(i) $TB_X = B_Y$,
(ii) $T(\partial B_X) = \partial B_Y$,
(iii) if F is an extreme subset of B_X then $T(F)$ is an extreme subset of B_Y.

Proof. Since a linear map sends 0 in X to 0 in Y, (i) and (ii) are evident from the preceding definitions.

Now suppose that F is an extreme subset of B_X and suppose $y \in T(F)$ with $y = \alpha y_1 + (1 - \alpha) y_2$ for some $y_1, y_2 \in B_Y$. Then $y = Tx$ for $x \in F$ and, by (i), $y_i = Tx_i$ for some $x_i \in B_X$ ($i = 1, 2$). Since T^{-1} is linear these equations imply $x = \alpha x_1 + (1 - \alpha) x_2$ and therefore, by Definition 0.2.6, $x_i \in F$ and hence $y_i \in T(F)$ as required. ∎

Proposition 1.1.5 *If $(X, \|.\|)$ is a Minkowski space then*

  (i) *each translation is an isometry of $(X, \|.\|)$;*
  (ii) *the mappings $\mathbf{1}$ and $-\mathbf{1}$ are isometries of $(X, \|.\|)$.*

Proof

  (i) For all x, y in X we have
  $$\delta(x, y) = \|x - y\| = \|(x + z) - (y + z)\| = \delta(x + z, y + z).$$

  (ii) That $\mathbf{1}$ is an isometry is clear and for $-\mathbf{1}$ we have
  $$\delta(-x, -y) = \|-x + y\| = \|(-1)(x - y)\| = \|x - y\| = \delta(x, y). \quad \blacksquare$$

It is possible that the mappings listed in Proposition 1.1.5 (and compositions of $-\mathbf{1}$ with a translation) are the only isometries of $(X, \|.\|)$. For example, this is so if $\mathbb{R}^2$ is given the norm

$$\|x\| = \|(\xi_1, \xi_2)^t\| := \begin{cases} |\xi_1| + 2^{-1}|\xi_2| & \text{if } \operatorname{sgn}\xi_1 = \operatorname{sgn}\xi_2 \text{ and } |\xi_2| \le 2|\xi_1|, \\ |\xi_2| & \text{if } \operatorname{sgn}\xi_1 = \operatorname{sgn}\xi_2 \text{ and } |\xi_2| > 2|\xi_1|, \\ (\xi_1^2 + \xi_2^2)^{1/2} & \text{if } \operatorname{sgn}\xi_1 = -\operatorname{sgn}\xi_2. \end{cases}$$

The unit ball for this norm is shown in Figure 1.1. If $x_1 := (1, 0)^t$, $x_2 := (\frac{1}{2}, 1)^t$ and $x_3 := (0, 1)^t$ then ∂B consists of the line segments $[x_1, x_2]$, $[x_2, x_3]$ and the circular arc from x_3 to $-x_1$ and the images of these under $-\mathbf{1}$.

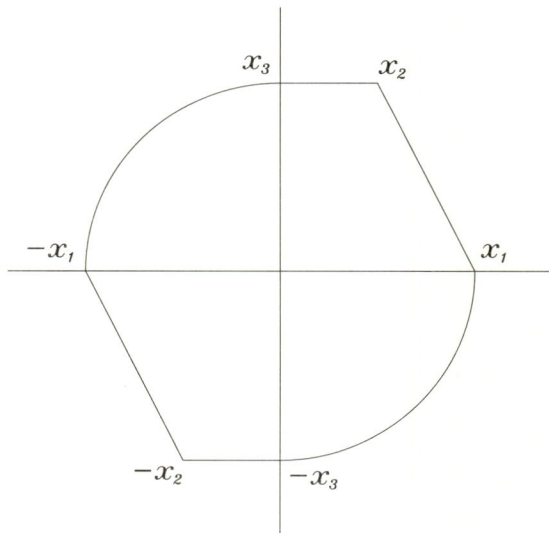

Figure 1.1

To prove that the only isometries are compositions of those listed in Proposition 1.1.5, we first need the Mazur–Ulam theorem (see 3.1.2 below), which states that any isometry of X onto itself that leaves the origin fixed is necessarily linear. Hence any isometry of a Minkowski space is a composition of a translation and a linear isometry. Proposition 1.1.4 implies that a linear isometry T maps line segments in ∂B to line segments of the same length. Since $\|x_1 - x_2\| = \sqrt{5}/2$ and $\|x_2 - x_3\| = \frac{1}{2}$ it follows that $Tx_i = \pm x_i$ and $T = \pm \mathbf{1}$.

Proposition 1.1.6 *The unit ball B in a Minkowski space $(X, \|.\|)$ is symmetric and convex. Moreover, with respect to the norm topology, B is closed and 0 is an interior point.*

Proof. The symmetry of B is evident from Definition 1.1.1(ii) with $\alpha = -1$. If $x, y \in B$ and $0 \leq \alpha \leq 1$ then

$$\|\alpha x + (1 - \alpha) y\| \leq \alpha \|x\| + (1 - \alpha) \|y\| \leq \alpha + (1 - \alpha) = 1.$$

This follows from Definition 1.1.1(i) and (ii) and the facts that $\|x\| \leq 1$, $\|y\| \leq 1$. Thus $\alpha x + (1 - \alpha) y \in B$ and B is convex.

The triangle inequality implies that $|\|x\| - \|y\|| \leq \|x - y\|$ and hence the norm is continuous as a mapping from $(X, \|.\|)$ to $\mathbb{R}$. Since B is the inverse image of a closed set, it is closed. Likewise $\{x : \|x\| < 1\}$ is open and so 0 is an interior point of B. ∎

Conversely, if B is a symmetric convex set in X with the property that each line through 0 meets B in a closed, bounded segment of positive length then we may define a functional $\|.\|_B$ on X by

$$\|x\|_B := \inf\{\xi \in \mathbb{R}^+ : x \in \xi B\}.$$

Definition 1.1.7 *The functional $\|.\|_B$ is called the **Minkowski functional** of B.*

Proposition 1.1.8 *If B is a symmetric convex set in X such that each line through 0 meets B in a non-trivial, closed, bounded segment then $\|.\|_B$ is a norm on X.*

Proof. From the definition, $\|x\|_B \geq 0$. If $x \neq 0$ then the ray joining 0 to x meets B in a segment $[0, \alpha x]$ with $0 < \alpha < +\infty$. Hence $\|x\|_B = \alpha^{-1}$ and so $\|x\|_B$ is a strictly positive real number. That $\|\beta x\|_B = \beta \|x\|_B$ for $\beta \geq 0$ follows readily from the definition and the symmetry of B implies that $\|-x\|_B = \|x\|_B$.

If x, y are non-zero elements of X then $x' := x/\|x\|_B$ and $y' := y/\|y\|_B$ are in B (because the line through 0 and x meets B in the *closed* segment

$[-x/\|x\|_B, x/\|x\|_B])$. Hence, by the convexity of B,

$$z := \frac{\|x\|_B}{\|x\|_B + \|y\|_B} x' + \frac{\|y\|_B}{\|x\|_B + \|y\|_B} y' = (\|x\|_B + \|y\|_B)^{-1}(x+y)$$

is in B, i.e. $\|z\|_B \leq 1$. But $\|z\|_B = (\|x\|_B + \|y\|_B)^{-1}\|x+y\|_B$ and hence

$$\|x+y\|_B \leq \|x\|_B + \|y\|_B.$$

This inequality is trivial if either x or y is zero. ∎

Remark. Propositions 1.1.6 and 1.1.8 show that there is a one–one correspondence between norms on X and symmetric, closed convex sets with non-empty interior in X. It is this correspondence which makes convexity an essential ingredient for the study of Minkowski spaces. Note that this correspondence is order reversing; i.e. $B_1 \subseteq B_2$ implies $\|.\|_{B_1} \geq \|.\|_{B_2}$ and $\|.\|_{\alpha B} = \alpha^{-1}\|.\|_B$ for $\alpha > 0$.

We now turn our attention to linear mappings. First we characterize the continuity of linear maps in terms of boundedness.

Definition 1.1.9 *If $(X, \|.\|_X)$ and $(Y, \|.\|_Y)$ are two normed spaces, a linear map T from X to Y is said to be **bounded** if $T(B)$, the image of the unit ball in X, is bounded in Y, i.e. if there exists a constant c such that*

$$\|Tx\|_Y \leq c \quad \text{whenever} \quad \|x\|_X \leq 1.$$

The least such c is called the *norm* of T, i.e.

$$\|T\| := \sup\{\|Tx\|_Y : \|x\|_X \leq 1\}. \qquad (1.1)$$

This terminology and notation are justified in Proposition 1.1.11 below.

Proposition 1.1.10 *If $(X, \|.\|_X)$ and $(Y, \|.\|_Y)$ are two normed spaces and if T is a linear map from X to Y then T is continuous if and only if T is bounded.*

Proof. Since T is linear and since translations are continuous in both X and Y, T is continuous if and only if it is continuous at 0. Since $T(0) = 0$, T is continuous at 0 if and only if given $\epsilon > 0$ there is a $\delta > 0$ such that

$$\|x\|_X \leq \delta \quad \text{implies} \quad \|Tx\|_Y \leq \epsilon.$$

But since T is linear and $\|.\|$ is positively homogeneous this is equivalent to saying $\|Tx\|_Y \leq \epsilon \delta^{-1}$ whenever $\|x\| \leq 1$. Hence the result. ∎

1.1 Norm topologies

Remark. We may also characterize $\|T\|$ as the smallest constant c such that $\|Tx\|_Y \leq c\|x\|_X$ for all x in X. This view of $\|T\|$ makes the proof of the following statements slightly easier.

Proposition 1.1.11 *The set of bounded linear maps from X to Y is a vector space under the usual pointwise operations and the functional $\|.\|$ defined by Equation (1.1) is a norm on that space. Furthermore, if $T : X \mapsto Y$ and $S : Y \mapsto Z$ are continuous linear maps then $\|ST\| \leq \|S\|\|T\|$.*

Proof. Rather than use a large number of subscripts, we will understand all the norms in the following calculations to refer to the appropriate spaces.

First, we have

$$\|(T_1 + T_2)x\| = \|T_1 x + T_2 x\| \leq \|T_1 x\| + \|T_2 x\| \leq (\|T_1\| + \|T_2\|)\|x\|$$

and hence, taking the suprema over $\|x\| \leq 1$, $\|T_1 + T_2\| \leq \|T_1\| + \|T_2\|$. This shows both that the set of bounded linear maps is closed under addition and that $\|.\|$ satisfies the triangle inequality.

Similarly, $\|\alpha T x\| = |\alpha|\|T x\|$ and so $\|\alpha T\| = |\alpha|\|T\|$. Again this shows both that the set of bounded linear maps is closed under scalar multiplication and that $\|.\|$ satisfies the second axiom for a norm.

It is clear that $\|T\| \geq 0$ and if $\|T\| = 0$ then, since $\|Tx\| \leq \|T\|\|x\|$, we have $\|Tx\| = 0$ for all x in X, i.e. $Tx = 0$ for all x in X and hence $T = 0$.

Finally, $\|(ST)x\| \leq \|S\|\|Tx\| \leq \|S\|\|T\|\|x\|$ and therefore $\|ST\| \leq \|S\|\|T\|$. ∎

In particular, if Y is the scalar field $\mathbb{R}$ then (1.1) defines a norm on the space of bounded linear functionals on X:

$$\|f\| := \sup\{|f(x)| : x \in B\}. \tag{1.2}$$

Looking ahead a little, we shall show in the next section that all linear maps defined on a Minkowski space are continuous. Thus (1.2) defines a norm on the algebraic dual space X^* of *all* linear functionals on X. In Chapter 2 we shall see that this norm on X^* coincides with the "support function" of B and that $\{f \in X^* : \|f\| \leq 1\}$ coincides with the "polar reciprocal" of B. For this reason we denote the unit ball in X^* by B°, i.e.

$$B^\circ := \{f \in X^* : \|f\| \leq 1\}. \tag{1.3}$$

This set is also called the *dual ball*.

The remainder of this section consists of several examples of norms on the space $\mathbb{R}^d$. In these examples we shall show that, for these particular norms, every linear functional is bounded.

Throughout, we suppose that $\mathbb{R}^d$ has the usual basis $(e_1, e_2, \ldots, e_d)$ and the dual space $(\mathbb{R}^d)^*$ has the dual (*i.e.* usual) basis $(e_1^*, e_2^*, \ldots, e_d^*)$. A linear functional f has coordinates (relative to the dual basis) $(\phi_1, \phi_2, \ldots, \phi_d)$ and, to save space, vectors x in $\mathbb{R}^d$ will be written $(\xi_1, \xi_2, \ldots, \xi_d)^t$.

Example 1.1.12 First, of course, is the usual Euclidean norm on $\mathbb{R}^d$:

$$\|x\|_2 := \left(\sum_{i=1}^{d} |\xi_i|^2 \right)^{1/2}. \tag{1.4}$$

Alternatively, we can write $\|x\|_2 = \langle x, x \rangle^{1/2}$, where $\langle , \rangle$ denotes the usual inner product on $\mathbb{R}^d$. That $\|.\|_2$ satisfies conditions (i) and (ii) of Definition 1.1.1 is evident. A standard argument using the Cauchy–Schwarz inequality shows that it also satisfies the triangle inequality. In Propositions 1.1.15 and 1.1.16 below, we shall prove generalizations of both of these inequalities.

The unit ball with respect to this norm is the usual Euclidean ball and is denoted by E (or by E_d to emphasize the dimension).

Consider the linear map from X^* to X, introduced in §0.1, that maps each e_i^* to e_i. Then a linear functional $f = (\phi_1, \phi_2, \ldots, \phi_d)$ is mapped onto the vector $y_f = \sum_{i=1}^d \phi_i e_i$. Hence $f(x) = \sum_{i=1}^d \phi_i \xi_i = \langle x, y_f \rangle$. Furthermore, the Cauchy–Schwarz inequality shows that if $x \in E$ then

$$|f(x)| = |\langle x, y_f \rangle| \leq \|y_f\|_2,$$

with equality if and only if $x = \|y_f\|_2^{-1} y_f$. Thus $\|f\|_{E^\circ} = \|y_f\|_2$ and this mapping of $(X^*, \|.\|_{E^\circ})$ onto $(X, \|.\|_E)$ is an isometry; *i.e.* the dual norm is given by the same formula (1.4) and the dual ball E° is mapped onto E.

Example 1.1.13 Let

$$\|x\|_\infty := \max\{|\xi_i| : i = 1, 2, \ldots, d\}. \tag{1.5}$$

This norm is called the ℓ_∞-norm (or the supremum norm, or the uniform norm) on $\mathbb{R}^d$. That $\|.\|_\infty$ satisfies the conditions of Definition 1.1.1 is straightforward (including the triangle inequality). The unit ball consists of all those vectors x such that $|\xi_i| \leq 1$ ($i = 1, 2, \ldots, d$). This set is called the *unit cube* and is denoted by C (C_d if the dimension is to be emphasized). It is the convex hull of its 2^d extreme points which are situated at $(\pm 1, \pm 1, \pm 1, \ldots, \pm 1)^t$. It has $2d$ facets which are cubes of dimension $(d-1)$ lying in the hyperplanes $\xi_i = \pm 1$ ($i = 1, 2, \ldots, d$).

Next consider $f = (\phi_1, \phi_2, \ldots, \phi_d)$ in X^*. We have $f(x) = \sum_{i=1}^d \phi_i \xi_i$. Hence if $x \in C$ then

$$|f(x)| = \left| \sum_{i=1}^d \phi_i \xi_i \right| \leq \sum_{i=1}^d |\phi_i \xi_i| \leq \max |\xi_i| \sum_{i=1}^d |\phi_i| \leq \sum_{i=1}^d |\phi_i|.$$

1.1 Norm topologies

This means that if $\sum |\phi_i| \le 1$ then $\sup\{|f(x)| : x \in C\} \le 1$ and so $\|f\| \le 1$. Conversely, if $f = (\phi_1, \phi_2, \ldots, \phi_d)$ is such that $\|f\| \le 1$ then choose $x_f \in C$ by setting

$$(x_f)_i := \operatorname{sgn} \phi_i = \begin{cases} 1 & \text{if } \phi_i > 0, \\ 0 & \text{if } \phi_i = 0, \\ -1 & \text{if } \phi_i < 0. \end{cases}$$

Then we have that $x_f \in C$ and therefore $\sum |\phi_i| = \sum \phi_i \operatorname{sgn} \phi_i = f(x_f) \le 1$. Thus $\|f\| \le 1$ if and only if $\sum |\phi_i| \le 1$. From this it follows, using the positive homogeneity of both expressions, that

$$\|f\| = \sum_{i=1}^d |\phi_i|.$$

This norm is called the ℓ_1-*norm* on X^*.

The dual ball in X^* is the d-dimensional cross-polytope

$$CP = CP_d := \left\{ f \in \mathbb{R}^d : \sum_{i=1}^d |\phi_i| \le 1 \right\}.$$

This convex body has $2d$ extreme points that are the dual basis vectors e_i^* and their negatives $-e_i^*$ ($i = 1, 2, \ldots, d$) and 2^d facets. Each facet is a simplex obtained by choosing one from each pair of opposite extreme points and then taking the convex hull of these points.

It is also possible to equip the original space with the ℓ_1-norm

$$\|x\|_1 := \sum_{i=1}^d |\xi_i| \tag{1.6}$$

and its dual is then the ℓ_∞-norm (see Theorem 1.3.8 below).

Example 1.1.14 Generalizing the preceding examples, for $p > 1$ we may define

$$\|x\|_p := \left(\sum_{i=1}^d |\xi_i|^p \right)^{1/p} \tag{1.7}$$

(which is called the ℓ_p-*norm*). As in the case when $p = 2$, conditions (i) and (ii) of Definition 1.1.1 are easy but the triangle inequality is non-trivial. For this we need to consider q defined by $p^{-1} + q^{-1} = 1$ and then we have the following basic inequality.

Proposition 1.1.15 (Hölder's inequality) *If $x \in \mathbb{R}^d$, $f \in (\mathbb{R}^d)^*$ and $p > 1$ then*

$$|f(x)| = \left| \sum_{i=1}^d \phi_i \xi_i \right| \le \left(\sum_{i=1}^d |\xi_i|^p \right)^{1/p} \left(\sum_{i=1}^d |\phi_i|^q \right)^{1/q}.$$

Proof. First we show that if $\alpha, \beta > 0$ and $p > 1$ then

$$\alpha\beta \leq p^{-1}(\alpha^p) + q^{-1}(\beta^q). \tag{1.8}$$

For this, consider the function $\Psi(t) = \beta t - p^{-1}t^p$. Elementary calculus shows that Ψ has an absolute maximum on $[0, \infty)$ at $t = \beta^{1/(p-1)}$. Hence

$$\alpha\beta - p^{-1}\alpha^p \leq \beta \cdot \beta^{1/(p-1)} - p^{-1}\beta^{p/(p-1)} = q^{-1}\beta^q,$$

which establishes (1.8). This inequality clearly holds also if α or β is zero.

For Hölder's inequality itself, first observe that both sides are positively homogeneous in both f and x and so it is sufficient to prove that if $\|x\|_p = 1$ and $\|f\|_q = 1$ then $|f(x)| \leq 1$.

But now

$$|f(x)| = \left|\sum_{i=1}^d \phi_i \xi_i\right| \leq \sum_{i=1}^d |\phi_i||\xi_i| \leq \sum_{i=1}^d \left(p^{-1}|\xi_i|^p + q^{-1}|\phi_i|^q\right),$$

where the last inequality comes from (1.8). Therefore,

$$|f(x)| \leq p^{-1}\sum_{i=1}^d |\xi_i|^p + q^{-1}\sum_{i=1}^d |\phi_i|^q$$
$$= p^{-1} + q^{-1} = 1,$$

which completes the proof. ∎

Proposition 1.1.16 (Minkowski's inequality) *If x, y are in $\mathbb{R}^d$ and $p > 1$ then $\|x + y\|_p \leq \|x\|_p + \|y\|_p$.*

Proof. Using Hölder's inequality twice for the crucial second inequality below, we have the following sequence of calculations:

$$\|x + y\|^p = \sum |\xi_i + \eta_i|^p$$
$$= \sum |\xi_i + \eta_i|^{p-1}|\xi_i + \eta_i|$$
$$\leq \sum |\xi_i + \eta_i|^{p-1}|\xi_i| + \sum |\xi_i + \eta_i|^{p-1}|\eta_i|$$
$$\leq \left(\sum |\xi_i + \eta_i|^{q(p-1)}\right)^{1/q} \left(\sum |\xi_i|^p\right)^{1/p}$$
$$+ \left(\sum |\xi_i + \eta_i|^{q(p-1)}\right)^{1/q} \left(\sum |\eta_i|^p\right)^{1/p}$$
$$= \left(\sum |\xi_i + \eta_i|^p\right)^{1/q} (\|x\|_p + \|y\|_p)$$
$$= \|x + y\|^{p/q}(\|x\|_p + \|y\|_p).$$

Hence $\|x + y\| = \|x + y\|^{p - \frac{p}{q}} \leq \|x\|_p + \|y\|_p$, as required. ∎

1.1 Norm topologies

This shows that the ℓ_p-norm defined by (1.7) satisfies the triangle inequality and so is a norm on $\mathbb{R}^d$.

Hölder's inequality shows, further, that if $f = (\phi_1, \phi_2, \ldots, \phi_d)$ satisfies the inequality $\left(\sum |\phi_i|^q\right)^{1/q} \leq 1$ then $|f(x)| \leq 1$ for all x with $\|x\|_p \leq 1$. In other words, $\|f\|_q \leq 1$ implies $\|f\| \leq 1$, where the last norm denotes the one dual to $\|\cdot\|_p$.

Conversely, suppose we have $\|f\| \leq 1$. Then we may define x by setting $\xi_i := \operatorname{sgn} \phi_i |\phi_i|^{q-1}$ and then let $\tilde{x} := \|x\|_p^{-1} x$. We have

$$\|x\|_p = \left(\sum |\xi_i|^p\right)^{1/p} = \left(\sum |\phi_i|^{p(q-1)}\right)^{1/p} = \left(\sum |\phi_i|^q\right)^{1/p} = \|f\|_q^{q/p}.$$

Hence $\tilde{x} = \|f\|_q^{-q/p} x$. Clearly $\|\tilde{x}\|_p = 1$ and, since $\|f\| \leq 1$, we have

$$1 \geq f(\tilde{x}) = \sum \phi_i \tilde{\xi}_i = \|f\|_q^{-q/p} \sum \phi_i \xi_i = \|f\|_q^{-q/p} \sum |\phi_i|^q$$
$$= \|f\|_q^{-q/p} \|f\|_q^q = \|f\|_q.$$

Thus $\|f\| \leq 1$ if and only if $\|f\|_q \leq 1$. Again, as in Example 1.1.13, this implies that $\|f\| = \|f\|_q$.

We have shown that the norm dual to the ℓ_p-norm is the ℓ_q-norm where $p^{-1} + q^{-1} = 1$. We interpret this to include the previous case where $p = 1$ and $q = \infty$ (the supremum norm) and vice versa. The calculations in Example 1.1.13 show that Hölder's inequality and Minkowski's inequality extend to these cases also.

In the case when $d = 2$, the boundaries of the unit ball for $p = 1, \frac{3}{2}, 2, 4, 10$ and ∞ are shown in Figure 1.2.

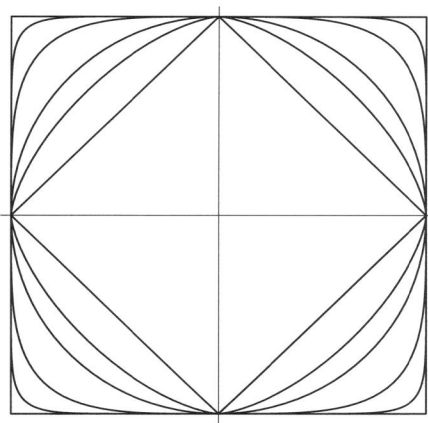

Figure 1.2

The curves for $p > 2$ have been called "super-circles" by Piet Hein (see, e.g., [167], Chap. 18).

Notation. The notation ℓ_p^d is used to denote the Minkowski space $(\mathbb{R}^d, \|.\|_p)$ for $1 \leq p \leq \infty$.

Example 1.1.17 The final example is a little more complicated than and also different from the others. The difference lies in the fact that in the previous examples we began with an expression for the norm. Here we begin with a description of the unit ball and derive a formula for the norm. First consider the d-dimensional simplex S_d with vertices at $e_0 = 0$ and e_i ($i = 1, 2, \ldots, d$) in $\mathbb{R}^d$. If $x = (\xi_1, \xi_2, \ldots, \xi_d)^t$ is in S_d then $\xi_i \geq 0$ and $\sum \xi_i \leq 1$ (and conversely). Now construct the ball $B = S_d - S_d$. This is a convex set which has $d(d+1)$ vertices at the points $e_i - e_j$ ($i, j = 0, 1, \ldots, d; i \neq j$).

In order to investigate the norm of a vector z relative to this ball we regard $\mathbb{R}^d$ as a lattice with the usual pointwise operations. For each $x \in \mathbb{R}^d$ we write $x^+ := x \vee 0$ and $x^- := -(x \wedge 0)$; *i.e.* in terms of coordinates we have

$$(x^+)_i = \begin{cases} \xi_i & \text{if } \xi_i \geq 0, \\ 0 & \text{otherwise} \end{cases} \quad \text{and} \quad (x^-)_i = \begin{cases} -\xi_i & \text{if } \xi_i < 0, \\ 0 & \text{otherwise.} \end{cases}$$

Let $z \in B$; then $z = x - y$ for some $x, y \in S_d$. Therefore, $(z)_i = \zeta_i = \xi_i - \eta_i$. Let $J_1 = \{i : \xi_i \geq \eta_i\}$, $J_2 = \{i : \xi_i < \eta_i\}$. Then

$$(z^+)_i = \begin{cases} \xi_i - \eta_i, & i \in J_1 \\ 0, & i \in J_2 \end{cases} \quad \text{and} \quad (z^-)_i = \begin{cases} \eta_i - \xi_i, & i \in J_2 \\ 0, & i \in J_1. \end{cases}$$

Hence

$$\|z^+\|_1 = \sum_{i \in J_1}(\xi_i - \eta_i) \leq \sum_{i \in J_1} \xi_i \leq \sum_{i=1}^d \xi_i \leq 1,$$

and

$$\|z^-\|_1 = \sum_{i \in J_2}(\eta_i - \xi_i) \leq \sum_{i \in J_2} \eta_i \leq \sum_{i=1}^d \eta_i \leq 1.$$

Thus $\max\{\|z^+\|_1, \|z^-\|_1\} \leq 1$.

Conversely, suppose $z = (\zeta_1, \zeta_2, \ldots, \zeta_d)^t$ is such that $\max\{\|z^+\|_1, \|z^-\|_1\} \leq 1$. Then $z^+ \in S_d$ and $z^- \in S_d$ and hence $z = z^+ - z^- \in B$. Thus we have shown that

$$\|z\|_B = \|z\|_{S_d - S_d} = \max\{\|z^+\|_1, \|z^-\|_1\}. \tag{1.9}$$

In the case when $d = 2$, the set $S_d - S_d$ is an affine regular hexagon with $6 = 2 \times 3$ vertices. This body is shown in Figure 1.3. In the case when $d = 3$, the set $S_d - S_d$ is an affine regular cubo-octahedron. This body has $12 = 3 \times 4$ vertices, 8 triangular facets and 6 square facets. It is illustrated in Figure 1.4.

In general, the facial structure of $S_d - S_d$ is complicated. For each non-empty subset J of $\{1, 2, \ldots, d\}$ there are two opposite facets of $S_d - S_d$ which we denote

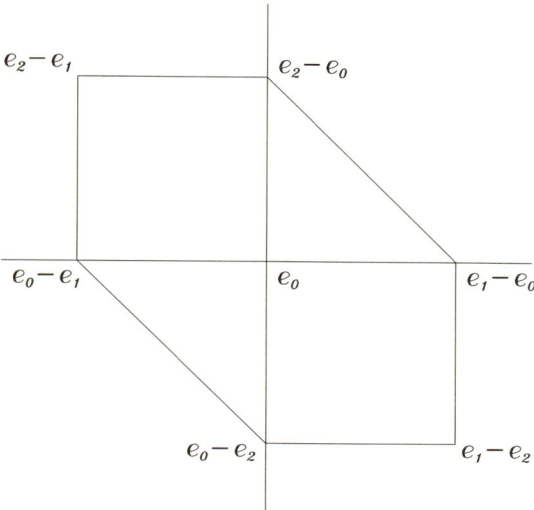

Figure 1.3

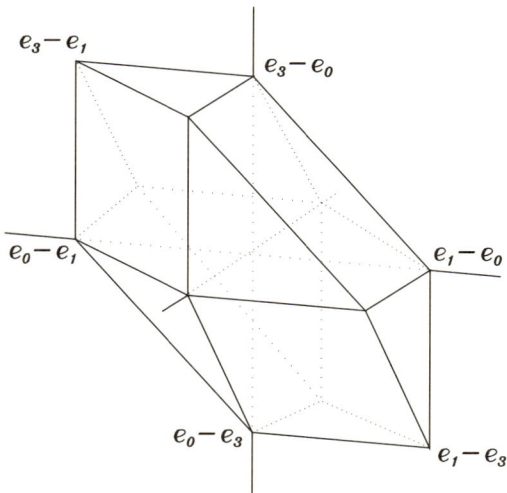

Figure 1.4

by F_J^+ and F_J^-. The facet F_J^+ consists of those vectors x such that $\xi_i \geq 0$ if and only if $i \in J$, $\|x^+\|_1 = 1$, $\|x^-\|_1 \leq 1$. Likewise, F_J^- consists of those x such that $\xi_i \leq 0$ if and only if $i \in J$, $\|x^-\|_1 = 1$, $\|x^+\|_1 \leq 1$. Thus there are $2^{d+1} - 2$ facets in total. When $d = 2$, $2^3 - 2 = 6$ and when $d = 3$, $2^4 - 2 = 14$.

The vertices of F_J^+ are $\{e_i : i \in J\}$ and $\{e_i - e_j : i \in J, j \notin J\}$ and this facet lies in the hyperplane $\{x : \sum_{i \in J} \xi_i = 1\}$. Similar statements apply to F_J^-.

Next we investigate the dual norm. If $f \in (\mathbb{R}^d)^*$ is such that $\|f^+\|_\infty + \|f^-\|_\infty \le 1$ and if $z \in B = S_d - S_d$ then

$$f(z) = \sum \phi_i \zeta_i \le \sum (f^+)_i (z^+)_i + \sum (f^-)_i (z^-)_i$$
$$\le \|f^+\|_\infty \sum (z^+)_i + \|f^-\|_\infty \sum (z^-)_i$$
$$\le \|f^+\|_\infty + \|f^-\|_\infty$$
$$\le 1.$$

Thus $f(z) \le 1$. Similarly

$$f(z) \ge -\sum (f^-)_i (z^+)_i - \sum (f^+)_i (z^-)_i \ge -1.$$

Therefore, $|f(z)| \le 1$ and hence $\|f\| \le 1$, where $\|f\|$ denotes the norm dual to $\|\cdot\|_B$.

Conversely, if $\|f\| \le 1$ then choose i_1 and i_2 such that $\phi_{i_1} = \|f^+\|_\infty$ and $\phi_{i_2} = -\|f^-\|_\infty$. Next define x and y in S_d by setting $\xi_j := \delta_{i_1 j}$ and $\eta_j := \delta_{i_2 j}$ (Kronecker δ's). Then $z := x - y \in B$. Therefore,

$$\|f^+\|_\infty + \|f^-\|_\infty = \phi_{i_1} - \phi_{i_2} = f(z) \le 1.$$

Thus the dual norm $\|f\|$ is defined by the formula

$$\|f\| := \|f^+\|_\infty + \|f^-\|_\infty. \tag{1.10}$$

To see what B° looks like we first recall the notation for line segments introduced in Example 0.2.3(iv). Let $u^* := \sum_{i=1}^d e_i^*$ and let Z_d be the zonotope

$$Z_d := [u^*] + \sum_{i=1}^d [e_i^*]. \tag{1.11}$$

Now suppose f is such that $\|f\| \le 1$ and, as above, let i_1 and i_2 be such that $\phi_{i_1} = \|f^+\|_\infty$ and $\phi_{i_2} = -\|f^-\|_\infty$ and let $t = 2^{-1}(\phi_{i_1} + \phi_{i_2})$. It is easy to see that $-\frac{1}{2} \le t \le \frac{1}{2}$ and also that $|\phi_i - t| \le \frac{1}{2}$ for all i. These inequalities show that $tu^* \in [u^*]$ and $f - tu^* \in \sum_{i=1}^d [e_i^*]$. Consequently $f \in [u^*] + \sum [e_i^*] = Z_d$. Thus $B^\circ \subseteq Z_d$. A similar computation proves the reverse inequality and hence $B^\circ = Z_d$.

The vertices of $B^\circ = Z_d$ are obtained by summing an arbitrary set of ends of its constituent line segments. Thus for each subset J of $\{1, 2, \ldots, d\}$ consider the vector

$$2^{-1} u^* + \sum_{i \in J} \frac{e_i^*}{2} + \sum_{i \notin J} \frac{(-e_i^*)}{2} = \sum_{i \in J} e_i^*$$

and its negative. These vectors are all the vertices of B° yielding $2^{d+1} - 2$ vertices which correspond to the facets of $B = S_d - S_d$. Likewise B° has $d(d+1)$ facets lying in the hyperplanes $\{f : \phi_i = \pm 1\}$ corresponding to the vertices of B. Each facet of Z_d is a parallelotope. One may express Z_d in the form $Z_d = [u^*] + 2^{-1} C_d^*$ (C_d^*

is the unit cube in X^*) and think of it as a cube stretched along an axis joining opposite vertices.

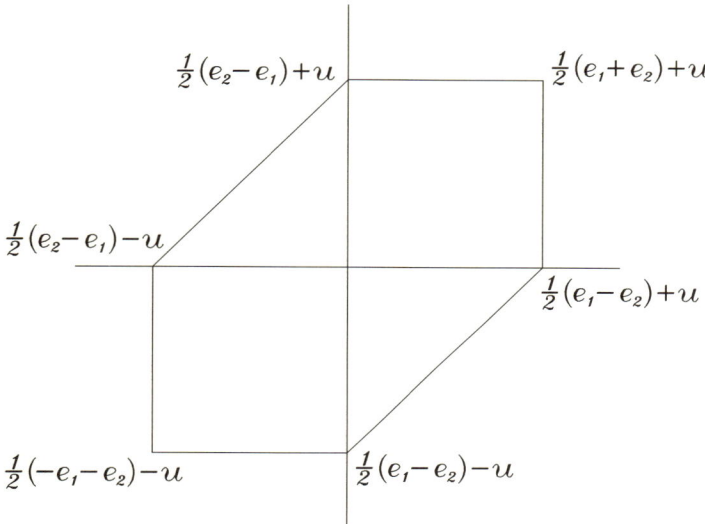

Figure 1.5

When $d = 2$, the body Z_d is an affine regular hexagon (dual to $S_d - S_d$) and when $d = 3$ the body Z_d is an affine regular rhombic dodecahedron; *i.e.* the 12 faces are affine rhombi. These convex bodies are shown in Figures 1.5. and 1.6. Both Z_d and $S_d - S_d$ may be regarded as generalizations of the regular hexagon to higher dimensions.

1.2 The unique linear topology on $\mathbb{R}^d$

In this section we develop the basic facts about the topological properties of a d-dimensional Minkowski space by showing that the topology coincides with the familiar Euclidean one.

Definition 1.2.1 *A topology τ on a linear space X is called a **linear topology** if the vector operations of addition and scalar multiplication (from the appropriate product spaces to X) are continuous.*

Proposition 1.2.2 *If $(X, \|.\|)$ is a normed linear space then the norm topology is a linear topology.*

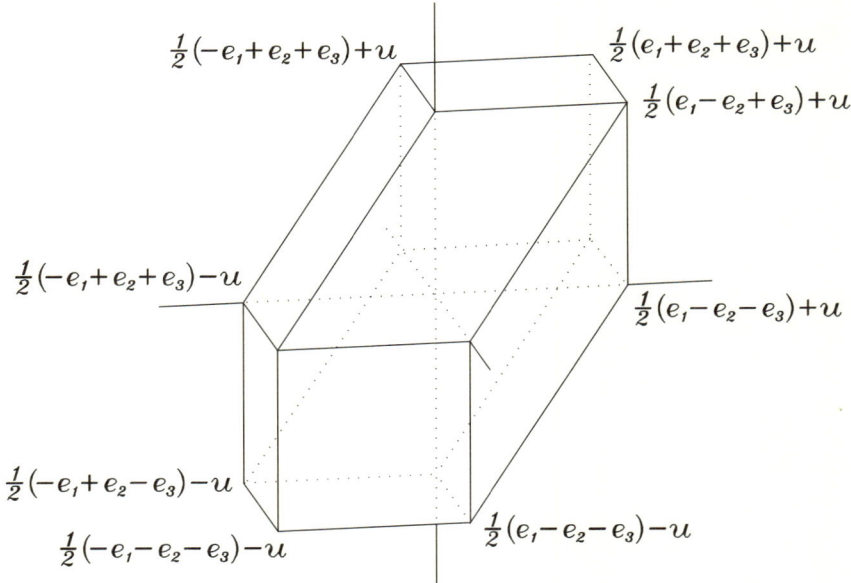

Figure 1.6

Proof. Firstly, if $\|x - x'\| < \epsilon/2$ and $\|y - y'\| < \epsilon/2$ then we have $\|(x + y) - (x' + y')\| = \|(x - x') + (y - y')\| \leq \|x - x'\| + \|y - y'\| < \epsilon$ and hence addition is continuous (even uniformly continuous). Secondly, given $\epsilon > 0$ and a pair (α, x) with $\alpha \neq 0$, $x \neq 0$, we have $\|\alpha x - \beta y\| = \|\alpha(x - y) + (\alpha - \beta)y\| \leq |\alpha|\|x - y\| + |\alpha - \beta|\,\|y\|$. Again, this is less than ϵ if $\|x - y\| < \min\{\epsilon/2\alpha, \|x\|\}$ and $|\alpha - \beta| < \epsilon/4\|x\|$. The inequalities are easier if either α or x is 0. ∎

Now we give the main theorem of this section.

Theorem 1.2.3 *Let ℓ_2^d be d-dimensional Euclidean space with elements $x = (\xi_1, \xi_2, \ldots, \xi_d)^t$ and norm $\|x\| = (\sum_{i=1}^d \xi_i^2)^{1/2}$. If (Y, τ) is a d-dimensional linear space with a Hausdorff linear topology τ then there is a linear homeomorphism from Y onto ℓ_2^d.*

Proof. Since Y is d-dimensional, it has a basis $(y_1, y_2, \ldots, y_d)$. Consider the map $T : \ell_2^d \mapsto Y$ defined by $Tx = T(\xi_1, \xi_2, \ldots, \xi_d)^t = \sum_{i=1}^d \xi_i y_i$. Elementary linear algebra shows that T is linear, injective and maps ℓ_2^d onto Y. Moreover, since the coordinates ξ_i depend continuously on x in ℓ_2^d and since τ is a linear topology, the map T is continuous. It remains to show that T^{-1} is continuous. Again because τ is a linear topology, it is sufficient to show continuity at 0. Suppose that $\{z_\alpha\}$ is a net in Y with $z_\alpha \xrightarrow{\tau} 0$ and $\|T^{-1} z_\alpha\| \geq \epsilon$ for all α. Let $z'_\alpha = \|T^{-1} z_\alpha\|^{-1} z_\alpha$. Then we have $z'_\alpha \to 0$ and $\|T^{-1}z'_\alpha\| = 1$. Because the unit ball in ℓ_2^d is compact, the net $\{x_\alpha\} = \{T^{-1}z'_\alpha\}$ has a convergent subnet $\{x_\beta\}$ with $x_\beta \to \bar{x}$ and $\|\bar{x}\| = 1$. But then $Tx_\beta \to T\bar{x} = \sum_{i=1}^d \bar{\xi}_i y_i$. On the other hand, $Tx_\beta = z'_\beta \to 0$. Because the topology τ is Hausdorff we have $T\bar{x} = 0$. Thus $\sum_{i=1}^d \bar{\xi}_i y_i = 0$ and hence $\bar{\xi}_i = 0$ for all i. Thus $\bar{x} = 0$, which contradicts the fact that $\|\bar{x}\| = 1$. ∎

It follows that in d-dimensional space the notions of open set, compact set and so on are independent of the norm being used. Consequently, more complex ideas such as the σ-algebra of Borel sets are also norm independent.

The core of the proof of Theorem 1.2.3 is to show that once a basis is chosen for a Minkowski space X then the map from X to $\mathbb{R}^d$ induced by that basis is continuous. The next theorem shows how this fact implies that every linear map between Minkowski spaces is bounded (*i.e.* continuous).

Theorem 1.2.4 *If $(X, \|.\|_X)$ and $(Y, \|.\|_Y)$ are two Minkowski spaces and if T is a linear map from X to Y then T is continuous (and hence bounded).*

Proof. Let $(b_1, b_2, \ldots, b_d)$ be a basis for X; each $x \in X$ has a unique representation as $x = \sum_{i=1}^d \xi_i b_i$. By Theorem 1.2.3, the map $x \mapsto (\xi_1, \xi_2, \ldots, \xi_d)^t$ from X to ℓ_2^d is continuous. Further, since $(Y, \|.\|_Y)$ is a topological linear space the map $(\xi_1, \xi_2, \ldots, \xi_d)^t \mapsto \sum_{i=1}^d \xi_i(Tb_i)$ from ℓ_2^d to Y is continuous. Since T is the composition of these two maps it is continuous. ∎

Corollary 1.2.5 *If a finite dimensional space X has two different norms $(X, \|.\|)$ and $(X, \|.\|')$ then those norms are equivalent; i.e. there are positive constants c_1 and c_2 such that*

$$c_1 \|x\| \le \|x\|' \le c_2 \|x\|, \quad \forall x \in X.$$

Proof. The identity map on X is linear and hence, by Theorem 1.2.4 and Proposition 1.1.10, is bounded as a map from $(X, \|.\|)$ to $(X, \|.\|')$. Hence $\|x\|' \le c_2 \|x\|$, $\forall x \in X$. Likewise we can interchange $\|.\|$ and $\|.\|'$. ∎

Norms that are equivalent in the sense of Corollary 1.2.5 generate not only the same topology but also the same uniformity. Therefore concepts such as "bounded", "totally bounded", "differentiability", "Lipschitz mapping" and "Cauchy sequence" which depend on the uniformity are all independent of which norm is used. It follows that in a Minkowski space a set is compact if and only if it is closed and bounded and, equivalently, if and only if it is complete and totally bounded. Since, as is well known, every Euclidean space is complete, so is each Minkowski space. Moreover, since any subspace is a Minkowski space with the relative norm, it must also be complete and hence closed. Alternatively, one can see this as a direct consequence of Theorem 1.2.4 because (as in Proposition 0.1.1) each subspace has an algebraic complement and hence is the kernel of a linear projection, and by Theorem 1.2.4, the kernel of a linear map is closed.

We now make explicit some of the earlier comments about compactness.

Theorem 1.2.6 *If $(X, \|.\|)$ is a Minkowski space then the unit ball is compact and hence X is a locally compact topological space.*

Proof. Let $(b_1, b_2, \ldots, b_d)$ be a basis for X; then each x in X has a unique representation $x = \sum_{i=1}^{d} \xi_i b_i$ and, as in Theorem 1.2.3, the map T defined by

$$Tx := (\xi_1, \xi_2, \ldots, \xi_d)^t$$

from X to $\mathbb{R}^d$ is a homeomorphism of $(X, \|.\|)$ onto ℓ_2^d. By Proposition 1.1.10, T is bounded and hence $T(B)$ is a bounded set in ℓ_2^d. Since T^{-1} is continuous and B is closed, $T(B)$ is also closed and hence compact. Again because T^{-1} is continuous, it follows that B is compact.

Since translations are isometries, the closed ball of radius 1 about each point of X is a compact neighbourhood of that point. ∎

The local compactness given by Theorem 1.2.6 is the basis for all the uniqueness theorems of this chapter. The last theorem of this section is the converse of Theorem 1.2.6. We shall give two proofs of this result, one here using a preliminary lemma of F. Riesz and one in the next section based on the Hahn–Banach theorem. The proof of the Riesz lemma is the first of many instances where we need to replace an easy orthogonal argument from Euclidean space with a surrogate of some sort.

Lemma 1.2.7 *If $(X, \|.\|)$ is a normed space with unit ball B and if L is a closed proper subspace of X then, for each $\epsilon > 0$, there exists $x \in B$ such that $\delta(x, L) := \inf\{\|x - y\| : y \in L\} > 1 - \epsilon$.*

Proof. Since $L \neq X$ there is an $x_0 \in X \setminus L$ and since L is closed $\delta_0 := \delta(x_0, L) > 0$. Choose $y_0 \in L$ such that $\delta_0 \leq \|x_0 - y_0\| < \delta_0 + \epsilon_0$, where $\epsilon_0 < \epsilon \delta_0/(1 - \epsilon)$. This choice is so that $\epsilon_0/(\delta_0 + \epsilon_0) < \epsilon$.

Consider $x_1 := (x_0 - y_0)\|x_0 - y_0\|^{-1}$. Then $x_1 \in B$ and, if $y \in L$, we have

$$\|x_1 - y\| = \|(x_0 - y_0)\|x_0 - y_0\|^{-1} - y\|$$
$$= \|x_0 - y_0\|^{-1}\|x_0 - (y_0 + \|x_0 - y_0\|y)\|$$
$$> (\delta_0 + \epsilon_0)^{-1} \delta_0,$$

since $\|x_0 - y_0\| < \delta_0 + \epsilon_0$ and $(y_0 + \|x_0 - y_0\|y) \in L$.

Thus $\|x_1 - y\| > 1 - \epsilon_0(\delta_0 + \epsilon_0)^{-1}$ and so $\delta(x_1, L) \geq 1 - \epsilon_0(\delta_0 + \epsilon_0)^{-1} > 1 - \epsilon$, as required. ∎

Theorem 1.2.8 *If $(X, \|.\|)$ is a normed linear space such that the unit ball B is compact then X is finite dimensional.*

Proof. Choose $x_1 \in B$ and let $L_1 := \text{span}\{x_1\}$. Then L_1 is finite dimensional and, by the remarks preceding Theorem 1.2.6, is complete and hence is a closed subspace of X.

If $L_1 = X$ the proof is complete. If not, then, by Lemma 1.2.7, there exists $x_2 \in B$ such that $\delta(x_2, L_1) \geq \frac{1}{2}$ and, in particular, $\|x_1 - x_2\| \geq \frac{1}{2}$. Let $L_2 := \text{span}\{x_1, x_2\}$, which is also a finite dimensional closed subspace. If $L_2 = X$ the proof is finished and if not the lemma yields $x_3 \in B$ such that $\|x_3 - x_2\| \geq \frac{1}{2}$ and $\|x_3 - x_1\| \geq \frac{1}{2}$. At some stage this process must stop because otherwise we obtain a sequence $(x_n) \subseteq B$ with $\|x_i - x_j\| \geq \frac{1}{2}$ for all i, j, which contradicts the compactness of B. However, the process stops only if $X = L_n$ for some n, i.e. only if X is finite dimensional. ∎

Returning to Theorem 1.2.4 we see that another consequence is that the space of all continuous linear functionals on a Minkowski space X is the same, as a set, as the space of all linear functionals on X; in other words, the topological dual coincides with the algebraic dual X^* and we may speak of "the dual space X^*". This justifies the assertion in the last section that (1.2) defines a norm on the whole of X^*. In §0.1 we saw that X and X^* have the same dimension and hence are isomorphic as linear spaces. The examples in §1.1 show that it is rare for there to be an isometry between X and X^* (though, as we have just seen, the norms are equivalent) and therefore X and X^* should not be identified as Minkowski spaces. Euclidean space is special in that X is isometrically isomorphic to X^*. However, that statement does not quite capture the essence of the speciality. The situation is rather like that with reflexivity in infinite dimensional spaces where, as James [267] showed, it is not a matter of X and X^{**} being isometrically isomorphic but of the natural map being such an isomorphism. There are non-Euclidean spaces where X and X^* are isometrically isomorphic. Easy examples are given by the plane with a regular $2n$-gon as unit ball (the case $n = 2$ being most familiar). Other examples that we shall meet again in dealing with the case $d = 2$, are the "Radon curves" discussed in detail by Day [124]. A less familiar example is to give $\mathbb{R}^3$ the following polyhedral unit ball B:

$$B = \text{conv}\{(0, \pm 1, \pm 1), (\pm 1, 0, \pm \tfrac{1}{2})\}.$$

This ball is shown in Figure 1.7.

The speciality of Euclidean space lies in the fact that (as was discussed in §0.1 and Example 1.1.10) the isometry is implemented by identifying a basis $(e_1, e_2, \ldots, e_d)$ in X with the dual basis $(e_1^*, e_2^*, \ldots, e_d^*)$ in X^*. This means that if $Te_i = e_i^*$ and if $x = \sum_{i=1}^d \alpha_i e_i$, $y = \sum_{i=1}^d \beta_i e_i$ then $(Tx)(y) = (\sum \alpha_i e_i^*)(\sum \beta_i e_i) = \sum \alpha_i \beta_i$.

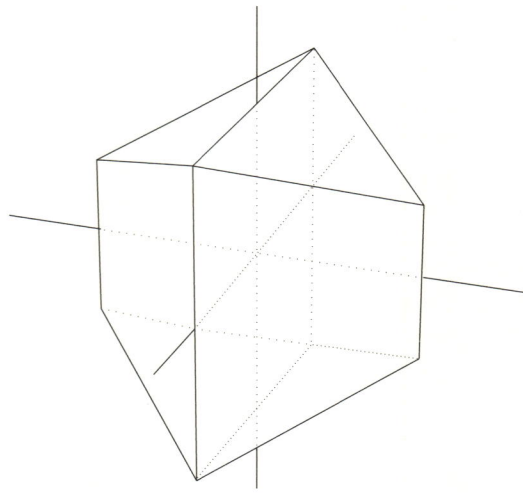

Figure 1.7

1.3 The Hahn–Banach theorem

The Hahn–Banach theorem comes in two forms. The first asserts that there is a continuous linear functional with certain properties and the second that there exists a closed hyperplane with certain properties. The connection, of course, is that a closed hyperplane is a translate of the kernel of a continuous linear functional. The first form is an *extension* theorem since the property required of the functional is that it extends a given functional (defined on a subspace) without increasing the norm. The second form is a *separation* theorem because the property required of the hyperplane is that it "separates" two given convex sets. We shall need both versions. The first one is dealt with here and the second in the next chapter.

In the development of this theorem in finite dimensional spaces it is possible to dispense with the topology and speak only of the "core" and "linear closure" of a convex set, which are both defined in terms of lines. This is the approach of R. B. Holmes [261] and of Valentine [516] in their respective books. This approach begins with a lemma of Stone that states: given disjoint convex sets A and B there exist disjoint convex sets C and D with $A \subseteq C$, $B \subseteq D$ and $C \cup D = X$. On the other hand, most accounts dealing with finite dimensional spaces rely not only on the topology but also on the Euclidean structure in the following way. If K is a closed convex set and x a point not in K then there is a unique nearest point y to x in K. The orthogonal complement of $(x - y)$ is then a hyperplane of the desired kind. A very good account using this approach is the one by Botts [52]. The intermediate approach following Banach's classical proof and using sublinear functionals seems the appropriate one in our context.

1.3 The Hahn–Banach theorem

Definition 1.3.1 *A functional h defined on a linear space X is said to be **sublinear** if*

(i) $h(x_1 + x_2) \leq h(x_1) + h(x_2), \forall x_1, x_2 \in X$;
(ii) $h(\alpha x) = \alpha h(x), \forall x \in X, \alpha \geq 0$.

A norm is a good example of a sublinear functional.

Theorem 1.3.2 (Hahn–Banach) *Suppose that Y is a (proper) subspace of a linear space X and that h is a sublinear functional defined on X. If f is a linear functional defined on Y such that*

$$f(y) \leq h(y), \qquad \forall y \text{ in } Y$$

then there is a linear functional $\tilde{f}$ defined on X such that

(i) $\tilde{f}(y) = f(y), \qquad \forall y \text{ in } Y$;   (ii) $\tilde{f}(x) \leq h(x), \qquad \forall x \text{ in } X$.

Proof. Suppose $x_0 \in X \setminus Y$ and let

$$Y_0 := \operatorname{span}\{Y, x_0\} = \{y + \alpha x_0 : y \in Y, \alpha \in \mathbb{R}\}.$$

If y_1 and y_2 are arbitrary points of Y we have

$$f(y_1) - f(y_2) = f(y_1 - y_2) \leq h(y_1 - y_2) = h((y_1 + x_0) + (-x_0 - y_2))$$
$$\leq h(y_1 + x_0) + h(-x_0 - y_2);$$

hence

$$f(y_1) - h(y_1 + x_0) \leq f(y_2) + h(-x_0 - y_2), \qquad \forall y_1, y_2 \in Y.$$

This inequality implies that

$$\sup\{f(y_1) - h(y_1 + x_0) : y_1 \in Y\} \leq \inf\{f(y_2) + h(-x_0 - y_2) : y_2 \in Y\},$$

and hence there is a real number γ for which

$$f(y_1) - h(y_1 + x_0) \leq \gamma, \qquad \forall y_1 \in Y \qquad (1.12)$$

and

$$\gamma \leq f(y_2) + h(-x_0 - y_2), \qquad \forall y_2 \in Y. \qquad (1.13)$$

Define $\tilde{f}$ on Y_0 by $\tilde{f}(y + \alpha x_0) := f(y) - \alpha \gamma$. With $y_1 = \alpha^{-1} y$, (1.12) yields

$$f(\alpha^{-1} y) - \gamma \leq h(\alpha^{-1} y + x_0)$$

and, therefore, if $\alpha > 0$, $f(y) - \alpha \gamma \leq h(y + \alpha x_0)$. Similarly from (1.13),

$$f(\alpha^{-1} y) - \gamma \geq -h(-\alpha^{-1} y - x_0)$$

and so, if $\alpha < 0$ then
$$f(y) - \alpha\gamma \leq -\alpha h((-\alpha)^{-1}y + (-\alpha)^{-1}(\alpha x_0))$$
$$= h(y + \alpha x_0).$$

In either case we get $\tilde{f}(y + \alpha x_0) \leq h(y + \alpha x_0)$ for all $y \in Y$.

In the case when X is finite dimensional a simple induction now shows that f can be extended to the whole of X as required. In the infinite dimensional case, which we use in Theorem 1.3.10 below, the proof is completed by a Zorn's lemma argument. ∎

Corollary 1.3.3 *If Y is a proper subspace of a Minkowski space X and if f is a linear functional on Y then there is a linear functional $\tilde{f}$ on X such that*

(i) $\tilde{f}(y) = f(y), \forall y$ in Y,
(ii) $\|\tilde{f}\|_{X^*} = \|f\|_{Y^*}$.

Proof. Let $h(x) = \|f\|_{Y^*}\|x\|$. Then h is a sublinear functional on X and $f(y) \leq h(y)$ for all y in Y. The required $\tilde{f}$ is now given by the theorem. ∎

Corollary 1.3.4 (Hahn) *If $x_0 \neq 0$ is a point in a Minkowski space $(X, \|.\|)$ then there exists a linear functional f_0 on X such that $f_0(x_0) = \|x_0\|$ and $f_0(x) \leq \|x\|$ for all x in X.*

Proof. Let Y be the one-dimensional subspace spanned by x_0 and define f on Y by $f(\alpha x_0) = \alpha\|x_0\|$. Then, as an element of Y^*, we have $\|f\| = 1$. Let f_0 be the extension of f guaranteed by Corollary 1.3.3. We have $f_0(x_0) = f(x_0) = \|x_0\|$ and $f_0(x) \leq \|f_0\|\|x\| = \|x\|$ for all x in X. ∎

Recall from Example 0.2.4(ii) the notation for half-spaces: $H[f, \alpha]^+ := \{x : f(x) \geq \alpha\}$. Similarly, $H[f, \alpha]^- := \{x : f(x) \leq \alpha\}$.

Definition 1.3.5 *A hyperplane $H := \{x : f(x) = \alpha\}$ is said to be a **support hyperplane** for a closed convex set K if $K \cap H \neq \emptyset$ and either $K \subseteq H[f, \alpha]^+$ or $K \subseteq H[f, \alpha]^-$. In this case we say that the linear functional f **supports** the convex set K at each point x in $K \cap H$.*

Corollary 1.3.6 *If $(X, \|.\|)$ is a Minkowski space and if $\|x_0\| = 1$ (i.e. $x_0 \in \partial B$) then there exists a linear functional f_0 with $\|f_0\| = 1$ and which supports B at x_0.*

Proof. With f_0 as in Corollary 1.3.4 and $H = \{x : f_0(x) = 1\}$ it is clear that H is a support hyperplane of B at x_0. ∎

1.3 The Hahn–Banach theorem

Corollary 1.3.7 *If x is a point in a Minkowski space $(X, \|.\|)$ then*
$$\|x\| = \sup\{|f(x)| : f \in B^\circ\}.$$

Proof. If $\|f\| \leq 1$ then $|f(x)| \leq \|f\| \|x\| \leq \|x\|$. On the other hand, by Corollary 1.3.4, there exists $f_0 \in B^\circ$ with $f_0(x) = \|x\|$. ∎

Just as Equation (1.2) defines a norm on X^*, in the same way we can define a norm on X^{**} by
$$\|\Phi\| := \sup\{|\Phi(f)| : f \in B^\circ\}, \qquad \forall \Phi \in X^{**}. \tag{1.14}$$

We have already seen in §0.1 that the mapping J defined by (0.4) is an algebraic isomorphism of X onto X^{**}. We show now that it is an isometry.

Theorem 1.3.8 *If $(X, \|.\|)$ is a Minkowski space and if X^{**} is given its natural norm defined by (1.14), then the mapping J of X onto X^{**} is an isometry.*

Proof. Using first Equation (1.14), then (0.4), and then Corollary 1.3.7, we have
$$\|Jx\| = \sup\{|Jx(f)| : f \in B^\circ\} = \sup\{|f(x)| : f \in B^\circ\} = \|x\|. \qquad \blacksquare$$

This theorem might be called "the fundamental theorem of Minkowski geometry". It asserts that X and X^{**} are identical as Minkowski spaces and that there is genuinely no problem if we forget the mapping J. Another way to look at it is that Minkowski spaces come in *unordered* pairs $(X, \|.\|_B)$, $(X^*, \|.\|_{B^\circ})$ in which either one is the dual space of the other. In the language of functional analysis the theorem says that all Minkowski spaces are reflexive. But it is stronger than that because the dual space X^* is the space of all linear functionals on X.

Equation (0.5) defined the dual T^* of a linear transformation T. Theorem 1.2.4 shows that both T and T^* are bounded. Corollary 1.3.4 can be used to show that T and T^* have the same norm.

Theorem 1.3.9 *If $(X, \|.\|_X)$ and $(Y, \|.\|_Y)$ are two Minkowski spaces and if T is a linear transformation from X to Y with dual transformation T^* then $\|T\| = \|T^*\|$.*

Proof. For all $g \in Y^*$ and all $x \in X$, we have
$$|(T^*g)(x)| = |g(Tx)| \leq \|g\| \|Tx\|_Y \leq \|g\| \|T\| \|x\|_X$$
and hence $\|T^*g\| \leq \|T\| \|g\|$, which implies that $\|T^*\| \leq \|T\|$.

Conversely, there exists $x_0 \in X$ with $\|x_0\|_X = 1$ and $\|Tx_0\|_Y = \|T\|$. Moreover, by Corollary 1.3.4 there is $g_0 \in Y^*$ such that $\|g_0\| = 1$ and $g_0(Tx_0) = \|Tx_0\|_Y = \|T\|$. Hence
$$\|T^*\| \geq |T^*g_0(x_0)| = |g_0(Tx_0)| = \|T\|. \qquad \blacksquare$$

Reference has been made to various characterizations of finite dimensional spaces. The first, purely algebraic one is that the mapping J of X into X^{**} (the second algebraic dual) is onto if and only if X is finite dimensional. The second is that there are no discontinuous linear functions on $(X, \|.\|)$ if and only if X is finite dimensional. The third is that the unit ball B is compact (and hence all closed and bounded sets are compact) if and only if $(X, \|.\|)$ is a Minkowski space (Theorems 1.2.6 and 1.2.8 above). As an application of the Hahn–Banach theorem we give a second proof of Theorem 1.2.8.

Theorem 1.3.10 *If $(X, \|.\|)$ is a normed linear space and if B is compact then X is finite dimensional.*

Proof. For each $f \in X^*$ (the set of all continuous linear functionals) consider the closed set $N_f := \{x \in X : \|x\| = 1 \text{ and } f(x) = 0\} = f^\perp \cap \partial B$.

By Corollary 1.3.4 (true for arbitrary normed spaces), for each $x \in X$ with $\|x\| = 1$ there is an $f_x \in X^*$ with $f_x(x) = 1$. Hence $\bigcap_{f \in X^*} N_f = \emptyset$. Since B (and hence ∂B) is compact, there is a finite set $\{f_1, f_2, \ldots, f_d\}$ in X^* such that $\bigcap_{i=1}^{d} N_{f_i} = \emptyset$. Define the map $T : X \mapsto \mathbb{R}^d$ by

$$Tx := (f_1(x), f_2(x), \ldots, f_d(x))^t.$$

Clearly T is linear. Moreover, if $y \neq 0$ then $x = \|y\|^{-1} y \in \partial B$ and so there exists i such that $f_i(x) \neq 0$ and hence $Ty \neq 0$ so that T is one–one. Hence $\dim X \leq d$. ∎

Finally, as a sequel to Corollary 1.3.6, we give the following definition.

Definition 1.3.11 *The unit ball B is said to be **smooth** at $x_0 \in \partial B$ if there exists precisely one linear functional f_0 such that $f_0(x_0) = \|f_0\| = 1$.*

1.4 The existence and uniqueness of Haar measure

Another consequence of local compactness is the existence of a translation-invariant, regular Borel measure that is finite on compact sets and positive on open sets. Moreover, this measure is unique up to a scalar factor. The existence and essential uniqueness of such a Haar measure plays a fundamental role in all the later discussions of area and volume in Minkowski spaces. One can think of this Haar measure as being defined on the group of translations of the Minkowski space. It is also possible to construct Haar measure on the non-abelian group of rotations of the Euclidean space $(\mathbb{R}^d, \|.\|_2)$. Haar [234] pointed out that the underlying structure is that of a locally compact topological group and that given such a group one can always construct a unique (up to scalar multiplication) measure that is invariant under the action of the group, is finite on compact sets and is strictly positive on open sets.

1.4 The existence and uniqueness of Haar measure

An account of this theory is found in books dealing with harmonic analysis and there the emphasis is often on the integral as a linear functional. The book on measure theory by Cohn [114] contains an account that emphasizes the more geometric idea of the measure. Because of its central importance in the sequel, and because it should be stressed that the fundamental concepts of Minkowski geometry are all independent of any Euclidean structure, we give a construction of Haar measure. We closely follow Cohn's treatment. The main difference from Cohn is that we will assume the group to be abelian and use additive notation. This allows a reader unfamiliar with topological groups to think of the group as a Minkowski space and to regard the neighbourhoods U of 0 as balls $B(0, \delta)$.

Definition 1.4.1 *Let* **G** *be an abelian topological group. Then a regular Borel measure μ is called a* **Haar measure** *on* **G** *if it has the following additional properties:*

(i) $\mu(K) < \infty$ *for each compact subset K of* **G**;
(ii) $\mu(U) > 0$ *for each open set U of* **G**;
(iii) μ *is invariant, i.e.* $\mu(a + A) = \mu(A)$ *for all $a \in$* **G**, *and all Borel sets $A \subseteq$* **G**.

Theorem 1.4.2 *If* **G** *is a locally compact abelian topological group then there exists a Haar measure on* **G**.

Proof. Our main aim is to construct a non-trivial additive function on the compact subsets of **G**. Once this is achieved there are standard procedures for obtaining from it an outer measure with respect to which Borel sets are measurable.

Given an open set U, we get an idea of the size of a compact set K (relative to U) by how many translates of U are needed to cover K. But this measure of size is crude if U is large, much better if U is small. This suggests taking "a limit" over the filter of open neighbourhoods of the identity 0. Let $\mathcal{K}$ denote the family of all compact subsets of **G** (including $\emptyset$) and let $\mathcal{U}$ denote the family of open neighbourhoods of 0. Since **G** is locally compact there is an element K_0 of $\mathcal{K}$ which has non-empty interior. Such an element K_0 is chosen and kept fixed for the rest of the discussion. This set will be used as a measuring "standard".

Let $K \in \mathcal{K}$. Since, for each $U \in \mathcal{U}$, $\bigcup_{x \in \mathbf{G}}(x + U) = \mathbf{G}$ we certainly have $K \subseteq \bigcup_{x \in \mathbf{G}}(x + U)$. Hence there are finite subsets F of **G** such that $K \subseteq \bigcup_{x \in F}(x + U)$. Let $n(K, U)$ be the minimal cardinal of such an F and let $q_U(K) := n(K, U)/n(K_0, U)$. Thus q_U is a function from $\mathcal{K}$ into $\mathbb{Q}$ which is 0 only for $K = \emptyset$. Before we list several other properties of q_U, note that $n(K, V)$ is also defined (as above) for any set V with non-empty interior, in particular for K_0.

(1) $0 \leq q_U(K) \leq n(K, K_0)$,
(2) $q_U(K_0) = 1$,

(3) $q_U(x+K) = q_U(K)$,
(4) if $K_1 \subseteq K_2$, then $q_U(K_1) \leq q_U(K_2)$,
(5) $q_U(K_1 \cup K_2) \leq q_U(K_1) + q_U(K_2)$,
(6) if $(K_1 - U) \cap (K_2 - U) = \emptyset$ then $q_U(K_1 \cup K_2) = q_U(K_1) + q_U(K_2)$.

The first inequality in (1) and properties (2), (3), (4) and (5) are all straightforward. If $K \subseteq \bigcup_i x_i + K_0$ and $K_0 \subseteq \bigcup_j y_j + U$ then $K \subseteq \bigcup_{i,j} x_i + y_j + U$ and so

$$n(K, U) \leq n(K, K_0) n(K_0, U),$$

which proves the second inequality in (1). For (6), suppose that $(K_1 - U) \cap (K_2 - U) = \emptyset$ and let $\{x_i + U\}_{i=1}^n$ be a minimal cover for $K_1 \cup K_2$. Each set $x_i + U$ meets either K_1 or K_2 but not both. So this cover can be partitioned into two, one piece covering K_1 and the other K_2 and so $n(K_1, U) + n(K_2, U) \leq n(K_1 \cup K_2, U)$, which, with (5), completes the proof of (6).

Now comes the limit step. Since the dependence of $q_U(K)$ on U is complicated, this step is an abstract existence one. For each $K \in \mathcal{K}$, let I_K be the interval $[0, n(K, K_0)]$ and let $C_\mathcal{K} := \prod_{K \in \mathcal{K}} I_K$ with the product topology. By the Tychonoff theorem $C_\mathcal{K}$ is compact. Moreover, (1) states that each of the functions q_U is an element of this product space.

For each $V \in \mathcal{U}$ let $Q(V) := c\ell\{q_U : U \in \mathcal{U} \text{ and } U \subseteq V\}$ (closure in $C_\mathcal{K}$). If $V_1, V_2, \ldots, V_n \in \mathcal{U}$ then $V_0 := \bigcap_{i=1}^n V_i$ is a non-empty neighbourhood of 0 and so $q_{V_0} \in Q(V_i)$ for all i; i.e. $\bigcap_{i=1}^n Q(V_i)$ is not empty. Hence, by the compactness of $C_\mathcal{K}$, $\bigcap_{V \in \mathcal{U}} Q(V) \neq \emptyset$. Choose an element q from this intersection. This q is the "limit" of q_U over $\mathcal{U}$.

Corresponding to the properties (1)–(6) of q_U we have the following properties of q:

(1') $q(K) \geq 0$,
(2') $q(K_0) = 1$,
(3') $q(x + K) = q(K)$,
(4') if $K_1 \subseteq K_2$ then $q(K_1) \leq q(K_2)$,
(5') $q(K_1 \cup K_2) \leq q(K_1) + q(K_2)$,
(6') if $K_1 \cap K_2 = \emptyset$ then $q(K_1 \cup K_2) = q(K_1) + q(K_2)$.

Since $q \in C_\mathcal{K}$, property (1') is clear from the definition of $C_\mathcal{K}$. Properties (2')–(5') all depend on the fact that evaluation maps on product spaces are continuous (by definition of the product topology). We illustrate the argument with (4'): the map $f \mapsto f(K_2) - f(K_1)$ from $C_\mathcal{K}$ to $\mathbb{R}$ is continuous and, by (4), if $K_1 \subseteq K_2$ it is positive for all q_U. Hence, by continuity, it is positive for all $f \in Q(V)$ and hence for q.

The proof of (6') needs a preliminary step. If $K_1 \cap K_2 = \emptyset$ there exist disjoint open sets G_1 and G_2 such that $K_i \subseteq G_i$. Hence there exist $U_i \in \mathcal{U}$ such that

1.4 The existence and uniqueness of Haar measure

$K_i + U_i \subseteq G_i$. Let $V := U_1 \cap U_2$. Then $(K_1 + V) \cap (K_2 + V) = \emptyset$. So for $U \in \mathcal{U}$ with $-U \subseteq V$ we have (by (6))

$$q_U(K_1 \cup K_2) = q_U(K_1) + q_U(K_2).$$

Again, by the continuity of the evaluation maps, this holds for all $f \in Q(-V)$ and hence for q.

Now we are ready for measure-theoretic arguments, which, because they are fairly standard, will not be presented in detail.

Define μ^* on all open subsets of **G** by

$$\mu^*(G) := \sup\{q(K) : K \subseteq G, K \in \mathcal{K}\} \tag{1.15}$$

and extend to all subsets of **G** by

$$\mu^*(S) := \inf\{\mu^*(G) : S \subseteq G, G \text{ open}\}. \tag{1.16}$$

It is clear that μ^* is non-negative and monotone and that $\mu^*(\emptyset) = 0$. To show that μ^* is an outer measure we need only establish the countable subadditivity. For this (by a standard argument) it is sufficient to prove subadditivity on the open sets.

Let $\{G_i\}$ be a countable collection of open subsets of **G** and let K be a compact subset of $\bigcup G_i$. Then there is a finite subcollection such that

$$K \subseteq \bigcup_{i=1}^{n} G_1$$

and hence compact subsets $K_1, K_2, \ldots, K_n$ such that $K_i \subseteq G_i$ and $K = \bigcup_{i=1}^{n} K_i$. Then, by (5'), (1.15) and the non-negativity of μ^*, we get

$$q(K) \leq \sum_{i=1}^{n} q(K_i) \leq \sum_{i=1}^{n} \mu^*(G_i) \leq \sum_{i=1}^{\infty} \mu^*(G_i).$$

Now use (1.15) again to get $\mu^*(\bigcup G_i) \leq \sum_{i=1}^{\infty} \mu^*(G_i)$. Thus μ^* is an outer measure.

We next show that each Borel set is measurable. By using the Carathéodory condition and (1.16) it is sufficient to show that

$$\mu^*(G_1) = \mu^*(G_1 \cap G_2) + \mu^*(G_1 \setminus G_2), \qquad \forall G_1, G_2 \text{ open and } \mu^*(G_1) < \infty.$$

Let $\epsilon > 0$ be given. Choose a compact subset K of $G_1 \cap G_2$ with

$$q(K) > \mu^*(G_1 \cap G_2) - \epsilon \qquad \text{(Equation (1.15))},$$

and a compact subset K' of $G_1 \setminus K$ with

$$q(K') > \mu^*(G_1 \setminus K) - \epsilon.$$

Then we have $K' \cap K = \emptyset$ and, because $G_1 \setminus G_2 \subseteq G_1 \setminus K$, we also have

$$q(K') > \mu^*(G_1 \setminus G_2) - \epsilon.$$

From these inequalities and (6') we get

$$\mu^*(G_1 \cap G_2) + \mu^*(G_1 \setminus G_2) - 2\epsilon < q(K') + q(K) = q(K' \cup K) \le \mu^*(G_1).$$

Since ϵ is arbitrary, this is the required inequality. Thus the restriction of μ^* to the Borel sets $\mathcal{B}$ is a Borel measure μ.

If G is open and $K \subseteq G$ then $q(K) \le \mu(G)$ and so, by (1.16), we have $q(K) \le \mu(K)$. Also, if G is open with $\overline{G} := c\ell G$ compact then if $K \subseteq G \subseteq \overline{G}$ we have, from (4'),

$$q(K) \le q(\overline{G})$$

and so, from (1.15),

$$\mu(G) \le q(\overline{G}).$$

Combining these, we get

$$q(K) \le \mu(K) \le \mu(G) \le q(\overline{G}).$$

Since $q(\overline{G})$ is finite, this shows that $\mu(K)$ is finite for all compact sets K.

The regularity of μ also follows from these inequalities and Equations (1.15) and (1.16). The invariance of q (i.e. property (3')) also easily carries over to μ via (1.15) and (1.16). Since $\mu(K_0) \ge q(K_0) = 1$, μ is non-trivial. But, more than this, if G is any open set, a finite number of translates $\{x_i + G\}$ covers K_0. Since all these sets have the same measure and their union has positive measure, each one has positive measure, i.e. $\mu(G) > 0$ for all open sets G.

This concludes the existence part of the proof. ∎

The proof of uniqueness, up to a scalar factor, is easier in the commutative than in the non-commutative case. In the latter case, the principles are the same but the technical details are more complicated.

Theorem 1.4.3 *If* **G** *is a locally compact abelian topological group and if μ and ν are two Haar measures on* **G** *then there is a constant c such that $\mu = c\nu$.*

Proof. We need three basic facts. Firstly, if g is continuous and non-negative on **G** with compact support and $g \ne 0$, then $\int g \, d\nu > 0$. This is because there is an ϵ and an open set U such that $g \ge \epsilon \chi_U$ and $\mu(U) > 0$ (χ_U denotes the characteristic function of U). Secondly, if λ is a regular Borel measure ($\lambda \ne 0$) then there is a continuous function f with compact support such that $\int f \, d\lambda \ne 0$. This follows from the regularity (there is a compact set K with $\lambda(K) \ne 0$) and Urysohn's lemma to construct a suitable continuous f with compact support. Thirdly, we need Fubini's theorem on interchange of order of integration.

With these preliminaries, let f be an arbitrary continuous function with compact support and g a fixed, non-negative continuous function with compact support

1.4 The existence and uniqueness of Haar measure

($g \neq 0$). Then consider $\int f \, d\mu \int g \, d\nu$. First write this product as an iterated integral and then use several changes of variable linked to the translation invariance of μ and ν as follows:

$$\int f \, d\mu \int g \, d\nu = \int\int f(x)g(y)\mu(dx)\nu(dy)$$
$$= \int\int f(x+y)g(y)\mu(dx)\nu(dy)$$
$$= \int\int f(y)g(y-x)\nu(dy)\mu(dx)$$
$$= \int\int f(y)g(-x)\mu(dx)\nu(dy)$$
$$= \int f \, d\nu \int \tilde{g}(d\mu),$$

where $\tilde{g}(x) := g(-x)$. In the second equation we used the substitution $x \mapsto x + y$ and the translation invariance of μ; in the third we first interchanged the order of integration to allow the substitution $y \mapsto y - x$; in the fourth we again changed the order and used the substitution $x \mapsto x + y$ again; finally we replaced the iterated integral by a product once more.

Thus if we define c by $c := \int \tilde{g} \, d\mu / \int g \, d\nu$ ($\int g \, d\nu \neq 0$ by the first basic fact) then $\int f \, d\mu = c \int f \, d\nu$, i.e. $\int f \, d(\mu - c\nu) = 0$ for all f. Hence, by the second of our basic facts, $\mu - c\nu = 0$ and $\mu = c\nu$. ∎

Following the "fundamental theorem" (Theorem 1.3.8), it was pointed out that Minkowski spaces occur in dual pairs. In some ways it is more natural to consider a product measure on $X \times X^*$ than a single Haar measure on X (see Saint Raymond [453]).

Recall from §0.1 that if T is a non-singular linear transformation on X then there is a dual transformation T^* on X^* given by Equation (0.5). Moreover, if T transforms a basis $(b_1, b_2, \ldots, b_d)$ in X to $(Tb_1, Tb_2, \ldots, Tb_d)$ then the corresponding dual bases on X^* are $(b_1^*, b_2^*, \ldots, b_d^*)$ and $(T^{*-1}b_1^*, T^{*-1}b_2^*, \ldots, T^{*-1}b_d^*)$ respectively.

Let λ (for Lebesgue) denote a member of the family of Haar measures on X. If $(b_1, b_2, \ldots, b_d)$ is a basis for X let $\sum[0, b_i]$ denote the parallelotope spanned by the basis vectors, i.e.

$$\sum[0, b_i] := \left\{ x \in X : x = \sum_{i=1}^{d} \alpha_i b_i, \ 0 \leq \alpha_i \leq 1 \right\}.$$

If $(b_1^*, b_2^*, \ldots, b_d^*)$ is the dual basis in X^*, then there is a unique Haar measure λ^* on X^* such that

$$\lambda\left(\sum[0, b_i]\right) \lambda^*\left(\sum[0, b_i^*]\right) = 1. \tag{1.17}$$

Proposition 1.4.4 *The measure λ^* is independent of the choice of basis $(b_1, b_2, \ldots, b_d)$ in X.*

Proof. If T is any non-singular linear transformation on X then
$$\lambda\left(\sum [0, Tb_i]\right) = \det T \, \lambda\left(\sum [0, b_i]\right).$$
Since $(T^{*-1} b_i : i = 1, 2, \ldots, d)$ is the corresponding dual basis, we have
$$\lambda^*\left(\sum [0, T^{*-1} b_i^*]\right) = \det(T^{*-1}) \lambda^*\left(\sum [0, b_i^*]\right)$$
$$= (\det T)^{-1} \lambda^*\left(\sum [0, b_i^*]\right).$$
Thus, the product $\lambda(\sum [0, Tb_i]) \lambda^*(\sum [0, T^{*-1} b_i^*]) = 1$ also. ∎

It is also clear that if λ is changed to some other Haar measure by multiplication by $\alpha \neq 0$, then to preserve Equation (1.17) the measure λ^* is multiplied by α^{-1}. Hence the product measure $\lambda \times \lambda^*$ on $X \times X^*$ is a *uniquely* defined measure satisfying (1.17) for *any* choice of basis in X and corresponding dual basis in X^*.

1.5 Notes

A good source for the early history of functional analysis is the book by Monna [391]. In the development of the idea of a normed linear space he traces two main strands ([391], p. 133):

There is the first line, leading from Fredholm via Hilbert, Schmidt, Riesz, Helly to Hahn and Banach and the Polish school where the explosive development of functional analysis started. ... These developments lie mainly in the period 1900 to 1930, culminating in 1932, the year that Banach's famous book was published.

There is the second approach from the Italian mathematicians Peano, Pincherle, Volterra, who refer in their work to Laguerre and Grassmann. Their work lies – at least as far as the introduction to functional analysis is concerned – mainly before 1900. ... Then follow in the first years of the 20th century the French mathematicians Hadamard and Fréchet whose work – as far as functionals and general analysis are concerned – is based on that of the Italian school.

In the development outlined by Monna above, the focus was on infinite dimensional spaces. The interest was in systems of linear equations with infinitely many unknowns and in the various function spaces. Early instances of the latter occur in the work of F. Riesz, who discusses "die Klasse [L^p] ..." in [441] and then introduces these spaces more specifically in [442]. Dunford and Schwartz [130] remark (p. 372) that on *finite* dimensional spaces "the study of norms other than Euclidean is primarily due to H. Minkowski". S. Mazur [361] uses the terms

Minkowski functional and *Minkowski space* perhaps coining both of them. It is, however, in Banach's work (his thesis published in 1922 [19] and his book of 1932 [21]) that these ideas achieve their modern form.

The axioms for a norm may be weakened and the resulting structures investigated. A *semi-norm* is a functional which satisfies Definition 1.1.1 except for the "only if" part of (i). In infinite dimensional spaces a Hausdorff topology generated by a family of semi-norms is an important type of topology (the *locally convex* topologies). See, *e.g.*, Köthe [301]. In finite dimensional spaces Theorem 1.2.3 shows that such a topology is equivalent to the usual one and it is, in fact, generated by a single norm. More significantly for finite dimensional spaces, axiom (ii) may be weakened to only require the positive homogeneity of $\|.\|$; *i.e.* for $\alpha \geq 0$ we have $\|\alpha x\| = \alpha \|x\|$. The unit ball then need not be symmetric and the "metric" generated by the norm need not be symmetric. Spaces with non-symmetric distance have been investigated by Zaustinsky [539]. The special case arising from a non-symmetric norm in the plane has been studied by Guggenheimer [226, 227]. On page 2 of [539] Zaustinsky comments: "One would have expected that the suppression of this hypothesis would entail the loss of most of the theory. We have the very surprising result, however, that nearly every theorem (so far attempted) of the symmetric theory which can be formulated at all without the assumption of symmetry holds without this assumption".

The relationship between continuity and boundedness is one of the themes of Banach's thesis [19]. Although he speaks of the smallest number M such that $\|F(x)\| \leq M\|x\|$ he does not (in [19]) explicitly define the norm of an operator. The notion of dual space ("polare Raum") and dual norm ("polare Maß Bestimmung") are introduced by Hahn [237], although in a footnote he refers to Helly [254]. In Banach's book [21] there is a systematic study of isometries between linear spaces, including the key fact (due to Mazur and Ulam [362]) that an isometry onto the range space is affine.

Norms on vector spaces induce, via Definition 1.1.9, norms on the spaces of linear operators between vector spaces, *i.e.* norms on spaces of matrices. While vector spaces of matrices may properly be regarded as a subclass of Minkowski spaces, the special considerations which enter the discussion because of the multiplicative structure are considerably different from the spirit of the present work. The reader interested in these topics might begin by looking at the recent survey of C.-K. Li [323].

The inequality due to Hölder [260] dates from 1889 and was generalized to integrals by F. Riesz [441]. A thorough survey of these and many related inequalities is given by Hardy, Littlewood and Pólya [244]. The idea used in Example 1.1.17 to look at $\mathbb{R}^d$ as a lattice and consider positive and negative parts and then take combinations of various norms on each part was exploited fully by Bynum [88] to construct examples of normed spaces with certain deep and surprising properties. Norms whose unit balls are affine rhombic dodecahedra have been studied by Ruiz

[449]. Leichtweiss [317] and Grünbaum [222] discuss properties of balls which are "close" to $S_d - S_d$.

The main result of §1.2 that there is only one Hausdorff linear topology on $\mathbb{R}^d$ was stated explicitly by Tychonoff [514] but the equivalence of two norms on $\mathbb{R}^d$ was given by Mazur [361]. The properties of a space which is isomorphic to its dual space were investigated by Leichtweiss [319] (without assuming the ball is symmetric).

The Hahn–Banach theorem was the final product of a long development. It originates from the study of systems of linear equations in infinitely many unknowns and from the "moment problem". In both areas the need to prove the existence of linear functionals with certain properties arose (see Monna [391] for more details of this development). The formulation of Corollary 1.3.4 and the statement that if Y is a proper closed subspace of X then there is a non-zero functional on X that vanishes on Y are due to Hahn [237]. In that paper Hahn also noted Corollary 1.3.7 and its consequence that the natural mapping of X into its second dual is an isometry. However, the generality of Theorem 1.3.2 and its formulation in terms of a linear functional dominated by a sublinear functional (thereby avoiding the immediate use of continuity) is due to Banach [20]. The proof in §1.3 is from [21].

The more geometric view of the Hahn–Banach theorem as a theorem asserting the existence of a hyperplane separating two convex sets also has a lengthy history which is somewhat disjoint from that of the analytic version. Since the account of this version is in the next chapter, the notes on its development will also be given there.

In 1933 Haar [234] proved the existence of a left invariant measure on a locally compact group with a countable basis. His proof (reformulated in terms of the integral as a linear functional) was extended to arbitrary locally compact groups by Weil [527] and by Kakutani [280], who also established the uniqueness. Von Neumann also gave proofs of both existence and uniqueness for compact groups [524] and then of uniqueness for locally compact groups with a countable basis [525]. All of these proofs use the axiom of choice to establish the existence, although the subsequent proof of uniqueness shows that its use cannot be crucial for the proof. H. Cartan [92] gave a simultaneous proof of existence and uniqueness that avoids the use of the axiom of choice. The proof given in §1.4 is Weil's proof, further modified by Halmos [239] and Cohn [114]; see also Husain [264]. Cartan's proof is given by Hewitt and Ross [259]. Since §1.4 is concerned only with the abelian case, the uniqueness part of the proof is quite elementary. All the references just given have proofs of uniqueness for the non-commutative case. A particularly strong version of the uniqueness was given by Loomis [331]. The idea of considering a pair of Haar measures, one in X and the other in X^*, occurs in Saint Raymond [453]. It is a particularly appropriate idea for Saint Raymond's purpose, which is to consider the volume product of convex bodies. That concept will be defined in Chapter 2 and play a major part in Chapter 6.

2
Convex bodies

As was stated in the introduction to Chapter 0, it is difficult to discuss the theory of convex sets without using such terms as *closed*, *bounded* and *interior point*. It is clear from Theorem 1.2.3 that, for a finite dimensional space X, these terms refer to the unique Hausdorff linear topology on X. One may think of this topology as coming from a Euclidean structure on X or from an arbitrary norm on X.

In the subsequent chapters there will be occasion to use a wide assortment of material from the theory of convex bodies. While it is not assumed that the reader is familiar with this material, to include a complete discussion of all the results that we shall need (especially the various inequalities relating to volumes and mixed volumes) would require a separate book. In this chapter we shall therefore only summarize the material, stating the main results, giving outlines for the proofs of some of them and indicating where complete proofs can be found. A full account of the material is contained in several places. The first systematic account is in Bonnesen and Fenchel, of which there is a relatively recent English translation [51]. The book of Eggleston [138] contains most of what we need. The best and most up-to-date reference is the book by Schneider [479], which also contains a wealth of references to the original sources.

The chapter begins with a section dealing with the various geometric versions of the Hahn–Banach theorem which were alluded to in §1.3. This is followed by a discussion of support functions and polar reciprocals for convex sets. Next comes a section on volumes and mixed volumes which will be heavily used but where few proofs are given. The fourth section deals with what are sometimes called "hyperspaces", *i.e.* with the way in which a metric on a set A generates a metric (the Hausdorff metric) on various spaces of subsets of A. In particular, we are interested in the notion of distance between convex sets. This naturally leads to the idea of distance between two different unit balls and hence to a distance between normed spaces. In the final section we look at the consequences which the local compactness of the linear space X has for the topology generated by the Hausdorff metric on the space of compact subsets of X. We show first that, although the Hausdorff metrics arising from different norms are different, they generate the same topology. Next there is the question of the approximation of convex bodies by special classes of convex sets (polytopes and smooth bodies). The main theorem of the section is a compactness result which is called the Blaschke selection theorem. The reason for using the term *selection* is that compactness in a metric space is equivalent to sequential compactness; *i.e.* any bounded sequence contains a convergent subsequence. It is the "selection" of such a subsequence which is referred to and which is crucial for many arguments.

2.1 Separation and support theorems

Throughout this chapter we shall refer to various classes of convex sets. It will be convenient to have some agreed-upon notation for these classes and so we begin with some notation and definitions. The substance of the section is relatively short and follows closely the ideas of §1.3. These results are often referred to as geometric versions of the Hahn–Banach theorem. As was mentioned in §1.3 alternative approaches can be found in Botts [52], Holmes [261] and Valentine [516].

Definition 2.1.1 *A **convex body** in a Minkowski space $(X, \|.\|)$ is a compact convex set with non-empty interior.*

As was remarked following Definition 0.2.5, when dealing with a single convex set in a finite dimensional space there is usually no loss of generality in supposing it has non-empty interior.

Next we make some notational conventions.

Definition 2.1.2 *If $(X, \|.\|)$ is a Minkowski space let*

  (i) *$\mathcal{C}$ denote the collection of non-empty compact, convex sets in X;*

  (ii) *$\mathcal{C}_b$ denote the collection of convex bodies in X;*

  (iii) *$\mathcal{C}_0$ denote the collection of closed convex sets which contain 0;*

  (iv) *$\mathcal{C}_i$ denote the collection of closed convex sets which have 0 as an interior point.*

The idea of *separating* sets by means of a linear functional has a hierarchy of levels. The only one that will be needed in the sequel is the strongest of these. This means that the following definition of *separate* is much stronger than the usual one. However, since we do not need the weaker versions, extra adjectives seem superfluous.

Definition 2.1.3 *A linear functional f is said to **separate** two sets A and B if there exist scalars α and β with $\alpha < \beta$ and such that*

$$A \subseteq H[f,\alpha]^- \quad \text{and} \quad B \subseteq H[f,\beta]^+.$$

In other words

$$\sup\{f(x) : x \in A\} \leq \alpha < \beta \leq \inf\{f(x) : x \in B\}.$$

Recall Definition 1.1.7 for the definition of the Minkowski functional of certain convex sets and Proposition 1.1.8 that such a functional is a norm. If we require only that K be in $\mathcal{C}_i$ then we may still define the Minkowski functional p_K by

$$p_K(x) := \inf\{\xi \in \mathbb{R}^+ : x \in \xi K\}$$

and, although p_K is no longer a norm, it is a sublinear functional (Definition 1.3.1) with the property that $p_K(x) \leq 1$ if and only if $x \in K$. The proof is exactly like Proposition 1.1.8.

Theorem 2.1.4 (support theorem) *If K is a convex set in $\mathcal{C}_i$ and if $x_0 \in \partial K$ then there exists a linear functional f which supports K at x_0.*

Proof. As was just observed, the Minkowski functional p_K of K is sublinear and we have $p_K(x_0) = 1$. Define f on the one-dimensional space spanned by x_0 by the equation $f(\alpha x_0) = \alpha$. This implies that $f(\alpha x_0) \leq p_K(\alpha x_0)$ and hence, by the Hahn–Banach theorem, 1.3.2, f can be extended to a linear functional, also denoted by f, on the whole space. The hyperplane $\{x : f(x) = 1\}$ supports K at x_0. ∎

Theorem 2.1.5 *If K is a convex set in $\mathcal{C}_i$ and if x_0 is not in K then there is a linear functional f which separates K and $\{x_0\}$.*

Proof. Since $x_0 \notin K$ we have $p_K(x_0) > 1$. Because $y = p_K(x_0)^{-1} x_0 \in \partial K$, the previous theorem implies that there is a hyperplane $\{x : f(x) = 1\}$ which supports K at y. Because $f(y) = 1$ we have $f(x_0) = p_K(x_0) > 1$ and so f separates K and $\{x_0\}$. ∎

The proof of the following proposition is straightforward.

Proposition 2.1.6 *A linear functional f separates two sets A and B if and only if f separates the sets $A - B$ and $\{0\}$.*

Theorem 2.1.7 (separation theorem) *If K_1 is a compact convex set and if K_2 is a closed convex set with $K_1 \cap K_2 = \emptyset$ then there is a linear functional which separates K_1 and K_2.*

Proof. First we show that $K_1 - K_2$ is closed. If $\{x_n\}$ is a sequence in $K_1 - K_2$ with x_n converging to $\bar{x}$ then $x_n = y_n - z_n$ with $y_n \in K_1$, $z_n \in K_2$. Since K_1 is compact, $\{y_n\}$ has a convergent subsequence, *i.e.* $y_{n_i} \to \bar{y} \in K_1$. Hence $z_{n_i} \to \bar{y} - \bar{x}$. But K_2 is closed, so $\bar{y} - \bar{x} \in K_2$ and hence $\bar{x} \in K_1 - K_2$. Since X is finite dimensional we may also suppose that $K_1 - K_2$ has an interior point x_0. Then $C := K_1 - K_2 - x_0$ is a convex set in $\mathcal{C}_i$. Since $K_1 \cap K_2 = \emptyset$, $0 \notin K_1 - K_2$ and so $-x_0 \notin C$. Hence there exists a linear functional f which separates $-x_0$ and C. Then, by Proposition 2.1.6 (used twice) f separates K_1 and K_2. ∎

It is clear that any intersection of closed half-spaces is a closed convex set. We are now in a position to prove the converse.

Theorem 2.1.8 *If K is a closed convex set in a finite dimensional space then K is the intersection of those closed half-spaces which contain it.*

Proof. Certainly K is contained in the stated intersection. Suppose $x \notin K$. Then (by Theorem 2.1.5) there is a linear functional f which separates $\{x\}$ and K. Hence x is not in the given intersection and the proof is complete. ∎

2.2 Support functions and polar reciprocals

This section deals with the important correspondence between convex sets in X and certain functions defined on X^*. Each convex set K induces an (extended real-valued) function

$$h_K(f) := \sup\{f(x) : x \in K\}. \tag{2.1}$$

Definition 2.2.1 *The function h_K defined by (2.1) is called the **support function** of K.*

This idea will play a crucial role in later chapters. We explore the relationship between K and h_K further.

The function h_K is sublinear on X^*, *i.e.*

$$h_K(\alpha f) = \alpha h_K(f) \text{ for } \alpha \geq 0,$$
$$h_K(f_1 + f_2) \leq h_K(f_1) + h_K(f_2).$$

If K is bounded then, because each f is continuous and hence bounded, h_K is a real-valued function on X^*. If $0 \in K$ then $h_K(f) \geq 0$ and if 0 is an interior point of K then $h_K(f) > 0$ for all $f \neq 0$. Moreover, if K is symmetric then h_K is an even function (*i.e.* $h_K(f) = h_K(-f)$) and, in this case, $h_K(f) = \sup\{|f(x)| : x \in K\}$. In the special case when K is the unit ball in a Minkowski space, h_K is the dual norm on X^*.

2.2 Support functions and polar reciprocals

The correspondence between convex sets and support functions is not one–one because if K_1 and K_2 have the same closure then $h_{K_1} = h_{K_2}$. However, it is a consequence of Theorem 2.1.7 that this is the extent of the problem.

Proposition 2.2.2 *If K_1 and K_2 are closed convex sets and if $h_{K_1} = h_{K_2}$ then $K_1 = K_2$.*

Proof. If $K_1 \neq K_2$ then there is a point x of K_1 which is not in K_2 (or vice versa). Then, by the separation theorem (2.1.7), there is a linear functional f for which $f(x) > \sup\{f(y) : y \in K_2\} = h_{K_2}(f)$. Since we have $f(x) \leq \sup\{f(z) : z \in K_1\}$ it follows that $h_{K_1} \neq h_{K_2}$. ∎

Example 2.2.3

(i) If $K = X$ then $h_X(f) = \begin{cases} +\infty, & f \neq 0, \\ 0, & f = 0. \end{cases}$

(ii) If $K = f_0^\perp = \{x : f_0(x) = 0\}$ then $h_K(f) = \begin{cases} 0, & \text{if } f = \alpha f_0, \\ +\infty, & \text{otherwise.} \end{cases}$

(iii) If $K = \{x\}$ then $h_K(f) = f(x)$.

The routine proof of the following proposition is omitted.

Proposition 2.2.4 *With respect to the algebraic structure on the collection of closed convex sets, the support function satisfies the following:*

(i) $h_{\alpha K} = \alpha h_K$ ($\alpha \geq 0$);
(ii) $h_{K_1 + K_2} = h_{K_1} + h_{K_2}$;
(iii) *if $K_1 \subseteq K_2$ then $h_{K_1} \leq h_{K_2}$.*

It follows from this proposition and Example 2.2.3(iii) that if $K' = K + \{x\}$ then $h_{K'}(f) = h_K(f) + f(x)$.

Next we look at this correspondence from the other side; *i.e.* we begin with a sublinear functional h on X^*. Eventually (see Theorem 2.2.8) we will show that h is the support function of a closed convex set in X so that the correspondence is a bijection between $\mathcal{C}$ and the sublinear functionals on X^*. The route to that result covers some other important ground first.

Given such a functional h, one can associate with it a set K_h in X^* where

$$K_h := \{f \in X^* : h(f) \leq 1\}. \tag{2.2}$$

If f and g are in K_h and if $0 \leq \alpha \leq 1$ then

$$h(\alpha f + (1-\alpha)g) \leq \alpha h(f) + (1-\alpha)h(g) \leq \alpha + (1-\alpha) = 1$$

and so K_h is convex. Since $0 \in K_h$, K_h is non-empty. Moreover, using the convexity of h it can be shown that h is continuous and hence that K_h is closed. If we begin with a closed convex set K in X and construct first the support function h_K and then the corresponding set for h_K via (2.2) we obtain a closed convex set in X^* called the polar reciprocal of K and denoted by $K°$.

Definition 2.2.5 *If K is a closed convex set in X, the **polar reciprocal** $K°$ of K is defined by*

$$K° := \{f \in X^* : f(x) \leq 1 \text{ for all } x \text{ in } K\}. \tag{2.3}$$

Note that if K is symmetric we have $K° = \{f \in X^* : |f(x)| \leq 1 \ \forall x \in K\}$. Thus if K is the unit ball in a Minkowski space then $K°$ is the dual ball as defined by Equation (1.3). As was pointed out there, this is the reason for denoting the dual ball by $B°$.

One can also see, directly from (2.3), that $K°$ is an intersection of closed half-spaces in X^* and hence it is a closed convex set containing 0. It is also evident that if $K_1 \subseteq K_2$ then $K_2° \subseteq K_1°$.

Since X^* is again a vector space and $K°$ is a closed convex set, we may repeat the procedure:

$$(K°)° := \{\Phi \in X^{**} : \Phi(f) \leq 1 \text{ for all } f \text{ in } K°\}.$$

But since X^{**} is isomorphic to X via the mapping J we have

$$K^{°°} := (K°)° = \{Jx \in X^{**} : (Jx)(f) \leq 1 \text{ for all } f \text{ in } K°\}$$
$$= \{x \in X : f(x) \leq 1 \text{ for all } f \text{ in } K°\}. \tag{2.4}$$

Comparing Equations (2.3) and (2.4) we see that $K \subseteq K^{°°}$.

Suppose that $0 \in K$ so that $h_K(f) \geq 0$ for all f in X^*. If $x \notin K$ then there is a (non-zero) linear functional f which separates K and $\{x\}$, i.e.

$$f(x) > \sup\{f(y) : y \in K\} = h_K(f).$$

If $h_K(f) > 0$ there is a positive scalar multiple βf of f such that $h_K(\beta f) = 1$ and $(\beta f)(x) > 1$. Thus $\beta f \in K°$ and $x \notin K^{°°}$. If $h_K(f) = 0$ then $\beta f \in K°$ for all $\beta > 0$ and, for some $\beta > 0$, $(\beta f)(x) > 1$. Hence, again, $x \notin K^{°°}$. Thus in both cases $x \notin K$ implies $x \notin K^{°°}$, which proves the following theorem.

Theorem 2.2.6 *If K is a closed convex set in X with $0 \in K$ then $K^{°°} = K$.*

Remarks

(i) In the special case when K is the unit ball in a Minkowski space this is a restatement of the "fundamental theorem" (1.3.8) that J is an isometry since it says that the ball B coincides with the ball $B^{°°}$ in X^{**}.

2.2 Support functions and polar reciprocals 51

(ii) It is understood that the equality sign in Theorem 2.2.6 refers to the identification of X^{**} with X via the mapping J as in (2.4).

(iii) If $0 \notin K$ then $K \neq K^{\circ\circ}$ because $0 \in K^{\circ\circ}$. In fact, in this case, $K^{\circ\circ} = \text{conv}(K \cup \{0\})$.

Recall from Definition 2.1.2(iii) that the collection of closed convex sets which contain 0 is denoted by $\mathcal{C}_0$. Let $\mathcal{C}_0^*$ denote the corresponding set of subsets in X^*.

Proposition 2.2.7 *The mapping $K \mapsto K^\circ$ is a one–one mapping of $\mathcal{C}_0$ onto $\mathcal{C}_0^*$, which reverses inclusions.*

Proof. That the mapping is a surjection follows from Theorem 2.2.6. The proof that it is injective is another application of the separation theorem exactly as in the proof of Proposition 2.2.2. ∎

We now prove the result advertised earlier.

Theorem 2.2.8 *If h is a sublinear functional on X^* then h is the support function of a closed convex set in X.*

Proof. The fact that h is convex implies that h is continuous (see Schneider [479], p. 23). Therefore, the set K_h defined by (2.2) is closed and belongs to $\mathcal{C}_0^*$. Hence, by Theorem 2.2.6, $K_h = K_h^{\circ\circ}$ and so h is the support function of K_h°. ∎

There is one rather more technical fact about the mapping $K \mapsto K^\circ$ which we shall need and which relates to Proposition 0.1.1. Suppose, as in 0.1.1, that $X = L_1 \oplus L_2$ and hence that $X^* = L_2^\perp \oplus L_1^\perp$ and that $L_1^* \cong L_2^\perp, L_2^* \cong L_1^\perp$. Suppose that K is in $\mathcal{C}_0 \cap \mathcal{C}$. Then, if Proj_1 is the canonical projection of X onto L_1, $\text{Proj}_1(K)$ is a closed convex set in L_1 and so has a polar $\text{Proj}_1(K)^\circ$ in $L_1^* \cong L_2^\perp$. In this situation we have the following result.

Theorem 2.2.9 *With the above notation, if $X = L_1 \oplus L_2$ and if $K \in \mathcal{C}_0 \cap \mathcal{C}$ then $\text{Proj}_1(K)^\circ = K^\circ \cap L_2^\perp$, where the polar on the left is with respect to the subspace L_1 and that on the right is with respect to X.*

Proof. Let $H[x, 1]^- := \{f \in X^* : f(x) \leq 1\}$ so that

$$K^\circ = \bigcap_{x \in K} H[x, 1]^-.$$

Then $K^\circ \cap L_2^\perp = \bigcap_{x \in K}(H[x, 1]^- \cap L_2^\perp)$. Each x in X has a unique representation as $x = x_1 + x_2$ with $x_i \in L_i$. We have $g \in H[x, 1]^- \cap L_2^\perp$ if and only if $g \in$

$L_2^\perp$ and $g(x) = g(x_1) + g(x_2) \le 1$. But if $g \in L_2^\perp$ then $g(x_2) = 0$ and so $g \in H[x, 1]^- \cap L_2^\perp$ if and only if $g \in L_2^\perp = L_1^*$ and $g(x_1) \le 1$, i.e. if and only if g is in $\{x_1\}^\circ = \{\text{Proj}_1 x\}^\circ$, where the polar is with respect to L_1. Hence $K^\circ \cap L_2^\perp = \bigcap_{x \in K} \{\text{Proj}_1 x\}^\circ = (\text{Proj}_1(K))^\circ$. ∎

Corollary 2.2.10 *If $0 \in \text{int } K$ we also have $(K \cap L_1)^\circ = \text{Proj}_1(K^\circ)$, where Proj_1 this time denotes the canonical projection of $X^* = L_2^\perp \oplus L_1^\perp$ onto its first component.*

Proof. In Theorem 2.2.9 replace K by K°, $L_2^\perp$ by $L_1 = (L_2^\perp)^*$ and then use Theorem 2.2.6; take polars of both sides (in L_1) and use Theorem 2.2.6 once more. ∎

There is an alternative way to describe a convex body K which has 0 as an interior point.

Definition 2.2.11 *If K is a convex body with 0 as an interior point then for each $x \ne 0$ in X we define $\rho_K(x)$ to be that positive number such that $\rho_K(x)x \in \partial K$. The function ρ_K is called the **radial function** of K.*

Proposition 2.2.12 *The radial function ρ has the following properties:*

  (i) $\rho_K(\alpha x) = \alpha^{-1} \rho_K(x)$ $(\alpha > 0)$;
  (ii) if $K_1 \subseteq K_2$ then $\rho_{K_1} \le \rho_{K_2}$;
  (iii) $\rho_{\alpha K} = \alpha \rho_K$ $(\alpha > 0)$.

Proof. Each of these is clear from the definition. ∎

More importantly, there is a close relationship between ρ, h and the mapping $K \mapsto K^\circ$.

Theorem 2.2.13 *If K is a convex body in X with 0 as an interior point and K° its polar then $\rho_{K^\circ}(f) = (h_K(f))^{-1}$ and $\rho_K(x) = (h_{K^\circ}(x))^{-1}$.*

Proof. It is sufficient to prove the first equation. Since $K^\circ = \{f \in X^* : h_K(f) \le 1\}$ we have that $f \in \partial K^\circ$ if and only if $h_K(f) = 1$. But since h_K is positively homogeneous this means that $(h_K(f))^{-1} f \in \partial K^\circ$ and hence the desired equation. ∎

Remark. In terms of our previous notation, we have $\rho_K(x) = 1/p_K(x)$, where p_K is the Minkowski functional of K.

2.3 Volumes and mixed volumes

Since, for a finite dimensional space X, there is a single Hausdorff linear topology, the concept of a Borel set is intrinsic to the space. In §1.4, it was shown how to construct a translation-invariant measure, Haar measure, on the Borel σ-algebra. Throughout this section we suppose that a Haar measure has been chosen and we shall refer to this measure as *volume*. As at the end of §1.4 we shall use $\lambda(.)$ to denote volume. Since closed convex sets are Borel sets, λ may be applied, in particular, to such sets. An alternative and conceptually simpler view of λ is to suppose that X comes equipped with an auxiliary Euclidean structure and that λ is the Lebesgue measure induced by that structure. The particular choice of Haar measure is immaterial; a scalar multiple corresponds to a change of basis for the Euclidean structure. Throughout the section the dimension of the ambient space X is assumed to be d. This notation is awkward in integral formulas like those in Theorem 2.3.13 where confusion with the differential symbol is possible.

Except for the most straightforward items, we shall not give proofs for the statements in this section. The reader is referred to Bonnesen and Fenchel [51], Eggleston [138] and Schneider [479] for proofs and fuller discussions.

The functional λ has the following properties (the arguments are always closed convex sets).

Proposition 2.3.1

  (i) $\lambda(K) \geq 0$ *with equality if and only if* $\dim K \leq d - 1$.
  (ii) *If* $K \subseteq K'$ *then* $\lambda(K) \leq \lambda(K')$.
  (iii) *If* $K' = K + \{x\}$ *then* $\lambda(K) = \lambda(K')$ (λ *is translation invariant*).
  (iv) *If* $\alpha \geq 0$ *then* $\lambda(\alpha K) = \alpha^d \lambda(K)$.
  (v) $\lambda(-K) = \lambda(K)$.
  (vi) *The volume λ "depends continuously" on the set* K.

Remarks

(i) Another way of phrasing Proposition 2.3.1(i) is to say that $\lambda(K) > 0$ if and only if K contains line segments $[x_i, x_i']$ such that the vectors $y_i = x_i' - x_i$ span X.

(ii) The reason for the quotation marks in (vi) is that continuity is somewhat meaningless until we have defined a topology on the collection of convex sets. That definition is the purpose of the next section.

At the end of §1.4 we described how the volume λ gives rise to a dual volume λ^* on the convex subsets of X^*. Furthermore, if K is a closed convex set in X, then K° is a closed convex set in X^*.

Definition 2.3.2 *For each closed convex body K in X the* **volume product** *of K is defined to be $\lambda(K)\lambda^*(K^\circ)$.*

If K is bounded then 0 is an interior point of K° and, conversely, if 0 is an interior point of K then K° is bounded. In dealing with volume products we shall suppose that K is a symmetric convex body (which implies that 0 is an interior point).

As was noted at the end of §1.4, this volume product is independent of which Haar measure is used for λ. Alternatively, it is independent of the choice of orthonormal basis for the Euclidean structure. Hence if T is a non-singular linear transformation then $\lambda(T(K))\lambda^*((TK)^\circ) = \lambda(K)\lambda^*(K^\circ)$.

One can also see this directly because $(TK)^\circ = T^{*-1}K^\circ$ and so

$$\lambda(T(K)) = (\det T)\lambda(K) \quad \text{and} \quad \lambda^*((TK)^\circ) = (\det T)^{-1}\lambda^*(K^\circ).$$

We describe this briefly by saying that the volume product is an affine invariant of K. That the volume product is an affine invariant of the convex body K is an important fact in many contexts but especially so in Chapter 6, where the whole development is based on this fact.

Two important inequalities give upper and lower bounds for the volume product of a symmetric convex body. The first, giving the upper bound, is the Blaschke–Santaló inequality (see §2.6 for the history and references).

Theorem 2.3.3 (Blaschke–Santaló) *If K is a symmetric convex body then*

$$\lambda(K)\lambda^*(K^\circ) \leq \lambda(E)\lambda^*(E^\circ), \tag{2.5}$$

where E is any ellipsoid centred at 0. Moreover, equality occurs in (2.5) if and only if K is an ellipsoid.

The second inequality, giving the lower bound, is a conjecture of Mahler's. It has been proved by Mahler [346] in the case when $d = 2$ and, more recently, by Reisner [438] if K is restricted to the smaller class of zonoids. A zonotope was defined in Example 0.2.3(iv) as a sum of line segments. The set of *zonoids* in a Minkowski space is the closure (with respect to the Hausdorff metric; Definitions 2.4.1 and 2.4.3 below) of the set of zonotopes. More will be said about zonoids in subsequent chapters.

Theorem 2.3.4 (Mahler–Reisner) *If K is a zonoid then*

$$\lambda(P)\lambda^*(P^\circ) \leq \lambda(K)\lambda^*(K^\circ), \tag{2.6}$$

where P is a parallelotope centred at 0. Moreover, equality occurs in (2.6) if and only if K is a parallelotope.

Remark. In the case when $d = 2$, all symmetric convex bodies are zonoids and hence Theorem 2.3.4 includes Mahler's result.

The value of $\lambda(E)\lambda^*(E^\circ)$ is ϵ_d^2, where $\epsilon_d := \pi^{2/d}/\Gamma((d/2) + 1)$ and is the volume of a d-dimensional Euclidean unit ball. The value of $\lambda(P)\lambda^*(P^\circ)$ can be calculated by considering the case when P is a cube and P° is a cross-polytope. Then $\lambda(P) = 2^d$ and P° consists of 2^d simplices each congruent to the one with vertices at 0 and the orthonormal basis vectors e_i. Hence $\lambda^*(P^\circ) = 2^d/d!$ and $\lambda(P)\lambda^*(P^\circ) = 4^d/d!$.

We now return to the discussion of the single functional λ. Proposition 2.3.1(iv) describes how λ behaves with respect to scalar multiplication. Its behaviour with respect to linear combinations is complex. The explanation of this behavior is due largely to Minkowski.

Theorem 2.3.5 (Brunn–Minkowski inequality) *The dth root of the volume is a concave function on the class of convex bodies, i.e.*

$$\lambda^{1/d}(\alpha K_1 + (1-\alpha)K_2) \geq \alpha \lambda^{1/d}(K_1) + (1-\alpha)\lambda^{1/d}(K_2). \qquad (2.7)$$

It follows from Proposition 2.3.1(iv) that if $\alpha_1, \alpha_2, \ldots, \alpha_r$ are non-negative scalars then $\lambda((\alpha_1 + \alpha_2 + \cdots + \alpha_r)K) = (\alpha_1 + \alpha_2 + \cdots + \alpha_r)^d \lambda(K)$, and this last expression may be expanded as a homogeneous polynomial of degree d in the α's, whose coefficients are certain multinomial coefficients multiplied by $\lambda(K)$. Perhaps surprisingly, this can be extended to arbitrary non-negative linear combinations of convex sets.

Theorem 2.3.6 *If $K = \alpha_1 K_1 + \alpha_2 K_2 + \cdots + \alpha_r K_r$ with $\alpha_i \geq 0$ then*

$$\begin{aligned}\lambda(K) &= \lambda(\alpha_1 K_1 + \alpha_2 K_2 + \cdots + \alpha_r K_r) \\ &= \sum V(K_{i_1}, K_{i_2}, \ldots, K_{i_d})\alpha_{i_1}\alpha_{i_2}\cdots\alpha_{i_d};\end{aligned} \qquad (2.8)$$

i.e. $\lambda(K)$ is a homogeneous polynomial of degree d in the α's.

In Equation (2.8) the sum is taken over all possible ordered sets of indices $i_i, i_2, \ldots, i_d$ chosen from the set $\{1, 2, \ldots, r\}$. However, we may (and do) assume that the coefficients $V(K_{i_1}, K_{i_2}, \ldots, K_{i_d})$ are symmetric in the sense that if $i'_1 i'_2 \cdots i'_d$ is a permutation of $i_1 i_2 \cdots i_d$ then $V(K_{i'_1}, K_{i'_2}, \ldots, K_{i'_d}) = V(K_{i_1}, K_{i_2}, \ldots, K_{i_d})$. The coefficients $V(K_{i_1}, K_{i_2}, \ldots, K_{i_d})$ are known collectively as *mixed volumes* and they have the following properties.

Proposition 2.3.7

(0) If $K_1 = K_2 = \cdots = K_d = K$ then
$$V(K_1, K_2, \ldots, K_d) = \lambda(K).$$

(i) $V(K_1, K_2, \ldots, K_d) \geq 0$ with strict inequality if and only if for each $i = 1, 2, \ldots, d$ one can find a line segment $[x_i, x_i']$ in K_i such that the vectors $y_i = x_i' - x_i$ span X.

(ii) If $K_i \subseteq K_i'$ then
$$V(K_1, K_2, \ldots, K_i, \ldots, K_d) \leq V(K_1, K_2, \ldots, K_i', \ldots, K_d).$$

(iii) If $K_i' = K_i + \{x\}$ then
$$V(K_1, K_2, \ldots, K_i, \ldots, K_d) = V(K_1, K_2, \ldots, K_i', \ldots, K_d).$$

(iv) If $\alpha \geq 0$ then
$$V(K_1, K_2, \ldots, \alpha K_i, \ldots, K_d) = \alpha V(K_1, K_2, \ldots, K_i, \ldots, K_d).$$

(v) If $K_i = K_i' + K_i''$ then
$$V(K_1, K_2, \ldots, K_i, \ldots, K_d) = V(K_1, K_2, \ldots, K_i', \ldots, K_d) \\ + V(K_1, K_2, \ldots, K_i'', \ldots, K_d).$$

(vi) The mixed volumes $V(K_1, K_2, \ldots, K_d)$ "depend continuously" on the bodies $K_1, K_2, \ldots, K_d$.

Proof. Property (0) follows from the remark preceding Theorem 2.3.6. Properties (iii) and (vi) follow from the corresponding properties for λ (Proposition 2.3.1). As before, property (vi) depends on a topology for the collection of convex sets. The linearity in each variable ((iv) and (v)) follows by representing K by
$$K = \sum \alpha_j K_j = \sum_{j \neq i} \alpha_j K_j + \alpha_i \alpha' K_i' + \alpha_i \alpha'' K_i''$$
and then expressing $\lambda(K)$ as two polynomials using (2.8) and finally equating coefficients in the two expressions. The seemingly innocent properties (i) and (ii) are not at all obvious. We refer to our standard references for proofs. ∎

As an example of the way in which 2.3.7(i) works in $\mathbb{R}^3$, take K_1 to be the parallelogram $[0, x_1] + [0, x_2]$ and K_2 to be the line segment $[0, x_3]$ with $\{x_1, x_2, x_3\}$ a linearly independent set. Then we have
$$\lambda(\alpha_1 K_1 + \alpha_2 K_2) = \alpha_1^2 \alpha_2 V(K_1, K_1, K_2),$$
where $V(K_1, K_1, K_2) = \lambda(\sum_{i=1}^{3}[0, x_i]) > 0$ and all other mixed volumes are zero.

It is a little awkward, in expressions like these, to have λ on one side and V on the other. The use of V for mixed volumes is widespread and the use of λ for measures is not uncommon. Unfortunately, it is not always possible to have accepted usage and consistency in matters of notation.

The linearity of mixed volumes in each variable and the fact that they are, by definition, symmetric (rather than antisymmetric like determinants) mean that they behave somewhat like permanents.

We shall be concerned almost exclusively with the case when $r = 2$ in Theorem 2.3.6. Then if $K = \alpha_1 K_1 + \alpha_2 K_2$ we have

$$\lambda(K) = \lambda(\alpha_1 K_1 + \alpha_2 K_2) = \sum_{i=1}^{d} \binom{d}{i} V(K_1[i], K_2[d-i]) \alpha_1^i \alpha_2^{d-i}. \quad (2.9)$$

In this expression we have used the symmetry of the coefficients to collect like terms (i.e. those for which i of the indices $j_1, j_2, \ldots, j_d$ are equal to 1 and the remaining $d - i$ of them are equal to 2; there are $\binom{d}{i}$ such terms). Since these coefficients now depend only on i we write them in the abbreviated form $V(K_1[i], K_2[d-i]) := V(K_1, K_1, \ldots, K_1, K_2, K_2, \ldots, K_2)$, where K_1 appears i times and K_2 appears $d - i$ times. Especially important, for the three reasons below, are the functionals $V(K_1[d-1], K_2)$ (the "1" for the last term is omitted).

Firstly, because $dV(K_1[d-1], K_2)$ is the coefficient of α in the polynomial $\lambda(K_1 + \alpha K_2)$ it can be calculated as the first derivative of this polynomial, i.e.

$$dV(K_1[d-1], K_2) = \lim_{\alpha \to 0} \alpha^{-1} \{\lambda(K_1 + \alpha K_2) - \lambda(K_1)\}. \quad (2.10)$$

Secondly, the Brunn–Minkowski inequality (2.3.5) implies the following relationship between $V(K_1[d-1], K_2)$ and $\lambda(K_1)$ and $\lambda(K_2)$.

Theorem 2.3.8 (Minkowski inequality for mixed volumes) *If K_1 and K_2 are convex bodies then*

$$V^d(K_1[d-1], K_2) \geq \lambda(K_1)^{d-1} \lambda(K_2). \quad (2.11)$$

Thirdly, properties (ii), (iv) and (v) of mixed volumes have the following consequence. If $u, v \in X$ then $[u] + [v]$ is a parallelogram and $[u + v]$ is one of its diagonals. Hence $[u+v] \subseteq [u]+[v]$. This implies that if K is a fixed convex body then $V(K[d-1], [u+v]) \leq V(K[d-1], [u]) + V(K[d-1], [v])$. Furthermore, $V(K[d-1], [\alpha u]) = \alpha V(K[d-1], [u])$ for $\alpha \geq 0$.

Definition 2.3.9 *For a fixed convex body K and for $u \in X$, the functional $\sigma_K : X \mapsto \mathbb{R}$ is defined by*

$$\sigma_K(u) := dV(K[d-1], [u]).$$

The preceding discussion proves the following proposition.

Proposition 2.3.10 *The functional $\sigma_K : X \mapsto \mathbb{R}$ is positively homogeneous and subadditive and hence is the support function of a convex set.*

Definition 2.3.11 *The convex set of which σ_K is the support function is called the **projection body** of K and is denoted by $\Pi(K)$.*

Remarks

(i) Since $[u] = [-u]$ we have that σ_K is an even function and $\Pi(K)$ is symmetric.

(ii) Note that, since σ_K is defined on X, it is the support function of a body in X^*, i.e. $\Pi(K) \subseteq X^*$.

(iii) The body $\Pi(K)$ will play a major role in Chapter 6. Observe that $\sigma_K(u)$ can be viewed as the area of the projection of K onto a hyperplane perpendicular to u.

When we need to make specific calculations of volumes and mixed volumes, there are so many formulas available for the Euclidean case that it is almost always convenient and expedient to introduce an auxiliary Euclidean metric. We list a number of formulas which depend on such a structure but, first, some background.

The Euclidean structure on X induces one on each $(d-1)$-dimensional subspace and hence there is a Lebesgue measure on each hyperplane. We call this measure *area* and use λ^{d-1} to denote it. If the context makes clear which dimension is intended the superscript may be omitted. Similarly, there is Lebesgue measure λ^k on subspaces of dimension k. Moreover, these Lebesgue measures are linked by the requirement that if Y and Z are orthogonal subspaces of X of dimensions j and k respectively then λ^{j+k} on $Y \oplus Z$ is equal to the product measure $\lambda^j \times \lambda^k$. Thus the Euclidean structure gives an easy solution to what will be the central problem in Chapter 5: how to choose a suitable Haar measure in each $(d-1)$-dimensional subspace of X.

The measure λ^{d-1} can be extended from regions contained in a hyperplane to convex surfaces in one of two ways. Either Equation (2.15) below can be used as a definition or, alternatively, surface area can be defined first for polytopes (using the area in each hyperplane to determine the area of each facet) and then extended to convex surfaces by an approximation argument. This will be discussed further in the next section.

The rest of the notation for Theorem 2.3.13 is as follows. A Euclidean unit vector in X will be denoted by $\hat{u}$ and in X^* by $\hat{f}$. The surface ∂K of a convex body K may be thought of as parametrized by a typical point on the surface. Thus

$$\int_{\partial K} \phi(x)\, d\lambda^{d-1}(x)$$

indicates that the function ϕ defined on ∂K is to be integrated over the surface ∂K

with respect to $(d-1)$-dimensional Lebesgue measure. Alternatively, ∂K may be thought of as parametrized by the unit normal; *i.e.* for each unit vector $\hat{f}$, we may consider the set of points of ∂K where the outward normal is $\hat{f}$. If K is smooth (so that each point of ∂K has a unique outward normal) and strictly convex (so that each $\hat{f}$ is the outward normal at a unique point) then this correspondence is straightforward. Otherwise, we first need a theorem of Reidemeister [435] that states that an arbitrary convex body has a unique normal λ^{d-1}-almost-everywhere and then the following construction. Let S^{d-1} denote the Euclidean unit sphere in X^* (we use the same notation for the sphere in X). If U is a Borel subset of S^{d-1} let $\Gamma^{-1}(U)$ be the set of points in ∂K at which there is an outward normal in U. Define ν_K by

$$\nu_K(U) := \lambda^{d-1}(\Gamma^{-1}(U)).$$

It follows (see Schneider [479], §4.2) that ν_K is a Borel measure on S^{d-1}. The measure ν_K is commonly denoted by $S(K, \cdot)$.

Definition 2.3.12 *The measure ν_K is called the* **surface area measure** *of K.*

Now if ψ is a function defined on S^{d-1} we may integrate it with respect to ν_K, for which we write

$$\int_{S^{d-1}} \psi(\hat{f}) \, d\nu_K(\hat{f}).$$

However, by a slight abuse of notation, this same integral is sometimes written

$$\int_{\partial K} \psi(\hat{f}) \, d\lambda^{d-1}(\hat{f}).$$

With these notational conventions we have the following formulas for volumes and surface areas. As stated in the Preface, the use of d for the dimension is awkward here where it is easily confused with the differential symbol.

Theorem 2.3.13 *If K is a convex body in X then:*

$$\lambda(K) = d^{-1} \int_{S^{d-1}} h_K(\hat{f}) \, d\nu_K(\hat{f})$$

$$= d^{-1} \int_{\partial K} h_K(\hat{f}) \, d\lambda^{d-1}(\hat{f}), \quad (2.12)$$

$$\lambda(K) = d^{-1} \int_{S^{d-1}} \rho_K(\hat{u})^d \, d\lambda^{d-1}(\hat{u}), \quad (2.13)$$

$$V(K_1[d-1], K_2) = d^{-1} \int_{S^{d-1}} h_{K_2}(\hat{f}) \, d\nu_{K_1}(\hat{f})$$

$$= d^{-1} \int_{\partial K_1} h_{K_2}(\hat{f}) \, d\lambda^{d-1}(\hat{f}), \quad (2.14)$$

$$\lambda(\partial K) = \int_{\partial K} 1 \, d\lambda^{d-1}(\hat{f})$$
$$= \int_{S^{d-1}} 1 \, dv_K(\hat{f}) = dV(K[d-1], E), \qquad (2.15)$$
$$\lambda(\partial K) = \epsilon_{d-1}^{-1} \int_{S^{d-1}} \sigma_K(\hat{u}) \, d\lambda^{d-1}(\hat{u}). \qquad (2.16)$$

Remark. Formula (2.16) is called Cauchy's surface area formula. It states that the surface area of K can be expressed as an average of the projections of K over all directions $\hat{u}$. For proofs of these formulas see either Schneider [479] or Bonnesen and Fenchel [51]. In particular, Equation (2.16) is discussed in §32 of [51].

We conclude this section with a uniqueness result which will be needed later both on its own and in conjunction with Equation (2.14).

Proposition 2.3.14 *If B_1 and B_2 are two symmetric convex bodies in a Minkowski space X and if $V(B_1[d-1], K) = V(B_2[d-1], K)$ or if $V(K[d-1], B_1) = V(K[d-1], B_2)$ for all symmetric convex sets K, then $B_1 = B_2$.*

Proof. We prove the first implication; the proof of the second is the same. If $K = B_2$ then $V(B_1[d-1], B_2) = \lambda(B_2)$. But since, by Minkowski's inequality, $V^d(B_1[d-1], B_2) \geq \lambda^{d-1}(B_1)\lambda(B_2)$, we have $\lambda(B_2) \geq \lambda(B_1)$. Interchanging B_1 and B_2 we get

$$\lambda(B_2) = \lambda(B_1).$$

This means that $V^d(B_1[d-1], B_2) = \lambda^d(B_2) = \lambda^{d-1}(B_1)\lambda(B_2)$, and hence B_1 and B_2 are homothetic. But they have the same volume and are centred at 0 so $B_1 = B_2$. ∎

2.4 Various derived metrics

Once there is a metric on X we can use it to consider distances between sets rather than between points. Although the construction which follows can be made very general (for a fuller discussion see [293]), we restrict ourselves to the setting of finite dimensional vector spaces and make use of the extra structure which is available there. We suppose $\|.\|$ is a norm on X with corresponding unit ball B.

For $\epsilon > 0$, ϵB is the ball of radius ϵ, $\epsilon B = \{x : \|x\| \leq \epsilon\}$. For bounded, non-empty subsets C and D of X, let

$$\delta'(C, D) := \inf\{\epsilon > 0 : C \subseteq D + \epsilon B\}.$$

2.4 Various derived metrics

We need C to be bounded and D to be non-empty to ensure that $\delta'(C, D)$ is finite. We have $\delta'(C, D) = 0$ if and only if $C \subseteq c\ell D$. Furthermore, if A, C and D are three subsets of X and if $A \subseteq C + \epsilon B, C \subseteq D + \eta B$ then $A \subseteq D + \epsilon B + \eta B = D + (\epsilon + \eta)B$. From this inclusion we get

$$\delta'(A, D) \leq \delta'(A, C) + \delta'(C, D). \tag{2.17}$$

Definition 2.4.1 *For bounded, non-empty subsets C, D of X let*

$$\delta(C, D) := \max\{\delta'(C, D), \delta'(D, C)\}.$$

Proposition 2.4.2 *The functional δ is a metric on the collection of closed, bounded, non-empty subsets of $(X, \|.\|)$.*

Proof. Since $\delta'(C, D) = 0$ if and only if $C \subseteq c\ell D$, it follows that $\delta(C, D) = 0$ if and only if $c\ell C = c\ell D$ and hence, for closed sets, if and only if $C = D$. The symmetry of δ is clear from the definition and the triangle inequality is a consequence of (2.17). ∎

Definition 2.4.3 *The metric δ is called the **Hausdorff metric** on the closed and bounded (i.e. compact) non-empty subsets of $(X, \|.\|)$.*

Let $\mathcal{K}$ denote the collection of non-empty compact subsets of $(X, \|.\|)$. (This is slightly at variance with the notation in §1.4, where $\mathcal{K}$ included the empty set.) By the results in §1.3, the set $\mathcal{K}$ is independent of which norm we impose on X. It is clear, however, that different norms give rise to different Hausdorff metrics δ on $\mathcal{K}$. In this direction, Kelley [284], p. 131, gives an example of a set with two topologically equivalent bounded metrics which generate different Hausdorff metric topologies on the set of closed subsets; a similar example is given in [293]. Here, however, because the norms are uniformly equivalent (Corollary 1.2.5) the topology on $\mathcal{K}$ is also independent of the norm.

Proposition 2.4.4 *If $\|.\|_1$ and $\|.\|_2$ are equivalent norms on X then the associated Hausdorff metrics δ_1 and δ_2 are uniformly equivalent on $\mathcal{K}$.*

Proof. If there are constants c_1 and c_2 such that

$$c_1\|x\|_2 \leq \|x\|_1 \leq c_2\|x\|_2 \quad \text{for all } x \text{ in } X$$

then $c_1 B_1 \subseteq B_2 \subseteq c_2 B_1$ and hence, from the definition of δ, it follows that $c_1\delta_2(C, D) \leq \delta_1(C, D) \leq c_2\delta_2(C, D)$ for all C, D in $\mathcal{K}$. ∎

Remark. Not only the topologies, but also the uniformities, on $\mathcal{K}$ coincide. Hence if $\mathcal{K}$ is complete in one Hausdorff metric, then it is complete in all of them.

The next step is to restrict δ to the convex sets in $\mathcal{K}$ (*i.e.* to the set $\mathcal{C}$ of Definition 2.1.2(i)). To each compact convex set K there corresponds its support function h_K defined on X^*. Also, $h_B(f) = 1$ if and only if $f \in \partial B°$. Hence, if we restrict the support functions to the boundary of $B°$ then the balls αB correspond to the constant functions $h_{\alpha B} \equiv \alpha$ (for $\alpha \geq 0$). Further, since Minkowski addition corresponds to sums of support functions,

$$h_{K+\epsilon B} = h_K + \epsilon.$$

Since $K_1 \subseteq K_2$ if and only if $h_{K_1} \leq h_{K_2}$, we have $K_1 \subseteq K_2 + \epsilon B$ if and only if $h_{K_1} \leq h_{K_2} + \epsilon$. Thus

$$\delta'(K_1, K_2) = \sup\{h_{K_1}(f) - h_{K_2}(f) : f \in \partial B°\}$$
$$= \|(h_{K_1} - h_{K_2})^+\|_\infty,$$

where $\|.\|_\infty$ denotes the uniform norm for functions on $\partial B°$. Now Definition 2.4.1 implies that

$$\delta(K_1, K_2) = \|h_{K_1} - h_{K_2}\|_\infty. \tag{2.18}$$

Thus we have proved the following proposition.

Proposition 2.4.5 *If $(X, \|.\|)$ is a Minkowski space and if K_1 and K_2 are compact, convex sets in X then the Hausdorff distance $\delta(K_1, K_2)$ is equal to the uniform norm $\|h_{K_1} - h_{K_2}\|_\infty$ between their support functions regarded as functions on $\partial B°$.*

We now turn our attention to the consideration of various norms or, equivalently, various unit balls on a vector space X. The purpose is to measure the distance between such norms (unit balls). In this situation there is no single metric to use for the measurements and the multiplicative structure works better than the additive structure.

Let B and B' be two symmetric convex bodies in X. Let

$$\Delta'(B, B') := \inf\{\alpha : B \subseteq \alpha B'\} = \min\{\alpha : B \subseteq \alpha B'\}.$$

Then $\Delta'(B, B') \leq 1$ if and only if $B \subseteq B'$. Also, if B'' is a third such body with $B' \subseteq \beta B''$ then $B \subseteq (\alpha\beta) B''$ and so

$$\Delta'(B, B'') \leq \Delta'(B, B')\Delta'(B', B''). \tag{2.19}$$

We can revert to the additive structure of $\mathbb{R}$ by taking logarithms and also symmetrize Δ' in either one of two ways:

$$\Delta_1(B, B') := \log[\Delta'(B, B')\Delta'(B', B)] = \log \Delta'(B, B') + \log \Delta'(B', B) \tag{2.20}$$

$$\Delta_2(B, B') := \log[\max\{(\Delta'(B, B'), \Delta'(B', B)\}]. \tag{2.21}$$

Then Δ_2 is a metric on the symmetric convex bodies (unit balls) in X.

Remark. Proposition 2.2.7 can now be extended to say that the mapping $K \mapsto K°$ is an isometry with respect to Δ_2 on the sets $\mathcal{C}_0$ and $\mathcal{C}_0^*$. This statement follows directly from the fact that $B_1 \subseteq B_2 \subseteq \alpha B_1$ if and only if $\alpha^{-1} B_1° \subseteq B_2° \subseteq B_1°$.

Although Δ_1 is not quite a metric, the way in which it fails can be turned into a useful feature. We have $\Delta_1(B, B') = 0$ if and only $B = \alpha B'$ for some $\alpha > 0$; *i.e.* the balls B and B' are homothetic. This relationship is an equivalence relation on the collection of balls and Δ_1 is a metric on the equivalence classes.

We can also express Δ' and hence Δ_1 in terms of the corresponding norms $\|.\|$ and $\|.\|'$. These norms are equivalent in the sense of Corollary 1.2.5 and if c_1, c_2 are the best constants such that

$$c_1 \|x\|' \leq \|x\| \leq c_2 \|x\|'$$

then $\Delta_1(B, B') = \log(c_2/c_1)$. Alternatively, $c_2 = \|\mathbf{1}\|$ and $c_1^{-1} = \|\mathbf{1}^{-1}\|$, where $\mathbf{1}$ denotes the identity map between $(X, \|.\|')$ and $(X, \|.\|)$. Thus

$$\Delta_1(B, B') = \log(\|\mathbf{1}\| \, \|\mathbf{1}^{-1}\|). \tag{2.22}$$

Also, $\Delta_1(B, B') = 0$ if and only if some multiple of the identity is an isometry between $(X, \|.\|)$ and $(X, \|.\|')$.

Identifying two spaces if the balls are homothetic is not especially relevant. We wish to think of two Minkowski spaces as identical if there is a linear isometry T of X onto Y. Therefore, we require a metric to measure the distance between spaces in such a way that it is zero if and only if the spaces are isometrically isomorphic. The final formula (2.22) for Δ_1 shows the direction to take.

If $(X, \|.\|_X)$ and $(Y, \|.\|_Y)$ are finite dimensional spaces of the same dimension then there exists a linear map T from X onto Y. One of the basic theorems of linear algebra shows that T is invertible and Theorem 1.2.4 shows that it is an isomorphism.

Definition 2.4.6 *The **Banach–Mazur distance** between isomorphic normed spaces X and Y is defined by*

$$\Delta(X, Y) := \inf\{\log(\|T\| \, \|T^{-1}\|) : T \text{ an isomorphism of } X \text{ onto } Y\}.$$

Remarks

(i) The definition is phrased in a way that is also applicable to infinite dimensional spaces.

(ii) One can restrict T in the definition to a uniformly bounded set, *e.g.* the set of T with $\|T\| = 1$, because $\|\alpha T\| \, \|(\alpha T)^{-1}\| = \|T\| \, \|T^{-1}\|$ and one could take $\alpha = \|T\|^{-1}$.

The triangle inequality for Δ is a straightforward consequence of the formula $\|ST\| \le \|S\| \|T\|$ proved in Proposition 1.1.11. This inequality also implies that

$$1 = \|\mathbf{1}_X\| = \|T^{-1}T\| \le \|T\| \|T^{-1}\|$$

so that $\Delta(X, Y) \ge 0$.

In finite dimensional spaces, $\|T\| = \min\{\alpha : T(B_X) \subseteq \alpha B_Y\}$ and $\|T^{-1}\| = \min\{\beta : T^{-1}(B_Y) \subseteq \beta B_X\} = \min\{\beta : \beta^{-1}B_Y \subseteq T(B_X)\}$. Therefore, $\|T\|^{-1} = \|T^{-1}\|$ if and only if $T(B_X) = \|T\|B_Y$ and $\|T\|^{-1}T$ is an isometry of X onto Y. Thus $\log(\|T\| \|T^{-1}\|) = 0$ if and only if X and Y are isometric.

The problem now arises as to whether the infimum in Definition 2.4.6 is attained or not. For infinite dimensional spaces this is a difficult problem. For finite dimensional spaces, however, the space of linear transformations between X and Y is again finite dimensional and hence, by Theorem 1.2.3, is locally compact. If (T_n) is a sequence of isomorphisms with $\|T_n\| \|T_n^{-1}\| \to \exp(\Delta(X, Y))$ and (using Remark (ii)) $\|T_n\| = 1$ then (T_n) has a convergent subsequence with limit T_0, say, and also $\|T_n^{-1}\|$ is uniformly bounded. From this it follows that T_0 is an isomorphism and $\log(\|T_0\| \|T_0^{-1}\|) = \Delta(X, Y)$. Thus, for finite dimensional spaces, $\Delta(X, Y) = 0$ if and only if X and Y are isometrically equivalent and Δ is a metric on the isometric equivalence classes of Minkowski spaces of a given dimension.

Let $\mathcal{D}_d$ denote the collection of isometric equivalence classes of Minkowski spaces of dimension d. We have shown that $(\mathcal{D}_d, \Delta)$ is a metric space. It can be shown that $(\mathcal{D}_d, \Delta)$ is both compact and connected.

2.5 Approximation of convex sets and the Blaschke selection theorem

This section is concerned with the topological properties of various metric spaces defined in the preceding section. We begin with the space $(\mathcal{C}_b, \delta)$, *i.e.* the Hausdorff metric on the set of convex bodies in X. As we saw above, it does not matter which norm we consider on X but, for definiteness, we shall suppose some fixed norm $\|.\|$ throughout with corresponding ball B.

First, we shall show that if $K \in \mathcal{C}_b$ then K can be approximated arbitrarily closely by either polytopes or by convex bodies with analytic boundary, *i.e.* these subclasses are dense in $(\mathcal{C}_b, \delta)$. Next we show – after some other technical preliminaries – that (Euclidean) volume and area are continuous with respect to δ (*i.e.* we legitimize 2.3.1(vi) and 2.3.7(vi)). Finally, we consider the space $(\mathcal{K}, \delta)$, *i.e.* the space of non-empty compact subsets of X with the Hausdorff metric. Our aim is to prove the Blaschke selection theorem, which states that $(\mathcal{K}, \delta)$ is locally sequentially compact.

Theorem 2.5.1 *The collection of polytopes is dense in* $(\mathcal{C}_b, \delta)$.

Proof. Let $K \in \mathcal{C}_b$ and let $\epsilon > 0$ be given. The family of open balls $B(x, \epsilon)$ with centres x in K and radius ϵ cover K and hence there is a finite set $F = \{x_1, x_2, \ldots, x_n\}$ in K such that $K \subseteq \bigcup_{i=1}^{n} B(x_i, \epsilon)$, i.e. $K \subseteq F + \epsilon B$. Note that since K has non-empty interior, we can, if necessary, add points to F so that it spans the whole space. Let $P := \text{conv}(F)$. Since F is finite, P is a polytope, $P \in \mathcal{C}_b$, and we have $P \subseteq K \subseteq F + \epsilon B \subseteq P + \epsilon B$ so that $\delta(K, P) \leq \epsilon$, as required. ∎

The next theorem is more complicated. We give only the outline of the proof from Bonnesen and Fenchel [51]. There is also a very elegant one-page proof by Firey [161].

Theorem 2.5.2 *The collection of strictly convex sets with analytic boundary is dense in $(\mathcal{C}_b, \delta)$.*

Proof. If we establish that each polytope can be approximated arbitrarily closely by sets of the required kind, then the previous theorem shows that the same is true of all convex bodies. Let P be a polytope in X. Then P has a finite number N of facets each of which lies in a hyperplane of the form $\{x : f_i(x) = \alpha_i\}$. Therefore,

$$P = \{x : f_i(x) - \alpha_i \leq 0; i = 1, 2, \ldots, N\}.$$

Let $\varphi(x) := N^{-1} \sum_{i=1}^{N} \exp(\omega(f_i(x) - \alpha_i))$, where ω is some arbitrary (but large) positive real number. Let $Q := \{x : \varphi(x) \leq 1\}$. It is clear that $P \subseteq Q$ and not difficult to show that, for sufficiently large ω, $Q \subseteq P + \epsilon B$. Elementary calculus (see [51] for the details) shows that Q is convex, is strictly convex and has analytic boundary. ∎

Thus in many arguments dealing with convex bodies it is sufficient to show the result for all polytopes or for all convex bodies with "sufficiently smooth" boundary and then argue that "by continuity" the result holds for all convex bodies. A typical case is the theorem we have just proved where polytopes are much easier to deal with than arbitrary convex bodies because they are defined by a finite number of linear inequalities.

In Equation (2.21) the metric Δ_2 was defined for centrally symmetric bodies. It is equally well defined on the whole of $\mathcal{C}_i$ (Definition 2.1.2(iv)). Since volume and area behave simply with respect to scalar multiplication, for purposes involving these concepts the metric space $(\mathcal{C}_i, \Delta_2)$ is easier to deal with than $(\mathcal{C}_b, \delta)$. The next proposition shows, however, that topologically they are the same space.

Proposition 2.5.3 *The metrics Δ_2 and δ are topologically equivalent on $\mathcal{C}_i$.*

Proof. Let $K \in \mathcal{C}_i$. We need to show that a sequence K_n converges to K with respect to δ if and only if it does so with respect to Δ_2.

Since $K \in \mathcal{C}_i$ there are scalars β_1 and β_2 such that $\beta_1 B \subseteq K \subseteq \beta_2 B$. Suppose, also, that K_n is such that $\Delta_2(K, K_n) \leq \log \xi_n$. Then $K \subseteq \xi_n K_n$ and $K_n \subseteq \xi_n K \subseteq \xi_n \beta_2 B$. Hence $K_n \subseteq K + (\xi_n - 1)K \subseteq K + (\xi_n - 1)\beta_2 B$ and $K \subseteq K_n + (\xi_n - 1)K_n \subseteq K_n + \xi_n(\xi_n - 1)\beta_2 B$. Thus $\delta(K, K_n) \leq \xi_n(\xi_n - 1)\beta_2$ and if $\xi_n \to 1$ then $\delta(K, K_n) \to 0$, as required.

Conversely, suppose $\delta(K, K_n) < \eta_n$ with $\eta_n < \beta_1/2$. Then $K_n \subseteq K + \eta_n B \subseteq (1 + \beta_1^{-1}\eta_n)K$. Also $\beta_1 B \subseteq K \subseteq K_n + \eta_n B$ and since $\eta_n < \beta_1/2$ we have $\beta_1/2 B \subseteq K_n$. Hence $K \subseteq K_n + \eta_n B \subseteq K_n + 2\eta_n \beta_1^{-1} K_n = (1 + 2\eta_n \beta_1)K$. Consequently, $\Delta_2(K, K_n) \leq \log(1 + 2\eta_n\beta_1) \to 0$ as $\eta_n \to 0$. ∎

Corollary 2.5.4 *The collection of polytopes is dense in $(\mathcal{C}_i, \Delta_2)$; hence for $K \in \mathcal{C}_i$ and arbitrary $\xi > 1$ there is a polytope P such that $P \subseteq K \subseteq \xi P$.*

Proof. The first statement follows directly from Proposition 2.5.3 and Theorem 2.5.1. This gives us a polytope P' such that $K \subseteq \xi^{1/2} P'$ and $P' \subseteq \xi^{1/2} K$ but then $P := \xi^{-1/2} P'$ has the desired properties. ∎

Next suppose that X has a Euclidean structure. In §2.3, just prior to Theorem 2.3.13, reference was made to the fact that (Euclidean) surface area for arbitrary convex bodies could be obtained from that for polytopes "by an approximation argument". We make that more precise. Recall (Example 0.2.4(i)) that throughout the book *polytope* means convex polytope. The distinction is vital here because it is well known that a cylinder (for example) can be approximated arbitrarily closely (in the Hausdorff metric) by non-convex polytopes of arbitrarily large surface area.

Definition 2.5.5 *The (Euclidean) surface area of a convex body K, $\lambda(\partial K)$, is defined by*

$$\lambda(\partial K) := \inf\{\lambda(\partial P) : P \text{ a polytope with } K \subseteq P\}.$$

Proposition 2.5.6 *If K_1 and K_2 are convex bodies with $K_1 \subseteq K_2$ then $\lambda(\partial K_1) \leq \lambda(\partial K_2)$. Also $\lambda(\partial(\alpha K_1)) = \alpha^{d-1} \lambda(\partial K_1)$ and $\lambda(\partial(K_1 + x_0)) = \lambda(\partial K_1)$.*

Proof. The first statement is immediate from the definition. Since $K_1 \subseteq P$ if and only if $\alpha K_1 \subseteq \alpha P$ ($\alpha > 0$) and $K_1 \subseteq P$ if and only if $K_1 + x_0 \subseteq P + x_0$, the remaining statements now also follow from the definition. ∎

Proposition 2.5.7 *If K is a convex body, then*

$$\lambda(\partial K) = \sup\{\lambda(\partial P') : P' \text{ a polytope with } P' \subseteq K\}.$$

2.5 Approximation of convex sets and the Blaschke selection theorem 67

Proof. We may suppose $K \in \mathcal{C}_i$. Choose $\xi > 1$. By Corollary 2.5.4 there is a polytope P with $P \subseteq K \subseteq \xi P$. Then we have

$$\lambda(\partial P) \leq \sup\{\lambda(\partial P') : P' \subseteq K\} \leq \inf\{\lambda(\partial P'') : K \subseteq P''\} = \lambda(\partial K).$$

(The second inequality follows from the monotonicity, Proposition 2.5.6.) Also $\lambda(\partial K) = \inf\{\lambda(\partial P'') : K \subseteq P''\} \leq \lambda(\partial \xi P) = \xi^{d-1}\lambda(\partial P)$. Since ξ can be chosen as close to 1 as we please, the result follows. ∎

Theorem 2.5.8 *Both volume and surface area are continuous functions on $(\mathcal{C}_i, \Delta_2)$ and hence on $(\mathcal{C}_b, \delta)$.*

Proof. If $K \in \mathcal{C}_i$ and $\epsilon > 0$ is given, choose η so close to 1 that $\eta^d(\eta^d - 1)\lambda(K) < \epsilon$. If $\Delta_2(K, K') < \log \eta$ we have $K' \subseteq \eta K$ and so $\lambda(K') \leq \eta^d \lambda(K)$, i.e. $\lambda(K') - \lambda(K) \leq (\eta^d - 1)\lambda(K) < \epsilon$. Also $K \subseteq \eta K'$ and so $\lambda(K) - \lambda(K') \leq (\eta^d - 1)\lambda(K') \leq \eta^d(\eta^d - 1)\lambda(K) < \epsilon$. This completes the proof for λ. The proof for surface area is exactly the same except that factors of η^{d-1} are involved. ∎

We now come to the Blaschke selection theorem and first consider the metric space $(\mathcal{K}, \delta)$. For metric spaces "sequentially compact", "compact" and "complete and totally bounded" are all equivalent. We approach the problem via the last of these and follow the development in [293].

Theorem 2.5.9 *If $\{K_n\}$ is a Cauchy sequence in $(\mathcal{K}, \delta)$ then $\{K_n\}$ converges to K_0, where*

$$K_0 := \bigcap_{i=1}^{\infty} c\ell \left(\bigcup_{j=1}^{\infty} K_j \right). \tag{2.23}$$

Moreover, K_0 is a non-empty compact set and so $(\mathcal{K}, \delta)$ is complete.

Proof. It is clear from (2.23) that K_0 is closed. Let $\epsilon > 0$ be given. There exists $n_0(\epsilon)$ such that $\delta(K_i, K_j) < \epsilon$ for all $i, j \geq n_0(\epsilon)$, i.e. $K_i \subseteq K_j + \epsilon B$ and $K_j \subseteq K_i + \epsilon B$ for all $i, j \geq n_0(\epsilon)$. From the second of these containments we get

$$\bigcup_{j \geq i} K_j \subseteq K_i + \epsilon B \qquad (i \geq n_0(\epsilon))$$

and hence

$$c\ell \left(\bigcup_{j \geq i} K_j \right) \subseteq K_i + 2\epsilon B \qquad (i \geq n_0(\epsilon))$$

so that $K_0 \subseteq K_i + 2\epsilon B$ ($i \geq n_0(\epsilon)$). This containment also shows that K_0 is bounded and hence compact.

For an inequality in the opposite direction we must work a little harder. For each natural number k, there exists $n_k(\epsilon)$ such that

$$\delta(K_i, K_j) < 2^{-k}\epsilon, \qquad \forall i, j \geq n_k(\epsilon).$$

Now begin an inductive construction by choosing $m_0 \geq n_0(\epsilon)$ and $x_0 \in K_{m_0}$; choose $m_1 > \max\{m_0, n_1(\epsilon)\}$ and $x_1 \in K_{m_1}$ with $\|x_0 - x_1\| < \epsilon$. If $m_0, m_1, \ldots, m_{k-1}$ and $x_0, x_1, \ldots, x_{k-1}$ have been chosen, choose $m_k > \max\{m_{k-1}, n_k(\epsilon)\}$ and $x_k \in K_{m_k}$ with $\|x_k - x_{k-1}\| < 2^{-(k-1)}\epsilon$. Then $\{x_k\}$ is a Cauchy sequence in $(X, \|.\|)$ and so converges to $\bar{x}$. For any natural number n, there exists $m_{n'}$ with $m_{n'} > n$ (the sequence m_k is strictly increasing). Thus $x_k \in \cup_{j \geq n} K_j$ for all $k \geq n'$. This implies that $\bar{x} \in c\ell(\cup_{j \geq n} K_j)$. Since n was arbitrary it follows that $\bar{x} \in K_0$. Thus $K_0 \neq \emptyset$.

We also have

$$\|\bar{x} - x_0\| = \lim_{n \to \infty} \|x_n - x_0\| \leq \lim_{n \to \infty} \sum_{k=1}^{n} \|x_k - x_{k-1}\| \leq \lim_{n \to \infty} \sum_{k=1}^{n} 2^{-(k-1)}\epsilon < 2\epsilon.$$

For each $x_0 \in K_{m_0}$ we have constructed a point $\bar{x}$ in K_0 with $\|\bar{x} - x_0\| \leq 2\epsilon$. Hence $K_{m_0} \subseteq K_0 + 2\epsilon B$. However, m_0 is an arbitrary integer $\geq n_0(\epsilon)$. Thus for all $i \geq n_0(\epsilon)$ we have $K_0 \subseteq K_i + 2\epsilon B$ and $K_i \subseteq K_0 + 2\epsilon B$, i.e. $\delta(K_i, K_0) \to 0$ as $i \to \infty$. ∎

Next we show that if we restrict the compact sets K to a bounded region of $(X, \|.\|)$, then $(\mathcal{K}, \delta)$ is totally bounded. As a corollary to the two results we get the compactness theorem for uniformly bounded compact sets. As a matter of notation, let $\mathcal{K}_a$ denote the set of non-empty, compact subsets of X which are contained in the ball $B[0, a]$.

Theorem 2.5.10 *The metric space $(\mathcal{K}_a, \delta)$ is totally bounded.*

Proof. Let $\epsilon > 0$ be given. The ball $B[0, a]$ is compact and so totally bounded. Hence there exists a finite ϵ-net for $B[0, a]$. Denote this by $F = \{x_1, x_2, \ldots, x_m\}$. Let J be the set of all non-empty subsets of F. Then J is a finite subset of $(\mathcal{K}_a, \delta)$. We show that it is an ϵ-net for $(\mathcal{K}_a, \delta)$. Let $K \in \mathcal{K}_a$ and let $F_K := \{x_i \in F : \exists x \in K$ with $\|x_i - x\| < \epsilon\}$. The definition of F_K implies $F_K \subseteq K + \epsilon B$. On the other hand, since F is an ϵ-net for $B[0, a]$, for each $x \in K$, there exists $x_i \in F$ with $\|x - x_i\| < \epsilon$. Hence $x_i \in F_K$ and $x \in F_K + \epsilon B$, i.e. $K \subseteq F_K + \epsilon B$. Thus $\delta(K, F_K) \leq \epsilon$ and the proof is complete. ∎

Corollary 2.5.11 *The space $(\mathcal{K}_a, \delta)$ is compact.*

Proof. All we need to know in addition to Theorems 2.5.9 and 2.5.10 is that $\mathcal{K}_a$ is a closed subset of $\mathcal{K}$. But this follows from the representation (2.23) for K_0, for if each $K_i \subseteq B[0, a]$ we have $K_0 \subseteq B[0, a]$. ∎

2.5 Approximation of convex sets and the Blaschke selection theorem

The purpose of the next, technical lemma is to enable us to deduce that the compact convex sets in X form a closed subset of $(\mathcal{K}, \delta)$.

Lemma 2.5.12 *If $\{K_n\}$ is a sequence in $(\mathcal{K}, \delta)$ such that $K_n \to K_0$ then*

$$K_0 = \bigcap_{i=1}^{\infty} c\ell \left(\bigcup_{j=1}^{\infty} K_j \right) = \bigcap_{\epsilon > 0} \bigcup_{i=1}^{\infty} \bigcap_{j \geq i} (K_j + \epsilon B).$$

Proof. The first equation is already known from Theorem 2.5.9. For the second, suppose first that $x \in \bigcap_{\epsilon > 0} \bigcup_{i=1}^{\infty} \bigcap_{j \geq i}(K_j + \epsilon B)$; then for all $\epsilon > 0$, there exists $i(\epsilon)$ such that $x \in K_j + \epsilon B$ for all $j \geq i(\epsilon)$. If now m is given, there is an n with $n \geq m$ and $n \geq i(\epsilon)$. Consequently,

$$x \in K_n + \epsilon B \subseteq \epsilon B + \bigcup_{n \geq m} K_n.$$

But ϵ is arbitrary (*i.e.* independent of m) and so $x \in c\ell(\bigcup_{n \geq m} K_n)$ and hence $x \in K_0$.

Secondly, since $\{K_n\}$ converges to K_0, if $\epsilon > 0$ is given there exists $i(\epsilon)$ such that $K_0 \subseteq K_j + \epsilon B$ for all $j \geq i(\epsilon)$. In other words,

$$K_0 \subseteq \bigcap_{\epsilon > 0} \bigcup_{i=1}^{\infty} \bigcap_{j \geq i} (K_j + \epsilon B). \qquad \blacksquare$$

Theorem 2.5.13 *The space $(\mathcal{C}, \delta)$ of non-empty, compact, convex sets in $(X, \|.\|)$ is a closed subset of $(\mathcal{K}, \delta)$.*

Proof. If $\{C_n\}$ is a sequence in $(\mathcal{C}, \delta)$ which converges to $K_0 \in \mathcal{K}$ then, by Lemma 2.5.12,

$$K_0 = \bigcap_{\epsilon > 0} \bigcup_{i=1}^{\infty} \bigcap_{j \geq i} (C_j + \epsilon B).$$

Because each $C_j + \epsilon B$ is convex and the union is taken over a nested family of convex sets, this representation shows that K_0 is convex. $\blacksquare$

Now we can put all these pieces together and derive the Blaschke selection theorem.

Theorem 2.5.14 *The space $(\mathcal{C}_a, \delta)$ of non-empty, compact, convex sets in $(X, \|.\|)$ which are uniformly bounded by a, and equipped with the Hausdorff metric, is sequentially compact.*

As a typical example of the way in which this theorem is used we prove the existence of a solution to the isoperimetric problem among compact convex sets. Another important and typical example of its use comes in the proof of the existence of the Löwner ellipsoid (Theorem 3.3.1).

Theorem 2.5.15 *Let C_V be the family of convex bodies which are uniformly bounded by some fixed radius a and which all have the same fixed volume V. Then there exists a body in C_V with minimal surface area.*

Proof. Since λ is a continuous function on (C_b, δ) and since any convex set with volume $V > 0$ is necessarily a convex body it follows that C_V is a closed subset of (C_a, δ) and hence is compact. Because surface area is also continuous, it follows that it attains its infimum on this set. ∎

In many of the applications of the Blaschke theorem the convex bodies under consideration are all of a particular type. We then need to know that the limit body is of the same type; *i.e.* these bodies form a closed subset. The final results of this section deal with this situation.

Proposition 2.5.16 *If K is a fixed convex body and if $\{T_i\}$ is a sequence of linear transformations such that the sequence of convex bodies $K_i := T_i K$ converges in the metric space (C, δ) to $\hat{K}$ then $\hat{K} = TK$ for some linear transformation T.*

Proof. Let E be an ellipsoid generating an auxiliary Euclidean metric. Because K is a body and $\hat{K}$ is in C (and so is compact), there exist $r_1, r_2 > 0$ such that $r_1 E \subseteq K$ and $\hat{K} \subseteq r_2 E$. If $\epsilon > 0$ then for all sufficiently large i we have $K_i \subseteq \hat{K} + \epsilon E \subseteq (r_2 + \epsilon)E$ and hence $r_1 T_i(E) \subseteq T_i K = K_i \subseteq (r_2 + \epsilon)E$. Therefore, $\|T_i\| \leq r_1^{-1}(r_2 + \epsilon)$ and so the sequence $\{T_i\}$ is uniformly bounded in the finite dimensional normed space of linear transformations on $\mathbb{R}^d$. Thus this sequence has a subsequence $\{T_{i_j}\}$ converging in norm to a linear transformation T. It follows that $T_{i_j} K$ converges to TK in the Hausdorff metric δ and hence that $TK = \hat{K}$. ∎

Remark. It does not follow that T is invertible and so $\hat{K}$ need not be a body. For example, it is possible to have a sequence of ellipses converging to a line segment.

Corollary 2.5.17 *If $\{K_i\}$ is a sequence of ellipsoids (resp. parallelotopes) in $\mathbb{R}^d$ which converges in (C, δ) to a convex set K then K is a (possibly degenerate) ellipsoid (resp. parallelotope).*

Proof. Any ellipsoid (resp. parallelotope) is the image under an affine map (linear composed with translation) of the Euclidean ball E (resp. the unit cube C). The convergence of $\{K_i\}$ implies that the translations are bounded and hence have a convergent subsequence. The result now follows from Proposition 2.5.16. ∎

Definition 2.5.18 *A **position** of K is the image of K under a linear transformation T with $\det T = 1$.*

Corollary 2.5.19 *If $\{K_i\}$ is a sequence of positions of a fixed body K and if $\{K_i\}$ converges in $(\mathcal{C}, \delta)$ to $\hat{K}$ then $\hat{K}$ is a position of K.*

Proof. Since $K_i = T_i K$ with $\det T_i = 1$ we have $\lambda(K_i) = \lambda(K)$ for all i. The continuity of λ implies that $\lambda(\hat{K}) = \lambda(K)$ and so, with T as in Proposition 2.5.16, $\det T = 1$. ∎

2.6 Notes

General references for most of the material in this chapter are the books by Schneider [479] (which deals with recent work as well as the background material), Bonnesen and Fenchel [51] and Eggleston [138]. An inviting introduction to the subject from a modern viewpoint and including many open problems is the survey article of Berger [31].

Support and separation theorems are geometric versions of the Hahn–Banach theorem. In finite dimensional spaces such theorems go back to Minkowski. In infinite dimensional spaces Mazur [361] proved the basic support theorem that through each boundary point of a convex body there passes at least one support hyperplane. He also proved that the set of points at which the supporting hyperplane is unique forms a dense G_δ subset of the boundary. Further major steps in the development of separation theorems were made by Eidelheit [141] and Kakutani [279]. In Euclidean spaces a straightforward approach to separation theorems is afforded by the notions of shortest distance and perpendicularity. This approach is well expounded by Botts [52]. However, in the same way that topology is not essential for the extension theorems neither is it for the separation theorem. Expositions of these non-topological versions usually begin with Stone's lemma [496]; see also Hammer [241], which states that two disjoint convex sets are contained, respectively, in two complementary disjoint convex sets. A comprehensive treatment of geometric forms of the Hahn–Banach theorem is contained in the book by Holmes [261] and there is a survey article by Klee [291]. Usually, support and separation theorems are proved in the context of closed or compact convex sets. Klee [290], however, studied separation theorems for convex sets which are intersections of *open* half-spaces, *i.e.* determined by systems of *strict* linear inequalities.

The results in §2.2 are all standard. The important fact (Theorem 2.2.8) that any sublinear functional is the support function of a convex set is given three different proofs by Schneider [479], pp. 38–39. The reason for the brevity of our proof is that we assume the continuity of the sublinear functional. The fact that convex functions are continuous is Theorem 1.5.1, p. 23 in [479]. Theorem 2.2.9 and its corollary are part of the folk-lore but an explicit statement is given by McMullen and Shephard [372], p. 70. These two results may be paraphrased by saying "the dual of a cross-section is the projection of the dual and *vice versa*". This statement is one example of the duality between projections and cross-sections which will play an important part in subsequent chapters and which is the basis for the relationship between Chapters 6 and 7.

Bounds for the volume product $\lambda(K)\lambda^*(K^\circ)$ appear to have arisen first in connection with the geometry of numbers (see, *e.g.*, Cassels [93], Gruber and Lekkerkerker [219] or the survey article by Gruber [216]) and were first investigated by Mahler [346]. He conjectured that the exact lower bound for centrally symmetric K is $4^d/d!$, which is attained by bodies (including parallelotopes and cross-polytopes) that are built up from line segments by Cartesian product and suspension operations. He proved this conjecture for $d = 2$ and was able to establish a very much smaller lower bound for general d. This lower bound was improved by Dvoretzky and Rogers [134] and further by Bambah [18] (to $4^d/(d!)^2$ for general K and to $\epsilon_d^2/d^{d/2}$ for centrally symmetric K). The strongest present result is due to Bourgain and Milman [55, 56], who proved that there is a universal constant $c > 0$ such that $\lambda(K)\lambda^*(K^\circ) \geq c^d \epsilon_d^2$. Mahler's conjecture for the restricted class of zonoids (and their duals) has been proved by Reisner [437, 438] and a shorter proof was given by Gordon, Meyer and Reisner [196]. Among zonoids, Reisner [438] showed that equality holds only for parallelotopes. Saint Raymond [453] has also established Mahler's conjecture for convex bodies with d linearly independent hyperplanes of symmetry (which can then be chosen as the coordinate hyperplanes). For equality in this last case see Meyer [375] and Reisner [439].

The exact upper bound was established by Blaschke [39] (see also [40] and [41]) for the cases $d = 2, 3$ and by Santaló [456] for higher dimensions. The fact that equality occurs only for ellipsoids was completely established by Saint Raymond [453] for symmetric bodies and by Petty [421] for arbitrary bodies. Simpler proofs of the Blaschke–Santaló inequality (based on Steiner symmetrization and the Brunn–Minkowski inequality) for centrally symmetric K have been given by Saint Raymond [453] and by Meyer and Pajor [376, 377]. A surprising use for this upper bound was found by Heil [251] in the estimation of the eigenvalues of a differential equation.

The Brunn–Minkowski inequality was first established for dimensions 2 and 3 by Brunn [60, 61] and for general d by Minkowski [389]. Schneider [479] gives two proofs (pp. 310–314), the first due to Kneser and Süss [294] and the second to Knothe [295]. Another proof based on the divergence theorem has been given by Gromov [207]. Schneider [479] gives generalizations, subsequent developments and applications of this fundamental inequality.

Schneider [479], p. 281, states baldly that "the theory of mixed volumes was created by Minkowski" [388, 390]. Inequality 2.3.8 was proved in [388]. As usual we refer to [479] both for proofs of these results and for subsequent developments. The book by Burago and Zalgaller [62] offers a wide-ranging account of the inequalities given in this chapter.

The formulas in Theorem 2.3.13 are the classical ones for volume and surface area. Cauchy's formula, which expresses the surface area as an average of the projected areas, is a special case of Kubota's integral recursion formulas (see [51] as well as [479]). References for the proofs of these formulas are [51] and [138].

The study of topologies on collections of subsets of a metric space was initiated by Hausdorff [248] (see also [250]) with the introduction of what is now called the Hausdorff metric on collections of closed and bounded subsets. This idea was extended to the idea of a topology on the closed subsets of a topological space by Vietoris [520–522]. Notions of convergence for sequences of sets were investigated by Kuratowski [307] (see also [308]) and the Polish school. In general these notions are distinct, although all share a "two-sided" feature: on the one hand a set is close to another if it is nearby and not much larger, and on the other hand it is close if it is nearby and not much smaller. The topology results from combining the two. On collections of compact sets, however, all these concepts coincide so that, for many purposes, compact sets are the natural ones to consider. A general reference for these topics is [293]. A survey of results about the metric space $(\mathcal{C}, \delta)$ has been made by Gruber [214]. In addition to the metric structure, he discusses the lattice structure and the semi-group structure on this space. The isometries of $(\mathcal{C}, \delta)$ have been investigated by Schneider [474], who showed that an isometry that is surjective must be one induced by an isometry of $\mathbb{R}^d$, and by Gruber and Lettl [220], who showed that an isometry is the composition of an isometry of $\mathbb{R}^d$ with Minkowski addition by some fixed compact convex set. See [214] for additional references. The "multiplicative" distances Δ_1 and Δ_2 are special cases of a construction for partially ordered linear spaces given in [507]. The Banach–Mazur distance is defined in Banach's book [21].

The metric space $(\mathcal{D}_d, \Delta)$ is compact and connected. The compactness (although using a different metric) was shown by Macbeath [343]; see also Dvoretzky [133], Shephard [485] and Schäffer [459]. The connectedness was discussed by McGuigan [363]. Natural questions to consider are the size of its diameter and, if that is known to be η say, whether there exist "central" spaces X in the sense that, for all other spaces Y, $\Delta(X, Y) \leq \eta/2$. The actual diameter is known only for $d = 2$; Stromquist [497] showed that it is $\log(\frac{3}{2})$ attained uniquely between the Euclidean ball and the regular hexagon. A closely related result is that of Lassak [311], which shows that if K is any convex set (not necessarily symmetric) and if B is an arbitrary Minkowski unit ball in the plane, then $\Delta(K, B) \leq \log(\sqrt{2} + 1)$. In higher dimensions, Gluskin [180] has shown that there is an absolute constant c such that $\operatorname{diam} \mathcal{D}_d \geq \log(cd)$. An upper bound on the diameter is given by John's result [272] on the distance to an ellipsoid, which is discussed in the next chapter. Related results will be given in the notes to that chapter.

The metric Δ is the starting point for much of the modern local theory of Banach spaces that considers various asymptotic properties with respect to Δ. There are several books dealing with topics from this local theory, *e.g.* Tomczak-Jaegermann [513], Pisier [430] and Milman and Schechtman [383]. A brief survey of some of these asymptotic properties is contained in Pełczyński [406]. There are more extensive surveys by Figiel [155] and by Lindenstrauss and Milman [327]. Some aspects of the local theory will be dealt with briefly in Chapter 9.

There are many results dealing with the approximation of convex sets by convex sets of a special type. Gruber has written two good survey articles on this topic [211, 215]. The fact that a convex body can be approximated by ones with analytic boundary is due to Minkowski [388]. Proofs which yield additional properties of the approximating sets can be found in Firey [161], Schneider [479], p. 158, [477] and recent articles of Schmuckenschlaeger [469] and of Grinberg and Zhang [203].

The Blaschke selection theorem is a very important tool in convexity theory. It was first given in [38]. The view presented here that the theorem asserts the sequential compactness of a metric space is due to Hausdorff [249]; see [293] for a full discussion of related results. The proof that $(\mathcal{K}, \delta)$ is complete is due to Castaing and Valadier [94]; see also Schneider for smoother proofs of 2.5.9 and 2.5.13 (Theorems 1.8.2 and 1.8.5 in [479]).

There are a variety of metrics possible on spaces of convex sets. A comparison of four of these was made by Shephard and Webster [488]. The idea, inherent in the Banach–Mazur distance, of considering affine equivalence classes of convex sets and the quotient topology (coming from the space $(\mathcal{C}_b, \delta)$) on those classes was also considered by Macbeath [343], who showed that the quotient topology is metrizable (his metric is different from the Banach–Mazur metric) and compact. See also Asplund [9]. Ewald and Shephard [146] have shown that by using equivalence classes of convex sets it is also possible to extend the semi-linear structure of Minkowski addition and scalar multiplication to a vector space structure and impose a norm on that space. Gruber's survey article [214] deals with these matters (among other things).

3
Comparisons and contrasts with Euclidean space

Euclidean space is, both historically and psychologically, our original idea of how a finite dimensional normed space behaves. The results of this chapter express in a variety of ways to what extent a Minkowski space is similar to, and how it differs from, a Euclidean space. There are three basic themes: (a) the group of isometries of the Minkowski space; (b) the notion of "perpendicularity" or "normality"; (c) the distance (in the sense of Banach–Mazur, Definition 2.4.6) to Euclidean space.

The chapter is divided into four sections. The first one is brief and presents the theorem of Mazur and Ulam, which states that any isometry of one normed space onto another one is necessarily linear. This is not true if the isometry is not surjective. It is not only a finite dimensional theorem but not usually prominent in books on functional analysis. The second section takes up the concept of normality. There is one main theorem but a variety of corollaries which indicate different ways of viewing the theorem. One is a geometric view stating that the unit ball in a Minkowski space (X, B) can be circumscribed by a parallelotope with certain properties. Another is an algebraic view that a Minkowski space (X, B) and its dual (X^*, B°) have a pair of dual "orthonormal" bases. The third section is an important one. It contains the central theorem of Löwner that the unit ball in a Minkowski space has both an inscribed ellipsoid of maximal volume and a circumscribed ellipsoid of minimal volume and, more importantly, that these ellipsoids are unique. This has far-reaching consequences both technically (as a means of proving theorems) and conceptually. For example, it follows that the group of isometries of a Minkowski space is a subgroup of that of a Euclidean space. It is also the Löwner ellipsoid which is needed to prove John's result that the Banach–Mazur distance of any Minkowski space to Euclidean space is at most $(\log d)/2$ (d is the dimension). The final section presents a small selection from the very large literature on characterizations of Euclidean space among Minkowski spaces. On the one hand, this shows what not to expect in an arbitrary Minkowski space and, on the other, gives a variety of possible approaches to try if one endeavours to show that a particular property of Minkowski space is, in fact, characteristic of Euclidean space.

3.1 The Mazur–Ulam theorem

The proof of this basic theorem about isometries of normed spaces depends on the concept of the metric centre of a set in a metric space.

Definition 3.1.1 *If (X, d) is a metric space and if A is a bounded subset of X then we say that x_0 is a centre of A of the first order if $d(x_0, a) \leq \operatorname{diam} A/2$ for all a in A. Further, x_0 is a centre of A of the nth order if it is a centre of the first order of C_{n-1}, the set of all centres of the $(n-1)$st order which belong to A. A point x_0 is a **metric centre** of A if, for all n, it is a centre of A of the nth order.*

Remarks

(i) It is clear from the definition that $\operatorname{diam} C_1 \leq \operatorname{diam} A/2$ and, in general, $\operatorname{diam} C_n \leq (\operatorname{diam} C_{n-1})/2$, so that $\operatorname{diam} C_n \to 0$ and A can have at most one metric centre.

(ii) In a normed linear space, routine calculations show that if the set A is symmetric about x_0 (i.e. $x \in A$ implies $2x_0 - x \in A$) then x_0 is the metric centre of A.

(iii) Since Definitions 3.1.1 are entirely in terms of the metric, it follows that if T is an isometry of X into another metric space Y and if x_0 is the metric centre of A then Tx_0 is the metric centre of $T(A)$.

With these preliminaries we can now give the elegant proof of the theorem of Mazur and Ulam [362].

Theorem 3.1.2 (Mazur–Ulam) *If X and Y are normed linear spaces and if T is an isometry of X onto Y with $T(0) = 0$ then T is linear.*

Proof. We shall show that T is additive. Since $T(0) = 0$ and T is certainly continuous, it follows by a standard argument using dyadic rationals that T is linear.

Let $x_1, x_2 \in X$ and let $A := \{x \in X : \|x_1 - x\| = \|x_2 - x\| = \|x_1 - x_2\|/2\}$. Simple calculations with the norm show that A is symmetric about $(x_1 + x_2)/2$ and so $(x_1 + x_2)/2$ is the metric centre of A. It follows that $T((x_1 + x_2)/2)$ is the metric centre of $T(A)$. However, because T maps X onto Y (and T is an isometry),

$$T(A) = \{Tx \in Y : \|x_1 - x\| = \|x_2 - x\| = \|x_1 - x_2\|/2\}$$
$$= \{y \in Y : \|Tx_1 - y\| = \|Tx_2 - y\| = \|Tx_1 - Tx_2\|/2\}.$$

Thus $T(A)$ is symmetric about $(Tx_1 + Tx_2)/2$, which is, therefore, the metric centre of $T(A)$. Hence

$$T((x_1 + x_2)/2) = (Tx_1 + Tx_2)/2, \qquad \forall x_1, x_2 \in X.$$

Setting $x_1 = 0$ (resp. $x_2 = 0$) in this equation and using $T(0) = 0$, we get

$$T(x_i/2) = T(x_i)/2 \qquad (i = 1, 2).$$

Thus $T((x_1 + x_2)/2) = T(x_1/2) + T(x_2/2)$. ∎

It is important to emphasize that surjectivity is crucial to this theorem. Also it is a theorem about spaces over the real field: complex conjugation is an isometry of $\mathbb{C}$ which is not complex linear.

Example 3.1.3 Let X be the real line $\mathbb{R}$ with the usual metric and let Y be $\mathbb{R}^2$ with the l_∞-norm. If $T(x) := (x, \sin x)$ then T is a non-linear, isometric embedding of X into Y. In fact, any function with Lipschitz constant ≤ 1 will do in place of sine.

3.2 Normality in Minkowski space

There are two key ideas in this section. One is Birkhoff's definition of normality given in Definition 3.2.2. Gruber [217], however, points out that the concept is older. In dealing with a problem in the calculus of variations, Carathéodory posed the problem of when normality (in the sense of Definition 3.2.2) is symmetric. Blaschke [36] showed that for $d \geq 3$ this property characterizes Euclidean space. This problem of Carathéodory was also one of the motives for the original construction of Radon curves [432]. The second is Taylor's idea of considering circumscribed parallelotopes and inscribed simplices to the unit ball B of a Minkowski space X. Taylor's theorem (3.2.1) is extremely general. For the purposes of Minkowski geometry the relevant statements are the corollaries which follow, especially 3.2.4 and 3.2.5.

Theorem 3.2.1 (Taylor [502]) *Let A be a closed, bounded set in $\mathbb{R}^d$ that spans $\mathbb{R}^d$. Then there exist points $x_1, x_2, \ldots, x_d$ in A and hyperplanes $H_1, H_2, \ldots, H_d$ such that*

(i) *for each i, $x_i \in H_i$;*
(ii) *for each i, H_i is parallel to the span of $\{x_1, \ldots, x_{i-1}, x_{i+1}, \ldots, x_d\}$;*
(iii) *for each i, H_i supports A at x_i.*

An illustration of the statement of this theorem with $d = 2$ is shown in Figure 3.1.

Proof. Consider any d points $\{y_1, y_2, \ldots, y_d\} \subseteq A$ and let $\lambda[y_1, y_2, \ldots, y_d]$ denote the volume of the parallelotope spanned by these vectors. Since A^d is compact

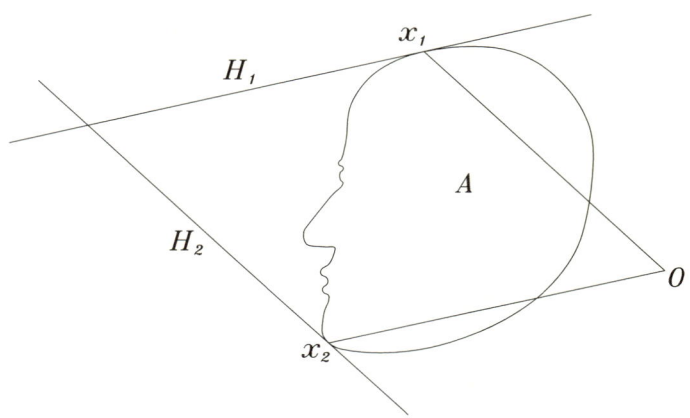

Figure 3.1

and λ is continuous, $\lambda[y_1, \ldots, y_d]$ attains a maximum at some set $\{x_1, x_2, \ldots, x_d\}$. Since A spans $\mathbb{R}^d$, λ takes on positive values and so the maximum is positive and the set $\{x_1, x_2, \ldots, x_d\}$ is linearly independent. Let H_i be the hyperplane $x_i + \text{span}\{x_1, x_2, \ldots, x_{i-1}, x_{i+1}, \ldots, x_d\}$; i.e. H_i is the hyperplane through x_i and parallel to the subspace spanned by the remainder. Then the sets $\{x_1, \ldots, x_d\}$ and $\{H_1, \ldots, H_d\}$ certainly satisfy (i) and (ii).

To complete the proof, suppose that (iii) is not satisfied for some i. Then there exists a point $x_i' \in A$ strictly on the opposite side of the hyperplane H_i from $x_1, x_2, \ldots, x_{i-1}, x_{i+1}, \ldots, x_d$. If we replace x_i by x_i' then we obtain a set $\{x_1, x_2, \ldots, x_i', \ldots, x_d\}$ with

$$\lambda[x_1, x_2, \ldots, x_i', \ldots, x_d] > \lambda[x_1, x_2, \ldots, x_i, \ldots, x_d],$$

which contradicts the maximality. ∎

Remarks

(i) In this theorem, the position of the origin is irrelevant.

(ii) Since $\lambda[x_1, x_2, \ldots, x_d] = d!\lambda(\text{conv}\{0, x_1, x_2, \ldots, x_d\})$, it follows that $\text{conv}\{0, x_1, x_2, \ldots, x_d\}$ is a simplex of maximal volume among all those with one vertex at 0 and the other vertices in A.

In Minkowski space, if we take for A the unit ball B then this theorem has a variety of reformulations. But first an important definition.

Definition 3.2.2 *If $(X, \|.\|)$ is a normed linear space and if $x, y \in X$ then we say that x is **normal** to y and write $x \dashv y$ if $\|x + \alpha y\| \geq \|x\|$ for all α in $\mathbb{R}$.*

3.2 Normality in Minkowski space

Geometrically this means that $x \dashv y$ if and only if the line $x + \alpha y$ ($\alpha \in \mathbb{R}$) supports the ball $B[0, \|x\|]$ at x (see Figure 3.2). The Hahn–Banach theorem then implies that $x + \alpha y$ lies in a hyperplane which supports $B[0, \|x\|]$ at x. It should be emphasized that $\dashv$ is not a symmetric relation; if $x \dashv y$ it is not usually true that $y \dashv x$.

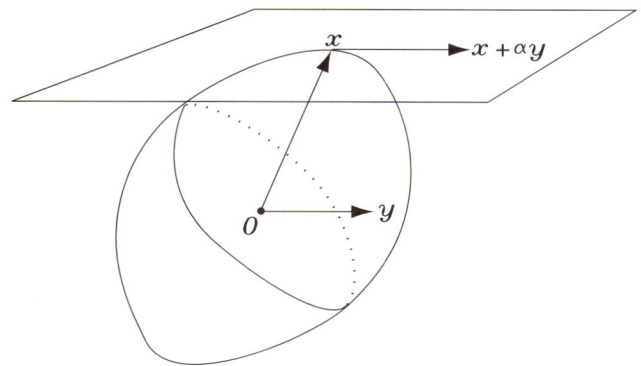

Figure 3.2

Corollary 3.2.3 *If B is the unit ball in a Minkowski space then there exists a basis $(x_1, x_2, \ldots, x_d)$ such that $\|x_i\| = 1$ and $x_i \dashv x_j$ for all i and j with $i \neq j$; i.e. each pair of basis vectors is mutually normal.*

Proof. Let $(x_1, x_2, \ldots, x_d)$ be an ordered set given by Theorem 3.2.1 when we take $A = B$. Then, since H_i supports B at x_i, we have $\|x_i\| = 1$ and also, since $x_i + x_j \in H_i$, we have $x_i \dashv x_j$ for all $j \neq i$. ∎

Corollary 3.2.4 *If B and B° are the unit balls in a Minkowski space X and its dual X^* then there are bases $(x_1, x_2, \ldots, x_d)$ in X and $(f_1, f_2, \ldots, f_d)$ in X^* such that $\|x_i\| = \|f_i\| = 1$ for all i and $f_i(x_j) = \delta_{ij}$ for all i, j.*

Proof. Let $(x_1, x_2, \ldots, x_d)$ be a basis for X as given in Corollary 3.2.3. Then define f_i on X by $f_i(\sum_{j=1}^{d} \alpha_j x_j) := \alpha_i$. Then we have $\|x_i\| = 1$ and $f_i(x_j) = \delta_{ij}$. That $\|f_i\| = 1$ follows from the fact that, in the notation of Theorem 3.2.1, $H_i = \{x : f_i(x) = 1\}$ and this hyperplane supports B at x_i. ∎

Corollary 3.2.5 *If B is the unit ball in a Minkowski space then there is a parallelotope C with 2^d vertices at $\{\pm x_1 \pm x_2 \cdots \pm x_d\}$ such that $B \subseteq C$ and each face of C supports B at the centre of that face.*

Proof. With $(x_1, x_2, \ldots, x_d)$ as in Corollary 3.2.3, consider the parallelotope C defined by

$$C := \text{conv}\{\pm x_1 \pm x_2 \pm \cdots \pm x_d\}.$$

Then the face of C that contains x_i lies in the hyperplane H_i and so supports B at x_i, which is the centre of that face. ∎

Corollary 3.2.3 allows one to introduce a Euclidean structure into the Minkowski space (X, B) in a more or less canonical way (the problem is that the basis $(x_1, x_2, \ldots, x_d)$ is not unique). Thus if we take $(x_1, x_2, \ldots, x_d)$ as the basis for a coordinate system then with the corresponding inner product (see §0.1) this becomes a genuine orthonormal basis. With this Euclidean structure the circumscribing parallelotope C is a d-dimensional cube of side 2. Since B is convex, the d-dimensional cross-polytope

$$CP := \text{conv}\{\pm x_1, \pm x_2, \ldots, \pm x_d\}$$

is inscribed to B and meets ∂B at each vertex. We summarize this in the following corollary.

Corollary 3.2.6 *If (X, B) is a Minkowski space then there is a Euclidean structure on X with orthonormal basis $(x_1, x_2, \ldots, x_d)$ such that the unit cross-polytope* $\text{conv}\{\pm x_1, \pm x_2, \ldots, \pm x_d\}$ *is inscribed to B and the cube* $\text{conv}\{\pm x_1 \pm x_2 \cdots \pm x_d\}$ *is circumscribed to B.*

We already know from Chapter 1 that the Minkowski norm is equivalent to a Euclidean norm. The construction in Corollary 3.2.6 gives us an estimate of the distance between the two norms. A better bound on the Banach–Mazur distance will be given in the next section.

Corollary 3.2.7 *If (X, B) is a Minkowski space then there is an ellipsoid E in X such that $d^{-1/2}E \subseteq B \subseteq d^{1/2}E$. Hence $\Delta(B, E) \leq \log d$.*

Proof. Let E be the unit ball of the Euclidean structure given in Corollary 3.2.6. Then we have the following sequence of inclusions:

$$d^{-1/2}E \subseteq \text{conv}\{\pm x_1, \pm x_2, \ldots, \pm x_d\} \subseteq B \subseteq \text{conv}\{\pm x_1 \pm x_2 \cdots \pm x_d\} \subseteq d^{1/2}E,$$

from which the result follows. ∎

3.3 The Löwner ellipsoid

Given a Minkowski space (X, B) we would like to be able to introduce a Euclidean structure in a more canonical way than in the last section. In other words, we would like the associated ellipsoid to be unique. Löwner's theorem asserts the existence and uniqueness of both an inscribed ellipsoid of maximal volume and a circumscribed ellipsoid of minimal volume. Either one of these is referred to in the literature as "the" Löwner ellipsoid. Löwner himself did not publish a proof of this result. We give here the proof due to Danzer, Laugwitz and Lenz [121], which is the most complete in that they do not assume central symmetry of the set to be circumscribed (inscribed) nor do they consider only ellipsoids with a common centre.

In the latter part of the section we use the Löwner ellipsoid to prove John's result that if (X, B) is a Minkowski space then there is an ellipsoid E such that $\Delta(B, E) \leq (\log d)/2$. The improvement from Corollary 3.2.7 well illustrates the extra strength of the uniqueness of the Löwner ellipsoid compared with the "orthonormal basis" results of the preceding section.

Theorem 3.3.1 (Löwner) *Let A be a convex body in $\mathbb{R}^d$. There exists a unique ellipsoid E_1 of minimal volume containing A and a unique ellipsoid E_2 of maximal volume contained in A.*

Proof (Danzer, Laugwitz and Lenz). We shall prove the existence and uniqueness of the minimal circumscribed ellipsoid. The proof of the existence of a maximal inscribed ellipsoid is the same; the proof that it is unique is similar in outline though different in detail (see [121]). In the case when the convex body A is centrally symmetric one need only consider ellipsoids with the same centre and then a duality argument will give the results for the inscribed ellipsoid from those of the circumscribed one.

To prove the existence of a minimal circumscribing ellipsoid we observe that, since A is bounded, it is sufficient to consider ellipsoids E for which $A \subseteq E \subseteq E_0$, where E_0 is some fixed ellipsoid containing A. The volume is a continuous function defined on these ellipsoids and so has a positive infimum $V \geq \lambda(A)$. Hence there is a sequence $\{E'_n\}$ of ellipsoids that contain A and with $\lim_{n \to \infty} \lambda(E'_n) = V$. By Blaschke's theorem (2.5.14), this sequence has a convergent subsequence whose limit (by Corollary 2.5.17) is also an ellipsoid E_1 which contains A and such that, by the continuity of λ, $\lambda(E_1) = V$.

In order to prove that E_1 is unique, we suppose that E and E' are two ellipsoids with the same volume V. We show that if E and E' are distinct then there is a third ellipsoid E'', obtained as the "average" of E and E', whose volume is strictly less than V. Given E and E' it is possible to choose a coordinate system so that both have equations in diagonal form but, possibly, with different centres:

$$E = \left\{ (x_1, x_2, \ldots, x_d) : \sum_{i=1}^{d} (x_i - \xi_i)^2 \leq 1 \right\},$$

$$E' = \left\{ (x_1, x_2, \ldots, x_d) : \sum_{i=1}^{d} \alpha_i^{-2} x_i^2 \leq 1 \right\}.$$

Then, since $\lambda(E) = \lambda(E') = \epsilon_d$ (in this normalized coordinate system), we have $\prod_{i=1}^d \alpha_i = 1$. Consider the "average" ellipsoid E'' defined by

$$E'' := \left\{ (x_1, x_2, \ldots, x_d) : 2^{-1} \left[\sum_{i=1}^d (x_i - \xi_i)^2 + \alpha_i^{-2} x_i^2 \right] \leq 1 \right\}.$$

Since any point of A is in both E and E' it follows that $A \subseteq E''$. Next we put the defining inequality for E'' in standard form:

$$E'' = \{(x_1, x_2, \ldots, x_d) : \sum_{i=1}^d \beta_i^{-2}(x_i - \eta_i)^2 \leq \gamma\},$$

where simple algebra shows that

$$\beta_i^{-2} = (1 + \alpha_i^{-2})/2,$$
$$\eta_i = \beta_i^2 \xi_i / 2 = \alpha_i^2 \xi_i / (1 + \alpha_i^2)$$

and

$$\gamma = 1 - 2^{-1} \sum_{i=1}^d \xi_i^2 / (1 + \alpha_i^2).$$

We see that $\gamma \leq 1$ with equality if and only if each $\xi_i = 0$, which means that E has the same centre as E'. Moreover, the arithmetic–geometric mean inequality implies that $\beta_i^{-2} \geq \alpha_i^{-1}$; i.e. $\beta_i \leq \sqrt{\alpha_i}$ with equality if and only if $\alpha_i = 1$.

Hence we have $\lambda(E'') = \epsilon_d \gamma^{d/2} \prod_{i=1}^d \beta_i \leq \epsilon_d \prod_{i=1}^d \sqrt{\alpha_i} \leq \epsilon_d$. Also, one of these inequalities is strict unless each $\xi_i = 0$ and each $\alpha_i = 1$. Thus $\lambda(E) = \lambda(E') > \lambda(E'')$ unless E and E' coincide and hence a minimal ellipsoid must be unique. ∎

Definition 3.3.2 *If (X, B) is a Minkowski space then the unique ellipsoid E of minimal volume containing B is called the **Löwner ellipsoid** of B.*

Theorem 3.3.1 has immediate consequences for the isometries of a Minkowski space (X, B). The next three corollaries deal with the isometries that leave the origin fixed, *i.e.* excluding the translations. By Theorem 3.1.2 these are the linear isometries.

Corollary 3.3.3 *If (X, B) is a Minkowski space, if E is the Löwner ellipsoid of B and if T is a linear isometry of X then $T(E) = E$.*

Proof. Since T is linear, T maps E into an ellipsoid. Moreover, since $T(B) = B$, T is volume preserving and so $T(E)$ is an ellipsoid with the same volume as E. The uniqueness of the Löwner ellipsoid completes the proof. ∎

3.3 The Löwner ellipsoid

The next corollary applies with "linear" either inserted or omitted in both places, the difference being whether or not translations and compositions of linear isometries with a translation are included.

Corollary 3.3.4 *If (X, B) is a Minkowski space then the group of (linear) isometries of X is a subgroup of the group of (linear) isometries of an associated Euclidean space.*

Proof. The Löwner ellipsoid of B is the unit ball for an associated Euclidean structure which, by Corollary 3.3.3, has the required property. ∎

A group of transformations is said to be *transitive* on a set A if, for every pair of points x and y in A, there is a member T of the group such that $T(x) = y$.

Corollary 3.3.5 *If (X, B) is a Minkowski space and if the group of linear isometries is transitive on ∂B then B is an ellipsoid and (X, B) is Euclidean.*

Proof. Since the Löwner ellipsoid E of B is minimal, it is clear that $\partial B \cap \partial E \neq \emptyset$. Let $x \in \partial B \cap \partial E$ and $y \in \partial B$. By the transitive property, there is a linear isometry T such that $Tx = y$. But since, by Corollary 3.3.3, $T(E) = E$, this means $y \in \partial B \cap \partial E$, i.e. $\partial B = \partial E$. Hence $B = E$. ∎

Remark. It is possible for the group of linear isometries to be infinite without B being Euclidean (*e.g.* if B is a circular cylinder in $\mathbb{R}^3$) but then $\partial B \cap \partial E$ must contain infinitely many points since it must contain the orbit of any one of its points under the group of isometries. Gruber [212] shows that for "most" Minkowski balls B (here "most" means a dense G_δ in the metric space $(\mathcal{D}_d, \Delta)$) this is not the case. He shows that in most cases a Minkowski ball B meets the boundary of its Löwner ellipsoid in $d(d+1)/2$ pairs of symmetric points and hence the group of isometries is finite.

In $\mathbb{R}^2$ a stronger statement is true, namely that if (X, B) is not Euclidean then the group of isometries is finite. To see this, let $\epsilon > 0$ be given. Consider the orbit $\{T_i x\}$ of some point $x \in B \cap \partial E$ (E is the Löwner ellipse) under the group of isometries $\{T_i\}$. If this orbit is infinite, it has an accumulation point. Hence, for all $\epsilon > 0$, there are isometries T_1 and T_2 such that $0 < \mu(T_1 x, T_2 x) < \epsilon$, where μ is the Minkowski arc length distance on ∂B. Thus $0 < \mu(T_2^{-1} T_1 x, x) < \epsilon$. Since we can choose T_1 and T_2 so that $T_2^{-1} T_1$ is orientation preserving, the sequence $(T_2^{-1} T_1)^n x$ is an ϵ-net for $(\partial B, \mu)$. This is true for all ϵ and so the orbit $\{T_i x\}$ is dense in ∂B. Since $T_i x \in \partial B \cap \partial E$, it follows that ∂B and ∂E coincide.

The two natural questions in this context are: Given a Minkowski unit ball B, how far is it from being an ellipsoid? What properties of B ensure that it is an ellipsoid? Corollary 3.3.5 gave one answer to the second question and we shall devote the next section to further possible answers. Corollary 3.2.7 gave a preliminary answer to the first question and we now give the precise answer due to F. John [272].

Theorem 3.3.6 (John) *If (X, B) is a Minkowski space of dimension d then there is a Euclidean ball E such that $E \subseteq B \subseteq \sqrt{d}E$, i.e. $\Delta(B, E) \leq (\log d)/2$.*

Proof (Amir [8]). Let E be the inscribed Löwner ellipsoid of maximal volume. The proof is complete if we show $B \subseteq \sqrt{d}E$. Let $x_0 \in \partial B$. Consider a coordinate system so that $E = \{(x_1, x_2, \ldots, x_d) : \sum_{i=1}^{d} x_i^2 \leq 1\}$ and $x_0 = \xi e_1$ ($\xi \geq 1$).

Then the situation in any two-dimensional subspace containing x_0 is given by Figure 3.3, where we have $E \subseteq \text{conv}\{\xi e_1, -\xi e_1, E\} \subseteq B$.

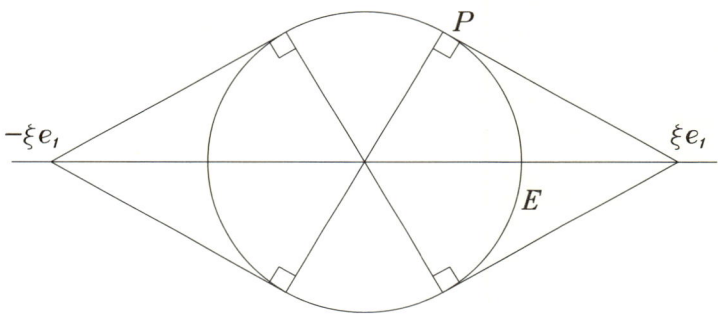

Figure 3.3

Next, for some α with $0 < \alpha < 1$, consider the transformation T_α defined by $T_\alpha(x_1, x_2, \ldots, x_d) := (\alpha x_1, x_2, \ldots, x_d)$. The preceding picture is now transformed to that shown in Figure 3.4 (using the original coordinate system).

The coordinates of the point P (Figure 3.3), where the tangent from ξe_1 meets ∂E, are $(\xi^{-1}, (\xi^2 - 1)^{1/2}\xi^{-1})$ and hence the transformed point $T_\alpha(P)$ (Figure 3.4) has coordinates $(\alpha\xi^{-1}, (\xi^2 - 1)^{1/2}\xi^{-1})$. Further elementary calculations show that the perpendicular distance ρ_α from 0 to the line segment $[T_\alpha(P), T_\alpha(x_0)]$ is

$$\rho_\alpha = \alpha\xi(\alpha^2(\xi^2 - 1) + 1)^{-1/2}.$$

Thus the image $T_\alpha(B)$ contains the ellipsoid

$$x_1^2 + x_2^2 + \cdots + x_n^2 \leq \rho_\alpha^2$$

(shown dotted in Figure 3.4), and hence B contains the ellipsoid

$$\alpha^2 x_1^2 + x_2^2 + \cdots + x_n^2 \leq \rho_\alpha^2.$$

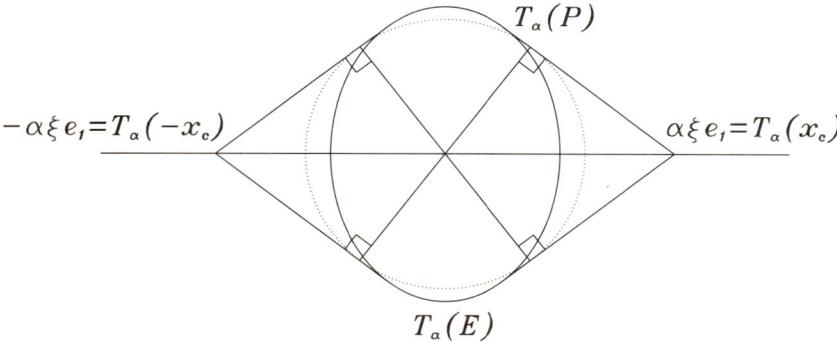

Figure 3.4

However, the volume $V(\alpha)$ of this new ellipsoid is given by

$$V(\alpha) = \alpha^{-1}\rho_\alpha^d \epsilon_d$$
$$= \xi^d \alpha^{d-1}(\alpha^2(\xi^2 - 1) + 1)^{-d/2}\epsilon_d.$$

Elementary calculus now shows that, as a function on $\mathbb{R}$, $V(\alpha)$ has a maximum value at $\alpha_0 = (d-1)^{1/2}(\xi^2 - 1)^{-1/2}$. If $0 < \alpha_0 < 1$ (i.e. if α_0 is in the range of permissible α's) the existence of this maximum contradicts the definition of E. Therefore, we have $\alpha_0 \geq 1$, which implies that $d - 1 \geq \xi^2 - 1$, i.e. $\xi \leq \sqrt{d}$. Hence $x_0 \in \sqrt{d}E$ and $B \subseteq \sqrt{d}E$, as required. ∎

Remark. This result of John shows that the Banach–Mazur distance of a Minkowski space from Euclidean space is at most $(\log d)/2$. In the case when B is the d-dimensional cross-polytope we have $\Delta(B, E) = (\log d)/2$ so that this bound is best possible. This implies that the diameter of the metric space $(\mathcal{D}_d, \Delta)$ is at most $\log d$. Szarek and Talagrand [500] have investigated the distance between an arbitrary d-dimensional Minkowski space and d-dimensional ℓ_∞-space.

3.4 Characterizations of Euclidean space

The very large literature on this subject is well covered and well organized in the book of Amir [8]. The purpose here is to present some of the more important geometric characterizations. Here *important* is a subjective term. The choice is made partly to show what cannot be expected in an arbitrary Minkowski space and partly to present criteria that may be relevant to establish some of the conjectured characterizations of Euclidean space which are mentioned in later chapters. The term *geometric* is used to exclude the many characterizations of ellipsoids via affine inequalities, *e.g.* the one given in Theorem 2.3.3. Although these characterizations are also highly relevant in Chapters 6 and 7 they would take us too far afield here.

The most well-known characterization is that of Jordan and von Neumann [276], which states that a normed linear space X is an inner product space if and only if the norm satisfies the parallelogram law:

$$\|x + y\|^2 + \|x - y\|^2 = 2\|x\|^2 + 2\|y\|^2, \qquad \forall x, y \in X.$$

As Day [125, 126] shows, this equation can be weakened significantly and still imply that X is an inner product space. The importance of the result is to show that being Euclidean is a "two-dimensional" property; *i.e.* if every two-dimensional subspace of X is Euclidean then so is X. Alternatively, if every two-dimensional cross-section of a centrally symmetric convex body B is an ellipse, then B is an ellipsoid. This fact is the starting point of many characterizations. Since we do not need to suppose X is finite dimensional, we will use the term *inner product space* rather than *Euclidean space*.

We begin in two dimensions with a very old characterization of ellipses, due to Bertrand [32]. The proof, which can be found in Grinberg [201], does equally well to prove an n-dimensional analogue of Bertrand's theorem due to Brunn [61].

Theorem 3.4.1 (Bertrand) *Let B be a convex body in the plane. For every vector $v \neq 0$, the locus of the mid-points of chords of B that are parallel to v is a straight line segment through 0 if and only if B is an ellipse with centre 0.*

Proof. We begin by remarking that the property is an affine one; *i.e.* if B has the property so does every affine image of B. Therefore, to prove the "if" part of the statement it is sufficient to point out that the statement is true for the Euclidean unit disc. We now prove the converse.

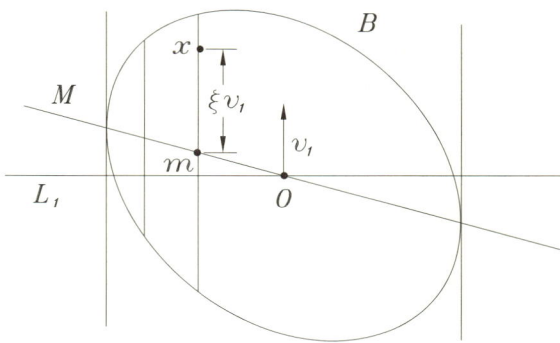

Figure 3.5

Let v_1 be a given direction, let M be the locus of mid-points of chords parallel to v_1 and let L_1 be the line through 0 perpendicular to v_1 (see Figure 3.5). We consider the Steiner symmetrization of B with respect to L_1. The main step is to

3.4 Characterizations of Euclidean space

show that the hypothesis on B means that this operation is *linear*. If $x \in B$ then x has a representation $x = m + \xi v_1$, with $m \in M$ and $\xi \in \mathbb{R}$. The image x_1 of x after symmetrization in L_1 is $x_1 = \text{Proj}\, m + \xi v_1$, where Proj is the orthogonal projection of M onto L_1. Thus the image B_1 of B under this symmetrization is given by $B_1 = T_1(B)$, where T_1 is the linear map that leaves v_1 fixed and maps M onto L_1 by projection along v_1. By our initial remark, B_1 has the same property with respect to chords as does B and so we can repeat the process. Moreover, the map T_1 preserves area. Therefore, B_1 is a position of B (Definition 2.5.18).

Now consider a sequence of successive symmetrizations, $B_1, B_2, \ldots, B_n, \ldots$ of B about lines $L_1, L_2, \ldots, L_n, \ldots$ perpendicular to directions $v_1, v_2, \ldots, v_n, \ldots$. With an appropriate choice of these directions, the sequence $\{B_n\}$ will converge in the Hausdorff metric to a circular disc. Because each B_i is a position of B it follows from Corollary 2.5.19 that the limit is also a position of B. Since the limit is a disc the original body B must be an ellipse. ∎

Remark. Although Grinberg remarks that one can use Corollary 2.5.19 in this way, he replaces it by an elegant argument involving the Löwner ellipse.

As we remarked above, this same proof will work in d-dimensional space to give Brunn's theorem. Alternatively, one can use Theorem 3.4.1 directly in conjunction with the Jordan–von Neumann result.

Theorem 3.4.2 (Brunn [61]) *Let B be a convex body in $\mathbb{R}^d$. For every vector $v \neq 0$, the locus of mid-points of chords of B parallel to v lies in a hyperplane through 0 if and only if B is an ellipsoid with centre 0.*

We now use Bertrand's theorem to give a characterization in terms of normality (Definition 3.2.2).

Theorem 3.4.3 *If for all $u, v \in X$ with $\|u\| = \|v\| = 1$ we have $(u+v) \dashv (u-v)$ then X is an inner product space.*

Proof. We first show that the condition implies that the unit ball B is strictly convex. If B is not strictly convex, let $[u, v]$ be a maximal line segment in ∂B. Consider the plane spanned by u and v (Figure 3.6).

Choose v' in this plane with $\|v'\| = 1$, v' outside the segment $[u, v]$ but sufficiently close to v so that $w := \|u + v'\|^{-1}(u + v')$ lies on $[u, v]$. Then $w \dashv (v-u)$. Since $(v' - u)$ is not parallel to $(v - u)$ but lies in the same plane, $(u + v')$ is not normal to $(v' - u)$, contradicting the hypothesis.

Next, let L be a two-dimensional subspace of X and let w be an arbitrary direction in L. Let z be one of the two points of $\partial B \cap L$ at which the tangent is parallel to w (Figure 3.7).

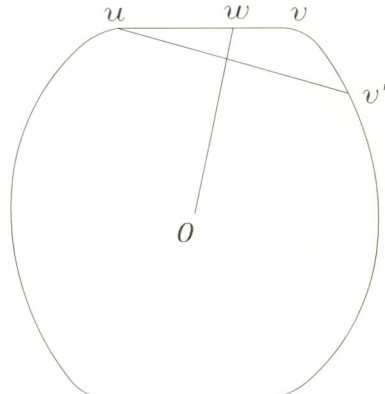

Figure 3.6

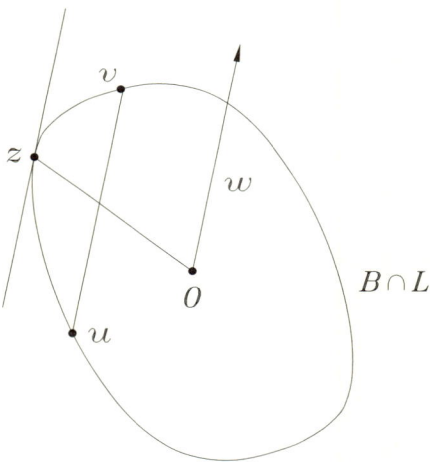

Figure 3.7

Consider a chord $[u, v]$ of $B \cap L$ that is parallel to w so that $(v-u) = \xi w$ ($\xi \in \mathbb{R}$). Because $z \dashv w$, we have $z \dashv (v-u)$. But since, by hypothesis, $(v+u) \dashv (v-u)$ and since the ball is strictly convex we must have $(u+v)/2 = \eta z$ for some $\eta \in \mathbb{R}$; i.e. the mid-points of chords parallel to w all lie on the line $0z$. Hence, by Theorem 3.4.1, $B \cap L$ is an ellipse. Finally, by the Jordan–von Neumann result, X is an inner product space. ∎

Corollary 3.4.4 *If for all $u, v \in X$ with $\|u\| = \|v\| = 1$ we have $\|u+v\|/2 \leq \|\xi u + (1-\xi)v\|$ for all $\xi \in [0, 1]$ then X is an inner product space.*

3.4 Characterizations of Euclidean space 89

Proof. Since $\|u\| = \|v\| = 1$, it is certainly true that $\|u + v\|/2 \leq 1 \leq \|\xi u + (1-\xi)v\|$ for all $\xi \in \mathbb{R} \setminus [0, 1]$. Hence the hypothesis implies that $\|u + v\|/2 \leq \|u + \xi(u - v)\|$ for all $\xi \in \mathbb{R}$. This inequality implies $(u + v) \dashv (u - v)$, and the result now follows from the theorem. ∎

In the next two theorems we follow the development given by Amir [8].

Theorem 3.4.5 (Joichi [275]) *Let M be a two-dimensional subspace of the normed space $(X, \|.\|)$ and let $x_0 \in X$ be such that $\|x_0\| \leq \|x_0 - y\|$ for all $y \in M$ (i.e. 0 is a nearest point in M to x_0). If, for all such M and all such x_0, $(x_0 + M) \cap B$ is centrally symmetric about x_0 (i.e. $\|x_0 + y\| = \|x_0 - y\|$ for all $y \in M$) then X is an inner product space.*

Proof. We show that the criterion of Corollary 3.4.4 is satisfied. First we observe that since $(x + M) \cap B$ is centrally symmetric about all points x which have 0 as their nearest point in M, x must be unique (up to scalar multiples) and hence B is strictly convex.

Let $u, v \in X$ with $\|u\| = \|v\| = 1$. Also let ξ_0 be the point at which the function $\phi(\xi) := \|\xi u + (1-\xi)v\|$ attains its minimum and let $x_0 := \xi_0 u + (1-\xi_0)v$. By the strict convexity of B, $\xi_0 \in (0, 1)$ and we must show that $\xi_0 = \frac{1}{2}$. If $r = \|x_0\|$ then the condition on ξ_0 means that the line segment $[u, v]$ supports the ball $B[0, r]$ at x_0. Consider any two-dimensional subspace M such that $x_0 + M$ contains the line segment $[u, v]$ and supports $B[0, r]$ at x_0.

Then 0 is the nearest point to x_0 in M. Also $y := (u - x_0) \in M$ and so, by hypothesis, $\|x_0 + y\| = \|x_0 - y\|$. But $\|x_0 + y\| = \|u\| = \|v\|$ and $\|x_0 - y\| = \|2x_0 - u\|$. However, since B is strictly convex, the line joining u and v cannot meet B in three distinct points and so either $2x_0 - u = u$ or $2x_0 - u = v$. The first implies that $x_0 = u$, which contradicts the fact that $\xi_0 \in (0, 1)$ and so $x_0 = (u + v)/2$, as required. ∎

Theorem 3.4.6 (Kakutani [281]) *If $\dim X \geq 3$ and if $(X, \|.\|)$ is such that for every two-dimensional subspace L of X there is a projection Proj_L of X onto L with $\|\text{Proj}_L\| = 1$ then X is an inner product space.*

Proof. We shall assume that B is smooth. For the general case one can either (see, *e.g.,* Phillips [428]) first show that the hypothesis implies that B is smooth or consider the non-smooth case separately (see, *e.g.,* Amir [8] or Bohnenblust [45]).

Let M be a two-dimensional subspace of X and let x_0 be such that $\|x_0\| \leq \|x_0 - y\|$ for all $y \in M$. If $S := B \cap (x_0 + M)$ we want to show that S is centrally symmetric about x_0. For this, consider any line $l := \{x_0 + \xi y : \xi \in \mathbb{R}, y \in M\}$ and let $[u, v] := l \cap S$. Let L be the two-dimensional subspace spanned by l. Then, by hypothesis, there is a projection Proj_L of X onto L with $\|\text{Proj}_L\| = 1$. In particular,

Proj_L projects the three-dimensional space $M + L$ onto L. Let N be the kernel of Proj_L in $M + L$ (see Figure 3.8).

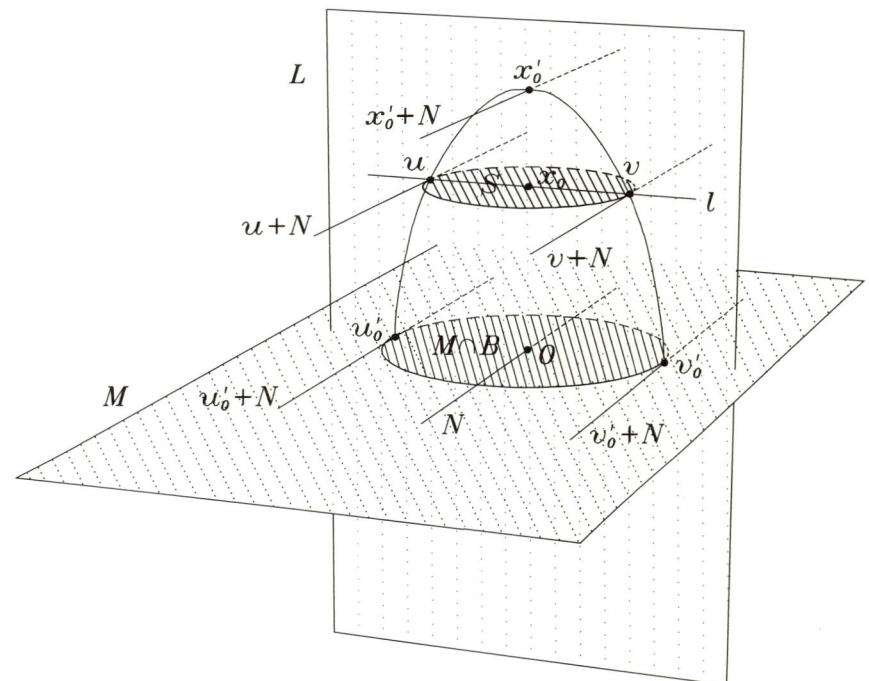

Figure 3.8

Since B is smooth it has a unique tangent hyperplane at $x_0' := x_0/\|x_0\|$ which contains M and so, since $\|\mathrm{Proj}_L\| = 1$, $x_0' + N$ is tangent to B at x_0' and so $N \subseteq M$. Again, since $\|\mathrm{Proj}_L\| = 1$, $u + N$ (resp. $v + N$) is tangent to B at u (resp. v). However, because $N \subseteq M$, $u + N$ (resp. $v + N$) is tangent to S at u (resp. v) in the plane $x_0 + M$. Let $u_0' := \|u - x_0\|^{-1}(u - x_0)$ and $v_0' := \|v - x_0\|^{-1}(v - x_0)$. Then the same argument shows that $u_0' + N$ and $v_0' + N$ are tangents to $M \cap B$ at u_0' and v_0' respectively; i.e. ∂S and $\partial(M \cap B)$ have parallel tangents at each pair of corresponding points u (on ∂S) and u_0' (on $\partial(M \cap B)$). It follows that the curves ∂S and $\partial(M \cap B)$ are similar. To see this, suppose that, in polar coordinates, the two curves have equations $r_1 = f_1(\theta)$ and $r_2 = f_2(\theta)$. Then, since for each θ the tangents are parallel, we have

$$r_1^{-1} \frac{dr_1}{d\theta} = r_2^{-1} \frac{dr_2}{d\theta},$$

i.e. $d(\log r_1 - \log r_2)/d\theta = 0$. Hence r_1/r_2 is constant. Since $M \cap B$ is centrally symmetric, it follows that S is centrally symmetric and the proof is completed by appealing to Theorem 3.4.3. ∎

Remark. It is clear from the proof of Theorem 3.4.6 that we need only suppose that there is a projection of norm 1 from each three-dimensional subspace M of X onto each two-dimensional subspace L of M.

The last remark enables Kakutani's theorem to be easily translated into geometric language. In that form it was proved earlier by Blaschke [36, 38] for $d = 3$. The translation needs the following definition.

Definition 3.4.7 *If K is a convex set in a Minkowski space X, if $x \in X$ and if L_x is the one-dimensional subspace spanned by x, then the **shadow boundary** of K in the direction x is the set $S_x := \partial(K + L_x) \cap \partial K$.*

Theorem 3.4.8 (Blaschke [36, 38]) *If (X, B) is a Minkowski space with $\dim X \geq 3$ and if, for each $x \in X$, there is a hyperplane H_x such that the shadow boundary of B in the direction x is contained in H_x then B is an ellipsoid and (X, B) is a Euclidean space.*

Proof. Suppose, first, that $\dim X = 3$ and let M be an arbitrary two-dimensional subspace of X. Then $M \cap \partial B$ contains two points x_1 and x_2 at which there are supporting hyperplanes to B, H_1 and H_2, respectively, that are distinct and not parallel. Therefore, $H_1 \cap H_2$ is a line with equation $x_3 + \alpha u$, say. Now both the points x_1, x_2 belong to the shadow boundary S_u of B in the direction u. By hypothesis S_u is planar, and hence $S_u \subseteq M$. It now follows that the projection along u onto M is of norm 1, *i.e.* each two-dimensional subspace has a norm 1 projection onto it.

In general, suppose that M is a three-dimensional subspace of X. If L is a line through 0 in M then, by hypothesis, $\partial(B+L) \cap \partial B$ lies in a hyperplane H. Hence,

$$\partial(B \cap M + L) \cap \partial(B \cap M) \subseteq M \cap H$$

and, since $L \not\subseteq H$, we have that $M \cap H$ is a two-dimensional subspace of M. Thus, $(M, B \cap M)$ satisfies the conditions of the theorem and so, by the first part, has a norm 1 projection onto each two-dimensional subspace. Therefore, by Theorem 3.4.6 and the remark following it, (X, B) is Euclidean. ∎

Gruber [217] regards this characterization of ellipsoids as "the most important one in convexity because of its many applications and generalizations". It can be thought of as dual to the Bertrand–Brunn characterization with which we began because that one deals with cross-sections by lines and this one deals with projections along lines.

In the final theorem of this section we follow the work of de Figueiredo and Karlovitz [127] and of Day [126] in order to show that three further geometric conditions are equivalent to the existence of projections of norm 1 and hence all imply that the space is an inner product space provided that the dimension is not less than 3.

If $(X, \|.\|)$ is a normed space the *radial projection* R of X onto the unit ball B is defined by

$$Rx := \begin{cases} x/\|x\|, & \text{if } \|x\| > 1, \\ x, & \text{if } \|x\| \leq 1. \end{cases}$$

The mapping R is said to be *non-expansive* if $\|Rx - Ry\| \leq \|x - y\|$ for all x, y in X. It is counter-intuitive that if $\dim X \geq 3$ then R is non-expansive only in the case of inner product spaces.

Example 3.4.9 In $(\mathbb{R}^3, \|.\|_\infty)$ consider the points $x := (1, 2\epsilon, 2\epsilon)$ $y := (\epsilon, 1 + \epsilon, 1 + \epsilon)$, where $\epsilon \leq \frac{1}{2}$. Then we have $\|x - y\|_\infty = \|(1-\epsilon, \epsilon-1, \epsilon-1)\|_\infty = 1-\epsilon$. However, $Rx = x$ and $Ry = (\epsilon/(1+\epsilon), 1, 1)$ so that

$$\|Rx - Ry\|_\infty = \|(1/1+\epsilon, 2\epsilon-1, 2\epsilon-1)\|_\infty = 1/1+\epsilon > 1-\epsilon.$$

In $(\mathbb{R}^3, \|.\|_1)$ if $x := (\frac{1}{3}, \frac{1}{3}, \frac{1}{3})$ and $y := x+(\epsilon, 0, 0)$ then $\|x-y\|_1 = \|(-\epsilon, 0, 0)\|_1 = \epsilon$. However, $Rx = x$, $Ry = 3^{-1}(1+\epsilon)^{-1}(1+3\epsilon, 1, 1)$ and therefore

$$\|Rx - Ry\|_1 = [3(1+\epsilon)]^{-1}\|(-2\epsilon, \epsilon, \epsilon)\|_1 = 4\epsilon/3(1+\epsilon) > \epsilon,$$

provided that $\epsilon < \frac{1}{3}$.

In a normed linear space $(X, \|.\|)$, if $\|x\| = \|y\| = 1$ let $\phi(x, y; \xi) := \|x + \xi y\|$. Then ϕ is a convex function of ξ whose minimum is attained either at a single point if the norm is strictly convex or, possibly, on an interval otherwise.

Theorem 3.4.10 *If $(X, \|.\|)$ is a normed linear space with $\dim X \geq 3$ then the following conditions are equivalent:*

(i) the radial projection R is non-expansive;
(ii) for all x, y with $\|x\| = \|y\| = 1$, all minima of $\phi(x, y; \xi)$ occur in $[-1, 1]$;
(iii) normality is symmetric, i.e. if $x \dashv y$ then $y \dashv x$;
(iv) for every two-dimensional subspace L of X there is a projection Proj_L of X onto L with $\|\text{Proj}_L\| = 1$;
(v) the space is an inner product space.

Proof. The assertion that (iv) implies (v) is the content of Theorem 3.4.6. That (v) implies (i) is a straightforward calculation with inner products. We shall prove the implications (i) $\Rightarrow$ (ii) $\Rightarrow$ (iii) $\Rightarrow$ (iv).

(i) ⇒ (ii). If $\|x\| = \|y\| = 1$ and if $\xi \geq 1$ then $R(\xi y) = y$ and $R(-\xi y) = -y$ hence condition (i) implies that

$$\|x - y\| \leq \|x - \xi y\| \quad \text{and} \quad \|x + y\| \leq \|x + \xi y\|.$$

These inequalities, together with the convexity of ϕ, show that ϕ is increasing on $[1, +\infty)$ and decreasing on $(-\infty, -1]$. To complete the proof we need to show that ϕ is strictly monotonic on these two intervals. Suppose, on the contrary, that, for some $\delta > 0$, ϕ is constant on the interval $[1, 1 + \delta]$ (the proof for $[-1 - \delta, -1]$ is very similar), *i.e.* $\phi(x, y; \xi) = c$ for $\xi \in [1, 1 + \delta]$.

For small positive α consider $z := -\alpha x + (1 - \alpha)y = \alpha(-x) + (1 - \alpha)y$. As $\alpha \to 0$ we have $z \to y$ and so $\|z\| \to 1$. Also, since $z \in [-x, y]$ we have $\|z\| \leq 1$. Consequently, $(1 - \alpha)/(\|z\| - \alpha) \to 1^+$ as $\alpha \to 0$.

Therefore, we can choose $\alpha > 0$ and $\epsilon > 0$ so that if $\xi_1 := (1 - \alpha)/(\|z\| - \alpha)$ and $\xi_2 := (1 + \epsilon)(1 - \alpha)/(\|z\| - (1 + \epsilon)\alpha)$ then we have

$$1 \leq \xi_1 < \xi_2 < 1 + \delta. \tag{3.1}$$

Then

$$\begin{aligned}
\phi(x, z/\|z\|; 1) &= \|x + z/\|z\|\| \\
&= \|x + \|z\|^{-1}(-\alpha x + (1 - \alpha)y)\| \\
&= (1 - \alpha/\|z\|)\|x + \xi_1 y\| \\
&= (1 - \alpha/\|z\|)\phi(x, y; \xi_1) \\
&= (1 - \alpha/\|z\|)c
\end{aligned}$$

because (3.1) holds.

Also

$$\begin{aligned}
\phi(x, z/\|z\|; 1 + \epsilon) &= \|x + (1 + \epsilon)\|z\|^{-1}z\| \\
&= \|x + (1 + \epsilon)\|z\|^{-1}(-\alpha x + (1 - \alpha)y)\| \\
&= (1 - (1 + \epsilon)\alpha\|z\|^{-1})\|x + \xi_2 y\| \\
&= (1 - (1 + \epsilon)\alpha\|z\|^{-1})\phi(x, y; \xi_2) \\
&= (1 - (1 + \epsilon)\alpha\|z\|^{-1})c \\
&< (1 - \alpha/\|z\|)c \\
&= \phi(x, z/\|z\|; 1).
\end{aligned}$$

But this last inequality contradicts the initial statement that ϕ is increasing on $[1, +\infty)$ applied to the pair $x, z/\|z\|$, which in turn contradicts (i). This completes the proof of the first implication.

(ii) ⇒ (iii). Suppose that $x \dashv y$, *i.e.* $\|x\| \leq \|x + \xi y\|$ for all ξ, but that $\|y + \xi_1 x\| < \|y\|$ for some $\xi_1 \neq 0$. Let $\eta := \|y + \xi_1 x\| \|y\|^{-1} < 1$. Then

$\|y + \xi_1 x\| = \|\eta y\|$ and we may suppose these vectors scaled so that both have norm 1. Then (ii) implies that

$$\phi(y + \xi_1 x, \eta y; -1) < \phi(y + \xi_1 x, \eta y; -\eta^{-1}),$$

i.e. $\|y + \xi_1 - \eta y\| < \|y + \xi_1 x - y\|$; in other words $\|x + (1-\eta)\xi_1^{-1} y\| < \|x\|$, which contradicts the supposition that $x \dashv y$.

(iii) $\Rightarrow$ (iv). By the remark following Theorem 3.4.6, it is sufficient to consider a three-dimensional subspace M of X and show that there is a projection of norm 1 from M onto each plane L in M. Let L be a plane through the origin. Then there exists $x_0 \in M$, $\|x_0\| = 1$ such that $x_0 + L$ supports B at x_0, i.e. $x_0 \dashv y$ for all y in L. Hence, by the symmetry of $\dashv$, we have $y \dashv x_0$ for all y in L, i.e. the line parallel to x_0 supports B at $\|y\|^{-1} y$ for all y in L. This means that the projection along x_0 onto L is of norm 1 and the proof is complete. ∎

Remarks. In condition (iv) of Theorem 3.4.10 we may delete the phrase "two-dimensional" so that if $\dim X \geq 3$ then X is an inner product space if and only if there is a norm 1 projection onto every subspace L.

In the proof it was only in the step from (iv) to (v) that the hypothesis $\dim X \geq 3$ was needed. For two-dimensional spaces, conditions (i)–(iii) and the modified (iv) are equivalent to the condition that the boundary of B is a Radon curve; see Day [124] and §4.7 below.

3.5 Notes

There are a number of extensions of the Mazur–Ulam theorem in various directions. Example 3.1.3 shows that no linearity condition can be established in general. Surprisingly, Figiel [152] (followed by Holsztiński [263]) has shown that any isometry (linear or not) has a linear left inverse. More precisely, if T is an isometry of a normed space X into a normed space Y then there is a linear map S from Y to X such that ST is the identity on X; and, although S may not be bounded on the whole of Y, it has norm 1 on the linear span of the range of T. Mankiewicz [351] has considered isometries mapping a subset U of X onto a subset V of Y. He shows that if U and V are either open and connected or convex bodies then the isometry is necessarily affine. On the other hand, it is easy to give an example of a non-linear isometry that maps a convex body *into* itself; e.g. consider the space c_0 of all sequences which converge to 0 equipped with the uniform norm and the mapping T defined on the unit ball of c_0 by the equation

$$Tx = T(\xi_1, \xi_2, \ldots, \xi_n, \ldots) := (1 - \|x\|, \xi_1, \xi_2, \ldots, \xi_{n-1}, \ldots).$$

Motivated by the idea that what permits non-linearity is that the range of T has too few directions, Tingley [510] considered isometries of the unit sphere in X onto the unit sphere in Y. After considerable effort he was able to establish only that if X is finite dimensional and if T is such an isometry then $T(-x) = -T(x)$. Charzyński [107] and Rolewicz [448] have considered extending the Mazur–Ulam result to F-spaces and have proved that in certain cases surjective isometries between F-spaces must be affine.

Carathéodory introduced the notion of *transversalität* to deal with a problem in Finsler geometry arising from the calculus of variations. It was reinvented by Birkhoff [35]. This and other concepts of orthogonality were defined by James [265, 266]. In [265] James considers two types: isosceles and pythagorean. The first is defined by $x \perp_i y$ if and only if $\|x - y\| = \|x + y\|$ and the second by the pythagorean theorem ($x \perp_p y$ if and only if $\|x + y\|^2 = \|x\|^2 + \|y\|^2$). Clearly both of these concepts are symmetric relations. James shows that if $x \perp_i y$ then $\|x + \xi y\| \geq \|x\|$ for $\xi \geq 1$ and $\|x + \xi y\| > \|x\|/2$ for all ξ. Furthermore, if orthogonality is either additive (i.e. $x_1 \perp_i y_1$ and $x_2 \perp_i y_2$ imply $(x_1 + x_2) \perp_i (y_1 + y_2)$) or homogeneous (i.e. $x_1 \perp_i y_1$ implies $x_1 \perp_i \xi y_1$) then the space is an inner product space. In [266] James is mainly concerned with a detailed discussion of the Carathéodory–Birkhoff orthogonality. Ohira [401] has shown that (in many cases) if one definition of orthogonality implies another then the norm must come from an inner product. Four more notions of orthogonality and their relationships to each other and to an inner product structure are surveyed by Amir [8].

Dual bases for X and X^* that are "orthonormal" are also referred to as Auerbach systems. Day [123] gives two proofs of the existence of such systems in the two-dimensional case – an algebraic one and one based on minimal circumscribing parallelograms. The idea of maximal inscribed cross-polytopes (or, dually, of minimal circumscribing parallelotopes) arises in several contexts. One, as we shall mention in Chapter 5, is to define Minkowski content via such objects. They also give rise to affine invariants for the Minkowski space. For example, one can consider the ratio $\tau(B) := \lambda(B)/\lambda(P)$, where P is some minimal circumscribing parallelotope for the unit ball B and λ is some Haar measure. Then, in each dimension d, one can consider τ_d defined to be $\inf\{\tau(B)\}$, where the infimum is taken over all unit balls B. Estimates for τ_d have been given by Dvoretzky and Rogers [134] and more recently by Pelczyński and Szarek [407] and a "stability" theorem for this ratio was given by Ball [15]. Petty and McKinney [424] characterize those symmetric convex bodies whose circumscribing parallelotopes have constant volume. Chakerian [100] considers the minimal value of the average of the volumes of parallelotopes circumscribed to a convex body K and shows that this minimum is attained only if K is a ball. McKinney [364] has very interesting results on the volume of a maximal simplex inscribed to a symmetric convex body. He shows that this volume is decreased by Steiner symmetrization and, further, that if S is a maximal simplex inscribed to K then

$$\lambda(S)/\lambda(K) \geq 1/d!\epsilon_d$$

with equality if and only if K is an ellipsoid. Lassak [310] made good use of the idea of a maximal inscribed octahedron and the corresponding circumscribed cube in his proof of the Hadwiger conjecture for symmetric convex bodies in $\mathbb{R}^3$.

The first proof of the uniqueness of the Löwner ellipsoid was probably that of Behrend [29], but John [272] was the first to consider the Banach–Mazur distance between a convex body and its Löwner ellipsoid. Proofs of John's result and of the uniqueness of the Löwner ellipsoid are given by Bollobás [48]. An earlier related result [271] showed that if K is a convex body whose minimal and maximal widths are $w(K)$ and $W(K)$ then there is a position K' of K with $W(K')/w(K') \leq d^{1/2}$. Similar results dealing with the ratio of inradius to circumradius and of diameter to mean width were obtained by Leichtweiss [318], who also used them to obtain John's result. The conditions that John obtains under which the Löwner ellipsoid is the Euclidean ball are related to the results of Petty [414] and Clack [113] on the positions of a convex body that minimize its surface area. Groemer [205] considers the question of whether, if E is a circumscribing ellipsoid whose volume is close to minimal, E is close (in the metric Δ_1) to the Löwner ellipsoid. Just as the ratio of the volume of K to that of a minimal circumscribing parallelotope is an affine invariant so is the ratio of the volume of K to that of its Löwner ellipsoid. Ball [15] shows that this ratio is maximal for a cube.

The question of which bodies attain the maximal distance to the Euclidean ball has been investigated by Milman and Wolfson [384]. Lewis [320] gave an upper bound of $d^{|1/2-1/p|}$ for the distance from an arbitrary d-dimensional subspace of an L_p space to ℓ_2^d. Lewis [321] has also generalized John's result by considering the distance to balls other than the Euclidean one and by considering other norms for the linear maps T in the definition of Δ than the usual operator norm. Various authors have also considered the distance from a general ball to either the cube or the cross-polytope. For results of this kind see Szarek and Talagrand [500], Bourgain and Szarek [57], Szarek [499] and Giannopoulos [177]. The best result here is that of Giannopoulos, who proves that the Banach–Mazur distance from any d-dimensional ball to the d-dimensional cube lies between $c_1 d^{0.5} \log d$ and $c_2 d^{0.875}$, where c_1 and c_2 are absolute constants. Note that here the Banach–Mazur distance is without the logarithm.

Corollary 3.3.5 was first proved (by a quite different method) by Auerbach [11]. Gruber [212] has a number of results dealing with the Löwner ellipsoid. As stated in the text, he shows that "most" symmetric convex bodies touch the boundaries of both the maximal inscribed and minimal circumscribed ellipsoids in exactly $d(d+3)/2$ points. He also shows, in contrast to Auerbach's result just mentioned, that "most" convex bodies are left invariant by no affine map other than the identity.

The literature and history on the characterization of Euclidean space among Minkowski spaces (or of inner product spaces among normed spaces) are far too extensive to summarize adequately here. Some idea of the development has been given in the course of §3.4. Corollary 3.4.4 (in a slightly weaker form) is due to

Goryachev [197] and in the stated form to Gurarii and Sozonov [232]. Schoenberg's [484] characterization via the Ptolemaic inequality

$$\|a-b\|\,\|c-d\| + \|a-d\|\,\|b-c\| \geq \|a-c\|\,\|b-d\|$$

is interesting because, like the Jordan–von Neumann result, it is quadratic, because he uses the Löwner ellipsoid to prove the result and because he only needs to suppose that $\|.\|$ is a semi-norm. Closely related to the result of de Figueiredo and Karlovitz [127] is that of Schäffer [458], who shows that the length of a curve outside the unit ball shrinks under radial projection onto the surface of the unit ball if and only if normality is symmetric and hence, for $d \geq 3$, if and only if the norm is Euclidean. Mann [352] considered sets of the type illustrated in Figure 4.11; *i.e.* given points x and y in a Minkowski space consider those z such that $\|x-z\| \leq \|y-z\|$. He showed that the norm is Euclidean if and only if all such sets are convex. This was generalized to the case when the unit ball is only assumed to be star-shaped by Woods [537] and further sharpened by Gruber [208, 209]. A very recent characterization of ellipsoids is that of Meyer, Reisner and Schmuckenschläger [378], who show that if B is a convex body such that for some τ the volume $\lambda(B \cap (x + \tau B))$ is a function of $\|x\|_B$ only then B is an ellipsoid. We mention again Amir [8] as a very comprehensive survey of results of the type considered in §3.4, where the emphasis is on geometric properties of the normed space. There is a similarly large literature that uses geometric properties to characterize ellipsoids among convex or more general classes of sets; see, *e.g.*, Gruber [208–210] and the survey of Gruber and Höbinger [218]. The Bertrand–Brunn theorem is a geometric result of this type, as is a result of Aitchison, Petty and Rogers [2] that is related to Theorem 3.4.5. If K is a convex body and if $x \in \text{int } K$ then x is a *pseudo-centre* if every two-dimensional cross-section of K through x is centrally symmetric (not necessarily about x). Such a pseudo-centre is a *false* centre if K itself is *not* symmetric about that point. The result is that for $d \geq 3$ a convex body is an ellipsoid if and only if it has a false centre. Both the Brunn theorem (3.4.2) and Blaschke's result (3.4.8) have been improved by Gruber [208, 209]. The Blaschke–Santaló inequality and the McKinney inequality [364] mentioned above are two examples where ellipsoids are characterized by equality in an inequality involving volumes of convex bodies. In later chapters we shall mention one or two more examples of the same phenomenon. A very different type of characterization is that of Kottman [303]. A reasonable conjecture that ellipsoids are the only zonoids whose polars are also zonoids was shown to be false by Schneider [473], who constructed other examples of such bodies. A survey of various types of characterizations of ellipsoids was made by Petty [420]. The section on ellipsoids in the survey article of Heil and Martini [252] is another useful source of information.

Any discussion of the relationship between Minkowski spaces and Euclidean spaces is not complete without some mention of Dvoretzky's theorem [133]. This theorem states that if a natural number n and a positive real number ϵ are given, then every Minkowski space of sufficiently large dimension (depending on n and ϵ) has an n-dimensional subspace whose Banach–Mazur distance from a Euclidean space is less than ϵ. In other words, there are n-dimensional cross-sections which are close to ellipsoids. The statement for infinite dimensional spaces is perhaps more striking: each infinite-dimensional normed space has, for each n and ϵ, an n-dimensional subspace whose Banach–Mazur distance from a Euclidean space is less than ϵ. For proofs see also Milman [381], Figiel [153, 154] and Szankowski [498]. Dvoretzky's theorem has been extremely influential as a starting point for the local theory of Banach spaces; both Figiel [155] and Lindenstrauss and Milman [327] emphasize this role.

4
Two-dimensional Minkowski spaces

The case of two-dimensional normed spaces is, in some respects, quite special. In other ways, it affords a good introduction to higher-dimensional geometry. This was evident in the last chapter. The results in the earlier sections dealing with circumscribed parallelotopes and inscribed and circumscribed ellipsoids are no different in dimension 2 than in other dimensions. On the other hand, in §3.4 the characterizations of Euclidean space via norm 1 projections and via the symmetry of orthogonality are only valid in dimension $d \geq 3$. When $d = 2$ these conditions lead to an interesting class of convex bodies.

From the point of view of our main topic – the normalization of Haar measure in all the subspaces of a Minkowski space – the particularity of dimension 2 comes from the fact that the only non-trivial subspaces are lines and there the normalization is already specified by the metric, *i.e.* by the unit ball. Thus, except for the choice of the single normalizing factor for Haar measure in the whole space, which affects the numerical value to be assigned to certain important concepts but not any of the theoretical framework, the geometry is entirely determined by the unit ball.

That this is not so in higher dimensions, that one has a certain amount of freedom in the choice of the *isoperimetrix* (the convex body of minimal surface area for a given volume), forms the content of Chapter 5. It is important, therefore, to have the unambiguous solution in two dimensions clearly determined first. Whatever solution is offered in higher dimensions, one of the constraints must be that it specializes to the right object when $d = 2$.

The contents of this chapter are a discussion, in this special situation, of topics which will recur in the rest of the book. The first section contains one result from three-dimensional space. It is a generalization (due to Petty) of a two-dimensional result which we need in both §4.2 and §4.3. Since it appears to be an open question whether the result is true in dimensions $d \geq 4$, it seems appropriate to include both theorems here as "low-dimension special cases". The first section ends with work of Asplund and Grünbaum on Minkowski analogues of certain results from the Euclidean geometry of triangles.

The second section introduces bodies of constant Minkowski width and also deals briefly with equichordal sets. The main results are Theorem 4.2.8, which gives a minimal property of Minkowski–Reuleaux triangles, and Example 4.2.11 (due to Petty and Crotty) of a convex set with two equichordal points.

The third section is concerned with the perimeter of the unit ball, *i.e.* the length of its circumference. This section contains what is probably the oldest result in the subject (Golab's theorem 4.3.6) and one of the prettiest (Schäffer's theorem 4.3.9). This topic gets a fuller treatment in Chapter 9.

The fourth and fifth sections are the central ones and deal with the solution to the isoperimetric problem in two dimensions. This is the introductory material for Chapters 5, 6 and 7. As well as the central theorem (Theorem 4.4.1) of Busemann we present a number of related inequalities. It is here that the question of normalizing two-dimensional Haar measure is relevant and the consequences of two possibilities are discussed.

The definitions of length and area imply another view of perpendicularity via the formula "area = base × perpendicular height". In Minkowski space this concept, which we shall term *transversality*, is distinct from that of normality (Definition 3.2.2). Transversality is the subject of §4.6.

An important question is whether there are unit balls, other than Euclidean ones, for which the solution to the isoperimetric problem is (a multiple of) the unit ball itself. We shall see that in two-dimensions this leads to the Radon curves. The last section discusses the special case when the unit ball is a Radon curve.

4.1 Inscribed regular hexagons and other constructions

The first theorem in this chapter shows that the Euclidean construction of an equilateral triangle is, in fact, a Minkowski construction; in other words, it is possible to inscribe a *regular* hexagon in any two-dimensional Minkowski ball. Moreover, the choice of one vertex is arbitrary. Here *regular* has two meanings: the hexagon is regular in the sense that all edges have the same Minkowski length; it is also affine regular – an affine image of a (Euclidean) regular hexagon. Another view of this construction is that one can construct a regular 2-simplex S_2 with one vertex at 0 and the other two on ∂B so that the hexagon $S_2 - S_2$ is inscribed to B.

Petty [417] has shown that Theorem 4.1.1 can be extended to any three-dimensional space. Despite the chapter title, this seems the most appropriate place for his theorem. The proofs of Theorems 4.1.1 and 4.1.3 are topological in nature. The last part of this section is devoted to showing that much of the geometry of a triangle is not essentially Euclidean. The results here are due to Asplund and Grünbaum [10] and concern the *circumcentre*, the *orthocentre*, the *nine-point centre* and the *Euler line* of a triangle. The notation for triangles is as follows: given three non-collinear points a, b, c in Minkowski space

$$\triangle abc := \text{conv}\{a, b, c\}.$$

Theorem 4.1.1 *If $(X, \|.\|)$ is a two-dimensional Minkowski space and if $x_0 \in X$ then there exist x_1, x_2 in X such that $\|x_i - x_j\| = 1$ $i, j = 0, 1, 2$.*

Proof. We may suppose that $x_0 = 0$. Let x_1 be an arbitrary point of ∂B (i.e. $\|x_1\| = 1$). Consider the ball $B[x_1, 1]$ with centre x_1 and radius 1. The boundary of this ball is a continuous curve which passes through 0 inside ∂B and through $2x_1$, which is outside ∂B (see Figure 4.1). Hence there exists a point $x_2 \in \partial B[x_1, 1] \cap \partial B$

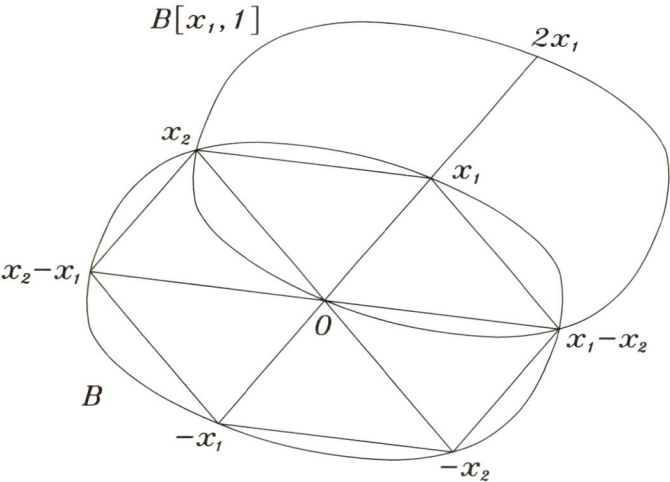

Figure 4.1

(and, if B is strictly convex, precisely two such points, one on either side of the line through $-x_1$ and x_1). Thus $\|x_2\| = \|x_2 - x_1\| = 1$ and so the six points $x_1, x_2, x_2 - x_1, -x_1, -x_2, x_1 - x_2$ form the vertices of a regular hexagon inscribed to B. ∎

One can also prove Theorem 4.1.1 in a slightly different way by showing that the distance $\varphi(z) := \|x_1 - z\|$ is a monotonic function as z moves round the boundary of B from x_1 ($\varphi(x_1) = 0$) to $-x_1$ ($\varphi(-x_1) = 2$) and so there is a point (or a subarc) where $\varphi(z) = 1$. The monotonicity of φ is needed for Petty's proof of the next theorem and it is also of some independent interest. A more straightforward proof would be preferable; an alternative one is given by Schäffer [466].

Lemma 4.1.2 *Let $(X, \|.\|)$ be a two-dimensional normed space and let $x \in \partial B$. If ∂B is given an orientation and if $x, z_1, z_2, -x$ lie in that order on ∂B with respect to the orientation then $\|x - z_1\| \leq \|x - z_2\|$.*

Proof. We will use the notation $c(z_1 z_2)$ to indicate the arc of ∂B from z_1 to z_2. We shall also consider the curve $c(x(-x))$ to be an ordered set, the order being induced by the orientation of ∂B. To be more precise one may regard ∂B as a curve parametrized by arc length, i.e. $\partial B = \{\psi(s) : -\ell < s \leq \ell\}$, where ℓ is the length of $c(x(-x))$ and $\psi(0) = x$, and then the order is that induced by the usual order on $(-\ell, \ell]$.

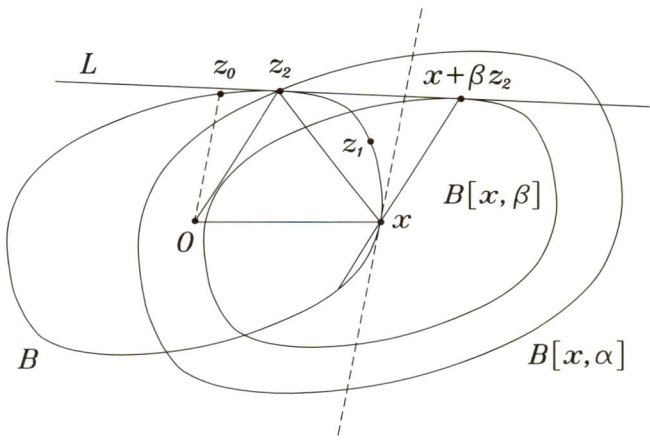

Figure 4.2

First suppose that $x < z_0 < -x$ and $x \dashv z_0$ in the sense of Definition 3.2.2 and that $z_1 \leq z_2 \leq z_0$. Let $\alpha := \|x - z_2\|$ and let L be a tangent to B at z_2 (see Figure 4.2). Then z_2 is a nearest point to 0 on L. Likewise, for a suitable β, $x + \beta z_2$ is a point of L nearest to x and since $\|(x + \beta z_2) - x\| = \beta$ we have $\beta \leq \alpha$. Hence $B[x, \alpha]$, the ball of radius α centred at x, contains the line segment $[x, z_2]$ and the line segment $[x, x + \beta z_2]$ and hence contains the triangle $T := \triangle x z_2 (x + \beta z_2)$. Since $z_2 \leq z_0$, the line $\{x + \zeta z_2 : \zeta \in \mathbb{R}\}$ is either tangent to B at x or it cuts ∂B at x and again at another point which precedes x on ∂B. Hence the triangle T contains the arc $c(xz_2)$ and hence $B[x, \alpha]$ contains this arc and so we have $\|x - z_1\| \leq \alpha$.

We have shown that the function $\varphi(z) := \|x - z\|$ is monotonic on the arc $c(xz_0)$. To complete the proof, let

$$\bar{z} := \sup\{z \in c(x(-x)) : \varphi \text{ is monotonic on } c(xz)\}.$$

Let $\tilde{z} := (\bar{z} - x)/\|\bar{z} - x\|$. Suppose, if possible, that $\bar{z} < -x$. Then we have $x < z_0 \leq \bar{z} < \tilde{z} < -x$. Consider z_1, z_2 with $z_0 \leq z_1 \leq z_2 \leq \tilde{z}$ (Figure 4.3).

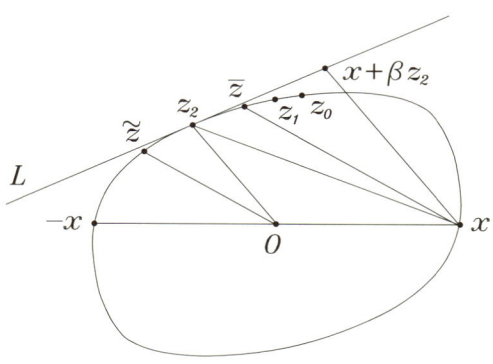

Figure 4.3

We repeat the argument of the first part and use the same notation. Now, however, since $z_2 \leq \tilde{z}$, the line $\{x + \xi z_2 : \xi \in \mathbb{R}\}$ either cuts ∂B at a second point $x' \leq \bar{z}$ or is tangent to B at x (in which case take $x' = x$). Hence the triangle $T := \triangle xz_2(x+\beta z_2)$ contains the arc $c(x'z_2)$ and $B[x, \alpha]$ contains this arc also. But since $x' \leq \bar{z}$ and φ is monotonic on $c(x\bar{z})$ it follows that $B[x, \alpha] \supseteq c(xx')$ and hence $B[x, \alpha] \supseteq c(xz_2)$. Again we get $\|x - z_1\| \leq \alpha = \|x - z_2\|$. Thus φ is monotonic on $c(x\tilde{z})$. This contradicts the definition of $\bar{z}$. Hence $\bar{z} = -x$, as required. ∎

We can now present Petty's generalization of Theorem 4.1.1 to three-dimensional space (see [417]).

Theorem 4.1.3 *If $(X, \|.\|)$ is a three-dimensional Minkowski space and if $x_0 \in X$ then there exist x_1, x_2, x_3 in X such that $\|x_i - x_j\| = 1$ $i, j = 0, 1, 2, 3$.*

Proof. We may suppose that $x_0 = 0$. Let x_1 be an arbitrary point of ∂B and let H be an arbitrary two-dimensional subspace containing x_1. By Theorem 4.1.1, there is an $x_2 \in H \cap \partial B$ such that $\|x_2\| = \|x_2 - x_1\| = \|x_1\| = 1$. The hyperplane H divides X into two closed half-spaces H^+ and H^-. Let $\partial B^+ := H^+ \cap \partial B$ ($\partial B^- := H^- \cap \partial B$) and let $B_H := \partial B \cap H = \partial B^+ \cap \partial B^-$. Both ∂B^+ and ∂B^- are compact, simply connected sets. Define $f : \partial B^+ \mapsto \mathbb{R}^2$ by $f(z) := (\|z - x_1\|, \|z - x_2\|)$ for all $z \in \partial B^+$. Since f is continuous, its image is a closed, bounded, simply connected subset of $\mathbb{R}^2$. Consider the image of B_H under f. We have $f(x_1) = (0, 1)$ and $f(x_2) = (1, 0)$. Further, by Lemma 4.1.2, there are real numbers $\alpha, \beta, \gamma \geq 1$ such that $f(x_2 - x_1) = (\alpha, 1)$, $f(-x_1) = (2, \beta)$, $f(-x_2) = (\beta, 2)$ and $f(x_1 - x_2) = (1, \gamma)$. It follows, again by using Lemma 4.1.2, that $f(B_H)$ is a simple closed curve in $\mathbb{R}^2$ which either passes through the point (1,1) or contains (1,1) in its interior. In either case we must have (1,1) in $f(\partial B^+)$ and so ∂B^+ (and similarly ∂B^-) contains a point x_3 with $\|x_3\| = 1 = \|x_3 - x_2\| = \|x_3 - x_1\|$. ∎

Remark. We can, however, say a little more than this. If $(1, 1) \in f(B_H)$ then it follows (by an argument similar to one presented in §4.3 below) that B_H is a parallelogram (with one of the points x_1, x_2, x_3 as a vertex and the other two at the mid-points of adjacent sides). Since, however, any three-dimensional ball has a plane cross-section which is not a parallelogram, it is always possible to begin with x_1 and x_2 in such a cross-section. Then $\{x_1, x_2, x_3\}$ is a linearly independent set. In this case $S_3 := \mathrm{conv}\{0, x_1, x_2, x_3\}$ is a non-degenerate, regular simplex and $S_3 - S_3$ is a *regular* cubo-octahedron inscribed to B (see Example 1.1.17).

Now we come to the work of Asplund and Grünbaum [10], who have shown that a good deal of the Euclidean geometry of the triangle carries over to Minkowski space.

In Euclidean geometry the orthocentre O of a triangle $\triangle abc$ is defined as the point at which the altitudes meet (which appears very dependent on orthogonality). However, the four points a, b, c, O are symmetrically situated in that each is the orthocentre of the other three. It is also well known that the circumradii of the four triangles $\triangle abc$, $\triangle abO$, $\triangle bcO$, $\triangle caO$ are all equal. This fact yields a metric definition and construction for the orthocentre. To avoid intersections of circles which are line segments rather than points we shall suppose that $(X, \|.\|)$ is a two-dimensional Minkowski space whose unit ball B is strictly convex. Then any two distinct circles intersect in at most two points. It follows that there is a unique circumcircle through three non-collinear points and that there are precisely two circles of a given radius ρ through two given points whose distance apart is less than 2ρ (the centres of the two circles lie one on each side of the line segment joining the points).

Theorem 4.1.4 *Suppose that $(X, \|.\|)$ is a two-dimensional Minkowski space with strictly convex unit ball B. Let $C := \partial B$ denote the unit circle. If three translates of C (C_1, C_2 and C_3) intersect in a single point p then the three further points p_{ij} ($i \neq j$) belonging to $C_i \cap C_j$ respectively lie on a fourth translate of C.*

Proof (MacKenzie[344]). We may suppose that p is the origin and let w_i be the centre of C_i. Then $\|w_i\| = 1$ and $C_i = w_i + C$. Let $p_{ij} := w_i + w_j$. Then $p_{ij} \in C_i \cap C_j$ because

$$\|w_i - (w_i + w_j)\| = \|-w_j\| = 1 = \|-w_i\| = \|w_j - (w_i + w_j)\|.$$

Finally, since $\|(w_1 + w_2 + w_3) - (w_i + w_j)\| = 1$, each point p_{ij} lies on the circle $C_0 := (w_1 + w_2 + w_3) + C$. ∎

4.1 Inscribed regular hexagons and other constructions

Remarks

(i) If B is not strictly convex then the points p_{ij} defined in the theorem are in $C_i \cap C_j$ and lie on a translate of C.

(i) If two circles touch at p then the corresponding point p_{ij} coincides with p.

That Theorem 4.1.4 is a generalization of Theorem 4.1.1 is seen as follows. If, in Figure 4.1, we draw a third circle centred at $x_2 - x_1$ then the three circles all pass through x_2. The points p_{ij} are then $x_1 - x_2$, 0 and $-x_1$, which all lie on the circle centred at $-x_2$.

According to MacKenzie [344] Theorem 4.1.4, for the Euclidean plane, was observed by Johnson [274] but, surely, it must have been known earlier.

Corollary 4.1.5 *If (under the conditions of Theorem 4.1.4) $\triangle abc$ is a triangle whose circumcircle has radius ρ then the three other circles which pass respectively through a and b, b and c, and c and a and have radius ρ have a common intersection point.*

Proof. Let C denote the circumcircle and C_1, C_2 and C_3 denote the other three circles. Then C, C_1 and C_2 all pass through b. The other intersection point of C and C_1 is a; that of C and C_2 is b. Let the second intersection point of C_1 and C_2 be O. Then by the theorem a, b and O all lie on a circle of the same radius which must, therefore, be C_3. ∎

Definition 4.1.6 *The common intersection point O described in Corollary 4.1.5 is called the **orthocentre** of the triangle $\triangle abc$.*

Remark. This definition is yet another way to use metric notions to introduce a concept of orthogonality. Note that the four points a, b, c, O are symmetric in that each is the orthocentre of the other three and the circumradii of the four triangles are all equal.

With the notation of Corollary 4.1.5 choose O to be the origin and let x_1, x_2 and x_3 be the centres of the circles C_1, C_2 and C_3. Then, as in the proof of Theorem 4.1.4, we have that $a = x_3 + x_1$, $b = x_1 + x_2$ and $c = x_2 + x_3$. Furthermore, again from Theorem 4.1.4, the centre O' of the circumcircle of $\triangle abc$ is given by $O' = x_1 + x_2 + x_3$. It is now clear that the centroid G of $\triangle abc$ satisfies the equation $G := (a + b + c)/3 = 2(x_1 + x_2 + x_3)/3$ and hence G lies on the line segment OO' and divides this segment (the Euler line) in the ratio $2 : 1$.

Observe also that the two triangles

$$\triangle x_1 x_2 x_3 \quad \text{and} \quad \triangle abc = \triangle(x_3 + x_1)(x_1 + x_2)(x_2 + x_3)$$

are related by reflection in the point $N := (x_1 + x_2 + x_3)/2$, which is the mid-point of OO'. Since this reflection interchanges O and O', O' is the orthocentre and O the circumcentre of $\triangle x_1 x_2 x_3$.

Finally note that the circle whose centre is N and whose radius is $\rho/2$ passes through the following six points: $(x_1 + x_2)/2$, $(x_2 + x_3)/2$ and $(x_2 + x_3)/2$ (which are the mid-points of the line segments Oa, Ob and Oc) and $x_1 + (x_2 + x_3)/2$, $x_2 + (x_3 + x_1)/2$ and $x_3 + (x_1 + x_2)/2$ (which are the mid-points of the sides of the $\triangle abc$). Thus the "nine-point circle" is reduced to a "six-point circle".

4.2 Sets of constant width and equichordal sets

In this section we give the Minkowski analogues of the Euclidean notions of *sets of constant width* and *equichordal points*. In the case of the second of these, the greater freedom afforded by certain norms enables one to construct examples more easily than in Euclidean space. In particular, we give Petty and Crotty's example [423] of a convex set in a two-dimensional normed space with two equichordal points (the non-existence of such a set in the Euclidean plane appears to have recently been settled; see Schäfke and Volkmer [468]).

Let $(X, \|.\|)$ be a Minkowski space with unit ball B. Then the dual space X^* has the dual norm induced by B°. We will denote this dual norm also by $\|.\|$. If K is a convex set, we may restrict the support function $h(K, .)$ to those f in X^* with $\|f\| = 1$. We shall call this restriction the Minkowski support function and denote it by $h_B(K, f)$, i.e. $h_B(K, f) = \sup\{f(x) : x \in K, \|f\| = 1\}$.

Definition 4.2.1 *Let K be a convex set. For each $f \in X^*$ with $\|f\| = 1$, the **Minkowski width** of K in the direction f, denoted by $w_B(K, f)$, is defined by*

$$w_B(K, f) := h_B(K, f) + h_B(K, -f).$$

Since the support function h is positively homogeneous, if g is a Euclidean unit linear functional, then $g/\|g\|$ is a Minkowski unit functional and so

$$\begin{aligned} w_B(K, g/\|g\|) &= h_B(K, g/\|g\|) + h_B(K, -g/\|g\|) \\ &= \{h(K, g) + h(K, -g)\}/\|g\| \\ &= w(K, g)/\|g\| \\ &= 2w(K, g)/w(B, g), \end{aligned}$$

where $w(K, g)$ is the usual Euclidean width of K in the direction g. The last equation follows from the fact that $\|g\| = \sup\{g(x) : x \in B\} = w(B, g)/2$.

Definition 4.2.2 *A convex set K is said to have **constant (Minkowski) width** if $w_B(K, f) = b$ for all $f \in X^*$, $\|f\| = 1$.*

4.2 Sets of constant width and equichordal sets

Theorem 4.2.3 *If K is a closed convex set with constant width 1, then $K + (-K) = B$.*

Proof. We have

$$1 = w_B(K, f) = h_B(K, f) + h_B(K, -f)$$
$$= h_B(K, f) + h_B(-K, f)$$
$$= h_B(K + (-K), f) \qquad \text{for all } f \in \partial B^\circ.$$

Therefore, $h(K + (-K), f) = h(B, f)$ for all $f \in X^*$, and hence, by Proposition 2.2.2, $B = K + (-K)$ since both are closed, convex sets. ∎

We next show, following an observation of Ohmann[402], that Theorem 4.1.1 allows us to construct Reuleaux triangles in Minkowski space. In what follows we refer to the notation of Theorem 4.1.1 and Figure 4.1.

Example 4.2.4 Consider the triangle $\triangle 0x_1 x_2$ of Figure 4.1. On the side $[x_1, x_2]$ construct the (translate of the) arc $x_1 x_2$ of B, on the side $[x_2, 0]$ construct the (translate of the) arc $(x_2 - x_1)(-x_1)$ of B and on the side $[0, x_1]$ construct the arc $(-x_2)(x_1 - x_2)$. Then the resulting convex body is a *Reuleaux triangle* of constant width 1. If B is smooth, the construction can be modified to make the resulting body smooth (see Figure 4.4).

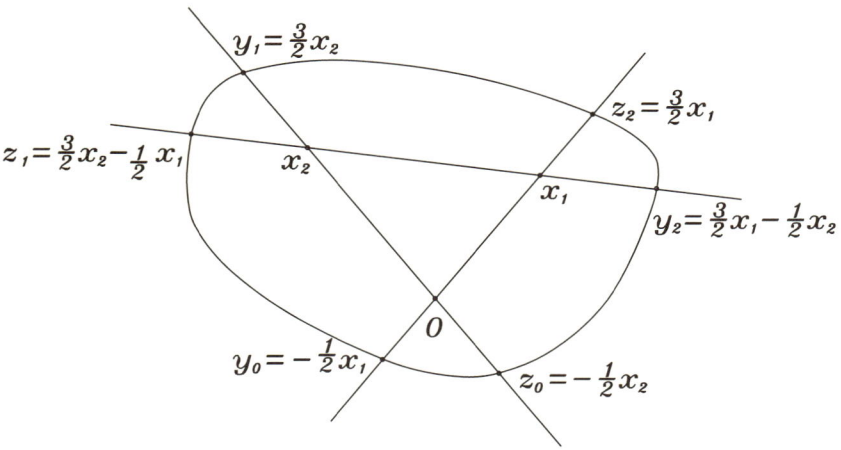

Figure 4.4

In Figure 4.4 the arc $z_2 y_1$ is homothetic to the arc $x_1 x_2$ of B with scale factor $\frac{3}{2}$; the arc $y_1 z_1$ is homothetic to the arc $x_2(x_2 - x_1)$ of B with scale factor $\frac{1}{2}$; likewise the arcs $z_1 y_0$, $y_0 z_0$, $z_0 y_2$ and $y_2 z_2$ are, respectively, translates of the arcs $3(x_2 - x_1)(-x_1)/2$, $(-x_1)(-x_2)/2$, $3(-x_2)(x_1 - x_2)/2$ and $(x_1 - x_2)(x_1)/2$.

Chakerian [98] gave a simple proof of the fact that, as in Euclidean space, Reuleaux triangles have minimum area among bodies of constant width. The result needs some preliminary lemmas.

Lemma 4.2.5 *If H is a regular hexagon inscribed to B as in Theorem 4.1.1 and if R is the Reuleaux triangle of constant width 1 that is based on H as in Example 4.2.4 then*

$$\lambda(R) = \lambda(B)/2 - \lambda(H)/3.$$

Proof. The Reuleaux triangle R is composed of $T := \triangle 0x_1x_2$ and three segments of B; half of B is made up of three triangles congruent to T and three segments each congruent to one in R; and H is composed of six triangles congruent to T. Therefore, the result is an elementary computation. Note that "congruent to" is an abbreviation for "is an isometric image of". ∎

The next lemma needs more definitions. If H is a regular hexagon inscribed to B as in Theorem 4.1.1 then a centrally symmetric hexagon H' circumscribed to B and whose sides support B at the vertices of H is called a *B-hexagon based on H*. The term *B-hexagon* denotes a circumscribed B-hexagon which is based on some H.

Lemma 4.2.6 *If (X, B) is a two-dimensional Minkowski space and if H is a regular hexagon inscribed to B as in Theorem 4.1.1 then $\lambda(B) \le 4\lambda(H)/3$.*

Proof. Let H' be a B-hexagon based on H. We will prove the stronger statement that $\lambda(H') \le 4\lambda(H)/3$. Let $x_1, x_2, x_3 := (x_2 - x_1)$ and $-x_1$ be successive vertices of H and let $-x_3'x_1'$, $x_1'x_2'$, $x_2'x_3'$ and $x_3'(-x_1')$ be successive sides of H' passing through x_1, x_2, x_3 and $-x_1$ respectively. Moreover, let sides x_1x_2 and $-x_1x_3$ of H meet at y. It is clear that $y = 2x_2 - x_1$, that $\|x_2 - y\| = \|x_3 - y\| = 1$ and that

$$\lambda(\triangle x_2 y x_3) = \lambda(\triangle x_2 0 x_3) = \lambda(H)/6.$$

Construct the line through y parallel to x_1x_1' (and hence also parallel to $-x_1x_3'$). Suppose that this line meets the side $x_2'x_3'$ of H' at z_2 and the side $x_1'x_2'$ produced at z_1. A symmetric argument covers the case when the line meets the side $x_1'x_2'$ at z_1 and the side $x_3'x_2'$ produced at z_2. Then $\triangle(-x_1)x_3'x_3$ is congruent to $\triangle yz_2x_3$ and $\triangle x_1x_1'x_2$ is congruent to $\triangle yz_1x_2$. Therefore,

$$\lambda(\triangle x_1x_1'x_2) + \lambda(\triangle x_2x_2'x_3) + \lambda(\triangle x_3x_3'(-x_1))$$
$$= \lambda(\triangle yz_1x_2) + \lambda(\triangle x_2x_2'x_3) + \lambda(\triangle yz_2x_3)$$
$$\le \lambda(\triangle x_2yx_3) = \lambda(H)/6,$$

which implies that $\lambda(H')/2 \le \lambda(H)/2 + \lambda(H)/6 = 2\lambda(H)/3$. ∎

If, now, K is an arbitrary convex body in X and if H' is a B-hexagon then it is possible to circumscribe about K a (possibly degenerate) hexagon whose sides are parallel to those of H'. Ohmann observed [402] that among all such circumscribed hexagons there is at least one which is symmetric.

Lemma 4.2.7 *If K is a convex body in a two-dimensional Minkowski space (X, B) then K admits a symmetric, circumscribed hexagon whose sides are parallel to those of some B-hexagon.*

Proof. Let x_1 be an arbitrary point of ∂B used as the initial vertex of an inscribed regular hexagon H. If H' is a B-hexagon based on H then the side through x_1 is called the *first* side of H'. The set of all B-hexagons can be parametrized continuously by the angle which the first side makes with some fixed direction in X. Let $H'(t)$, $0 \leq t \leq 2\pi$, denote such a parametrization. Let $C(t)$ denote the hexagon circumscribed about K with sides parallel to those of $H'(t)$ and, finally, let $s_i(t)$, $i = 1, 2, \ldots, 6$, denote the lengths of the sides of $C(t)$. Then $s_4(0) = s_1(\pi)$ and $s_1(0) = s_4(\pi)$. Hence $s_1(t) - s_4(t)$ changes sign on $[0, \pi]$ (or is zero at both ends). Therefore, for some t_0 we have $s_1(t_0) = s_4(t_0)$ and these two sides form a parallelogram with centre z (say). Elementary linear algebra then shows that, since the remaining four sides are in two parallel pairs, the hexagon $C(t_0)$ is symmetric about z. ∎

Now we can conclude Chakerian's proof [98] of the minimal property of Reuleaux triangles.

Theorem 4.2.8 *If (X, B) is a Minkowski space of dimension 2 and if K is a convex body of constant Minkowski width 2, then $\lambda(K) \geq \lambda(R)$, where R is some Reuleaux triangle of width 2.*

Proof. By Lemma 4.2.7 there is a centrally symmetric hexagon C circumscribed about K whose sides are parallel to a B-hexagon H'. Furthermore, since K has constant width 2, each pair of sides of C are distance 2 apart and so are those of H'. Thus C is a translate of H'. Let H be the inscribed hexagon on which H' is based, let R_1 be the Reuleaux triangle corresponding to H as in Example 4.2.4 and let $R := 2R_1$.

By Theorem 4.2.3, $K + (-K) = 2B$ and hence

$$\begin{aligned} 4\lambda(B) = \lambda(K+(-K)) &= \lambda(K) + 2V(K,-K) + \lambda(-K) \\ &\leq 2\lambda(K) + 2V(C,-C) \\ &= 2\lambda(K) + 2V(H',-H') \\ &= 2\lambda(K) + 2\lambda(H') \\ &\leq 2\lambda(K) + 8\lambda(H)/3. \end{aligned}$$

The first inequality follows since $K \subseteq C$, the next equality is because H' is a translate of C, the next because $H' = -H'$ and the last inequality is from the proof of Lemma 4.2.6. Thus $\lambda(K) \geq 2\lambda(B) - 4\lambda(H)/3 = \lambda(R)$. The final equality comes from Lemma 4.2.5 and the fact that $R = 2R_1$. ∎

Definition 4.2.9 *If K is a convex set and if $x \in K$, x is said to be an **equichordal point** of K if, for all lines L through x, the line segment $L \cap K$ has constant (Minkowski) length.*

Clearly the unit ball B has 0 as an equichordal point. If x is an equichordal point of K and, at the same time, a centre of symmetry of K, then K is homothetic to B. We give an easy example in $\mathbb{R}^2$ with the ℓ_∞ norm of a set other than B which has an equichordal point. This is followed by Petty and Crotty's remarkable example of a norm and a closed convex set which has two equichordal points.

Example 4.2.10 Consider the convex set K pictured in Figure 4.5 whose boundary consists of four line segments and two hyperbolic arcs with equations $x^2 \pm xy \mp 2x - 4y = 0$. It is readily checked that the set is convex.

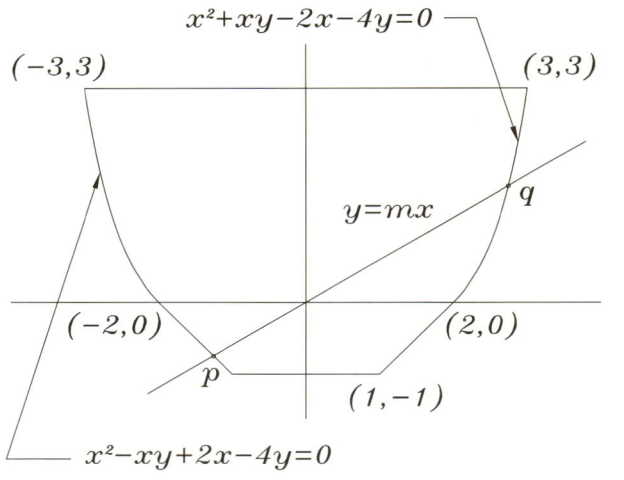

Figure 4.5

Lines through 0 with slope greater than 1 or less than -1 have length 4 in the ℓ_∞-norm. Consider a line $y = mx$ with $0 < m < 1$. Then, because $m < 1$, its ℓ_∞ length is the difference in the X-coordinates of its end points p, q. Elementary algebra shows that $p = (-2/(m+1), -2m/(m+1))$ and $q = (2(1+2m)/(1+m), 2m(1+2m)/(1+m))$ and hence that $\|p - q\|_\infty = 4$ as required.

Example 4.2.11 (Petty and Crotty [423]) We present only a particular case for the plane, not the full generality for a linear space of arbitrary dimension given in [423]. Consider the vector space $\mathbb{R}^2$ and for $x = (\xi_1, \xi_2) \in \mathbb{R}^2$ let

$$\|x\| := 2(\xi_1^2 + \xi_2^2)(4\xi_1^2 + 3\xi_2^2)^{-1/2}$$

(if $x \neq 0$) and $\|0\| = 0$. Except for the triangle inequality, the properties of a norm are clear. To verify that the triangle inequality holds, we check that the unit ball B is convex. The equation for the boundary of B in polar coordinates is $r = (1 - \sin^2 \theta/4)^{1/2}$, from which one can verify that the curvature is strictly positive for all θ and hence that B is convex.

Next, consider the convex set K which is the Euclidean unit disc. We show that $p := (2^{-1}, 0)$ is an equichordal point of K. If $x := (\cos \theta, \sin \theta)$ is a point of ∂K then the chord of K from x through p meets ∂K again at the point $y := [5 - 4\cos\theta]^{-1}(4 - 5\cos\theta, -3\sin\theta)$ and a further calculation gives us that $p - y = [5 - 4\cos\theta]^{-1}(x - p)$. Hence

$$\|x-p\| + \|p-y\| = \frac{8 - 4\cos\theta}{5 - 4\cos\theta}\|x-p\| = \frac{8 - 4\cos\theta}{5 - 4\cos\theta}\|((\cos\theta - 2^{-1}), \sin\theta)\| = 2.$$

Thus, with respect to this norm, p is an equichordal point of K. By the symmetry with respect to the Y-axis it follows that $-p$ is also an equichordal point.

Remark. If one is willing to consider non-closed convex sets K then it is easier to give an example. In the plane with the ℓ_∞-norm consider the rectangle with vertices $(0, -1)$, $(1, -1)$, $(1, 1)$ and $(0, 1)$ and with the two segments $\{(\xi_1, \xi_2) : \xi_1 = 1, -1 \leq \xi_2 < 0\}$ $\{(\xi_1 \xi_2) : \xi_1 = 0 \ 0 < \xi_2 \leq 1\}$ deleted from the boundary. Then $(0, 0)$ and $(0, 1)$ are equichordal points with all chords through them having length 1.

4.3 Lengths of curves, perimeter of the unit ball

This section and the following one contain the most important material of the chapter. Here we are concerned with the length (measured in its own metric) of the perimeter of the unit ball. This number is perhaps the most important affine invariant associated with a two-dimensional, symmetric, convex body. Certainly, in the Euclidean case it is the most interesting (and most precisely calculated) of all the irrational numbers.

We shall denote the length of a curve c (these concepts are defined below), measured with respect to a unit ball B, by $\mu_B(c)$. Thus the numbers we are concerned with in this section are $\mu_B(\partial B)$ for two-dimensional convex bodies B. There are two main theorems concerning this self-circumference. The first, Golab's theorem [181], is one of the oldest in Minkowski geometry and states that $6 \leq \mu_B(\partial B) \leq 8$.

The second is Schäffer's more recent result [465] that $\mu_B(\partial B) = \mu_{B^\circ}(\partial B^\circ)$. Before we come to these theorems we need preliminary material on curves in the plane and their lengths. These ideas will be discussed more generally in Chapter 9.

By a *curve* c in a Minkowski plane X we understand the range of a continuous function φ whose domain is a closed, bounded interval, *i.e.*

$$c = \{x : x = \varphi(t), \alpha \le t \le \beta\}.$$

For example, an interval $[x_1, x_2] = \{x : x = (1-t)x_1 + tx_2, 0 \le t \le 1\}$ is a curve. A finite union of adjacent intervals $[x_1, x_2] \cup [x_2, x_3] \cup \cdots \cup [x_{n-1}, x_n]$ is called a *polygon* and is also a curve. The points x_i are the *vertices* of the polygon and the segments $[x_{i-1}, x_i]$ are the *edges*. The polygon will be denoted by $[x_1, x_2, \ldots, x_n]$. More generally, two curves c_1 and c_2 for which the last point of c_1 coincides with the initial point of c_2 may be joined to form a new curve $c_1 \cup c_2$.

The *length* of a polygon $\wp := [x_0, x_1, \ldots, x_n]$ with respect to a unit ball B is $\mu_B(\wp) := \sum_{i=1}^n \|x_i - x_{i-1}\|_B$. If $c := \{x : x = \varphi(t) \alpha \le t \le \beta\}$ is a curve and if $\alpha = t_0 < t_1 < \cdots < t_n = \beta$ is a partition of $[\alpha, \beta]$ then we may consider the polygon $[\varphi(t_0), \varphi(t_1), \ldots, \varphi(t_n)]$ with length $\sum_{i=1}^n \|\varphi(t_i) - \varphi(t_{i-1})\|$. The triangle inequality implies that this length increases if the partition is refined. The curve is said to be *rectifiable* if the lengths of all such polygons are bounded above. The *length of c*, $\mu_B(c)$, is defined to be

$$\mu_B(c) := \sup\left\{\sum_{i=1}^n \|\varphi(t_i) - \varphi(t_{i-1})\| : \{t_0, t_1, \ldots, t_n\} \text{ a partition of } [\alpha, \beta]\right\}.$$

Note that, although the length of c depends on B, the property of being rectifiable does not.

A simple closed curve is said to be *convex* if its interior (in the Jordan curve sense) is a convex set. A curve c from x_0 to x_1 ($x_0 \ne x_1$) is said to be convex if $c \cup [x_1, x_0]$ is a simple, closed, convex curve. If c_1 and c_2 are two convex curves from x_0 to x_1 then c_1 is said to *lie inside c_2* if $c_1 \subseteq \operatorname{conv} c_2$.

Lemma 4.3.1 *If the polygon $\wp := [x_0, x_1, x_2, \ldots, x_n]$ is a convex curve from x_0 to x_n which lies inside the polygon $[x_0, y, x_n]$ then*

$$\sum_{i=1}^n \|x_i - x_{i-1}\| \le \|y - x_0\| + \|x_n - y\|.$$

Proof. We may suppose the vectors $u_1 := y - x_0$ and $u_2 := x_n - y$ to be linearly independent (otherwise the result is trivial). Express each $x_i - x_{i-1}$ in this basis as

$$x_i - x_{i-1} = \alpha_i u_1 + \beta_i u_2.$$

(Because $\wp$ is convex and lies inside $[x_0, y, x_1]$ each α_i and β_i is ≥ 0.) Then

$$x_n - x_0 = \sum_{i=1}^n (x_i - x_{i-1}) = \sum \alpha_i u_1 + \sum \beta_i u_2 = (y - x_0) + (x_n - y) = u_1 + u_2.$$

Hence $\sum_{i=1}^{n} \alpha_i = \sum_{i=1}^{n} \beta_i = 1$. Since $\|x_i - x_{i-1}\| \leq \alpha_i \|u_1\| + \beta_i \|u_2\|$,

$$\mu_B(\wp) = \sum_{i=1}^{n} \|x_i - x_{i-1}\| \leq \sum_{i=1}^{n} \alpha_i \|u_1\| + \sum \beta_i \|u_2\| = \|u_1\| + \|u_2\|$$

and the proof is complete. ∎

Proposition 4.3.2 *If c is a convex curve from x_0 to x_1 which lies inside the polygon $[x_0, y, x_1]$ then $\mu_B(c) \leq \|y - x_0\| + \|x_1 - y\|$.*

Proof. If $\wp = [z_0, z_1, \ldots, z_n]$ is any polygon inscribed to c then it also lies inside $[x_0, y, x_1]$. Hence, by Lemma 4.3.1, $\mu_B(\wp) \leq \|y - x_0\| + \|x_1 - y\|$. Since $\mu_B(c) = \sup\{\mu_B(\wp) : \wp$ inscribed to $c\}$, the result follows. ∎

Definition 4.3.3 *If c is a convex curve from x to y, a convex polygon from x to y is said to be **inscribed** to c if each vertex belongs to c. A convex polygon $\wp'$ from x to y is said to be **circumscribed** to c if c lies inside $\wp$ and each edge of $\wp$ meets c.*

Theorem 4.3.4 *If c is a convex curve from x to y and if $\wp_1$ and $\wp_2$ are, respectively, inscribed and circumscribed polygonal paths from x to y then*

$$\mu_B(\wp_1) \leq \mu_B(c) \leq \mu_B(\wp_2).$$

Proof. The first inequality is immediate from the definition of $\mu_B(c)$. Suppose $\wp_2 = [x, y_2, y_3, \ldots, y_{n-1}, y]$. Set $y_1 = y_1' = x$ and $y_n = y_{n-1}' = y$ and suppose each edge $[y_i, y_{i+1}]$ meets c at y_i', $i = 2, 3, \ldots, n-2$. Then apply Proposition 4.3.2 to each arc from y_i' to y_{i+1}' and to the corresponding line segments $[y_i', y_{i+1}, y_{i+1}']$, $i = 1, 2, \ldots, n-2$. ∎

Proposition 4.3.5 *If c_1 and c_2 are two convex curves from x to y with c_1 lying inside c_2 then $\mu_B(c_1) \leq \mu_B(c_2)$.*

If there is a polygon $\wp$ from x to y which is circumscribed to c_1 and inscribed to c_2 then Theorem 4.3.4 shows that $\mu_B(c_1) \leq \mu_B(\wp) \leq \mu_B c_2$. Unfortunately, such a polygon need not exist and it is not easy to complete an argument along these lines.

A formal and succinct proof using the monotonicity of mixed volumes will be given in the next section. Until we reach that point we will use only Theorem 4.3.4.

Note that the words *parallelogram* and *hexagon* (like *triangle*) are used to describe both a two-dimensional convex figure and its boundary curve.

Theorem 4.3.6 (Golab) *If B is the unit ball in a two-dimensional Minkowski space then $6 \leq \mu_B(\partial B) \leq 8$.*

Proof. By Theorem 4.1.1, B has an inscribed hexagon with each edge of length 1. The first inequality now follows from Theorem 4.3.4. For the second inequality we apply Corollary 3.2.5. This states that B has a circumscribed parallelogram $\wp := [x_1 + x_2, -x_1 + x_2, -x_1 - x_2, x_1 - x_2]$ with $\|x_1\| = \|x_2\| = 1$. Hence, $\mu_B(\wp) = 8$ and the result again follows from Theorem 4.3.4. ∎

Theorem 4.3.7 *If (X, B) is a two-dimensional Minkowski space then*

  (i) $\mu_B(\partial B) = 6$ *if and only if B is an affine regular hexagon;*
  (ii) $\mu_B(\partial B) = 8$ *if and only if B is a parallelogram.*

Proof. The "if" part of each statement is clear. We give Schäffer's proofs [459] of the "only if" parts.

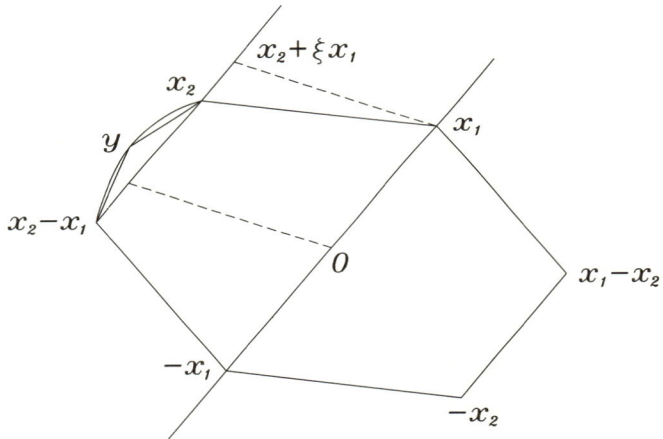

Figure 4.6

(i) Suppose that $\mu_B(\partial B) = 6$. First use Theorem 4.1.1 to inscribe a hexagon to B beginning at an extreme point x_1 of ∂B (Figure 4.6). Let y be the mid-point (in arc length) of the arc $c(x_2(x_2 - x_1))$ of ∂B. Then we have

$$3 = \|x_1 - x_2\| + \|x_2 - (x_2 - x_1)\| + \|(x_2 - x_1) - (-x_1)\|$$
$$\leq \|x_1 - x_2\| + \|x_2 - y\| + \|y - (x_2 - x_1)\| + \|(x_2 - x_1) - (-x_1)\|$$
$$\leq \|x_1 - x_2\| + \mu_B(c(x_2 y)) + \mu_B(c(y[x_2 - x_1])) + \|(x_2 - x_1) - (-x_1)\|$$
$$\leq \mu_B(\partial B)/2 = 3.$$

Thus we must have equality throughout and so

$$\|x_2 - y\| = \mu_B(c(x_2 y)) = \mu_B(c(y[x_2 - x_1])) = \|y - (x_2 - x_1)\|,$$

with the central equation coming from the choice of y. Also

$$\|x_2 - y\| + \|y - (x_2 - x_1)\| = \|x_2 - (x_2 - x_1)\| = 1,$$

i.e. $\|x_2-y\| = \|y-(x_2-x_1)\| = \frac{1}{2}$. Hence $2(x_2-y) \in \partial B$ and $2(y-(x_2-x_1)) \in \partial B$. But $x_1 = (x_2 - y) + (y - (x_2 - x_1))$ and x_1 is an extreme point of B. Hence $x_1 = 2(x_2 - y) = 2(y - (x_2 - x_1))$; i.e. $y = x_2 - (x_1/2)$ and y is the mid-point of the segment $[x_2, x_2 - x_1]$. Since $y, x_2, x_2 - x_1$ all have norm 1, $[x_2, x_2 - x_1] \subseteq \partial B$. We show, further, that x_2 is an extreme point of B. If not, since $[x_2, x_2 - x_1] \subseteq \partial B$ there must exist ξ with $0 < \xi < 1$ such that $x_2 + \xi x_1 \in \partial B$ (see Figure 4.6). Then $\|(x_2 + \xi x_1) - x_1\| = \|\xi x_2 + (1 - \xi)(x_2 - x_1)\| = 1$. Hence

$$\mu_B(\partial B)/2 \geq \|x_2 + \xi x_1 - x_1\| + \|(x_2 + \xi x_1) - (x_2 - x_1)\|$$
$$+ \|(x_2 - x_1) - (-x_1)\|$$
$$= 1 + (1 + \xi) + 1 > 3,$$

which is a contradiction.

Thus x_2 is also an extreme point. Hence we can apply the proof again beginning with x_2 and a third time beginning with $(x_2 - x_1)$ to show that the whole hexagon is contained in ∂B. Hence B is an affine regular hexagon.

(ii) Suppose that $\mu_B(\partial B) = 8$ and suppose, from Corollary 3.2.5, that $\wp = [x_1 + x_2, -x_1 + x_2, -x_1 - x_2, x_1 - x_2]$ is a circumscribed parallelogram with $\|x_1\| = \|x_2\| = 1$ (Figure 4.7). Since the mid-point of each side of $\wp$ is in ∂B, if each vertex also is in ∂B then $\partial B = \wp$ and we are finished. If not, then by suitable renaming, we may suppose that $x_1 + x_2 \notin \partial B$. Then there exists ξ with $0 \leq \xi < 1$ such that the line $\{(1 + \xi)x_1 + t(x_2 - x_1) : t \in \mathbb{R}\}$ is tangent to B. This line meets $\wp$ at $(x_1 + \xi x_2)$ and $(x_2 + \xi x_1)$ (see Figure 4.7). Hence, by

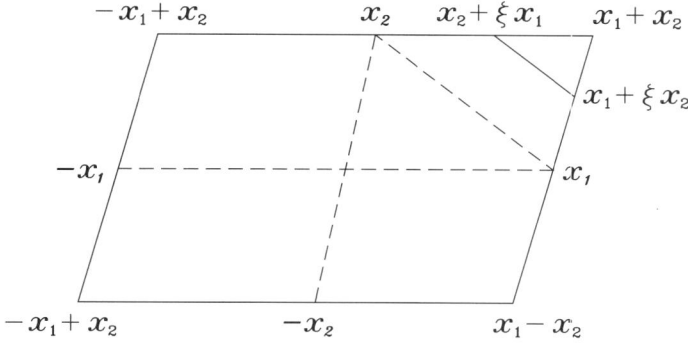

Figure 4.7

116 Two-dimensional Minkowski spaces

Theorem 4.3.4,

$$4 = \mu_B(\partial B)/2 \leq \|(x_1 + \xi x_2) - x_1\| + \|(x_2 + \xi x_1) - (x_1 + \xi x_2)\|$$
$$+ \|(-x_1 + x_2) - (x_2 + \xi x_1)\| + \|(-x_1 + x_2) - (-x_1)\|$$
$$= \xi + (1 - \xi)\|x_2 - x_1\| + (1 + \xi) + 1$$
$$= 2 + 2\xi + (1 - \xi)\|x_2 - x_1\|$$
$$\leq 4$$

(since $\|x_2 - x_1\| \leq \|x_2\| + \|x_1\| = 2$).

Again we must have equality throughout, but since $\xi < 1$ this implies that $\|x_2 - x_1\| = 2$, i.e. $(x_2 - x_1)/2 \in \partial B$. Since $-x_1, x_2 \in \partial B$ this means that the line segment $[x_2, -x_1] \subseteq \partial B$ and also that the vertex $x_2 - x_1$ of $\wp$ does not belong to ∂B. Hence we may repeat the argument with that vertex and conclude that $\partial B = [x_1, x_2, -x_1, -x_2]$; i.e. B is a parallelogram. ∎

One can check (see Figure 4.8) that if

$$B := \text{conv}\{(1, 1), (-\xi, 1), (-1, \xi), (-1, -1), (\xi, -1), (1, -\xi)\}$$

then

$$\mu_B(\partial B) = 6 + 2\xi \quad (0 \leq \xi \leq 1).$$

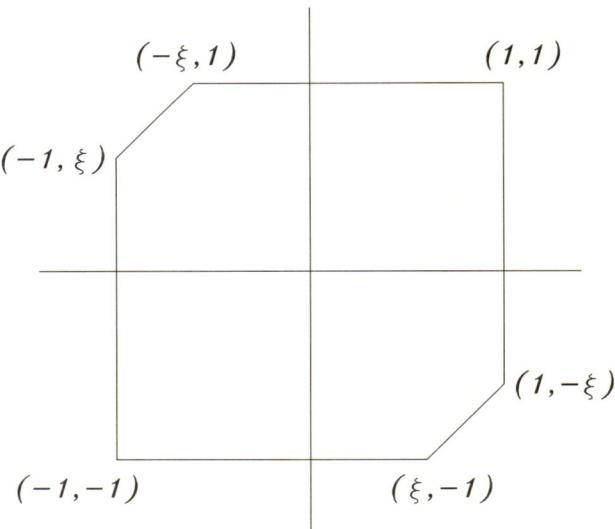

Figure 4.8

4.3 Lengths of curves, perimeter of the unit ball   117

Thus all values between 6 and 8 are assumed (and assumed by hexagons, accepting a parallelogram as a degenerate hexagon). Thus, in Minkowski geometry, the range of possible values for "π" is [3,4].

Theorem 4.3.8 *If X is a two-dimensional linear space and if X is equipped with two unit balls B_1 and B_2 then*

$$\mu_{B_2}(\partial B_1) = \mu_{B_1^\circ}(\partial B_2^\circ).$$

Proof (see [508]). Let $[x_0, x_1, x_2, \ldots, x_n = x_0]$ be a simple, closed, convex polygon inscribed to B_1 (see left-hand side of Figure 4.9) and consider its length $\sum_{i=1}^{n} \|x_i - x_{i-1}\|_2$ with respect to B_2. For each $i = 1, 2, \ldots, n$, let

$$y_i := (x_i - x_{i-1})/\|x_i - x_{i-1}\|_2$$

be the B_2-unit vector in the direction of $(x_i - x_{i-1})$. By the Hahn–Banach theorem, for each i there exists $f_i \in X^*$ such that $\|f_i\|_2 = 1$ and $f_i(y_i) = 1$ (i.e. f_i supports B_2 at y_i). Hence $\|x_i - x_{i-1}\|_2 = f_i(x_i - x_{i-1})$ ($i = 1, 2, \ldots, n$).

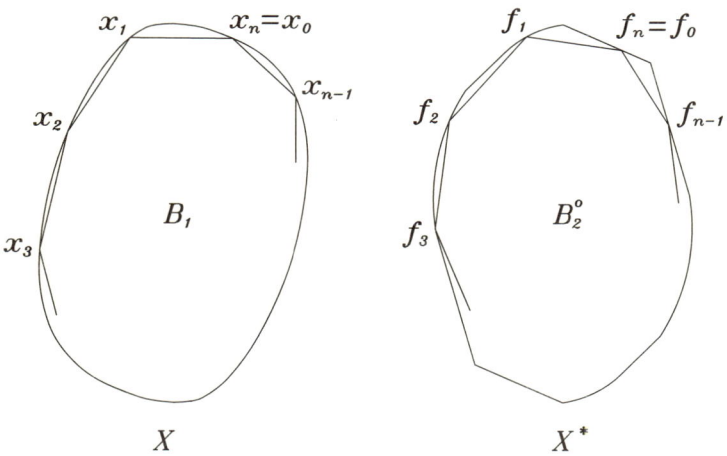

Figure 4.9

Because the x_i's come in order round ∂B_1, the unit vectors $\{y_i\}$ come in order round the curve ∂B_2. Moreover, since B_2 is convex, ∂B_2 has positive curvature and so the tangent lines $\{x : f_i(x) = 1\}$ to B_2 at y_i have slopes which are monotonically increasing. Hence the points $\{f_i\}$ come in order round the curve ∂B_2° (see right-hand side of Figure 4.9). Let $f_0 := f_n$. Therefore, $[f_0, f_1, \ldots, f_n]$ is a simple, closed, convex polygon inscribed to B_2°. Moreover,

$$\sum_{i=1}^{n} \|x_i - x_{i-1}\|_2 = \sum_{i=1}^{n} f_i(x_i - x_{i-1}) = \sum_{i=0}^{n-1}(f_i - f_{i+1})(x_i) \qquad (4.1)$$

$$\leq \sum_{i=1}^{n} \|x_i\|_1 \|f_i - f_{i+1}\|_1 = \sum_{i=1}^{n} \|f_i - f_{i+1}\|_1 \qquad (4.2)$$

$$\leq \mu_{B_1^\circ}(\partial B_2^\circ). \qquad (4.3)$$

Since this is true of every inscribed polygon we get

$$\mu_{B_2}(\partial B_1) \leq \mu_{B_1^\circ}(\partial B_2^\circ).$$

Using the reflexivity of both spaces, Theorem 1.3.8 (or by repeating the proof beginning with an inscribed polygon to B_2°), we get the reverse inequality. ∎

Corollary 4.3.9 (Schäffer) *If (X, B) is a two-dimensional Minkowski space with dual (X^*, B°) then $\mu_B(\partial B) = \mu_{B^\circ}(\partial B^\circ)$.*

Proof. Let $B_1 = B_2$ in Theorem 4.3.8. ∎

Remark. Schäffer proves a slightly more general result for non-symmetric norms; i.e. the balls B_1 and B_2 need not be centrally symmetric. In this case there are two perimeters depending on which orientation is used. By keeping track of the orientation one can adapt the preceding proof to obtain Schäffer's more general result. Note that in the "integration by parts" interchange (Equation (4.1)) there is a change of orientation between the measurement of ∂B_1 and that of ∂B_2°.

4.4 The isoperimetric problem in a Minkowski plane

One of the central problems in plane metric geometry is to investigate which (convex) bodies have maximal area for a given perimeter or minimal perimeter for a given area. In this section we present Busemann's [66, 68] elegant solution in the case of the Minkowski plane (see also Benson [30] for another account).

As usual, $(X, \|.\|)$ denotes a two-dimensional normed space whose unit ball is B. We are concerned with a general convex body K and its boundary ∂K whose Minkowski length is denoted by $\mu(\partial K)$. We also need to consider the area of K. Since all translations are isometries of X, we need area in X to be translation invariant; i.e. area must be Haar measure in X suitably normalized. We shall return to the question of a suitable normalization of two-dimensional Haar measure in the later part of this section. For now we suppose that some Haar measure has been agreed upon and is fixed for the discussion. Because there are so many well-known formulas for both area and arc length in Euclidean space, it is most convenient for the purposes of calculation to suppose that there is an auxiliary

Euclidean metric defined on X. We may also suppose that this is done so that the two-dimensional Lebesgue measure, λ, induced by the Euclidean metric coincides with the chosen Haar measure. We shall use $|.|$ to denote the Euclidean norm, λ^1 to denote Euclidean arc length and λ^2 to denote Euclidean plane measure (in most contexts the dimension is clear and the superscripts will be omitted).

To begin with, let us suppose that the convex set K has a smooth boundary; *i.e.* at each point $x \in \partial K$ there is a unique (up to scalar multiple) non-zero tangent vector y_x. Then the *unit outward normal* to K at x is that unique linear functional $\hat{f}_x$ such that $\hat{f}_x(y_x) = 0$, $|\hat{f}_x| = 1$ (a Euclidean unit normal) and $\hat{f}_x(z) \leq \hat{f}_x(x)$ for all $z \in K$. If ∂K is parametrized by

$$\partial K = \{x = x(t) : \alpha \leq t \leq \beta\}$$

then the *Euclidean element of arc length* at x is

$$d\lambda(x) := |x'(t)|dt.$$

The *Minkowski element of arc length* at x is, likewise, defined as

$$d\mu_B(x) := \|x'(t)\| \, dt.$$

If we choose y_x to be the Minkowski unit tangent, *i.e.* $y_x = x'(t)/\|x'(t)\|$, then in the direction $y_x = \hat{f}_x^\perp$ we have the ratio

$$\sigma(\hat{f}_x) := \frac{\text{Minkowski length in the direction } \hat{f}_x^\perp}{\text{Euclidean length in the direction } \hat{f}_x^\perp} = \frac{\|y_x\|}{|y_x|} = \frac{1}{|y_x|} = \frac{1}{\rho_B(y_x)}, \tag{4.4}$$

where ρ_B is the radial function of the unit ball B.

By means of approximating polygons whose lengths can be considered to be Riemann sums, one gets that the Minkowski length of ∂K is given by the following equation:

$$\mu_B(\partial K) = \int_{\partial K} d\mu_B(x). \tag{4.5}$$

However, from (4.4) we have a variety of other expressions for this length.

$$\mu_B(\partial K) = \int_\alpha^\beta \|x'(t)\| \, dt = \int_\alpha^\beta \frac{1}{\rho_B(y_x)} |x'(t)| \, dt$$

$$= \int_{\partial K} \sigma(\hat{f}_x) \, d\lambda(x) = \int_{\partial K} \frac{1}{\rho_B(y_x)} \, d\lambda(x). \tag{4.6}$$

Recall two familiar formulas from Chapter 2. The first (Equation (2.14)) is that if the function $\sigma(\hat{f}_x)$ is extended to all of X^* by positive homogeneity and if the extended function is convex then the integral

$$2^{-1} \int_{\partial K} \sigma(\hat{f}_x) d\lambda(x)$$

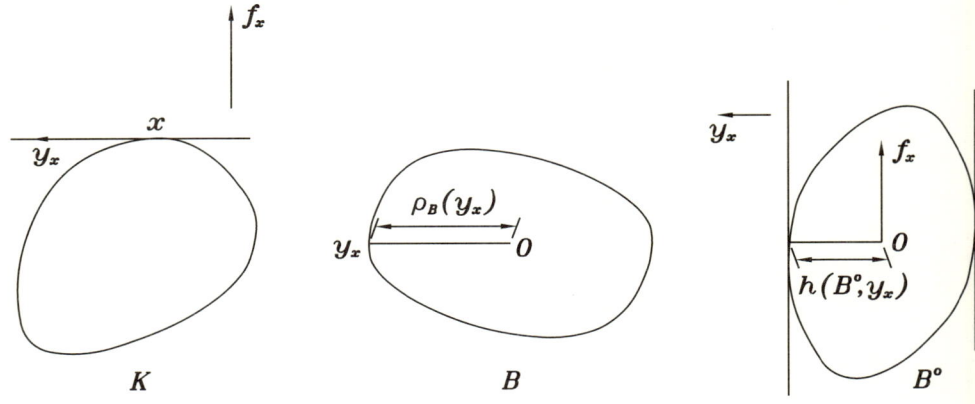

Figure 4.10

in (4.6) represents a two-dimensional mixed volume $V(K, \mathbf{I}_B)$, where $\mathbf{I}_B$ is that convex set of which the extended function σ is the support function. The second (Theorem 2.2.13) is that if X and X^* are identified then the reciprocal of the radial function of B is the support function of $B°$, i.e. $\rho_B(y_x)^{-1} = h_{B°}(y_x)$ (see Figure 4.10). Thus $\sigma(\hat{f}_x) = \rho_B(y_x)^{-1}$ is convex and is the support function, not of $B°$ because of the change from $\hat{f}_x$ to y_x in the argument, but of "$B°$ rotated through 90°". We emphasize, however, that $\mathbf{I}_B \subseteq X$ whereas $B° \subseteq X^*$ and the "rotation" depends on the usual identification of X and X^*. Another view of this identification is to consider the map $\Lambda : X \mapsto X^*$ defined by

$$(\Lambda x)(y) := \det[x, y]. \tag{4.7}$$

Then Λ is the unary operation in $\mathbb{R}^2$ which replaces the binary cross-product in $\mathbb{R}^3$ and $\mathbf{I}_B$ is the inverse image of $B°$ under Λ (see Guggenheimer [226, 227]).

The functional σ and the set $\mathbf{I}_B$ can be interpreted in two other ways, to which we shall have occasion to refer more explicitly in Chapters 6 and 7. Firstly,

$$\sigma(\hat{f}_x) = 2^{-1} \text{ (the projection of } B° \text{ along } \hat{f}_x)$$

and hence $\mathbf{I}_B$ is the projection body of $B°$ (see Figure 4.10); secondly,

$$\sigma(\hat{f}_x) = 2 \text{ (the cross-section of } B \text{ by } \hat{f}_x^\perp)^{-1}$$

and so $\mathbf{I}_B$ is the polar reciprocal of the intersection body of B (in both cases up to a scalar factor).

Returning to the formulas for $\mu(\partial K)$ we obtain

$$\mu(\partial K) = \int_{\partial K} \sigma(\hat{f}_x)\, d\lambda(x) = \int_{\partial K} h_{B°}(y_x)\, d\lambda(x) = 2V(K, \mathbf{I}_B). \tag{4.8}$$

4.4 The isoperimetric problem in a Minkowski plane

We have established these formulas for convex bodies K with smooth boundary. The set of points of ∂K at which ∂K is not differentiable is at most countable and this set does not affect the values of the integrals. Therefore, these formulas are valid for all convex bodies K.

Remark. Formula (4.8) provides the promised proof of Proposition 4.3.5. If c_1 and c_2 are convex curves let $K_i := \operatorname{conv} c_i$ for $i = 1, 2$. Since c_1 lies inside c_2 we have $K_1 \subseteq K_2$ and therefore, by Theorem 2.3.7(ii), $V(K_1, \mathbf{I}_B) \leq V(K_2, \mathbf{I}_B)$, which, with Equation (4.8), completes the proof of Proposition 4.3.5.

The proof of the isoperimetric theorem for Minkowski space is now a consequence of the Minkowski inequality for mixed volumes.

Theorem 4.4.1 *If K is a convex body in the Minkowski space (X, B) with area equal to that of $\mathbf{I}_B$ then $\mu_B(\partial K) \geq \mu_B(\partial \mathbf{I}_B)$ with equality if and only if K is a translate of $\mathbf{I}_B$.*

Proof. From Equation (4.8) we have $\mu_B(\partial K) = 2V(K, \mathbf{I}_B)$ and $\mu_B(\partial \mathbf{I}_B) = 2V(\mathbf{I}_B, \mathbf{I}_B) = 2\lambda(\mathbf{I}_B)$. Further, from Minkowski's inequality (Theorem 2.3.8) and since $\lambda(K) = \lambda(\mathbf{I}_B)$, we have $V(K, \mathbf{I}_B)^2 \geq \lambda(\mathbf{I}_B)\lambda(K) = \lambda(\mathbf{I}_B)^2$ with equality if and only if K is a translate of $\mathbf{I}_B$, i.e.

$$\mu_B(\partial K) \geq 2\lambda(\mathbf{I}_B) = \mu_B(\partial \mathbf{I}_B), \qquad (4.9)$$

as required. ∎

Thus, among convex bodies with a fixed value for the area, those with minimal perimeter are precisely the translates of some fixed multiple of $\mathbf{I}_B$. As the value for the area is varied we obtain all homothetic images of $\mathbf{I}_B$. The same applies, dually, for bodies of maximal area for a given perimeter. Among the homothetic images of $\mathbf{I}_B$ we wish to specify a unique one to be called *the isoperimetrix* $\tilde{\mathbf{I}}_B$. Since $\mathbf{I}_B$ is centrally symmetric, it is natural to specify that $\tilde{\mathbf{I}}_B$ be centred at 0. Once we have normalized Haar measure in a way to be called "Minkowski area" in X and denoted by μ (μ^2 if confusion with length is likely and μ_B to emphasize the dependence on B), we normalize $\tilde{\mathbf{I}}_B$ so that

$$\mu_B(\partial \tilde{\mathbf{I}}_B) = 2\mu(\tilde{\mathbf{I}}_B). \qquad (4.10)$$

As in (4.4) define the number σ_B by

$$\sigma_B := \frac{\text{Minkowski area in } X}{\text{Euclidean area in } X} = \frac{\mu}{\lambda}. \qquad (4.11)$$

If $\tilde{\mathbf{I}}_B = \alpha \mathbf{I}_B$ then from the definition and Equations (4.8) and (4.10) we get

$$\sigma_B = \frac{\mu^2(\tilde{\mathbf{I}}_B)}{\lambda^2(\tilde{\mathbf{I}}_B)} = \frac{\mu_B^1(\partial \tilde{\mathbf{I}}_B)}{2\alpha^2\lambda^2(\mathbf{I}_B)} = \frac{\mu_B^1(\partial \tilde{\mathbf{I}}_B)}{\alpha \mu_B^1(\partial \tilde{\mathbf{I}}_B)} = \alpha^{-1}. \quad (4.12)$$

Thus $\alpha = \sigma_B^{-1}$ and $\tilde{\mathbf{I}}_B = \sigma_B^{-1} \mathbf{I}_B$.

The choice of the normalizing factor σ_B begs the question of how area is to be defined in two-dimensional subspaces of three (or higher) dimensional spaces. This is the subject of the next chapter. Here we mention two possibilities, which are (respectively) those explored in Chapters 7 and 6.

Firstly, following Busemann [71], we may normalize area by specifying

$$\mu(B) := \pi. \quad (4.13)$$

In this case $\sigma_B = \mu(B)/\lambda(B) = \pi/\lambda(B)$. Thus $\tilde{\mathbf{I}}_B = \pi^{-1}\lambda(B)\mathbf{I}_B$ and

$$\mu(\tilde{\mathbf{I}}_B) = \sigma_B^{-2}\mu(\mathbf{I}_B) = \sigma_B^{-2}\sigma_B\lambda(\mathbf{I}_B) = \sigma_B^{-1}\lambda(\mathbf{I}_B).$$

If we regard $\mathbf{I}_B$ as B° rotated then $\lambda(\mathbf{I}_B) = \lambda^*(B^\circ)$ and

$$\mu(\tilde{\mathbf{I}}_B) = \sigma_B^{-1}\lambda^*(B^\circ) = \lambda(B)\lambda^*(B^\circ)/\pi;$$

i.e. the isoperimetrix has Minkowski area equal to the normalized volume product of B (Definition 2.3.2).

Secondly, as in [262], we may interchange the roles of B and $\tilde{\mathbf{I}}_B$ and normalize area by specifying

$$\mu(B) := \lambda(B)\lambda^*(B^\circ)/\pi. \quad (4.14)$$

Recall that this volume product is an affine invariant independent of which particular Euclidean structure is used. In this case, $\sigma_B = \mu(B)/\lambda(B) = \lambda^*(B^\circ)/\pi$. Thus $\tilde{\mathbf{I}}_B = \pi[\lambda^*(B^\circ)]^{-1}\mathbf{I}_B$. A repetition of the calculation above this time yields $\mu(\tilde{\mathbf{I}}_B) = \pi$. This second normalization is interesting because, by (4.10), we have

$$\mu_B(\partial \tilde{\mathbf{I}}_B) = 2\pi.$$

Thus we may measure an angle either by twice the area of the sector of $\tilde{\mathbf{I}}_B$ or by the length of the perimeter of $\tilde{\mathbf{I}}_B$ which it cuts off. These trigonometric questions are the subject of Chapter 8.

The bounds for the volume product $\lambda(B)\lambda^*(B^\circ)$ for centrally symmetric, two-dimensional bodies are given by the Blaschke–Santaló and the Mahler–Reisner inequalities (Theorems 2.3.3 and 2.3.4). Specifically, we have

$$8 \leq \lambda(B)\lambda^*(B^\circ) \leq \pi^2$$

with equality on the left if and only if B is a parallelogram and equality on the right if and only if B is an ellipse. We summarize this discussion in the following theorem.

Theorem 4.4.2 *Let (X, B) be a two-dimensional Minkowski space and let λ denote some Haar measure on X.*

(i) If Minkowski area μ_B on X is defined to be that multiple of λ such that $\mu_B(B) = \pi$ then $\tilde{\mathbf{I}}_B = \lambda(B)\pi^{-1}\mathbf{I}_B$, $\mu_B(\tilde{\mathbf{I}}_B) = \lambda(B)\lambda^*(B^\circ)/\pi$ and $8/\pi \leq \mu_B(\tilde{\mathbf{I}}_B) \leq \pi$.

(ii) If Minkowski area μ_B on X is defined to be that multiple of λ such that $\mu_B(B) = \lambda(B)\lambda^*(B^\circ)/\pi$ then $\tilde{\mathbf{I}}_B = \pi[\lambda^*(B^\circ)]^{-1}\mathbf{I}_B$, $\mu_B(\tilde{\mathbf{I}}_B) = \pi$, $\mu_B(\partial\tilde{\mathbf{I}}_B) = 2\pi$ and $8/\pi \leq \mu_B(B) \leq \pi$.

Remark. These formulas for $\mu(\tilde{\mathbf{I}}_B)$ depend on being able to identify $\lambda(\mathbf{I}_B)$ with $\lambda^*(B^\circ)$ because $\mathbf{I}_B$ is a rotation of B°. This is very special to dimension 2.

A more general calculation, valid in all dimensions, is the following. If $\mu_B = \sigma_B \lambda$ is a renormalization of Haar measure in X then the dual Haar measure in X^* is $\mu^* = \sigma_B^{-1}\lambda^*$. Therefore, $\mu_B^*(B^\circ) = \sigma_B^{-1}\lambda^*(B^\circ)$. With $\sigma_B = \epsilon_d/\lambda(B)$ we get $\mu_B^*(B^\circ) = \lambda(B)\lambda^*(B^\circ)/\pi$ and with $\sigma_B = \lambda^*(B^\circ)/\epsilon_d$ we get $\mu_B^*(B^\circ) = \epsilon_d$.

These last equations and those in Theorem 4.4.2 are evidence for two imprecise ideas. The first is the duality between the two normalizations of Haar measure given in Equations (4.13) and (4.14). The second is that $\tilde{\mathbf{I}} = \sigma_B^{-1}\mathbf{I}_B$ is the right normalization to be called *the* isoperimetrix. Section 4.6 provides further evidence for the second statement.

4.5 Isoperimetric inequalities

In this short section we give the Minkowski analogues of the classical isoperimetric inequality and Bonnesen's strengthening of it [50]. In all these formulas the value of $\mu_B(\tilde{\mathbf{I}}_B)$ appears. In order that the formulas appear most like the usual Euclidean case we use the normalization of Equation (4.14) throughout the section. Readers who prefer (4.13) may modify the formulas accordingly; they appear in Petty's fundamental paper [413] and some of them also in Guggenheimer [226].

First, since Minkowski area $\mu_B(K) = \sigma_B \lambda(K)$, where $\sigma_B = \lambda^*(B^\circ)/\pi$, we may also define Minkowski mixed volume $V_B(K_1, K_2)$ by

$$V_B(K_1, K_2) := \sigma_B V(K_1, K_2).$$

Then we have Minkowski's inequality

$$V_B(K_1, K_2)^2 \geq \mu_B(K_1)\mu_B(K_2), \quad (4.15)$$

and, by the linearity of V in the second variable,

$$V(K, \mathbf{I}_B) = V_B(K, \tilde{\mathbf{I}}_B). \quad (4.16)$$

Proposition 4.5.1 *If K is a convex body in a Minkowski space (X, B) then*

$$\mu(\partial K)^2 \geq 4\pi \mu(K).$$

Proof. Since $\mu(\partial K) = 2V(K, \mathbf{I}_B) = 2V_B(K, \tilde{\mathbf{I}}_B)$ (by 4.16) we have, using Equation (4.15) and Theorem 4.4.2(ii),

$$\mu(\partial K)^2 = 4V_B(K, \tilde{\mathbf{I}}_B)^2 \geq 4\mu_B(K)\mu_B(\tilde{\mathbf{I}}_B) = 4\pi\mu_B(K). \qquad \blacksquare$$

Definition 4.5.2 *If K is a convex body in a Minkowski space (X, B) then the **parallel body** K_α is defined by $K_\alpha := K + \alpha\tilde{\mathbf{I}}_B$, where $\tilde{\mathbf{I}}_B$ is the isoperimetrix.*

Proposition 4.5.3 *If K_α is the parallel body to the convex body K, then*

(i) $\mu_B(K_\alpha) = \mu_B(K) + \alpha\mu_B(\partial K) + \alpha^2\pi$,
(ii) $\mu_B(\partial K_\alpha) = \mu_B(\partial K) + 2\pi\alpha$.

Proof

(i) We have $\mu_B(K_\alpha) = \mu_B(K + \alpha\tilde{\mathbf{I}}_B) = \mu_B(K) + 2\alpha V_B(K, \tilde{\mathbf{I}}_B) + \mu_B(\alpha\tilde{\mathbf{I}}_B)$, from which the result is immediate.

(ii) $\mu_B(\partial K_\alpha) = 2V_B(K_\alpha, \tilde{\mathbf{I}}_B) = 2V_B(K + \alpha\tilde{\mathbf{I}}_B, \tilde{\mathbf{I}}_B)$. The result now follows from the linearity of V_B in the first variable. $\qquad \blacksquare$

Definition 4.5.4 *If K is a convex body, define the **inner radius** of K, $r(K)$ by*

$$r(K) := \max\{\alpha : \exists x \in X \text{ with } \alpha\tilde{\mathbf{I}}_B \subseteq K + x\}$$

*and the **outer radius** of K, $R(K)$ by*

$$R(K) := \min\{\alpha : \exists x \in X \text{ with } \alpha\tilde{\mathbf{I}}_B \supseteq K + x\}.$$

The following generalization of Bonnesen's inequality is also due to Petty [413].

Theorem 4.5.5 *If K is a convex body in a Minkowski space (X, B) with inner and outer radii r and R respectively, then*

$$\mu(\partial K)^2 - 4\pi\mu(K) \geq \pi^2(R^2 - r^2).$$

Proof. The proof is a translation of Eggleston's proof [138], p. 108, of the Bonnesen inequality. A standard argument shows that $r\tilde{\mathbf{I}}_B$ is inscribed to some translate K' of K in the sense of Eggleston and, likewise, that K'' (some other translate of K) is inscribed to $R\tilde{\mathbf{I}}(B)$. In the plane, if K_1 and K_2 are convex bodies with K_1 inscribed to K_2 then $2V(K_1, K_2) \geq \lambda(K_1) + \lambda(K_2)$ and hence

$2V_B(K_1, K_2) \geq \mu(K_1) + \mu(K_2)$. Thus we have

$$r\mu_B(\partial K) \geq \mu_B(K) + \pi r^2,$$
$$R\mu_B(\partial K) \geq \mu_B(K) + \pi R^2;$$

i.e. the quadratic form $\mu_B(K) + x\mu_B(\partial K) + \pi x^2$ is negative at both $x = -r$ and $x = -R$. Hence if x_1 and x_2 are the roots of this expression then we have $\mu_B(\partial K)^2 - 4\pi \mu_B(K) = \pi^2(x_1 - x_2)^2 \geq \pi^2(R - r)^2$. ∎

4.6 Transversality

In the same way that various roles of the unit ball in Euclidean space are divided among several players in Minkowski geometry, so are the roles of orthogonality. We have already looked at the idea of shortest distance from a point to a line or a hyperplane in Definition 3.2.2. We now turn to the idea that for parallelograms

area = base × perpendicular height.

Let P be the parallelogram with one vertex at 0 and spanned by the vectors x and y. If $\hat{f}$ is the Euclidean unit linear functional such that $\hat{f}(x) = 0$ and $\hat{f}(y) > 0$ then the perpendicular height of P is given by $\hat{f}(y)$ so that

$$\lambda(P) = |x|\hat{f}(y).$$

Using Equation (4.4), we get

$$\lambda(P) = \|x\|\hat{f}(y)/\sigma(\hat{f}) = \|x\|f'(y),$$

where $f' := \hat{f}/\sigma(\hat{f})$ is a unit vector relative to $\mathbf{I}_B^\circ$. Finally,

$$\mu_B(P) = \sigma_B \lambda(P) = \|x\|\sigma_B f'(y) = \|x\|\tilde{f}(y),$$

where $\tilde{f} := \sigma_B \hat{f}/\sigma(\hat{f})$ is a unit vector relative to $\tilde{\mathbf{I}}_B^\circ$. Thus we have proved the following proposition.

Proposition 4.6.1 *In a two-dimensional space (X, B) if P is the parallelogram defined above then*

$$\mu_B(P) = \|x\|\tilde{f}(y),$$

where $\tilde{f}$ is the linear functional such that $\tilde{f}(x) = 0$, $\tilde{f}(y) > 0$ and $\|\tilde{f}\|_{\tilde{\mathbf{1}}} = h_{\tilde{\mathbf{1}}}(\tilde{f}) = 1$.

Definition 4.6.2 *With $\tilde{f}$ as above we say that y is **transversal** to x and write $y \triangleleft x$ if $\tilde{f}(y) = \|y\|_{\tilde{\mathbf{1}}}$, i.e. if $\tilde{f}$ supports $\tilde{\mathbf{I}}$ at $y/\|y\|_{\tilde{\mathbf{1}}}$.*

The proof of the following corollary is immediate.

Corollary 4.6.3 *With P as above, if* $y \triangleleft x$ *then*

$$\mu_B(P) = \|x\|_B \|y\|_{\bar{1}}.$$

We return to these ideas in Chapters 5 and 8.

The final version of orthogonality that we consider is that of perpendicular bisector. In Euclidean space the locus of points equidistant from x and y is the perpendicular bisector of $[x, y]$. More generally, the locus of points z such that $|x - z| = \alpha |y - z|$ is a circle of Apollonius which has x and y as inverse points and which cuts each circle through x and y orthogonally. It is possible to consider these loci in Minkowski space but, as far as I know, their only value is as interesting (and attractive) pictures. Figures 4.11a and b were drawn by Brian Ingalls using MAPLE. They show loci analogous to circles of Apollonius and the locus of points equidistant from two given points when the norm is an ℓ_p-norm. It is interesting that the closed curves need not be convex and that there appears to be no family of orthogonal circles as there is in the Euclidean case.

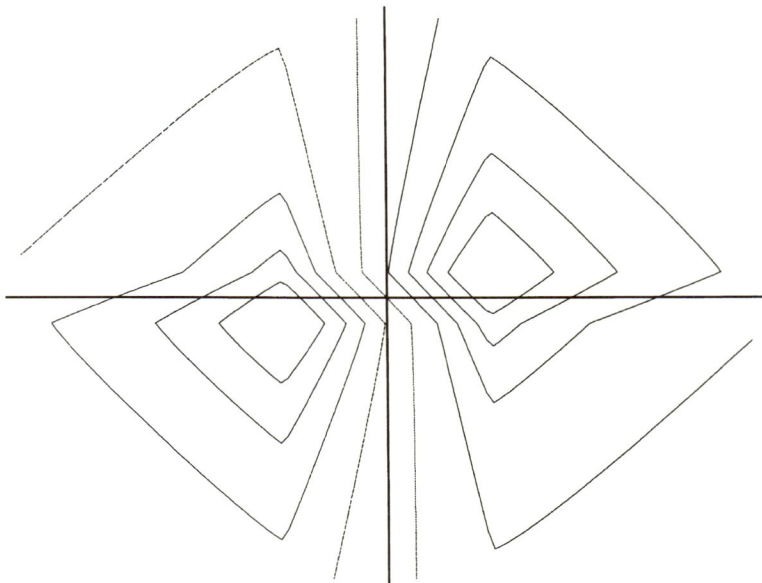

Figure 4.11a. "Circles" of Apollonius with respect to the points $a := (2, \frac{1}{2})$ and $b := (-2, -\frac{1}{2})$ when the norm is the ℓ_p-norm for $p = 20/19$. The curves are the loci $\|x - a\| = \alpha \|x - b\|$ for $\alpha = 1, 1.5, 2.1, 3, 5$ and their reciprocals.

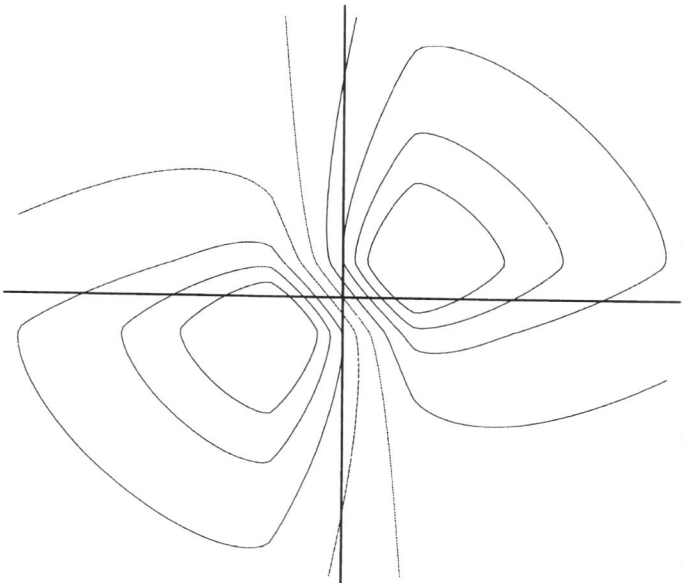

Figure 4.11b. "Circles" of Apollonius with respect to the points $a := (2, 1)$ and $b := (-2, -1)$ when the norm is the ℓ_p-norm for $p = \frac{4}{3}$. The curves are the loci $\|x-a\| = \alpha \|x-b\|$ for $\alpha = 1, 1.3, 1.7, 2.2, 3$ and their reciprocals.

4.7 Radon curves

In this section we shall write $\mathbf{I}(B)$ instead of $\mathbf{I}_B$ because we want to regard $\mathbf{I}$ as a mapping on the centrally symmetric convex sets in the plane. It is clear from the construction of $\mathbf{I}(B)$ in §4.4 that for every ball B we have $\mathbf{I}^2(B) = B$. In this section we consider the possibility $\mathbf{I}(B) = B$ and, more generally, $\mathbf{I}(B) = \alpha B$ for some multiple α. Neither formulation, however, captures the essence of the problem, because the size of $\mathbf{I}(B)$ depends on the Euclidean structure used for its construction. The questions should be: are there balls B such that for every Euclidean structure $\mathbf{I}(B) = \alpha B$ (α depending on the Euclidean structure), and are there balls B for which there is a Euclidean structure with respect to which $\mathbf{I}(B) = B$? The answer to both questions is the same; this happens precisely when the boundary of B is a Radon curve. These curves were introduced by Radon in [432]. Most of the section is devoted to the construction of such curves. Examples which we have seen already are the hexagons of Example 1.1.17.

To simplify the notation a little we will denote the oriented unit circle by c_0, i.e. $c_0 = \partial B$ together with an orientation. Now use Corollary 3.2.3 to construct a basis (x_1, x_2) whose vectors are mutually normal. We will use coordinates relative to this basis in X and relative to the dual basis in X^*.

Each linear functional $f = (\phi_1, \phi_2) \in \partial B^\circ$ supports B at some point $x \in c_0$. In

Euclidean terms (ϕ_1, ϕ_2) represents the outward normal to c_0 at x. On the other hand, we may consider the direction t of that tangent to c_0 at x for which $f(t) = 0$; the orientation determines the sign of t. The vector t is in X and is given by $t = -\phi_2 x_1 + \phi_1 x_2 = (-\phi_2, \phi_1)^t$. Denote the set of vectors constructed in this way by c_0'. Thus

$$c_0' := \{y = (-\phi_2, \phi_1)^t : (\phi_1, \phi_2) \in \partial B^\circ\}.$$

In terms of the discussion in §4.4, c_0' is the curve described there as "∂B° rotated through $90°$" and can also be written as

$$c_0' = \Lambda^{-1}(\partial B^\circ) = \partial \mathbf{I}(B),$$

where Λ is defined by (4.7).

Because we have chosen x_1 and x_2 to be mutually normal, there is a tangent to c_0 at x_1 parallel to x_2 and a tangent to c_0 at x_2 in the direction $-x_1$. Therefore, both c_0 and c_0' pass through the four points $x_1, x_2, -x_1, -x_2$ and the two curves have at least one common tangent at each of these points. This allows us to combine the two curves as follows: let $c_0'' := c_1 \cup c_1' \cup -c_1 \cup -c_1'$, where $c_1 := \{(\xi_1, \xi_2)^t \in \partial B : \xi_1 \geq 0, \xi_2 \geq 0\}$ and $c_1' := \{(-\phi_2, \phi_1) : (\phi_1, \phi_2) \in \partial B^\circ, \phi_1 \geq 0, \phi_2 \geq 0\}$. Then c_0'' is a symmetric *convex* curve passing through the basis vectors and sharing a common tangent with both c_0 and c_0' at those points.

The above process may be described by saying that c_0'' is equal to ∂B in the first and third quadrants and is equal to $\partial \mathbf{I}_B$ in the other two quadrants. A curve of this kind is called a *Radon curve* (Radon [432]). See, e.g., Figure 4.12 ($R = \text{conv } c_0''$).

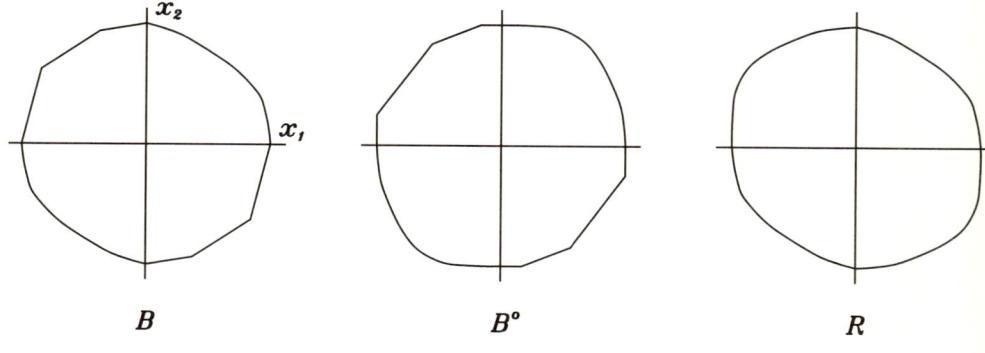

Figure 4.12

Since $f = (\phi_1, \phi_2)$ supports B at $x = (\xi_1, \xi_2)^t$ if and only if $J(x) = (\xi_1, \xi_2)^t$ supports B° at (ϕ_1, ϕ_2) it follows that $-\phi_2 x_1 + \phi_1 x_2$ is tangent to c_0 at $(\xi_1, \xi_2)^t$ if and only if $-\xi_1 x_1 - \xi_2 x_2$ is tangent to c_0' at $(-\phi_2, \phi_1)$. Hence, if R denotes the ball consisting of the vectors inside and on c_0'', we see that $(\xi_1, \xi_2)^t \in \partial R$ ($\xi_i \geq 0$)

is normal relative to R to $(-\phi_2, \phi_1)^t \in \partial R$ if and only if $(-\phi_2, \phi_1)^t$ is normal to $(\xi_1, \xi_2)^t$ relative to R, and similarly if $\xi_i < 0$. Thus normality is symmetric for the ball R. For more precise details of this calculation we refer to Day [124], who also proves the converse that if normality is symmetric for a ball R then ∂R must be a Radon curve.

It is also clear from the construction that ∂B is a Radon curve if and only if $c_0 = c_0'$ and hence, from the above description of c_0', if and only if $B = \mathbf{I}(B)$. Here, however, the Euclidean structure used for computing $\mathbf{I}(B)$ is the same as the one used in the construction of the Radon curve. Had we used a different Euclidean ball then we would only have $\mathbf{I}(B) = \alpha B$. The reason for this is that $\mathbf{I}(B)$ does not transform naturally when B is transformed by an invertible linear map T (there is a scale factor depending on $\det T$). In this regard, $\tilde{\mathbf{I}}(B)$ is the more natural object to use (see Equation (5.21) in the next chapter). In dimensions $d \geq 3$, symmetry of normality characterizes ellipsoids (Theorem 3.4.10). We conjecture that if $d \geq 3$ then the equation $B = \alpha \mathbf{I}(B)$ characterizes ellipsoids. Note that even in the case of Radon curves it is not true that $B = \tilde{\mathbf{I}}(B)$. In fact, we shall show in Chapter 6 that if we use the normalization given by Equation (4.14) then in all dimensions $B = \tilde{\mathbf{I}}(B)$ if and only if B is an ellipsoid.

4.8 Notes

The fact that the Euclidean construction of an equilateral triangle works in any Minkowski plane has often been rediscovered. One of the earlier references is Kelly [288]. The extension to $\mathbb{R}^3$ is due to Petty [417]. That an extension to higher dimensions is not straightforward is made clear by Petty's example [417] of the situation when the ball in $\mathbb{R}^4$ is the double cone over the Euclidean ball in $\mathbb{R}^3$; *i.e.* the ball B is given by $\xi_1^2 + \xi_2^2 + \xi_3^2 + |\xi_4| \leq 1$. With this ball, if we choose $x_0 := 0$, $x_1 := (0, 0, 0, 1)$ and $x_2 := (\frac{1}{2}, 0, 0, \frac{1}{2})$ then the only choice for x_3 is $x_3 := (-\frac{1}{2}, 0, 0, \frac{1}{2})$ and this planar equilateral set cannot be extended by a fifth point. Nevertheless, by starting at a different place it is possible to construct an equilateral 4-simplex in this ball, *e.g.*

$$x_0 := (0, 0, 0, 0), \quad x_1 := (1, 0, 0, 0), \quad x_2 := (\tfrac{1}{2}, \tfrac{\sqrt{3}}{2}, 0, 0),$$
$$x_3 := (\tfrac{1}{2}, \tfrac{1}{\sqrt{12}}, 0, \tfrac{2}{3}), \quad x_4 := (\tfrac{1}{2}, \tfrac{1}{\sqrt{12}}, \tfrac{1}{\sqrt{2}}, \tfrac{1}{6}).$$

Moreover, this last example can easily be extended to cones in higher dimensional spaces.

It is remarkable that it is still an open problem whether or not every Minkowski space (even if the ball is smooth and strictly convex) admits a regular simplex. This problem has been mentioned explicitly by Morgan [394]; see also Lawlor and Morgan [315].

Slightly related to the problems discussed in §4.1 are the Minkowski analogues of combinatorial problems of Erdös [143]. Given n points in the plane, what is the maximum number of unit distances between pairs of the points and what is the minimum number of distinct distances that there can be? Erdös [144] credits Ulam with suggesting that these problems also be investigated in the Minkowski plane. They received some attention from, *e.g.*, Lassak [312] and Chilakamarri [109] and have been thoroughly investigated recently by Brass [58]. Also related to a problem of Erdös is the work of Makai and Martini [348, 349] on the number of "antipodal" points in a finite set in $\mathbb{R}^d$. In [348] they point out connections with the number of diameters that a finite set in a Minkowski space can have.

I first became aware of the "three-circles" theorem from the article of MacKenzie [344] and was delighted to see that it was true in Minkowski planes. I am grateful to Horst Martini for giving me the reference to the beautiful results of Asplund and Grünbaum [10]. In addition to the results mentioned in §4.1 it is shown in [10] that the "six-point circle" is a "nine-point circle" only if the unit ball is an ellipse. Ceder [95] shows that both this three-circles theorem and the possibility of inscribing a regular hexagon do not depend on symmetry. In fact he shows that if K is an arbitrary convex body in the plane with 0 in its interior then there are three points x, y, z, on the boundary of K such that $x + y$, $y + z$, $z + x$ are also on the boundary and hence these six points form a "regular" hexagon. Furthermore, the three "circles" with centres at $-x, -y, -z$ pass through y and z, z and x, and x and y respectively, and 0 lies on all three so that 0 is the "orthocentre" of x, y and z. In this situation it is no longer possible to choose the first point arbitrarily. Golab and Tamássy [184] have considered altitudes of triangles using the Carathéodory–Birkhoff definition of normality. They show that if for all triangles the altitudes are concurrent then the ball is an ellipse. If, again using Definition 3.2.2, "area" of a triangle is defined as half the product of the length of a side and the corresponding altitude then (see Busemann [73]) the "area" is independent of the side chosen if and only if the ball is a Radon curve. This was generalized to the case when the ball is supposed only to be a star-shaped set by Tamássy [501]. Other Minkowski analogues of elementary Euclidean results on the geometry of the triangle have been given by Guggenheimer [227, 231].

Many authors have discussed convex sets of constant width in a Minkowski space. As early as 1911, Meissner [374] considered the problem of whether a set of constant width in a Minkowski space is diametrically maximal (*i.e.* the addition of any point to the set increases the diameter) and showed that this is the case for $d = 2, 3$. Eggleston [140] established the same result for arbitrary dimension and showed that for $d \geq 3$ the converse is false even if the ball B is both smooth and strictly convex; *i.e.* there are sets which are diametrically maximal but not of constant width. Consequently the necessary and sufficient statements by Meissner (for $d = 3$) and by Kelly [287] are erroneous. Eggleston [139] proved that if B is a parallelotope then any diametrically maximal set (and hence any set of constant width) is homothetic to B. Soltan [493] proved the converse. Note that the construction in Example 4.2.4 yields a ball if B is a parallelogram.

In his interesting paper [140], Eggleston considers several other known characterizations of sets of constant width in Euclidean space and determines to what extent these characterizations hold in Minkowski spaces. Makai and Martini [347] characterize bodies of constant width in $\mathbb{R}^d$ in the following way. If K is a convex body of diameter 1 then the sum of the lengths of every set of d mutually orthogonal (non-degenerate) chords is bounded below by 1 if and only if K has constant width. They observe that Soltan posed the corresponding question for Minkowski planes where "mutually orthogonal" is taken in the sense of Corollary 3.2.3. Further extensions of known Euclidean results to Minkowski spaces are given by Chakerian and Groemer in their excellent survey article [105]. While this article is concerned primarily with Euclidean spaces, the Minkowski generalizations are included where they are known. The very useful survey of Heil and Martini [252] contains a section on bodies of constant width which also refers to results in Minkowski spaces and which complements and updates the results in [105].

Reuleaux triangles in Minkowski spaces are considered by Ohmann [402], whose proof of Theorem 4.2.8, while more complicated than Chakerian's [98], is also based on inscribed and circumscribed hexagons. The use of hexagons also features in several proofs of the corresponding Euclidean result (see, e.g., Lebesgue [316] and Eggleston [137]) and also in other related minimum problems (e.g. Kubota and Hemmi [306] and Sholander [489]). Reimann [436] re-proves Lemma 4.2.6 in the following form: the ratio $\lambda(B)/\lambda(T)$ where T is an equilateral triangle of side 1 lies in the interval [6, 8]. Further results of this type are given by Wernicke [530].

Reuleaux polygons were investigated by Sallee [454] and Hammer [242]. However, Hammer [242] is primarily concerned with analytic representations of curves of constant width. He presents a specific formula (involving the definition of the unit ball) which always yields a curve of constant width. He also shows that if two convex bodies have the same width in every direction then they have identical arc length functions.

If K is a convex body of constant width w in the Euclidean plane and if r and R are the inradius and circumradius respectively then $r + R = w$. Chakerian [99] has proved that this result also holds in Minkowski planes, and Sallee [455] both extended it to arbitrary dimension and showed that it depends on the set having the "spherical intersection property" rather than being of constant width. Eggleston [140] had already shown that the spherical intersection property is equivalent to being diametrically maximal. More general questions about the relationships between the inradius, circumradius and diameter of a convex set in a Minkowski space are discussed in detail by Eggleston [139]; for related results see also Bohnenblust [44] and Leichtweiss [317]. Further results on bodies of constant width in normed spaces are given in §32 of the forthcoming book by Boltyanski, Martini and Soltan [49].

A good account of the problem of whether, in Euclidean space, there exist sets with two equichordal points together with some of the history of the problem is given in the short article by Klee [292]. There is a full discussion of the problem and much related theory in Gardner [172]. In the note added in proof at the end of their long analysis of the problem, Schäfke and Volkmer [468] claim that improved error bounds combined with results of Michelacci and Volčič finally settle the problem; they also refer to preprints by J. C. Rogers which, in their words, "might lead to a negative solution of the problem." Gardner [172] also refers to work of Rychlik that solves the problem. As stated in the text, the paper of Petty and Crotty [423] contains considerably more information than we have included.

The bounds on the length of the boundary of the unit ball are well known and have been rediscovered many times. They can be viewed as asserting that the value of π for a Minkowski space lies in the interval $[3, 4]$ and that all values in this interval are possible. Thus the various values ascribed to π by various cultures at various times are all valid if the correct geometry is used. The observation that, in fact, all values are attained by balls which are hexagons (with the parallelogram as a limiting case) was made by Laugwitz [313].

The best accounts of Golab's result and its relatives are those of Schäffer [459, 466] and Petty [413]. Less well known than these results for symmetric balls B are those dealing with non-symmetric K. In this case the problem is more difficult since one must consider the metric with respect to some arbitrary origin z inside K (see Hammer [243]). If $\ell(K, z)$ denotes the length of K (in one of the orientations) with respect to z as origin, then Golab [181] conjectured that $6 \leq \ell(K, z)$ and, if $\ell(K) := \inf\{\ell(K, z) : z \in K\}$, that $\ell(K) \leq 9$. He established that $4 + \sqrt{2} \leq \ell(K, z)$ and $\ell(K) \leq 24$. Grünbaum [224] (for the upper bound) and [225] (for the lower bound) confirmed these conjectures.

Schäffer [465] was the first to prove the fact that a two-dimensional ball and its dual have the same self-circumference (his proof was also for the non-symmetric case). However, it was the alternative proof in [508] which motivated Holmes to propose this property as a way of *defining* surface area in $\mathbb{R}^3$. It was this idea which eventually led to our work in [262] and the present Chapter 6. The more general result (Theorem 4.3.8) was shown also to hold in the non-symmetric case by Chakerian [102].

The solution of the isoperimetric problem for Minkowski spaces in general and for Minkowski planes in particular is due to Busemann [66, 68, 71]. Other accounts for the two-dimensional case can be found in Petty [413] and Barthel [22]; see also Guggenheimer [227] and Benson [30]. Recently, Wallen [526] has written a beautiful account from a very elementary point of view. In particular, the case when B is a polygon needs very little background.

Equation (4.10) can be generalized to say that if A is an arc of $\partial \tilde{\mathbf{I}}$ then its length is exactly twice the area which it subtends at the centre. Therefore, as Wallen [526] points out, if a particle travels along $\partial \tilde{\mathbf{I}}$ with constant Minkowski speed, then it "sweeps out equal areas in equal times". Wallen further shows that, if B is a Euclidean circle whose centre is not at 0, then $\tilde{\mathbf{I}}$ is an ellipse with one focus at 0.

The normalization of area via Equation (4.14) comes from [262] and the results dealing with the normalized isoperimetrix which this equation yields can mostly be found in [262]. Equivalent equations and inequalities for the Busemann normalization can be found in Petty [413]. However, there is no equivalent for the pleasing result that the area and circumference of the normalized isoperimetrix are π and 2π respectively.

Theorem 4.5.5 is due to Petty [413], although his proof differs from that given here. Other extensions of the Bonnesen isoperimetric inequalities to the Minkowski plane have been given by Guggenheimer [230] and by Peri, Wills and Zucco [411]. Some of the ideas in [411] were extended to spaces of arbitrary dimension by Peri [410]. Chakerian [96] proves another isoperimetric-type inequality: namely, that if K is a convex polygon in the plane and if K' is the polygon circumscribed about $\tilde{\mathbf{I}}$ whose sides are parallel to those of K then

$$(\mu(\partial K))^2 \geq 4\mu(K)\mu(K')$$

with equality if and only if K is itself circumscribed to a homothetic image of $\tilde{\mathbf{I}}$.

The notion of transversality is central to Minkowski geometry and is one of the underlying themes of Busemann's work; see, *e.g.*, [68], [71] and [87]. It is precisely for the purpose of defining this notion via support hyperplanes to the isoperimetrix that we shall require the area function σ in Chapter 5 to be convex. In two dimensions, however, since the isoperimetrix is determined by the norm, no essentially new ideas are involved. For this reason it is useful to understand the concept here before looking at the higher-dimensional generalizations. Phadke [425] (see also [426]) has considered the question of the loci of points equidistant from lines and planes in situations more general than the Minkowski plane.

Radon curves were introduced in [432]. They have reappeared in various contexts since then. For our purposes their significance is two-fold. Firstly, the two curves which determine the geometry, the unit ball B and the isoperimetrix $\mathbf{I}$ are homothetic exactly when ∂B is a Radon curve. Secondly, many conditions which characterize Euclidean spaces for dimensions $d \geq 3$ yield, instead, characterizations of Radon curves in dimension 2; see, *e.g.*, Day [124] and de Figueiredo and Karlovitz [127]. The combination of these two views suggests that in dimensions $d \geq 3$ the bodies B and $\mathbf{I}$ are homothetic only in Euclidean spaces. It is perhaps of interest that it is precisely these two aspects which Radon emphasizes [432].

Many other analogues of results from plane Euclidean geometry are possible. This is especially true, of course, for results which deal exclusively with distance. Chakerian and Ghandehari [104], for example, have considered Fermat's problem of minimizing the sum of the distances from a point to n given points in a Minkowski plane. See also Durier and Michelot [131] for a discussion of this problem and their [132] for related problems on optimizing sums of distances. A very complete study of Fermat's problem and the more general Steiner problem (to construct a network of minimal total length connecting n points) has been made by Cieslik, not only for the plane but also in higher dimensional spaces. See Cieslik [111] and the lengthy reference list therein for this work. Ghandehari [176] has considered the problem of minimizing the sum of the distances to a line from two given points on one side of the line. Grünbaum [221] looked at the Borsuk problem in Minkowki planes and showed that if the ball is not a parallelogram then any set of diameter 1 can be covered by three balls of diameter less than 1; see §33 in Boltyanski, Martini and Soltan [49] for a full discussion of this problem in normed spaces. For results on packing inequalities in a Minkowski plane see Graham, Witsenhausen and Zassenhaus [198]. The particular Minkowski metric on the plane that has received the most attention is the "taxicab" metric, *i.e.* the ℓ_1-norm, see, *e.g.*, the book by Krause [304].

5

The concept of area and content

In this chapter we deal with a d-dimensional Minkowski space $(X, \|.\|)$. The purpose of the chapter is to show how to introduce a "suitable" measure on each k-dimensional subspace of X ($k = 1, 2, \ldots, d - 1, d$) and hence on translates of such subspaces. Strictly speaking, these measures are defined not on the flat but on some σ-algebra of subsets of each flat. Here "suitable" means, roughly, that the measure should relate to the geometry. After the properties to be imposed are listed examples of measures that meet the requirements are given. The final part of the chapter is concerned with the immediate consequences of the relationship between the measure and the geometry. These consequences are restricted here to those which are common to all definitions that meet the requirements. Specific properties of specific definitions are postponed to later chapters.

The first requirement is that each measure μ should be invariant under all isometries of X, i.e.

$$\mu(T(U)) = \mu(U)$$

for all μ-measurable sets U and all isometries T. In particular, since all translations are isometries (and in §1.1 we saw that these may be the only ones other than ± 1), each measure must be translation invariant. This shows how to transfer a measure from a subspace to an arbitrary flat: if μ is defined for certain subsets U of a subspace M then we extend μ to the corresponding subsets of $M + y$ by $\mu(U + y) = \mu(U)$.

Next, since the topology of X is independent of the norm, the obvious requirement on the σ-algebra of μ measurable sets is that it contain the relative Borel σ-algebra in a subspace M. To be interesting, the measures ought to be non-trivial, i.e. non-zero on relatively open subsets of M and finite on compact subsets of M. Thus, in a subspace M of dimension k, we require a non-trivial, locally finite, translation-invariant measure defined on the relative Borel σ-algebra. By Theorem 1.4.3, we know that such a measure must be a k-dimensional Haar measure.

Since different Haar measures are related by a scalar multiple, the problem of introducing a suitable measure on each subspace of X is reduced to the problem of specifying a suitable normalization of Haar measure in each subspace. There is freedom here compared with Euclidean spaces. Since there are in general few rotational or reflectional isometries, the normalizing factor in subspaces unconnected by an isometry may be unrelated. The difficulty lies in choosing the factor in a way consistent with the geometry.

In the case of one-dimensional subspaces (*i.e.* lines through the origin) the metric provides the normalization of the one-dimensional Haar measure because if x and y are two points on a line through the origin then the Minkowski length of the segment $[x, y]$ is $\|x - y\|$. This situation indicates how to proceed in other dimensions. If L is a line through the origin, we normalize Haar measure μ_B^1 in L by specifying that $\mu_B^1(B \cap L) = 2$. Hence, another way of stating the problem is to say that for each k-dimensional subspace M of X we must assign a number $\mu_B^k(B \cap M)$ which is to be the k-dimensional measure of the relative unit ball in M. Because each centrally symmetric convex body, of each dimension k, can appear as a k-dimensional section of some larger ball we require a function γ defined on *all* centrally symmetric convex bodies, of all dimensions, with range in $\mathbb{R}^+$ so that we can set

$$\mu_B^k(B \cap M) := \gamma(B \cap M). \tag{5.1}$$

Yet another way of restating the problem is to first impose an auxiliary Euclidean metric on X so that we have the usual k-dimensional Lebesgue measure λ^k on each k-dimensional subspace. Now the problem is to specify the ratio

$$\sigma_B^k(M) := \frac{\text{Minkowski } k\text{-dimensional measure in } M}{\text{Lebesgue } k\text{-dimensional measure in } M}. \tag{5.2}$$

Then if U is a measurable subset of M we have

$$\mu_B^k(U) = \sigma_B^k(M) \lambda^k(U). \tag{5.3}$$

Clearly, σ_B^k and γ are related by the equation

$$\gamma(B \cap M) = \sigma_B^k(M) \lambda^k(B \cap M). \tag{5.4}$$

It is preferable to specify γ (in a coordinate-free way) than to specify σ_B^k, which requires λ^k first. However, because so many formulas and computational tools have been developed over many centuries for dealing with λ^k, the formula (5.3) is usually the most practical for calculating $\mu_B^k(U)$. It also turns out, at least in dimension $d - 1$, that the function σ_B^{d-1} is of considerable significance. We note therefore that σ_B^k must be a positive, real-valued function defined (for each pair (X, B)) on the Grassmann manifold of all k-dimensional subspaces of X ($k = 1, 2, \ldots, d$).

The case $k = d$ is not very important for the geometry of X. There is only one d-dimensional subspace and to specify a function on a singleton is usually not very interesting. Formulas involving d-dimensional content can always be adjusted by this single constant. It is, however, pleasant if this adjustment can be done in a way that retains some of the familiar formulas from Euclidean geometry. In Chapter 4 we saw how this question arises when X is two-dimensional.

5.1 Requirements and examples

After this lengthy introduction, we come to the geometric requirements for our measures, in other words the requirements for γ and (equivalently) σ (we usually omit the subscript B). With one exception it seems preferable to relate these requirements to γ.

Requirements 5.1.1

(a) *The function γ is an affine invariant; i.e. if K is a centrally symmetric convex body that spans a k-dimensional linear space and if T is an invertible linear map defined on that space then $\gamma(TK) = \gamma(K)$.*

(b) *The function γ is continuous with respect to the metric Δ_2 (Equation (2.21)); i.e. if two Minkowski spaces are nearly isometric then the measures of their unit balls should be close.*

(c) *If E_k is a k-dimensional ellipsoid (i.e. if the Minkowski space is Euclidean) then*

$$\gamma(E_k) := \epsilon_k, \tag{5.5}$$

where $\epsilon_k := \pi^{k/2} \Gamma(1 + k/2)^{-1}$. *In particular, for $k = 1$, $\gamma(E_1) = 2$ and for $k = 2$, $\gamma(E_2) = \pi$.*

(d) *The function σ^{d-1} is convex; see Remark (iv) below.*

The purpose of Requirement 5.1.1(a) is to establish the following proposition and its corollary. Remarks on the other requirements will follow the corollary.

Proposition 5.1.2 *If (X_1, B_1) and (X_2, B_2) are isometric Minkowski spaces then $\gamma(B_1) = \gamma(B_2)$.*

Proof. By the Mazur–Ulam theorem (3.1.2) there is a linear map T from X onto Y such that $T(B_1) = (B_2)$. Hence by Requirement 5.1.1(a) $\gamma(B_1) = \gamma(B_2)$. ∎

Corollary 5.1.3 *If (X_1, B_1) and (X_2, B_2) are two Minkowski spaces and if M_i is a subspace of X_i ($i = 1, 2$) such that M_1 and M_2 are isometric then $\gamma(B_1 \cap M_1) = \gamma(B_2 \cap M_2)$.*

Proof. This result follows by applying the proposition to the spaces $(M_1, B_1 \cap M_1)$ and $(M_2, B_2 \cap M_2)$. ∎

Remarks

(i) Corollary 5.1.3 implies that $\gamma(K)$ depends only on K and not on the larger ball of which K is a cross-section. For example, for an ellipse E, $\gamma(E) = \pi$ whether that ellipse is a cross-section of a cylinder or an ellipsoid; and for a parallelogram P, $\gamma(P)$ is independent of whether P is a cross-section of a cube, cylinder or octahedron, and all cross-sections of a cubic unit ball that are parallelograms have the same Minkowski area $\gamma(P)$.

(ii) The metric Δ_2 in Requirement 5.1.1(b) differs from the Hausdorff metric precisely in that the distance between sets that span spaces of different dimension is infinite (or not defined). In the Hausdorff metric it is possible, *e.g.*, to have a sequence of ellipsoids E_n converging to a line segment L and we do not want $\gamma(E_n) \to \gamma(L)$. To this extent, the statement in [262] requiring that γ be continuous with respect to the Hausdorff metric needs modification.

(iii) It would be possible to combine Requirements 5.1.1(a) and (b) by using the Banach–Mazur distance Δ (see Definition 2.4.6). If we regard γ as a function defined on the collection of all Minkowski spaces (X, B) by setting $\gamma(X, B) = \gamma(B)$ then 5.1.1(a) and (b) are combined in the single statement that γ is continuous with respect to Δ.

(iv) Requirement 5.1.1(d) needs explanation. The Grassmann manifold of (oriented) $(d-1)$-dimensional subspaces can be identified (via null spaces) with the set of (normalized) linear functions on X. Since σ^{d-1} already involves an auxiliary Euclidean metric we may suppose σ^{d-1} defined, in the first instance, on the Euclidean unit sphere in X^* and satisfying $\sigma^{d-1}(f) = \sigma^{d-1}(-f)$. We then extend σ^{d-1} to the whole of X^* by positive homogeneity and denote it simply by σ (σ_B when the dependence on B needs emphasis); *i.e.*

$$\sigma(f) := |f|\sigma^{d-1}(f/|f|) \tag{5.6}$$

if $f \neq 0$ and $\sigma(0) := 0$. It is this extended function which we require to be convex on X^*. We restate 5.1.1(d) in this form:

The positively homogeneous function σ defined on X^* by (5.6) is the support function of a convex set $\mathbf{I}$ in X.

This last requirement has a geometric justification and is also a great technical convenience. However, we shall give an example (5.1.4(v) below) of a plausible function γ for which the corresponding function σ is not convex on at least one Minkowski space. The geometric justification of the requirement is as follows. If $\mathcal{R}$ is a region in some hyperplane in a Minkowski space and if $\mathcal{S}$ is some other $(d-1)$-dimensional surface which has the same boundary as $\mathcal{R}$ then we would like the area of $\mathcal{R}$ not to exceed that of $\mathcal{S}$. In short, we say that "flat" regions should minimize area. We shall see below that the convexity of σ implies that this property does indeed hold in the same way that the convexity of B yields, via the triangle inequality, the fact that straight lines are geodesics.

One further comment on Requirement 5.1.1(d). It may seem odd to require only the convexity of σ^{d-1} and not that of σ^k for all k. First it is difficult to specify exactly what convexity of σ^k should mean. It is natural to ask that "flat" k-regions minimize k-measure; that is, if $\mathcal{R}$ is a region in some k-dimensional flat then its k-measure should not exceed that of any other k-dimensional surface with the same boundary. Busemann and others have investigated conditions on σ^k which would yield this property. However, it appears to be open whether any of the functions σ^k used in the literature satisfy such a requirement. Of course, σ^1 is the support function of B° and is certainly convex.

A natural question that now arises is: how much freedom is there within these requirements? It is clear that if γ_1 and γ_2 both satisfy Requirements 5.1.1(a)–(d) then so does $\alpha\gamma_1 + (1-\alpha)\gamma_2$ for $0 \leq \alpha \leq 1$. In this sense the set of all possible functions is a convex set. What are its extreme points? We give examples of some suitable functions γ. The first two will be examined in greater detail in subsequent chapters. We also give an example of a function that satisfies all but the convexity requirement showing that this one is independent of the others. Following these examples the rest of the chapter will be devoted to a study of the properties common to any example that satisfies Requirements 5.1.1. To a great extent this follows the pioneering work of H. Busemann [68, 71].

Example 5.1.4

(i) Certainly the simplest and most obvious way to satisfy Requirements 5.1.1(a)–(c) is to make γ constant in each dimension; *i.e.* if K spans a d-dimensional space then, as in Equation (4.13),

$$\gamma(K) := \epsilon_d = \pi^{d/2}\Gamma(1 + d/2)^{-1}. \tag{5.7}$$

Note that γ is continuous with respect to both Δ_2 and the Banach–Mazur distance but not the Hausdorff metric (see Remark (ii) above).

Using this γ we get from (5.4) and (5.6)

$$\sigma_B(f) = \frac{\epsilon_{d-1}|f|}{\lambda^{d-1}(B \cap f^\perp)}. \tag{5.8}$$

It is not obvious that σ_B defined by (5.8) is convex. The proof is a major result of Busemann [67]. From now on we shall refer to this definition of γ and σ as the *Busemann definition*. The idea had previously been considered by Bouligand and Choquet [53]. Barthel [24], *e.g.*, refers to it as the Choquet–Busemann definition. Its special characteristics are explored in Chapter 7.

(ii) Another affine invariant for a convex body K of dimension d is the volume product, Definition 2.3.2. Using this as in Equation (4.14), we have

$$\gamma(K) := \lambda(K)\lambda^*(K^\circ)/\epsilon_d. \tag{5.9}$$

The normalizing factor ϵ_d^{-1} is inserted to ensure that Requirement 5.1.1(c) is satisfied. The continuity of λ and λ^* (Theorem 2.5.8) and that of the polar map (see the remark following Equation (2.21)) imply that 5.1.1(b) is satisfied.

With this choice of γ we get from (5.4) and (5.6) (and replacing λ^* by the particular Haar measure λ^{d-1})

$$\sigma_B(f) = \frac{\lambda^{d-1}((B \cap f^\perp)^\circ)|f|}{\epsilon_{d-1}}. \tag{5.10}$$

As we saw in Chapter 4 there is a certain duality between (5.8) and (5.10). This has been explored extensively by Lutwak [337] and further evidence of it will be seen in subsequent chapters.

This definition also needs a name. With some diffidence I will use the one which seems to be becoming standard and refer to it as the Holmes–Thompson definition. Its special characteristics are explored in the next chapter.

(iii) A slightly different approach to the problem of normalizing Haar measure was used by Benson [30]. Given a centrally symmetric convex body K in a space X of dimension d one can circumscribe about K a parallelotope P of minimal volume (cf. Corollary 3.2.5). We can normalize Haar measure by setting $\mu^d(P) := 2^d$. For the Minkowski space (X, B) we can then write the formula for σ_B as

$$\sigma_B(f) = \frac{2^{d-1}|f|}{\lambda^{d-1}(P^f)}, \tag{5.11}$$

where P^f is a minimal box circumscribed about $B \cap f^\perp$. For more details of this example when $d = 2, 3$, see [30].

(iv) One might modify the approach in Example (iii) in a variety of ways. Firstly, instead of circumscribed parallelotopes one might assign the volume $2^d/d!$ to a maximal inscribed cross-polytope. Secondly, one might use the Löwner ellipsoid (Theorem 3.1.2) in either the minimal circumscribed or maximal inscribed form and, in either case, assign it the volume ϵ_d. I do not know of any investigation of these ideas. It appears that any calculations based on them are likely to be difficult. In particular, the question of convexity (Requirement 5.1.1(d)) seems difficult.

(v) Finally, we mention another possibility for which our main purpose is a negative one. If B is the unit ball in a two-dimensional space X then $\mu_B^1(\partial B)$ is an affine invariant. Hence we may set

$$\gamma(B) := \mu_B^1(\partial B)/2.$$

Having defined γ for all two-dimensional balls, it is then theoretically possible to calculate the surface area $\mu_B^2(\partial B)$ for all three-dimensional balls and, for such a B, set

$$\gamma(B) := \mu_B^2(\partial B)/3$$

and so on, inductively. Except in the simplest cases (*e.g.* the d-dimensional cube) this would be a difficult calculation. Using the three-dimensional cube, we show in the next section that the σ derived from this γ is not convex and, consequently, that Requirement 5.1.1(d) is not implied by the others.

5.2 The role of the function σ_B

In this section we extend the results of §4.4 to a general d-dimensional Minkowski space $(X, \|.\|)$. We suppose that X comes equipped with an auxiliary Euclidean norm $|.|$, with unit ball E. In each k-dimensional subspace there is an associated Lebesgue measure λ which, in the general case, can be scaled using the formula (5.3) and, in specific cases, by (5.8), (5.10) or (5.11). The resulting Minkowski measure will be denoted by μ with superscripts to indicate the dimension where this seems necessary. We will be almost exclusively concerned with dimensions $1, d-1$ and d. For simplicity of language the measures in these dimensions will be referred to as length, area and volume, respectively. When $k = 1$, we have two notations for the length of a line segment, $[x, y]$, $\mu^1([x, y])$ and $\|y - x\|$. We shall use the second on the grounds of common usage. But when, as in Chapter 4, we deal with the length of a curve c, we shall use $\mu^1(c)$.

Except in §5.5.2, the term *unit normal vector* will also have the same meaning as in Chapter 4; *i.e.* if H is a hyperplane then the unit normal to H is (up to a choice of sign) that unique linear functional $\hat{f}$ in X^* such that $\hat{f}(M) = 0$ and $|\hat{f}| = 1$, where M is the subspace of which H is a translate. If U is a measurable subset of a hyperplane in X and if $\hat{f}$ is the unit normal vector to that hyperplane, then following (5.3) and (5.6) we have

$$\mu_B(U) = \sigma(\hat{f})\lambda(U). \tag{5.12}$$

We are now able to state precisely what was meant by our previous comment that if σ is convex then flat surfaces minimize area. Recall from (5.6) that σ is positively homogeneous.

Theorem 5.2.1 *The function σ is convex if and only if for each d-polytope P with facets F_i ($i = 1, 2, \ldots, n$) we have $\mu_B(F_1) \leq \sum_{i=2}^{n} \mu_B(F_i)$.*

Proof. Suppose σ is convex. Let f_i be the unit outward normal to the facet F_i which has Euclidean area $\lambda(F_i)$. To avoid excessive clutter we omit the carets (^) in this argument. Since

$$\lambda(F_1)(-f_1) = \sum_{i=2}^{n} \lambda(F_i) f_i,$$

we have

$$\lambda(F_1)\sigma(f_1) = \lambda(F_1)\sigma(-f_1) = \sigma\left(\sum_{i=2}^{n}\lambda(F_i)f_i\right)$$

$$\leq \sum_{i=1}^{n}\lambda(F_i)\sigma(f_i).$$

But, from (5.12), this is exactly $\mu_B(F_1) \leq \sum_{i=2}^{n}\mu_B(F_i)$.

Conversely, by the homogeneity of σ, if it is not convex then it is not convex on the Euclidean unit ball. Hence there exist unit vectors $f_1, \ldots, f_d$ and non-negative scalars $\lambda_1, \ldots, \lambda_d$ such that

$$\sigma\left(\sum_{i=1}^{d}\lambda_i f_i\right) > \sum_{i=1}^{d}\lambda_i\sigma(f_i).$$

Let $\lambda_0 := |\sum_{i=1}^{d}\lambda_i f_i|$ and $f_0 := -\lambda_0^{-1}(\sum_{i=1}^{d}\lambda_i f_i)$. Then f_0 is a unit vector and $\sum_{i=0}^{d}\lambda_i f_i = 0$. Hence there is a (possibly degenerate) simplex with facets $F_0, F_1, \ldots, F_d$ whose outward normals are $f_0, f_1, \ldots, f_d$ and whose Euclidean areas are $\lambda_0, \lambda_1, \ldots, \lambda_d$. But since $\lambda_0\sigma(f_0) = \sigma(-f_0\lambda_0) = \sigma(\sum_{i=1}^{1}\lambda_i f_i) > \sum_{i=1}^{1}\lambda_i\sigma(f_i)$ we have (by 5.10)

$$\mu_B(F_0) > \sum_{i=1}^{d}\mu_B(F_i),$$

which completes the proof. ∎

Next we need to discuss briefly the type of hypersurfaces whose areas we will measure without getting into the intricacies of this complicated question. Its complexity is illustrated by the results of Besicovitch [33], who constructs a set A in $\mathbb{R}^3$ which is homeomorphic to a ball; $\lambda^3(A)$ can be made arbitrarily large and yet $\lambda^2(\partial A)$ is arbitrarily small. By the term *smooth hypersurface* in the d-dimensional space X we understand a $(d-1)$-dimensional manifold-with-boundary as defined, e.g., by Spivak [494], p. 113. The term *piecewise smooth hypersurface* denotes a finite union of smooth hypersurfaces which intersect only along boundaries and which, if two of them share a common boundary, are joined along that common boundary. This definition requires some technicalities about the meaning of the word *joined* and about the type of boundary. It is usual to use induction and require the boundaries to be piecewise smooth manifolds of one lower dimension. This class, however, is not large enough for our purposes since we certainly also wish to include the boundary of each convex body in X. We will use the union of these two classes. From now on *surface* will mean either a piecewise smooth hypersurface or the boundary of a convex body in X. Most frequently it denotes the latter. The important point about a surface so defined is that it has a unique tangent hyperplane

5.2 The role of the function σ_B

λ^{d-1}-almost-everywhere. For convex bodies this fact is a theorem of Reidemeister [435] (see also Schneider [479], Theorem 2.2.4).

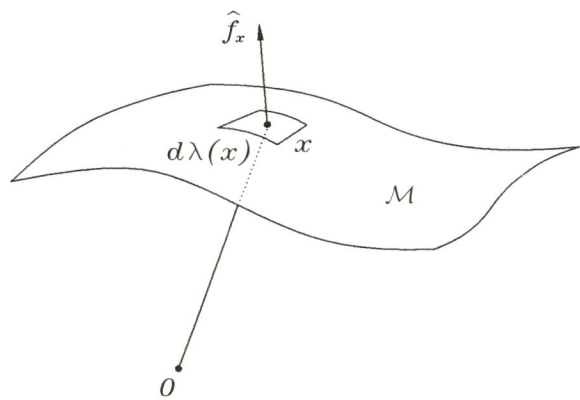

Figure 5.1

Let $\mathcal{M}$ be a surface in X. With the possible exception of a λ^{d-1} null set, at each point x of $\mathcal{M}$ the surface has a unique tangent hyperplane which, in turn, has a unit normal vector $\hat{f}_x$ (see Figure 5.1). We extend (5.12) by scaling λ^{d-1} at each point x of $\mathcal{M}$ by the factor $\sigma(\hat{f}_x)$ (regarded here as a function of x) and write

$$d\mu_B^{d-1}(x) := \sigma(\hat{f}_x)\, d\lambda^{d-1}(x).$$

This formula is understood to mean that the Minkowski surface area of $\mathcal{M}$ is defined by the following equation:

$$\mu_B^{d-1}(\mathcal{M}) := \int_{\mathcal{M}} d\mu_B^{d-1}(x) = \int_{\mathcal{M}} \sigma(\hat{f}_x)\, d\lambda^{d-1}(x). \tag{5.13}$$

If K is a convex body we may also scale the surface area measure ν_K (Definition 2.3.12) by the factor $\sigma_B(\hat{f})$. Thus if ψ is a function defined on S^{d-1} then the alternative notation established in §2.3, namely,

$$\int_{\partial K} \psi(\hat{f})\, d\lambda^{d-1}(\hat{f}) := \int_{S^{d-1}} \psi(\hat{f})\, d\nu_K(\hat{f}),$$

can be similarly extended to

$$\int_{\partial K} \psi(\hat{f})\, d\mu_B^{d-1}(\hat{f}) := s \int_{S^{d-1}} \psi(\hat{f})\sigma_B(\hat{f})\, d\nu_K(\hat{f}).$$

From now on we shall frequently omit the superscript $d-1$.

Recall (Equation (2.14)) that if K, K' are convex bodies then the mixed volume $V(K[d-1], K')$ satisfies the equation

$$dV(K[d-1], K') = \int_{\partial K} h_{K'}(\hat{f}_x) \, d\lambda(x). \tag{5.14}$$

Using Requirement 5.1.1.(d) and by comparing Equations (5.13) and (5.14), we obtain the following result of H. Busemann [68]:

Theorem 5.2.2 *If K is a convex body in the Minkowski space (X, B) then the Minkowski surface area $\mu_B(\partial K)$ of K is given by*

$$\mu_B(\partial K) = dV(K[d-1], \mathbf{I}_B), \tag{5.15}$$

where $\mathbf{I}_B$ is that convex body whose support function is σ_B.

Remarks

(i) Theorem 5.2.2 shows the technical reason for Requirement 5.1.1(d). It allows the Brunn–Minkowski theory of convex bodies to be applied.

(ii) Alternatively, to prove 5.2.2 one could first restrict attention to smooth convex bodies and then approximate a general convex body by smooth ones (Theorem 2.5.2).

As Busemann first showed [68], Theorem 5.2.2 and the Minkowski inequality (2.3.8) make clear the central role played by $\mathbf{I}_B$ as the solution to the isoperimetric problem.

Theorem 5.2.3 *In a Minkowski space (X, B), among all convex bodies with volume $V = \lambda(\mathbf{I}_B)$ those with minimum area are precisely the translates of $\mathbf{I}_B$. Likewise, among convex bodies with surface area $s = \mu_B(\partial \mathbf{I}_B)$ those with maximum volume are the translates of $\mathbf{I}_B$.*

Proof. If K is a convex body with volume V then, by Theorem 5.2.2 and the Minkowski inequality, we have

$$\mu_B(\partial K) = dV(K[d-1], \mathbf{I}_B) \geq dV^{(d-1)/d} \lambda(\mathbf{I}_B)^{1/d}$$

with equality if and only if K is homothetic to $\mathbf{I}_B$. If $V = \lambda(\mathbf{I}_B)$ then $\mu_B(\partial K) \geq d\lambda(\mathbf{I}_B)$ with equality if and only if K is a translate of $\mathbf{I}_B$. Likewise, if $\mu_B(\partial K) = \mu_B(\partial \mathbf{I}_B) = d\lambda(\mathbf{I}_B)$ then $V \leq \lambda(\mathbf{I}_B)$ with equality if and only if K is a translate of $\mathbf{I}_B$. ∎

If V (resp. s) is some other fixed constant then the convex bodies with minimal area (resp. maximal volume) are precisely the translates of an appropriate multiple of $\mathbf{I}_B$.

Note that in this theorem we have taken V to be Euclidean volume and hence the notation $\lambda(\mathbf{I}_B)$. If, however, we wish to consider the Minkowski volumes of K (and $\mathbf{I}_B$) then the extra factors σ_B^d make no difference to the proof. In this sense, Theorems 5.2.2 and 5.2.3 are hybrids, expressing Minkowski surface areas in terms of Euclidean mixed volumes. This hybrid nature is most clearly shown if we set $K = \mathbf{I}_B$ in Equation (5.15). For then we get

$$\mu_B(\partial \mathbf{I}_B) = d\lambda(\mathbf{I}_B);$$

i.e. the Minkowski surface area of $\mathbf{I}_B$ is d times its Euclidean volume.

More than this is true: the formula works locally as well as globally. Some notation is needed in order to describe precisely what we mean. Let K be a convex body with 0 as an interior point. Let A be a subset of ∂K. Then cone(A) denotes the cone subtended at 0 by A; i.e. cone(A) := $\{\xi x : x \in A,\ 0 \leq \xi \leq 1\}$.

Theorem 5.2.4 *If U is an arbitrary measurable subset of $\partial \mathbf{I}_B$ then*

$$\mu_B^{d-1}(U) = d\lambda^d(\text{cone}(U)).$$

Proof. Recall Theorem 2.3.13(i), which expresses volume in terms of the support function. This yields $\lambda^d(\text{cone}(U)) = d^{-1} \int_U \sigma(\hat{f}_x)\, d\lambda^{d-1}(x)$. On the other hand, from (5.13), $\mu_B^{d-1}(U) = \int_U \sigma(\hat{f}_x)\, d\lambda^{d-1}(x)$. ∎

The process of tidying up these formulas depends on the normalization of $\mathbf{I}$, which in turn is the same question as the normalization of μ^d. These topics and some of the properties of $\mathbf{I}$ form the content of the next section.

5.3 The properties and the normalization of I

The body $\mathbf{I}_B$ affords another way of approaching the problem of area in Minkowski space. Since what is required is a convex, positively homogeneous function σ_B, we may specify σ_B (and hence γ) by specifying the convex body $\mathbf{I}_B$ of which σ_B is the support function. For this point of view, we need to know the restrictions that Requirement 5.1.1(a) places on σ_B and $\mathbf{I}_B$. Therefore, we begin this section with some formulas that show how $\mathbf{I}_B$ and σ_B behave under linear transformations of B (or, alternatively, how they depend on the choice of auxiliary Euclidean metric). The first statement is an obvious but important consequence of 5.1.1(a).

Theorem 5.3.1 *Let (X, B) be a Minkowski space and let T be an invertible linear transformation on X. If $\mathcal{M}$ is a surface in X then*

$$\mu_{TB}(T\mathcal{M}) = \mu_B(\mathcal{M}); \qquad (5.16)$$

i.e. the Minkowski area of a surface is independent of the choice of Euclidean coordinate system used to represent it.

Proof. We give the proof for a measurable set U in a $(d-1)$-dimensional flat M. The extension to a surface $\mathcal{M}$ via (5.13) is then clear.

Suppose that M is a translate of $f^\perp$. Then $T(U)$ lies in $T(M)$, which is a translate of $T(f^\perp)$. Since T transforms all areas in $f^\perp$ by the same factor and since μ_B is translation invariant, we have

$$\frac{\mu_B(U)}{\mu_B(B \cap f^\perp)} = \frac{\mu_B(TU)}{\mu_B(T(B \cap f^\perp))}.$$

But since ratios of area in any fixed direction are independent of the norm,

$$\frac{\mu_B(TU)}{\mu_B(T(B \cap f^\perp))} = \frac{\mu_{TB}(TU)}{\mu_{TB}(T(B \cap f^\perp))}.$$

Finally, since $\mu_B(B \cap f^\perp) = \gamma(B \cap f^\perp)$ and $\mu_{TB}(T(B \cap f^\perp)) = \gamma(T(B \cap f^\perp))$ and, by Requirement 5.1.1(a), $\gamma(B \cap f^\perp) = \gamma(T(B \cap f^\perp))$ we have

$$\mu_B(U) = \mu_{TB}(TU). \qquad \blacksquare$$

The next theorem, showing the connection between Theorem 5.3.1 and the way the body $\mathbf{I}_B$ behaves under linear transformations, seems to have been first stated explicitly by R. Clack [112].

Theorem 5.3.2 *Let T be an invertible linear transformation on a Minkowski space (X, B). Then $\mu_B(\partial K) = \mu_{TB}(\partial TK)$ for all convex bodies K if and only if*

$$\mathbf{I}_{TB} = |\det T|^{-1} T(\mathbf{I}_B).$$

Proof. Using (5.15), we have

$$\begin{aligned}
d^{-1}\mu_{TB}(\partial TK) &= V(TK[d-1], \mathbf{I}_{TB}) \\
&= V(TK[d-1], TT^{-1}\mathbf{I}_{TB}) \\
&= |\det T| V(K[d-1], T^{-1}\mathbf{I}_{TB}) \\
&= V(K[d-1], |\det T| T^{-1}\mathbf{I}_{TB}).
\end{aligned}$$

The third equation expresses a general fact about mixed volumes and the last comes from the linearity of mixed volumes in each variable (Proposition 2.3.7(iv)). On the other hand, $d^{-1}\mu_B(\partial K) = V(K[d-1], \mathbf{I}_B)$. This completes one direction.

To reverse the argument we need to begin with Proposition 2.3.14, that if $V(K[d-1], C_1) = V(K[d-1], C_2)$ for centrally symmetric C_1 and C_2 and for all K, then $C_1 = C_2$. Thus, if $\mu_B(\partial K) = \mu_{TB}(\partial TK)$ for all K, then the previous sequence of equations shows that $\mathbf{I}_B = |\det T| T^{-1}\mathbf{I}_{TB}$, as required. $\blacksquare$

5.3 The properties and the normalization of **I**

Corollary 5.3.3 *Let T be an invertible linear transformation on a Minkowski space (X, B). Then*

$$T(\mathbf{I}_B) = |\det T| \mathbf{I}_{TB} \quad \text{and} \quad \sigma_B(f) = |\det T| \sigma_{TB}(T^{*-1}f).$$

Proof. The first equation follows directly from Theorems 5.3.1 and 5.3.2. The second is a consequence of the fact that σ_B (resp. σ_{TB}) is the support function of $\mathbf{I}_B$ (resp. $\mathbf{I}_{TB}$). ∎

Corollary 5.3.4 *The symmetry group of B is a subset of the symmetry group of $\mathbf{I}_B$.*

Proof. If S is a symmetry of B, i.e. a linear transformation that maps B onto B, then it is volume preserving and $|\det S| = 1$. Thus, by Corollary 5.3.3, $S(\mathbf{I}_B) = \mathbf{I}_B$. ∎

This shows how **I** must behave with respect to linear transformations of the space as a whole. For example, Corollary 5.3.4 shows that if B is a solid of revolution then $\mathbf{I}_B$ is one also. However, it is also possible for the space (X, B) to have two $(d-1)$-dimensional isometric subspaces without that isometry being realized as a motion of the whole space. Requirement 5.1.1(a) implies that such subspaces have the same value for γ. For example, if B is a cylinder (with axis vertical and regardless of the shape of the cross-section) then all subspaces that are not too steeply inclined to the horizontal cross-section (see Figure 5.2) are

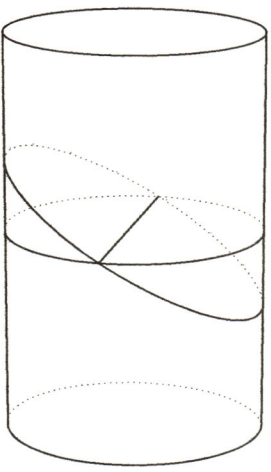

Figure 5.2

isometric. Similarly, both the cube and the octahedron have many cross-sections that are parallelograms which, by 5.1.1(a), must *all* have the same value for γ.

More precisely, if $\hat{f}_1$ and $\hat{f}_2$ are unit normals to subspaces $f_1^\perp$ and $f_2^\perp$ and if there is a linear map T of $f_1^\perp \cap B$ onto $f_2^\perp \cap B$ then $\gamma(f_1^\perp \cap B) = \gamma(f_2^\perp \cap B)$ and hence $\sigma(\hat{f}_1) = \lambda(f_2^\perp \cap B)(\lambda(f_1^\perp \cap B))^{-1}\sigma(\hat{f}_2) = |\det T| \sigma(\hat{f}_2)$.

We illustrate this process with an arbitrary cylinder and then with the cube and octahedron in $\mathbb{R}^3$.

Example 5.3.5

(i) Let B_0 be an arbitrary unit ball in a linear space X and consider the cylinder $B := B_0 \times [-1, 1]$ in $X \times \mathbb{R}$. If $f_0 := (0, 1)$, i.e. $f_0(x, \alpha) = \alpha$, then $B \cap f_0^\perp = B_0$. If $\hat{u}$ is an arbitrary linear functional of (Euclidean) norm 1 on X, let $f := (\hat{u} \sin\phi, \cos\phi)$, i.e. $f(x, \alpha) = \hat{u}(x)\sin\phi + \alpha\cos\phi$. Then, provided ϕ is sufficiently small ($|\cos\phi| \geq \sup\{|\hat{u}(x)| : x \in B_0\}$), $f^\perp$ intersects ∂B in the generators of the cylinder. Thus $T(f^\perp \cap B) = B_0$, where $T(x, \alpha) := (x, 0)$ and $|\det T| = |\cos\phi|$. So we have $\sigma_B(f) = |\cos\phi|\sigma_B(f_0)$. This implies that f_0 and the linear functionals f all support $\mathbf{I}_B$ at the same point $(0, 1)$ and hence that $\mathbf{I}_B$ has a vertex at this point. See Figures 5.3(b),(d) and 5.4(b),(d) below.

(ii) Let B be the cube in $\mathbb{R}^3$ with vertices at $(\pm 1, \pm 1, \pm 1)$. Since we may regard B as a cylinder over each coordinate plane, Example 5.3.5(i) shows that $\mathbf{I}_B$ has vertices on the coordinate axes. One can calculate the shape of these vertices and hence show that $\mathbf{I}_B$ (suitably scaled) is contained in the rhombic dodecahedron

$$\text{conv}\{(\pm 1, 0, 0), (0, \pm 1, 0), (0, 0, \pm 1), (\pm\tfrac{1}{2}, \pm\tfrac{1}{2}, \pm\tfrac{1}{2})\}.$$

Furthermore, if $f := (\cos\phi, \sin\phi, 0)$ then $f^\perp \cap B$ is a rectangle which is affinely equivalent to the coordinate plane cross-sections. Calculation of the areas of these cross-sections shows that the intersection of $\mathbf{I}_B$ with the plane $z = 0$ (and scaled as before) is $\text{conv}\{(\pm 1, 0, 0), (0, \pm 1, 0)\}$. Similarly for the other coordinate planes. Thus, no matter what definition of γ is used, $\mathbf{I}_B$ lies between the octahedron and the rhombic-dodecahedron. See Figures 5.3(b) and 5.4(b) below.

(iii) For this example, let B be the octahedron in $\mathbb{R}^3$ with vertices at $(\pm 1, 0, 0)$ $(0, \pm 1, 0)$ and $(0, 0, \pm 1)$. The same calculations with all the rhombic cross-sections of B (linear functionals that lie in one of the coordinate planes of the dual) show that $\partial \mathbf{I}_B$ contains the three squares $\text{conv}\{(\pm 1, \pm, 1, 0)\}$, $\text{conv}\{(\pm 1, 0, \pm 1)\}$ and $\text{conv}\{(0, \pm 1, \pm 1)\}$ in the three coordinate planes. Hence, since $\mathbf{I}_B$ is convex, it lies between the cubo-octahedron (the convex hull of the squares just listed) and the cube $\text{conv}\{(\pm 1, \pm 1, \pm 1)\}$. See Figures 5.3(a) and 5.4(a) below.

Remarks

(i) These examples illustrate how Requirement 5.1.1(a) restricts γ and hence $\mathbf{I}$. The last two examples, in particular, indicate how much freedom remains.

5.3 The properties and the normalization of I

(ii) These examples show that 5.1.1(a) precludes the possibility that one may specify $\mathbf{I}_B = B$ even for the most simple type of non-Euclidean ball.

(iii) In these simple cases, the general shape of $\mathbf{I}_B$ resembles that of B°, giving rise to the general philosophy that $\mathbf{I}_B$ is a surrogate of the dual ball in the original space. In two-dimensional spaces, as we saw in Chapter 4, this resemblance is an isometry.

(iv) In two-dimensional spaces we always have $\mathbf{I}^2(B) = B$. These examples show that $\mathbf{I}^2(B)$ resembles B.

Next comes the question of the normalization of $\mathbf{I}_B$. Firstly, if (X, B) is a d-dimensional Minkowski space then, in accordance with Equation (5.2) (see also Equation (4.11)), we denote the ratio of Minkowski volume μ^d to Lebesgue d-measure λ^d by σ_B^d. Secondly, recall from (2.8) the defining formula for mixed volumes: if $K = \sum_{i=1}^n \xi_i K_i$ then

$$\lambda(K) = \sum_{i_1, i_2, \ldots, i_d = 1}^n V(K_{i_1}, K_{i_2}, \ldots, K_{i_d}) \xi_{i_1} \xi_{i_2} \ldots \xi_{i_d},$$

where the coefficients $V(K_{i_1}, \ldots, K_{i_d})$ are invariant under permutations of the arguments. If we multiply this equation by σ_B^d and set the *Minkowski mixed volume* $V_B(K_{i_1}, \ldots, K_{i_d}) := \sigma_B^d V(K_{i_1}, \ldots, K_{i_d})$ then we have

$$\mu_B(K) = \sum V_B(K_{i_1}, \ldots, K_{i_d}) \xi_{i_1} \ldots \xi_{i_d}. \tag{5.17}$$

Using these numbers we can tidy up some of our previous equations starting with (5.15). We have

$$\mu_B(\partial K) = dV(K[d-1], \mathbf{I}_B)$$
$$= d(\sigma_B^d)^{-1} V_B(K[d-1], \mathbf{I}_B)$$
$$= dV_B(K[d-1], \tilde{\mathbf{I}}(B)), \tag{5.18}$$

where $\tilde{\mathbf{I}}(B) = (\sigma_B^d)^{-1} \mathbf{I}_B$.

Definition 5.3.6 *If (X, B) is a Minkowski space then $\tilde{\mathbf{I}}(B)$ defined by $\tilde{\mathbf{I}}(B) := (\sigma_B^d)^{-1} \mathbf{I}_B$ is called **the isoperimetrix** of the space.*

Setting $K = \tilde{\mathbf{I}}(B)$ in (5.18) we get, as in Chapter 4,

$$\mu_B^{d-1}(\partial \tilde{\mathbf{I}}) = d\mu_B^d(\tilde{\mathbf{I}}(B)); \tag{5.19}$$

i.e. the Minkowski surface area of $\tilde{\mathbf{I}}$ is d times the Minkowski volume of $\tilde{\mathbf{I}}$.

Corresponding to Theorem 5.2.4 we have that if U is an arbitrary measurable subset of $\partial \tilde{\mathbf{I}}(B)$ and if $\text{cone}(U) = \{\xi x : x \in U, 0 \leq \xi \leq 1\}$ then

$$\mu_B^{d-1}(U) = d\mu_B^d(\text{cone}(U)). \tag{5.20}$$

150   The concept of area and content

The final equation we need to modify is the first one of Corollary 5.3.3. Equation (5.4) and Requirement 5.1.1(a) give

$$\sigma_B^d = |\det T| \sigma_{TB}^d$$

for any non-singular linear transformation T. Therefore (using Corollary 5.3.3),

$$\tilde{\mathbf{I}}(TB) = (\sigma_{TB}^d)^{-1}\mathbf{I}_{TB} = (\sigma_B^d)^{-1}|\det T|\mathbf{I}_{TB} = (\sigma_B^d)^{-1}T\mathbf{I}_B = T((\sigma_B^d)^{-1}\mathbf{I}_B).$$

In other words,

$$\tilde{\mathbf{I}}(TB) = T(\tilde{\mathbf{I}}B) \tag{5.21}$$

and $\tilde{\mathbf{I}}(B)$ transforms properly under invertible linear maps.

Equations (5.15) and (5.18) are two instances of a general situation. It is possible to use one ball B for measuring area and another ball B' for measuring volume. In (5.15) $B' = E$ (E is a Euclidean ball) and in (5.18) $B' = B$. Since the choice of B' affects the measurement of volume by only a scalar factor, the shape of the isoperimetrix $\mathbf{I}$ is unaffected. However, for each B' there is an appropriate normalization $\mathbf{I}(B, B')$ of $\mathbf{I}$ so that

$$\mu_B(\partial K) = dV_{B'}(K[d-1], \mathbf{I}(B, B'))$$

and, in particular, $\mu_B(\partial \mathbf{I}(B, B')) = d\mu_{B'}(\mathbf{I}(B, B'))$. Thus $\mathbf{I}(B, E) = \mathbf{I}_B$ and $\mathbf{I}(B, B) = \tilde{\mathbf{I}}_B$. Corollary 5.3.3 shows how $\mathbf{I}(B, B')$ behaves when one of the two balls is transformed by T. Equation (5.21) is the special case when both balls are transformed in the same way.

5.4 The isoperimetrices that arise from Examples 5.1.4

In this section we explore the nature of the isoperimetrices that arise when we use the various definitions of γ given in Examples 5.1.4. The subsections are numbered to correspond to those examples. Throughout this section, linear functionals are used as directions for hyperplanes in X and are normalized to have Euclidean length 1. However, so that the formulas appear less cluttered, the caret will be omitted where it seems unessential.

(i) The Busemann definition

This is the one used most widely in the literature. From (5.8) we have

$$\sigma_B(f) = \epsilon_{d-1}/\lambda(f^\perp \cap B).$$

To see the nature of $\mathbf{I}_B$, it is preferable to consider

$$\rho(f) = \sigma_B(f)^{-1} = \epsilon_{d-1}^{-1}\lambda(f^\perp \cap B).$$

5.4 The isoperimetrices that arise from Examples 5.1.4

Since σ_B is the support function of $\mathbf{I}_B$, ρ is the radial function of $\mathbf{I}_B^\circ$. The body whose radial function in a given direction is equal to the area of the cross-section of K perpendicular to that direction was called the *intersection body* of K by Lutwak [337]. We shall denote this body by IK. Thus with σ given by (5.8) the solution to the isoperimetric problem is

$$\mathbf{I}_B = \epsilon_{d-1}(IB)^\circ. \tag{5.22}$$

There is, however, a difference between giving a name to $\mathbf{I}_B$ as the polar of the intersection body and knowing what it looks like. We know from Example 5.3.5(ii) that when B is the usual cube in $\mathbb{R}^3$, $\mathbf{I}_B = (IB)^\circ$ must have vertices corresponding to the facets of B and, indeed, that $(IB)^\circ$ lies between an octahedron and a rhombic dodecahedron. I am indebted to Jennifer Overington for help with the use of Mathematica to produce Figures 5.3a–d, which illustrate some intersection bodies and their duals. Note that in these figures (and the following ones of projection bodies and their duals) scale factors are ignored in order to fit the illustrations to the page.

(ii) The Holmes–Thompson definition

For this definition of Minkowski area we begin with (5.10):

$$\sigma_B(f) = \frac{\lambda^{d-1}((B \cap f^\perp)^\circ)}{\epsilon_{d-1}}.$$

Next we recall Corollary 2.2.10, which relates the dual of a cross-section to the projection of the dual. Using that corollary we have

$$\sigma_B(f) = \epsilon_{d-1}^{-1}\lambda^{d-1}(\text{Proj}_f(B^\circ)),$$

where Proj_f denotes the projection along f onto the orthogonal complement of f. This formula appears highly dependent on the Euclidean structure. To see that this is largely illusory, we first recall the definition of the shadow boundary (Definition 3.4.7).

Observe that the projection of ∂B° along f covers $\text{Proj}_f(B^\circ)$ exactly twice because the two parts of ∂B° on either side of its shadow boundary relative to f each cover the projection once. If $d\lambda(\hat{x})$ denotes the *Euclidean area-element* at that point of ∂B° whose unit normal is $\hat{x}$ then the projection of this area-element perpendicular to f is given by $|f(\hat{x})|d\lambda(\hat{x})$. Thus

$$\lambda(\text{Proj}_f(B^\circ)) = 2^{-1}\int_{\partial B^\circ}|f(\hat{x})|\,d\lambda(\hat{x}) \tag{5.23}$$

and hence

$$\sigma_B(f) = (2\epsilon_{d-1})^{-1}\int_{\partial B^\circ}|f(\hat{x})|\,d\lambda(\hat{x}). \tag{5.24}$$

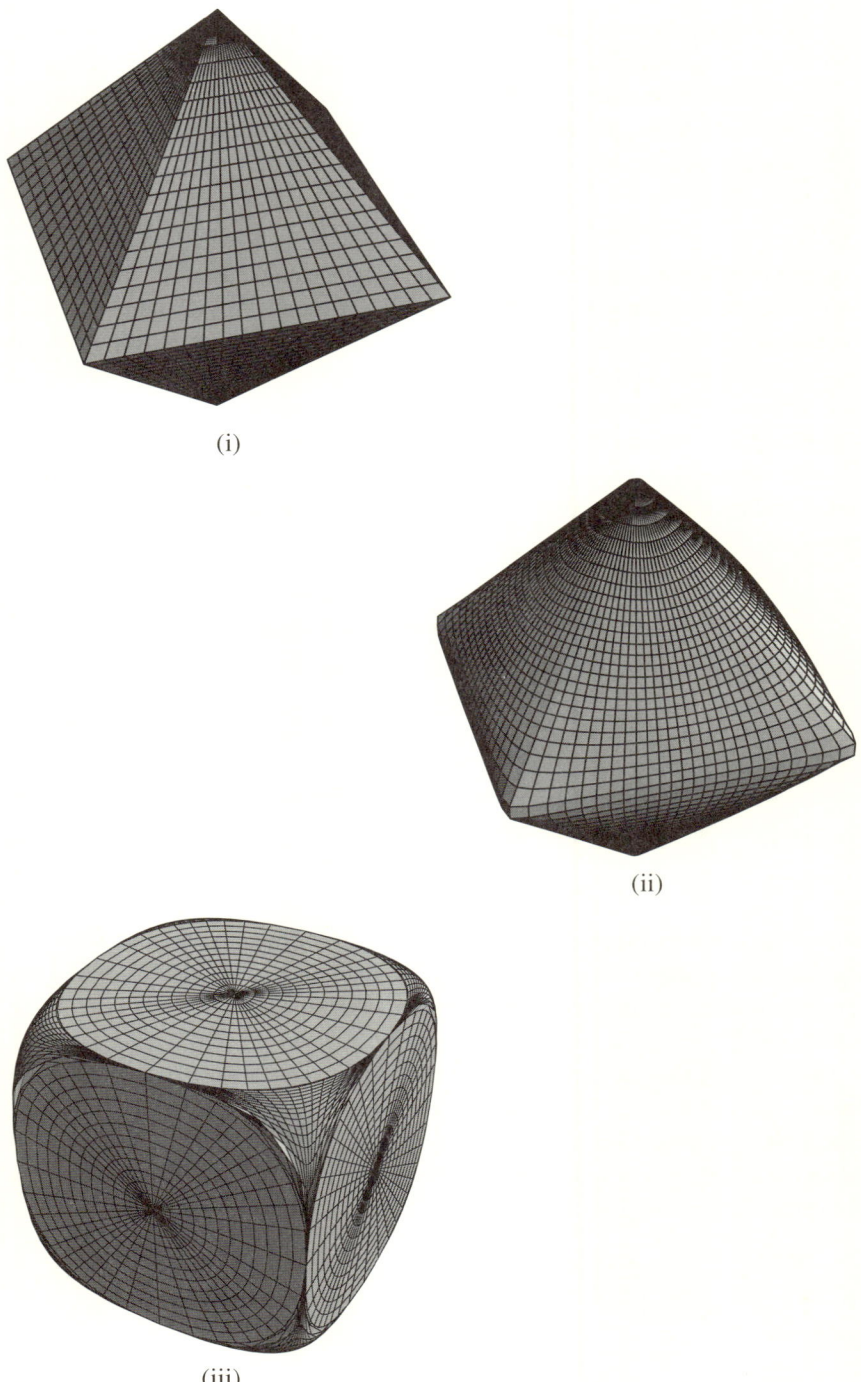

Figure 5.3a. (i) The ball B, (ii) the intersection body $I(B)$ and (iii) the isoperimetrix $\mathbf{I}(B) = (I(B))^\circ$ in the case when B is an octahedron. Note that the boundaries of the flat faces of $\mathbf{I}(B)$ are not circles but four parabolic arcs.

5.4 The isoperimetrices that arise from Examples 5.1.4   153

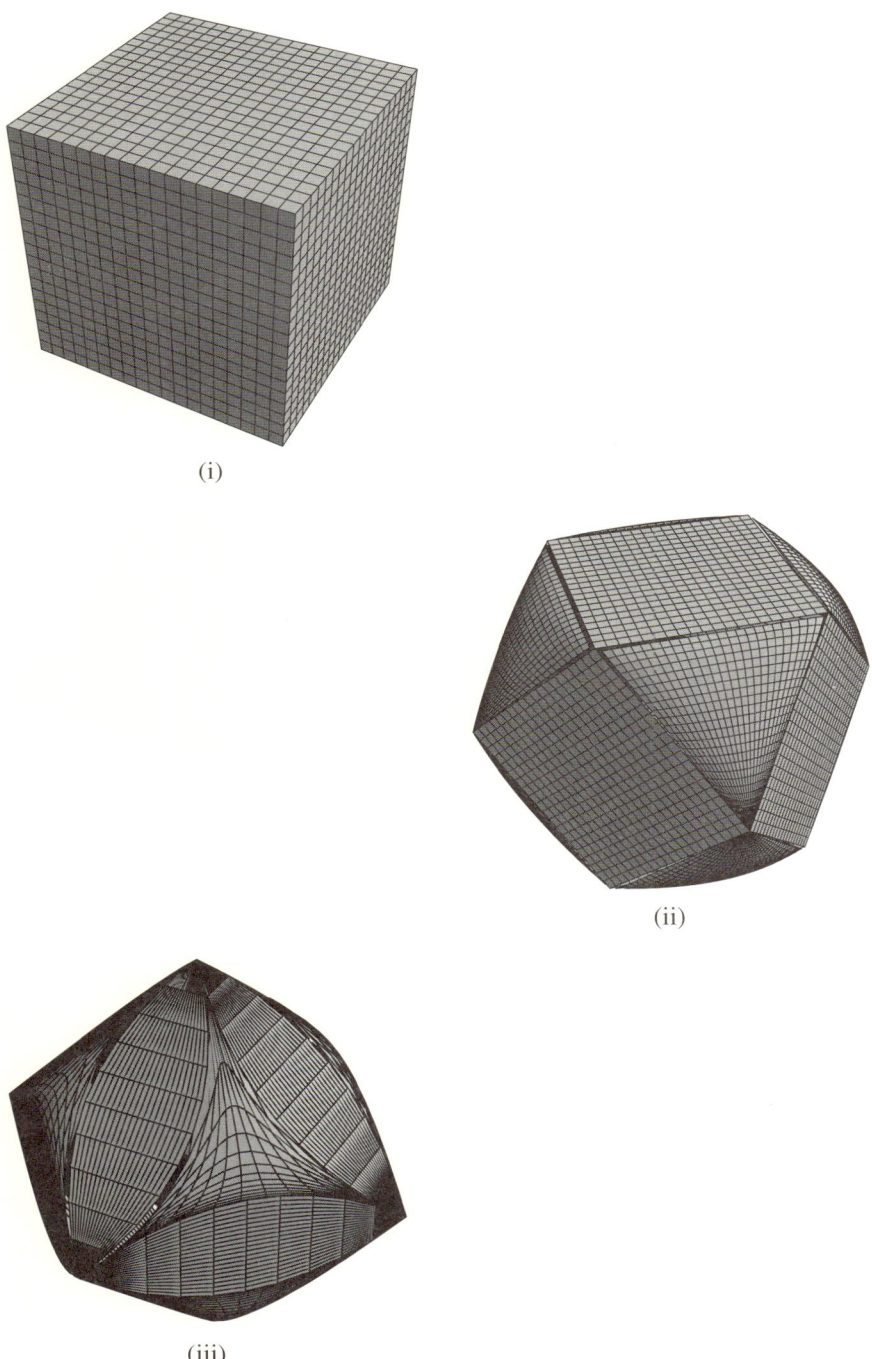

(i)

(ii)

(iii)

Figure 5.3b. (i) The ball B, (ii) the intersection body $I(B)$ and (iii) the isoperimetrix $\mathbf{I}(B) = (I(B))^\circ$ in the case when B is a cube.

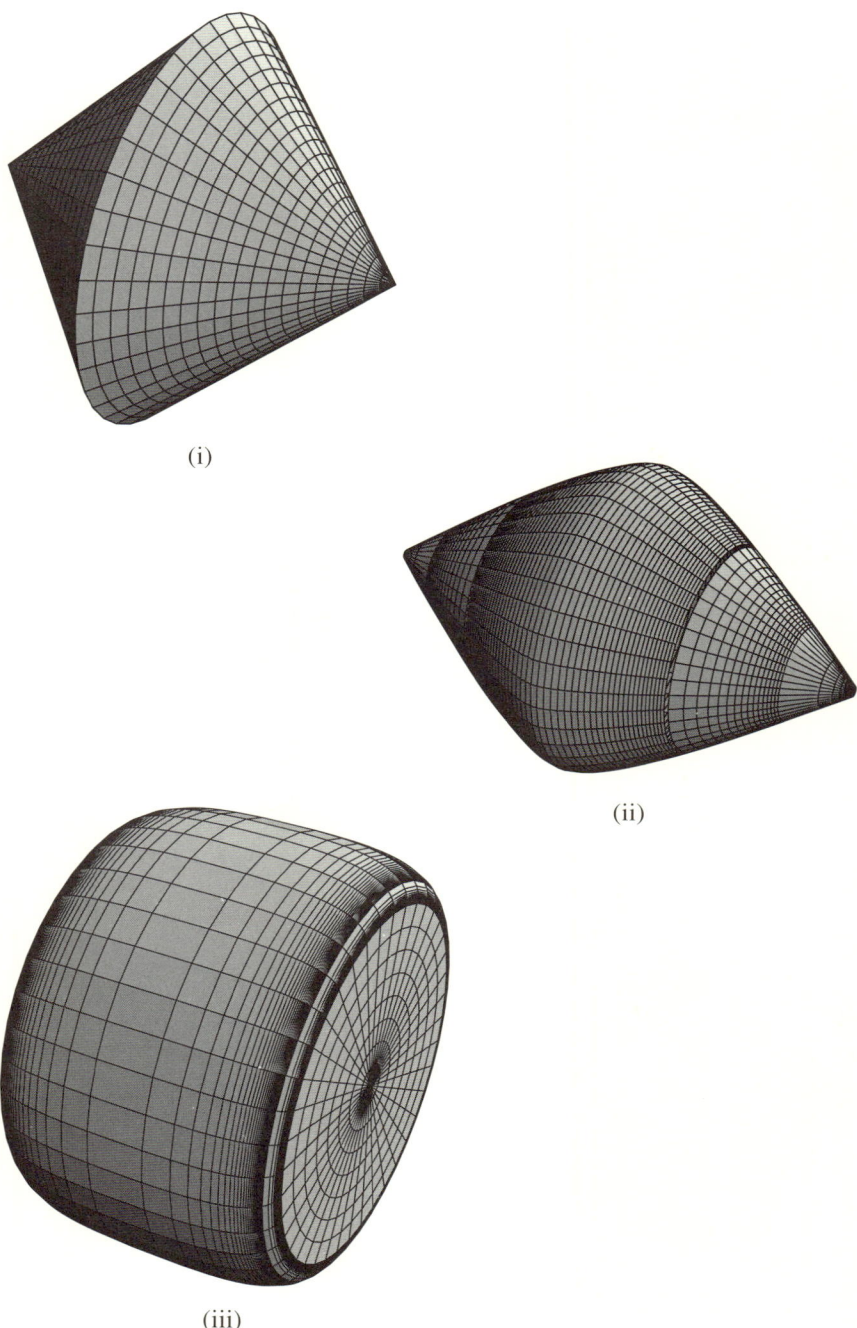

Figure 5.3c. (i) The ball B, (ii) the intersection body $I(B)$ and (iii) the isoperimetrix $\mathbf{I}(B) = (I(B))^\circ$ in the case when B is a double cone.

5.4 The isoperimetrices that arise from Examples 5.1.4

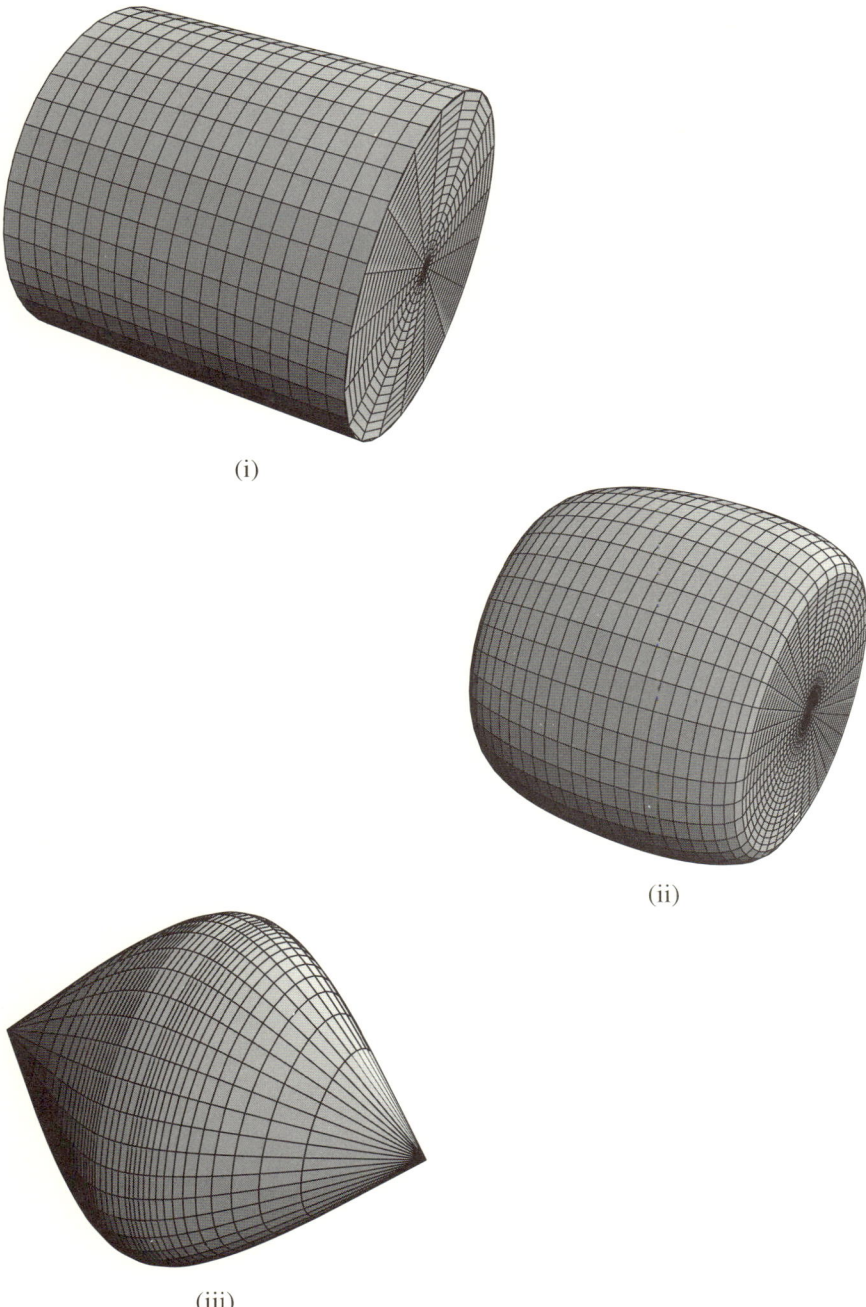

Figure 5.3d. (i) The ball B, (ii) the intersection body $I(B)$ and (iii) the isoperimetrix $\mathbf{I}(B) = (I(B))^\circ$ in the case when B is a cylinder.

As was explained in §2.3 (just before Theorem 2.3.13) the more precise formulation of this integral is as an integral over S^{d-1} in X with respect to the surface area measure $\nu_{B°}$ (Definition 2.3.12). Equation (5.24) avoids the need to speak of "orthogonal projections" and, since both sides are positively homogeneous in f, also avoids the restriction that $|f| = 1$. We shall make extensive use of (5.24) in the sequel.

Using the notation for segments from Example 0.2.3(iv), Example 2.2.3(iii) shows that $|f(\hat{x})|/2 = h_{[f]}(\hat{x})$. Comparison of (5.24) with Theorem 2.3.13(iii) now gives

$$\sigma_B(f) = dV(B°[d-1], [f])/\epsilon_{d-1}.$$

Finally, Definitions 2.3.9 and 2.3.11 show that if σ is given by (5.10) then

$$\mathbf{I}_B = \Pi(B°)/\epsilon_{d-1}. \tag{5.25}$$

If B is a polytope then $B°$ is also a polytope. Let the facets of $B°$ be F_i, with unit normals $\hat{x}_i (i = 1, 2, \ldots, n)$. Then (5.24) becomes

$$\sigma_B(f) = (2\epsilon_{d-1})^{-1} \sum_{i=1}^{n} |f(\hat{x}_i)| \lambda^{d-1}(F_i) \tag{5.26}$$

$$= (2\epsilon_{d-1})^{-1} \sum_{i=1}^{n} |f(\alpha_i \hat{x}_i)|, \tag{5.27}$$

where $\alpha_i = \lambda(F_i)$. Using Examples 0.2.3(iv) and 2.2.3(iii) again, we see that $\mathbf{I}_B$ is the zonotope $\sum_{i=1}^{n} [\alpha_i x_i]/\epsilon_{d-1}$. Thus, for this σ, when B is a polytope, $\mathbf{I}_B$ is not only a polytope but a zonotope. It follows by a continuity argument that, in general, $\mathbf{I}_B$ is a zonoid, *i.e.* a limit of zonotopes. These facts make it easier to construct examples of $\mathbf{I}_B$, Figure 5.4, than in the previous case (see, *e.g.,* [262]).

Indeed, the extra structure that zonotopes have make it possible to use $\mathbf{I}_B$ as a new ball and iterate the construction several times. The sequence of six iterations beginning with the octahedron was first drawn several years ago by Barbara Taylor using a FORTRAN programme. Recently, Brian Ingalls used Mathematica to reproduce the original sequence and to produce several more. I am grateful to these two students for the following computer-drawn illustrations using Mathematica. In Figures 5.5–5.8 the map $(.)°$ goes horizontally from left to right, the map $\Pi(.)$ goes diagonally from upper right to lower left and the composition $\mathbf{I}(.)$ goes vertically down on the left.

(iii) The Benson definition

As far as I know, the theory for $\mathbf{I}_B$ has not been investigated in this case. It is, however, worth reproducing one or two sample calculations. The purpose of these is to show that $\mathbf{I}_B = (\Pi B)°$ for some but, it seems, not all bodies B.

5.4 The isoperimetrices that arise from Examples 5.1.4

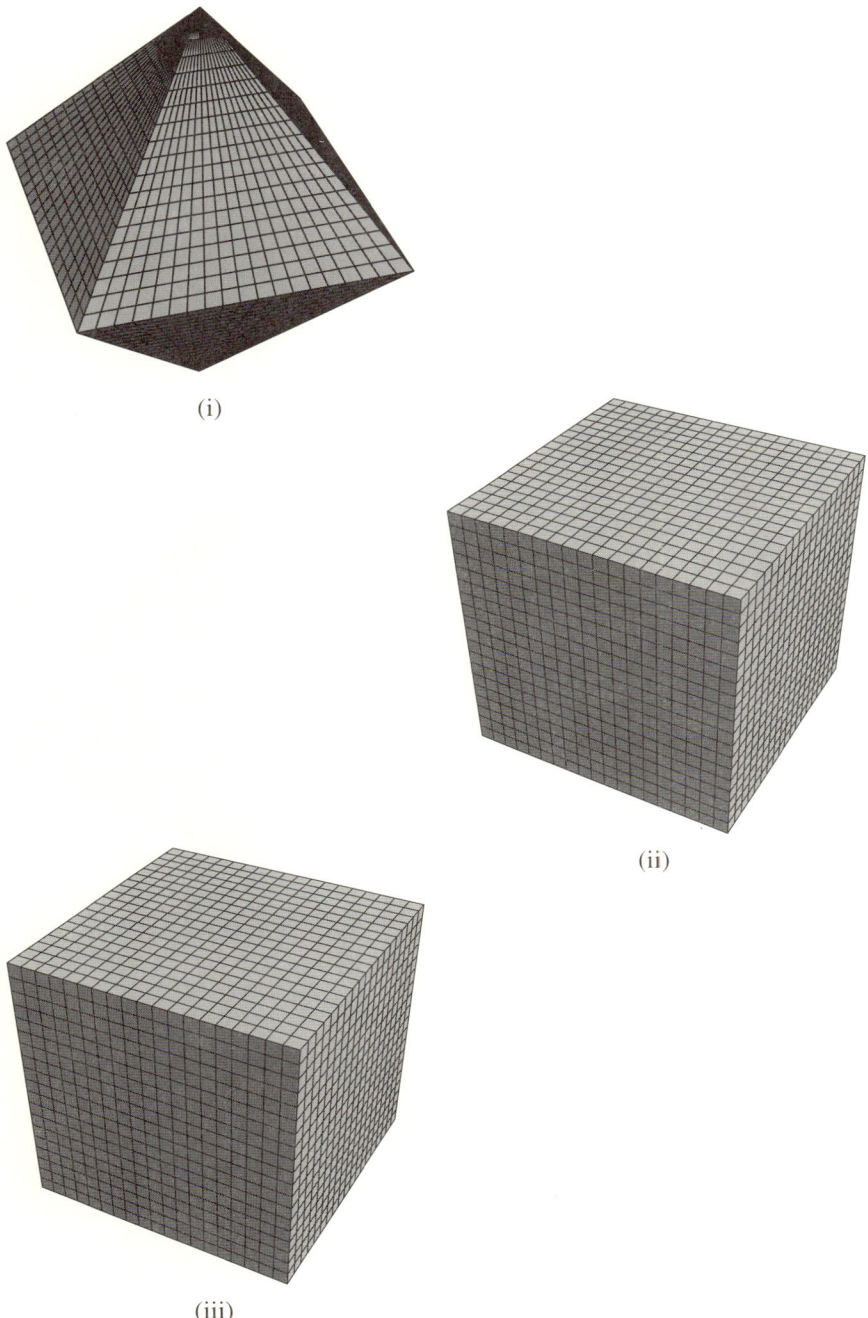

Figure 5.4a. (i) The ball B, (ii) the dual ball B° and (iii) the isoperimetrix $\mathbf{I}(B) = \Pi(B^\circ)$ in the case when B is an octahedron.

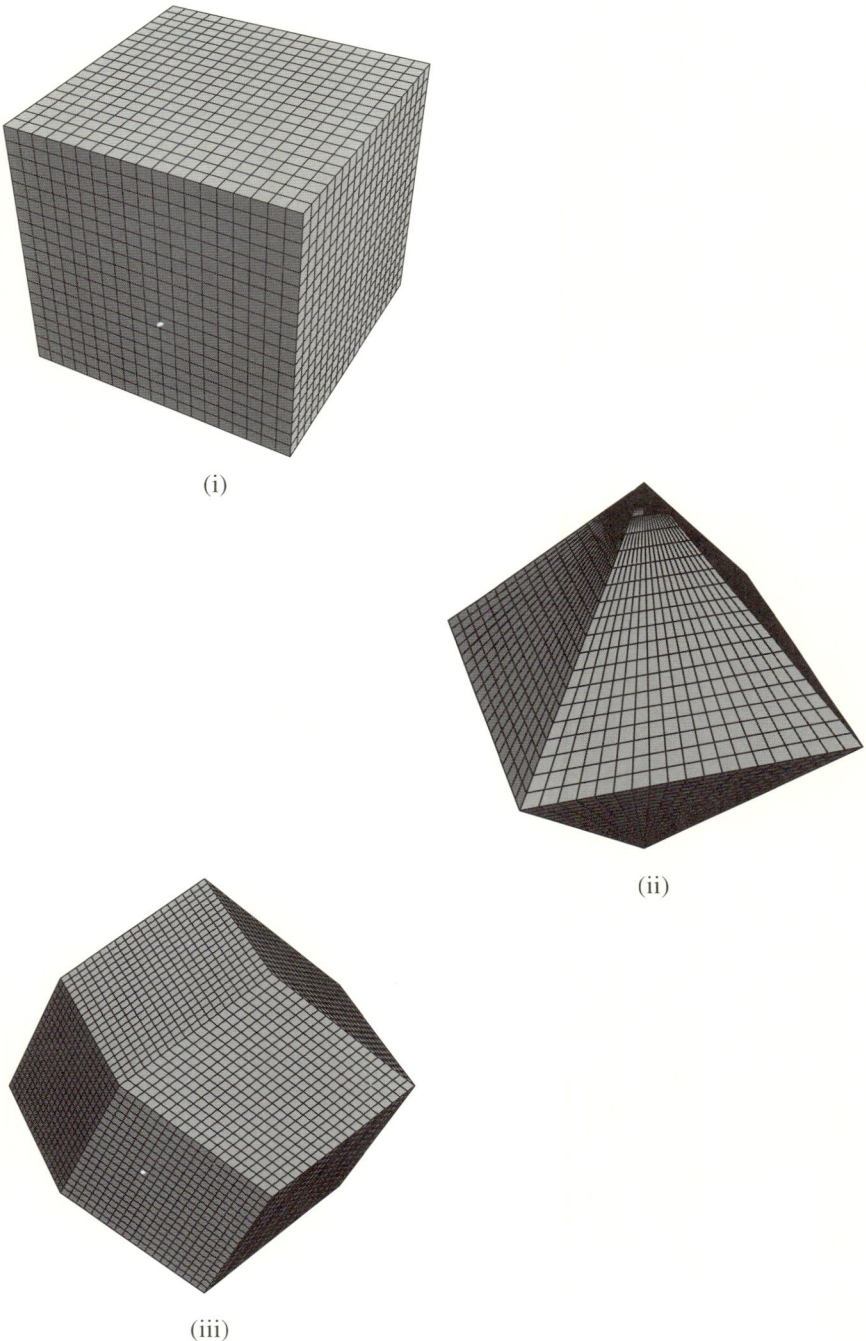

Figure 5.4b. (i) The ball B, (ii) the dual ball B° and (iii) the isoperimetrix $\mathbf{I}(B) = \Pi(B^\circ)$ in the case when B is a cube.

5.4 The isoperimetrices that arise from Examples 5.1.4 159

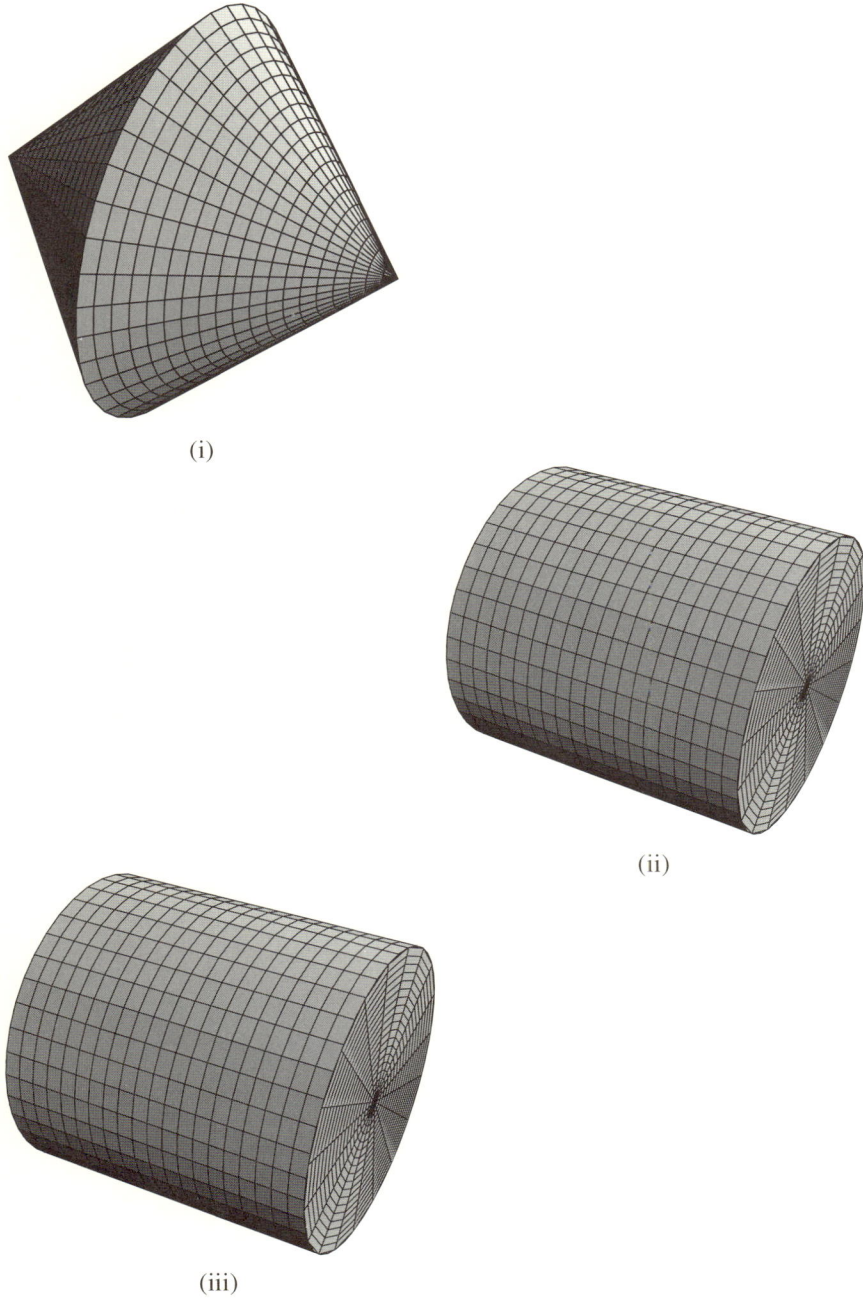

(i)

(ii)

(iii)

Figure 5.4c. (i) The ball B, (ii) the dual ball B° and (iii) the isoperimetrix $\mathbf{I}(B) = \Pi(B^\circ)$ in the case when B is a double cone.

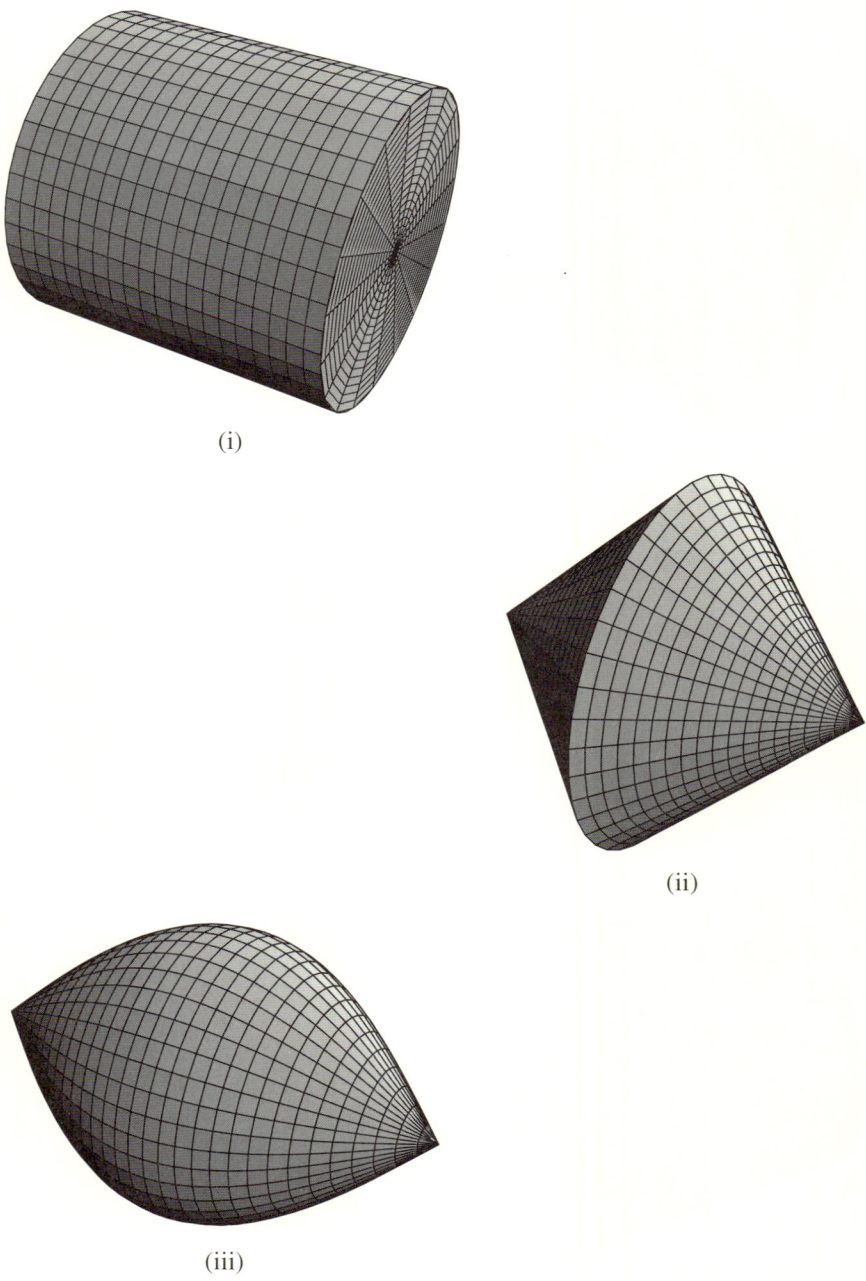

Figure 5.4d. (i) The ball B, (ii) the dual ball B° and (iii) the isoperimetrix $\mathbf{I}(B) = \Pi(B^\circ)$ in the case when B is a cylinder. Note that in this case $\mathbf{I}(B)$ is the solid of revolution of the cosine curve; see Chilton and Coxeter [110] and [262].

5.4 The isoperimetrices that arise from Examples 5.1.4   161

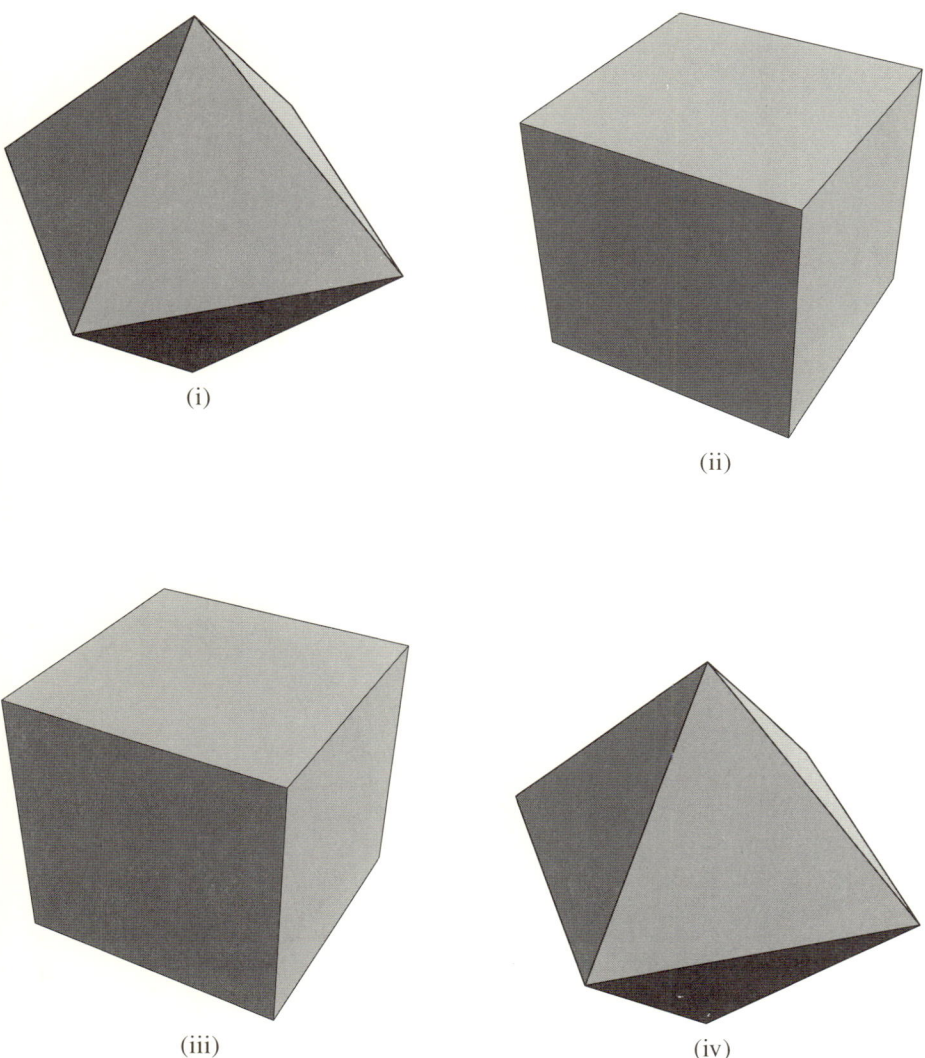

Figure 5.5. Iterations of the isoperimetric map $\mathbf{I}(.)$ when B is the octahedron. (i) The ball B, (ii) the dual ball B°, (iii) $\mathbf{I}(B) = \Pi(B^\circ)$, (iv) $(\mathbf{I}B)^\circ$.

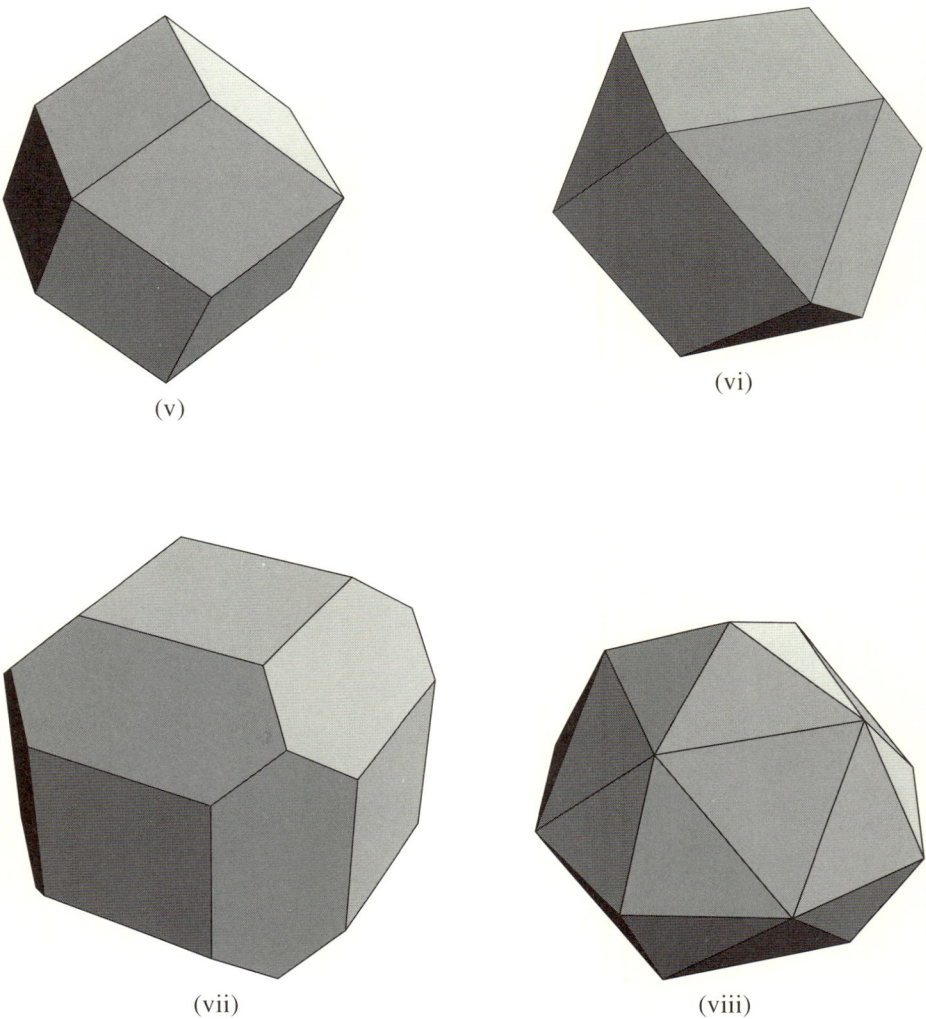

Figure 5.5 (*cont.*). Iterations of the isoperimetric map $\mathbf{I}(.)$ when B is the octahedron. (v) $\mathbf{I}^2 B = \Pi((\mathbf{I}B)^\circ)$, (vi) the dual $(\mathbf{I}^2 B)^\circ$, (vii) $\mathbf{I}^3 B = \Pi((\mathbf{I}^2 B)^\circ)$, (viii) the dual $(\mathbf{I}^3 B)^\circ$.

5.4 *The isoperimetrices that arise from Examples 5.1.4* 163

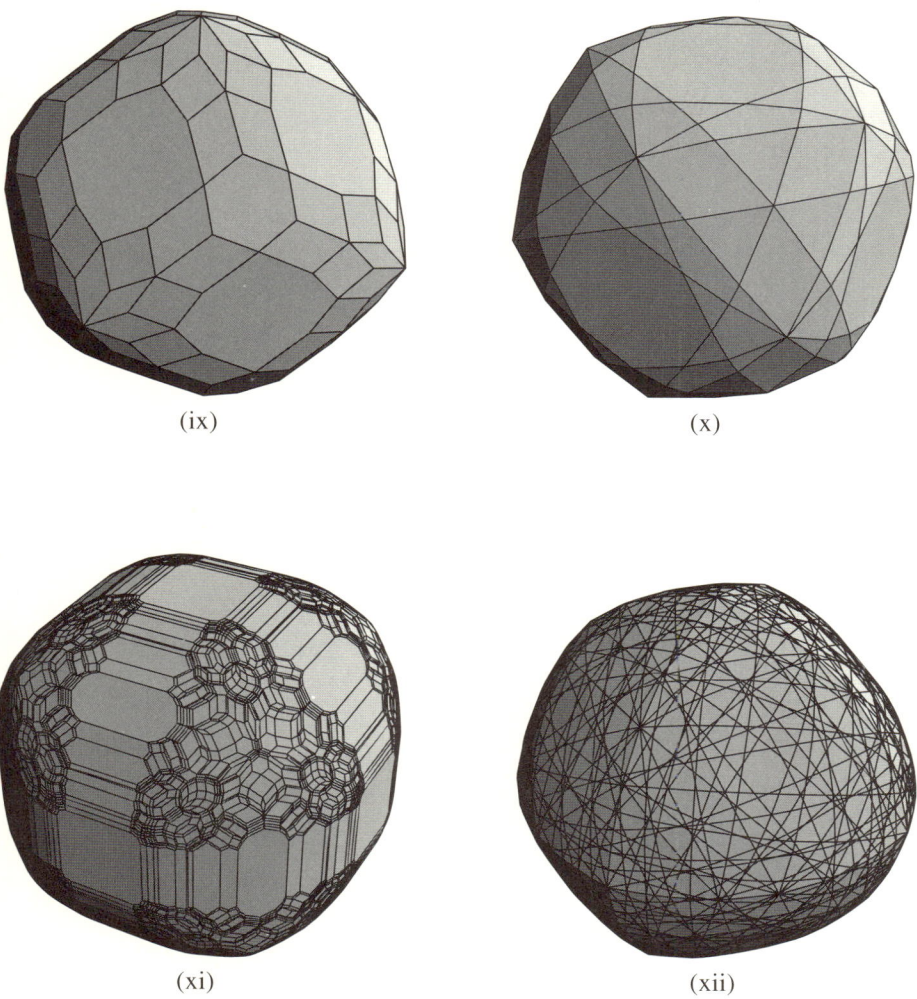

Figure 5.5 (*cont.*). Iterations of the isoperimetric map $\mathbf{I}(.)$ when B is the octahedron. (ix) $\mathbf{I}^4 B = \Pi((\mathbf{I}^3 B)^\circ)$, (x) the dual $(\mathbf{I}^4 B)^\circ$, (xi) $\mathbf{I}^5 B = \Pi((\mathbf{I}^4 B)^\circ)$, (xii) the dual $(\mathbf{I}^5 B)^\circ$.

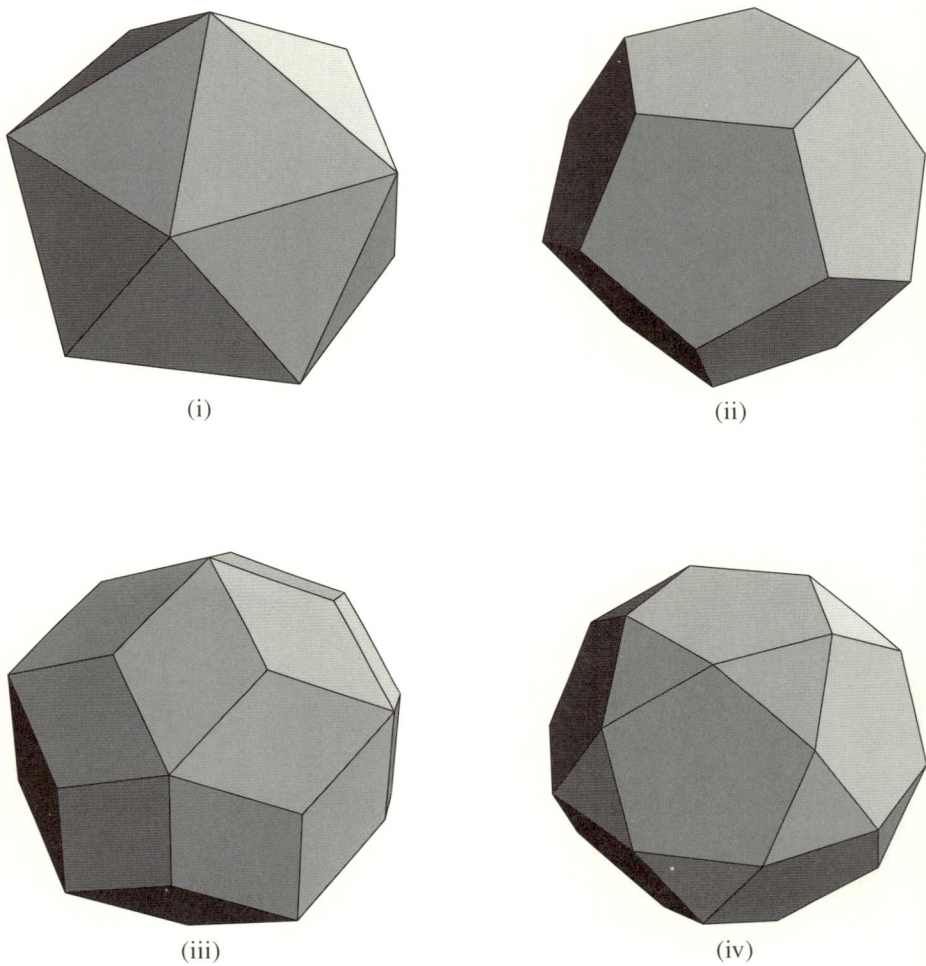

Figure 5.6. Iterations of the isoperimetric map $\mathbf{I}(.)$ when B is the icosahedron. (i) The ball B, (ii) the dual ball B°, (iii) $\mathbf{I}(B) = \Pi(B^\circ)$, (iv) $(\mathbf{I}B)^\circ$.

5.4 The isoperimetrices that arise from Examples 5.1.4 165

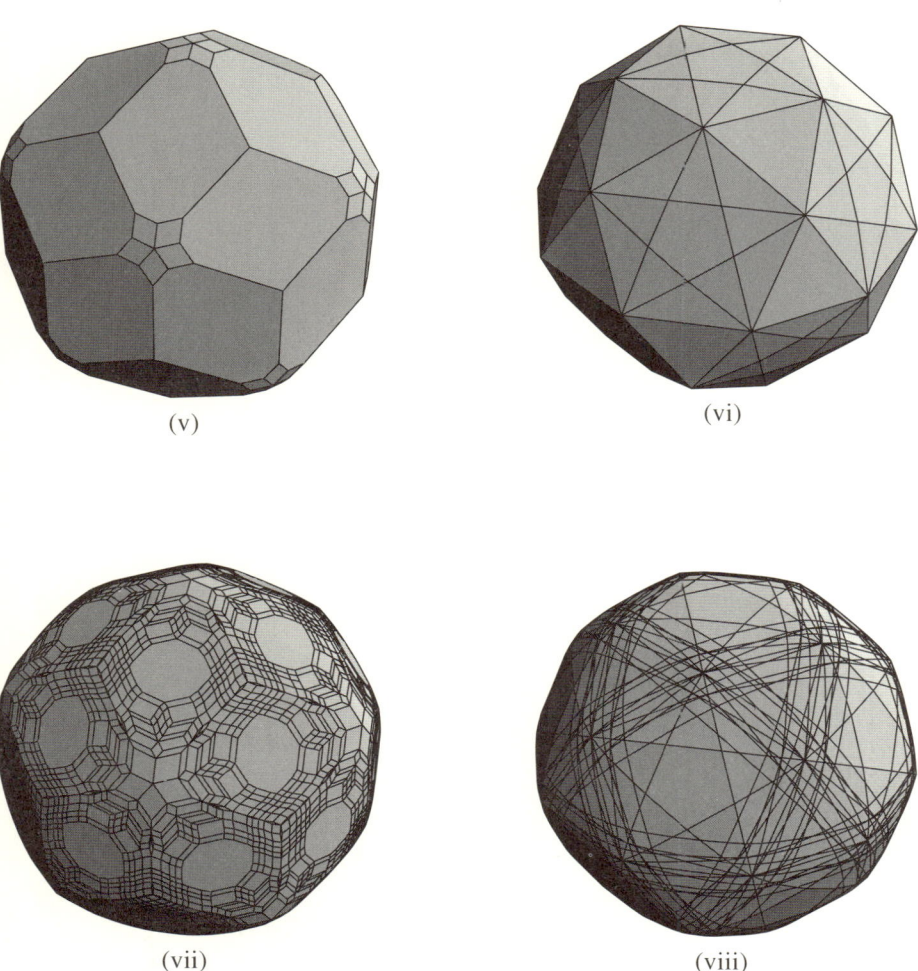

Figure 5.6 (*cont*.). Iterations of the isoperimetric map $\mathbf{I}(.)$ when B is the icosahedron. (v) $\mathbf{I}^2 B = \Pi((\mathbf{I}B)^\circ)$, (vi) the dual $(\mathbf{I}^2 B)^\circ$, (vii) $\mathbf{I}^3 B = \Pi((\mathbf{I}^2 B)^\circ)$, (viii) the dual $(\mathbf{I}^3 B)^\circ$.

166 The concept of area and content

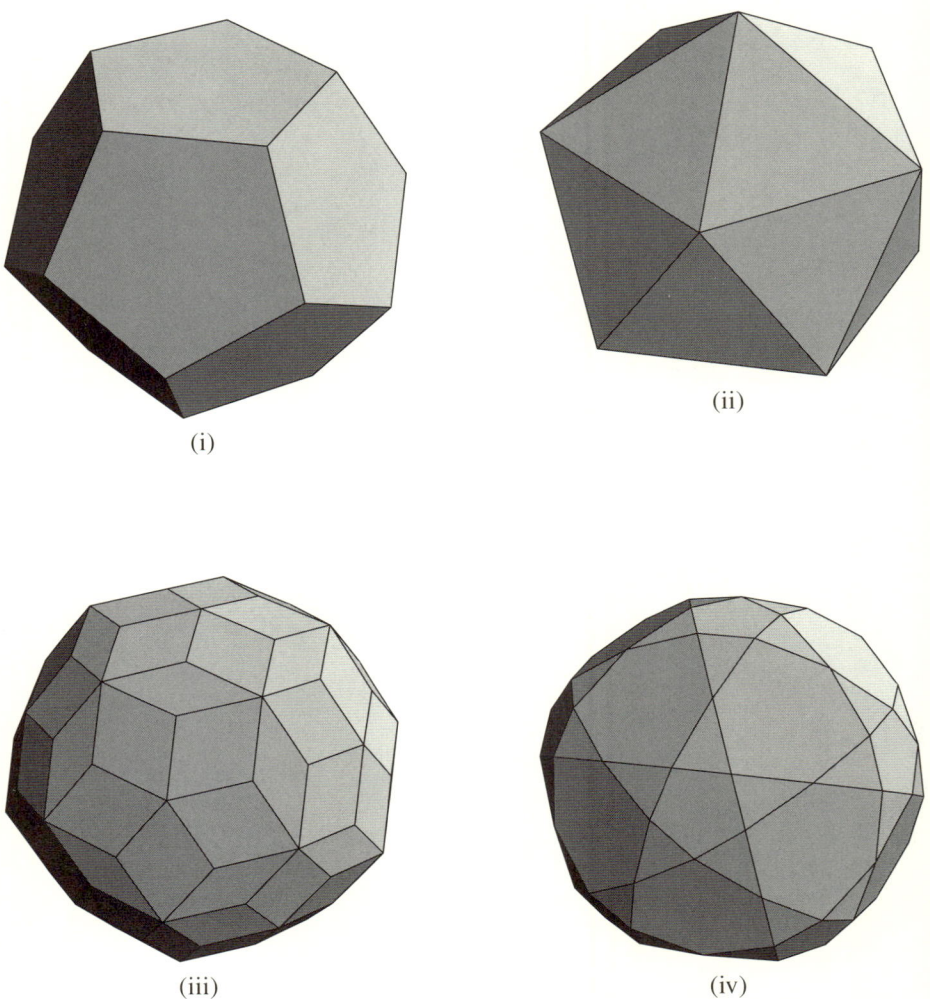

Figure 5.7. Iterations of the isoperimetric map $\mathbf{I}(.)$ when B is the dodecahedron. (i) The ball B, (ii) the dual ball B°, (iii) $\mathbf{I}(B) = \Pi(B^\circ)$, (iv) $(\mathbf{I}B)^\circ$.

5.4 The isoperimetrices that arise from Examples 5.1.4   167

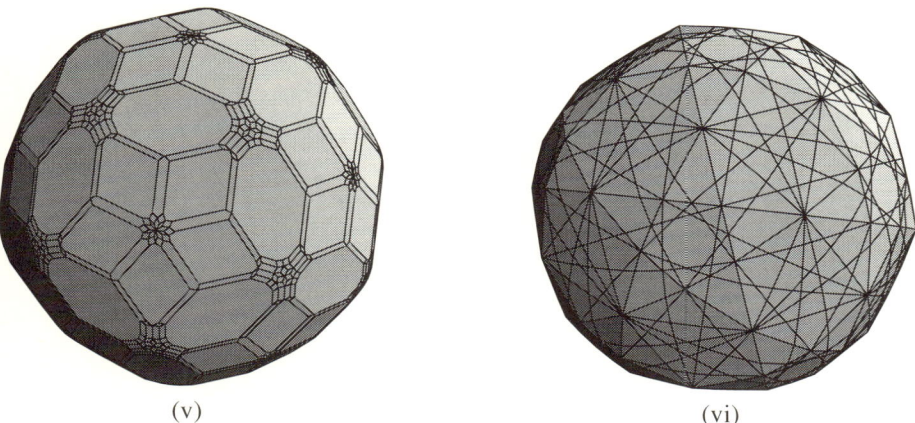

Figure 5.7 (*cont.*). Iterations of the isoperimetric map **I**(.) when B is the dodecahedron. (v) $\mathbf{I}^2 B = \Pi((\mathbf{I}B)^\circ)$, (vi) the dual $(\mathbf{I}^2 B)^\circ$.

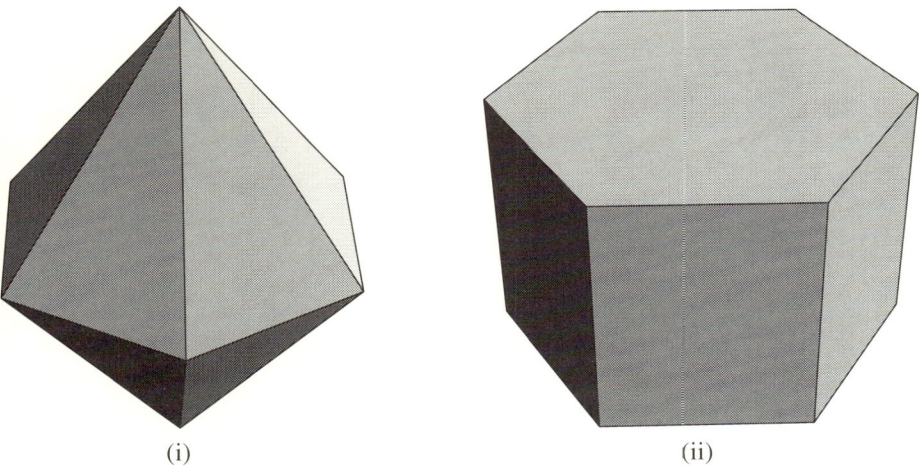

Figure 5.8. Iterations of the isoperimetric map **I**(.) when B is the double cone over a regular hexagon. (i) The ball B, (ii) the dual ball B°.

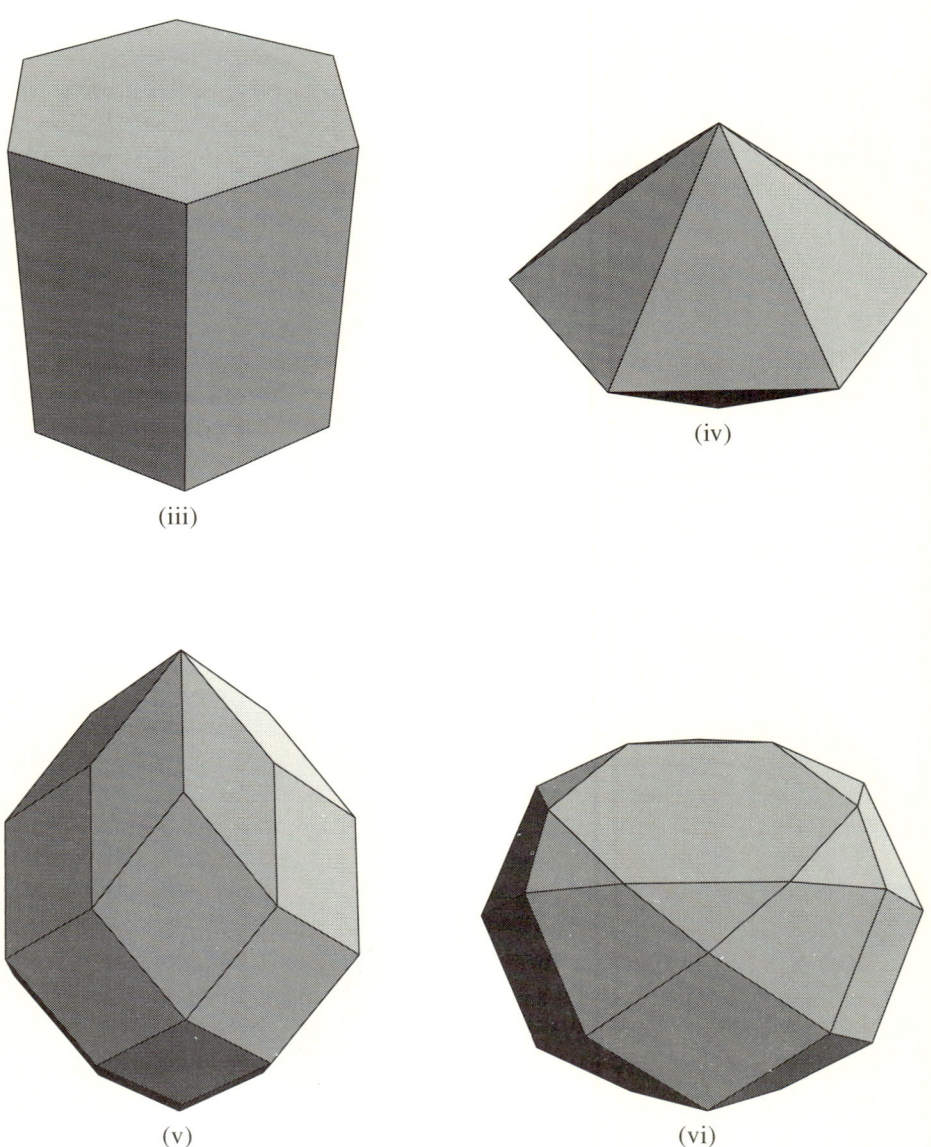

Figure 5.8 (*cont.*). Iterations of the isoperimetric map $\mathbf{I}(.)$ when B is the double cone over a regular hexagon. (iii) $\mathbf{I}B = \Pi(B^\circ)$, (iv) the dual $(\mathbf{I}B)^\circ$, (v) $\mathbf{I}^2 B = \Pi((\mathbf{I}B)^\circ)$, (vi) the dual $(\mathbf{I}^2 B)^\circ$.

5.4 The isoperimetrices that arise from Examples 5.1.4   169

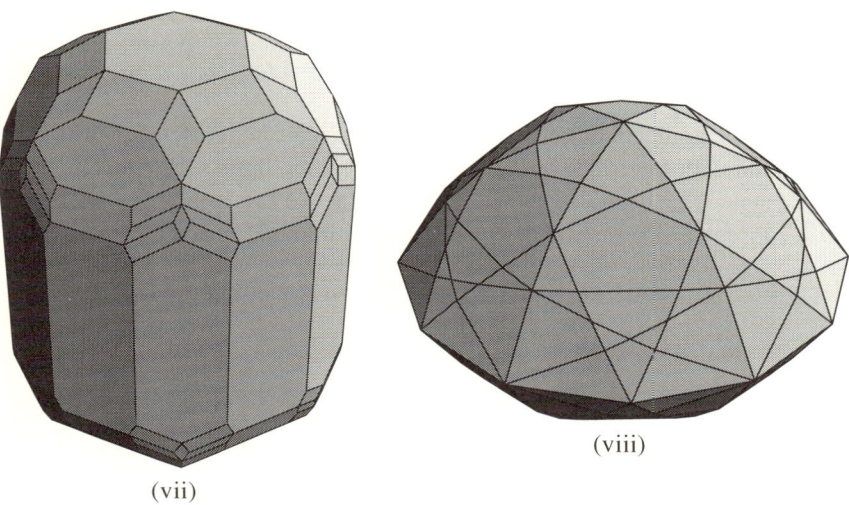

(vii)

(viii)

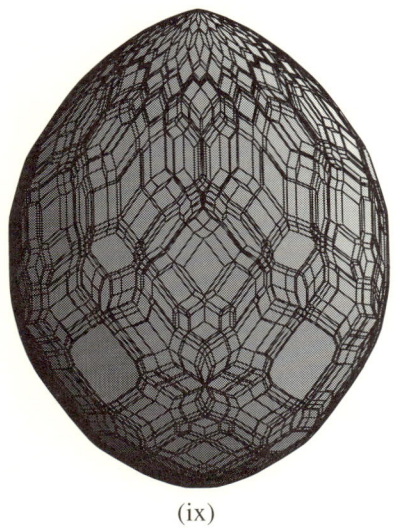

(ix)

Figure 5.8 (*cont.*). Iterations of the isoperimetric map $\mathbf{I}(.)$ when B is the double cone over a regular hexagon. (vii) $\mathbf{I}^3 B = \Pi((\mathbf{I}^2 B)^\circ)$, (viii) the dual $(\mathbf{I}^3 B)^\circ$, (ix) $\mathbf{I}^4 B = \Pi((\mathbf{I}^3 B)^\circ)$.

(a) The ball B is the unit octahedron: $\mathrm{conv}\{(\pm 1, 0, 0), (0, \pm 1, 0), (0, 0, \pm 1)\}$. Consider a linear functional (a, b, c) of Euclidean norm 1 and with $a, b, c > 0$. The vertices of the hexagon in which the plane $ax + by + cz = 0$ intersects the

boundary of the octahedron are

$$\pm(a+c)^{-1}(-c, 0, a), \ \pm(b+c)^{-1}(0, -c, b), \ \pm(a+b)^{-1}(b, -a, 0). \quad (5.28)$$

A parallelogram circumscribed to a hexagon and of minimal area has its sides in common with four sides of the hexagon. There are three such parallelograms and elementary calculations show that, for the above hexagon, these have areas $4(a+b)^{-1}$, $4(b+c)^{-1}$, $4(c+a)^{-1}$. Hence the minimal such area is

$$4\min\{(a+b)^{-1}, (b+c)^{-1}, (c+c)^{-1}\} = 4[\max\{(a+b), (b+c), (c+a)\}]^{-1}.$$

Using this in (5.11) we get

$$\sigma(a, b, c) = \max\{(a+b), (b+c), (c+a)\} \qquad (a, b, c \geq 0)$$

and, in general,

$$\sigma(a, b, c) = \max\{(|a|+|b|), (|b|+|c|), (|c|+|a|)\}.$$

This is precisely the formula for the norm given by the rhombic dodecahedron. Therefore, $\mathbf{I}(B)^\circ$ is the rhombic dodecahedron and $\mathbf{I}(B)$ is the cubo-octahedron; i.e. for this B, $\mathbf{I}(B) = (\Pi B)^\circ$.

(b) The ball B is the unit cube: conv $\{(\pm 1, \pm 1, \pm 1)\}$. Again consider a linear functional (a, b, c) of Euclidean unit norm, with $a, b, c > 0$ and suppose $\max\{a, b, c\} = c$. If $c \geq a+b$ then the corresponding hyperplane intersects the cube in a parallelogram of area $4/c$. On the other hand, if $c < a+b$ then the intersection is a hexagon with vertices

$$\pm(-1, (a-c)/b, 1), \quad \pm((b-c)/a, -1, 1), \quad \pm(1, -1, (b-a)/c). \quad (5.29)$$

The minimal circumscribing parallelogram is obtained by considering the intersection of the hyperplane with the unbounded cylinder over the square $(\pm 1, \pm 1, 0)$. The area of this parallelogram is also $4/c$.

Thus, in general, $\sigma(a, b, c) = \max\{|a|, |b|, |c|\}$ and $\mathbf{I}(B)^\circ$ is the cube which is ΠB. Therefore, $\mathbf{I}(B) = (\Pi B)^\circ$.

Calculations like the last one will work for an arbitrary cylinder. Unfortunately it is not always true that we have $\mathbf{I}(B) = (\Pi B)^\circ$. For example, rather lengthy calculations indicate that if B is a double cone then $\mathbf{I}(B) \neq (\Pi B)^\circ$.

(iv) The definition of area by means of the Löwner ellipsoid

We have no calculations for $\mathbf{I}$ corresponding to the functions σ suggested in Example 5.1.4(iv).

(v) The definition of area by means of the perimeter

Finally, we consider Example 5.1.4(v) and calculate σ_B in the case when B is an octahedron and a cube.

(a) The ball B is an octahedron. As in Example (iii), let $u = (a, b, c)$ with $0 \leq a \leq b \leq c$. Then the hyperplane $ax + by + cz = 0$ meets ∂B in the hexagon given by (5.28). Using the l_1 norm, it is a simple matter to calculate the self-circumference of this intersection. We get

$$2^{-1}\mu_1(\partial(B \cap u^\perp)) = 2(a/(a+b) + b/(a+b) + c/(a+c))$$
$$= 2 + 2c/(a+c).$$

On the other hand, the Euclidean area of this cross-section is

$$\frac{2(ab+bc+ca)}{(a+b)(b+c)(c+a)}$$

and hence

$$\sigma_B(a,b,c) = \frac{(a+b)(b+c)(a+2c)}{(ab+bc+ca)} \qquad (0 \leq a \leq b \leq c).$$

(b) The ball B is a cube. With the same notation, if $c \geq a+b$ then the hyperplane $u^\perp$ meets ∂B in a parallelogram with self-circumference 4 and Euclidean area $4/c$. If $c < a+b$ then the intersection is the hexagon given by (5.29). Using the l_∞ norm, the calculation of the self-circumference of this intersection yields

$$2^{-1}\mu_1(\partial(B \cap u^\perp)) = 3 + (c-a)/b \qquad (0 \leq a \leq b \leq c),$$

while $\lambda(\partial(B \cap u^\perp)) = [2(ab+bc+ca) - (a^2+b^2+c^2)]/abc$. Hence

$$\sigma_B(a,b,c) = \begin{cases} (3b+c-a)(a-c)/D & \text{if } a+b > c, \\ c & \text{if } a+b \leq c, \end{cases}$$

where $D = 2(ab+bc+ca) - (a^2+b^2+c^2)$. Thus we have $\sigma_B(\alpha, \alpha, 1) = 1$ if $0 \leq \alpha \leq \frac{1}{2}$, $\sigma_B(1,1,1) = 1$, $\sigma_B(\frac{3}{4}, \frac{3}{4}, 1) = \frac{15}{16}$ and $\sigma_B(\frac{2}{3}, \frac{2}{3}, 1) = \frac{14}{15}$. Thus σ_B and $\mathbf{I}_B$ are *not convex* in this case.

5.5 Further properties of I

As we have seen, in passing from Euclidean space to a Minkowski space the role of the unit ball is divided among the unit ball B, the isoperimetrix $\mathbf{I}_B$ and their duals. In this section we give three further examples of this division.

5.5.1 Transversality

The first example is the notion of transversality introduced in §4.6. We show that Proposition 4.6.1 can be extended to d dimensions. By a *cylinder* in a linear space X we mean a set of the form $U \times [0, x]$, where U is a measurable subset of a $(d-1)$-dimensional subspace $f^\perp$ and $x \notin f^\perp$. The cylinder is convex if U is convex.

Proposition 5.5.1 *If $C = U \times [0, x]$ is a cylinder in a Minkowski space (X, B) with $U \subseteq f^\perp$ then $\mu_B^d(C) = \mu_B^{d-1}(U)|\tilde{f}(x)|$, where $\tilde{f} = f/\|f\|_{\tilde{\mathbf{I}}_B^\circ}$.*

Proof. Recall from Equation (5.2) and Definition 5.3.6 the definitions of σ_B^d and $\tilde{\mathbf{I}}$. We may suppose that f is a Euclidean unit functional. Then $\lambda^d(C) = \lambda^{d-1}(U)|f(x)|$. Therefore,

$$\mu_B^d(C) = \sigma_B^d \lambda^d(C) = \sigma_B^d \mu_B^{d-1}(U)|f(x)|/\sigma(f) = \mu_B^{d-1}(U)|\tilde{f}(x)|. \qquad \blacksquare$$

We also extend Definition 4.6.2.

Definition 5.5.2 *We say that the vector y is **transversal** to the linear functional f and write $y \triangleleft f$ if $f(y) = \|y\|_{\tilde{\mathbf{I}}}\|f\|_{\tilde{\mathbf{I}}^\circ}$, i.e. iff f supports $\tilde{\mathbf{I}}$ at $y/\|y\|_{\tilde{\mathbf{I}}}$. If $y \triangleleft f$ and if $x \in f^\perp$ then we shall also say that y is **transversal** to x and write $y \triangleleft x$.*

Thus $y \triangleleft x$ with respect to B if and only if $y \dashv x$ with respect to $\tilde{\mathbf{I}}$. Therefore, the picture to illustrate this concept is the same as Figure 3.2 but with $\tilde{\mathbf{I}}$ in place of B. This relationship is another reason for requiring σ to be convex since it implies that for any $y \neq 0$ there is a linear functional f with $y \triangleleft f$. If, in Proposition 5.5.1, we have $x \triangleleft f$ then $\mu_B(C) = \mu_B(U)\|x\|_{\tilde{\mathbf{I}}}$. In other words, if "height" is transversal to the "base" then "volume = area of base × height" provided that each term is measured appropriately. We shall return to this question in Chapter 8 and modify the form of Proposition 5.5.1 after we have introduced the Minkowski sine function.

5.5.2 Divergence and the Laplacian

Next we consider the Minkowski analogue of the Gauss divergence theorem. The main ideas in this subsection are due to Alan Thompson; see [506]. In connection with Stokes' theorem, Spivak [494], p. 104, says that the proof is trivial once the concepts involved are properly defined. The same situation arises here. It is necessary to spend some time setting the stage before stating and then proving the theorem.

We make two overriding assumptions throughout this subsection. The first is that the unit ball B is smooth and strictly convex and the second is that the adjective *smooth* applied to a surface will mean that the coordinate functions defining the surface are infinitely differentiable. (Recall that differentiability and smoothness are independent of which norm is used on X.) The next step is to relate the concept of *unit normal* to the Minkowski geometry. If $\mathcal{M}$ is a smooth $(d-1)$-dimensional surface in the Minkowski space X then at each point x_0 of $\mathcal{M}$ there is a unique tangent hyperplane H given by $H := \{x : f(x) = \alpha\}$ for some $f \in X^*$ and $\alpha \in \mathbb{R}$. Because B is strictly convex there is a unique (up to sign) vector $u(x_0) \in X$ with $\|u(x_0)\|_B = 1$ and $u(x_0) \dashv f$. Throughout this subsection the phrase "the *unit normal* to $\mathcal{M}$ at x_0" will mean the vector $u(x_0)$. This is at variance with the usage in the rest of the book, where the unit normal to a surface in X is always an element of X^*.

The next requirement involves some of the language of differential forms. At each point x of X there is the tangent space $X_x = \{x\} \times X$ whose elements are pairs (x, y) with y in X and which can be thought of as a copy of X situated at x. Likewise, at each point x of X there is the cotangent space $X_x^* = \{x\} \times X^*$. A *smooth 1-form* ω is a mapping that assigns to each x in X an element ω_x in X_x^* in such a way that if a basis is chosen for X^* then the coordinate functions for ω are infinitely differentiable. Similarly, a *smooth vector field* $\mathcal{F}$ is a mapping that assigns to each x in X an element $\mathcal{F}_x$ of X_x and whose coordinate functions relative to some basis are infinitely differentiable. A 1-form ω can be evaluated on a vector field $\mathcal{F}$ to produce a real-valued function on X by the equation

$$\langle \omega, \mathcal{F} \rangle(x) := \omega_x(\mathcal{F}_x).$$

We often, as in this equation, omit the distinction between ω_x and its second coordinate, and similarly for both $\mathcal{F}_x$ and, in what follows, ϑ_x; *i.e.* we should write $\omega_x = (x, f_x) \in X_x^*$ and $\mathcal{F}_x = (x, y_x) \in X_x$ and then $\omega_x(\mathcal{F}_x) := f_x(y_x)$. If both ω and $\mathcal{F}$ are smooth then $\langle \omega, \mathcal{F} \rangle$ is a smooth function. The set of smooth 1-forms will be denoted by $\Omega^1(X)$.

More generally a smooth k-form ϑ is a mapping that assigns to each x in X an element ϑ_x that is an alternating, k-linear functional on $X_x^k := \{x\} \times X^k$; *i.e.* ϑ_x has, as argument, a k-tuple of vectors $(x_1, x_2, \ldots, x_k)$ and $\vartheta_x(x_1, x_2, \ldots, x_k)$ is linear in each variable separately and alternates in sign with each interchange of variables. Moreover, ϑ_x is infinitely differentiable with respect to x. The alternating, k-linear functionals on X_x^k form a vector space of dimension $\binom{d}{k}$. When $k = 1$ or $(d-1)$ the dimension of this space of forms is d and when $k = d$ the space is just one-dimensional. The last fact is the well-known statement that, up to scalar multiples, the only alternating d-linear functional is the determinant.

Definition 5.5.3 *The divergence is a mapping*, div, *from $\Omega^1(X)$ to $C^\infty(X)$ defined by the following equation:*

$$(\mathrm{div}\ \omega)(x) := \lim_{\epsilon \to 0} \left\{ V_\epsilon^{-1} \int_{\partial B[x,\epsilon]} \langle \omega, u \rangle(y)\, d\mu_B^{d-1} \right\}, \tag{5.30}$$

where (see Definition 1.1.2) $B[x, \epsilon] := \epsilon B + \{x\}$, $u(y)$ is chosen to be the outward unit normal at each point $y \in \partial B[x, \epsilon]$ and $V_\epsilon := \mu_B^d(B[x, \epsilon])$.

Theorem 5.5.4 *Let $(X, \|.\|)$ be a Minkowski space with unit ball B that is both smooth and strictly convex. Let $\mathcal{M} \subset X$ be the closure of a bounded open set in X whose boundary $\partial \mathcal{M}$ is a smooth $(d-1)$-dimensional surface. Then, for every smooth 1-form ω on X,*

$$\int_{\mathcal{M}} (\operatorname{div} \omega)(x)\, d\mu_B^d(x) = \int_{\partial \mathcal{M}} \langle \omega, u \rangle(y)\, d\mu_B^{d-1}(y), \tag{5.31}$$

where u is the unit outward normal to $\partial \mathcal{M}$ at x.

Proof. Choose an orientation of X and let ω be a given smooth 1-form on X. We need to construct a $(d-1)$-form ϑ on X such that

$$\int_{\partial \mathcal{M}} \vartheta = \int_{\partial \mathcal{M}} \langle \omega, u \rangle(y)\, d\mu_B^{d-1}(y).$$

For this purpose, let $x \in \mathcal{M}$ and $(v_1, v_2, \ldots, v_{d-1}) \in X_x^{d-1}$. Let P_{d-1} be the (possibly degenerate) parallelotope spanned by $v_1, v_2, \ldots, v_{d-1}$. If P_{d-1} has dimension $(d-1)$, then there is a linear functional f such that $f^\perp = \operatorname{span}(v_1, v_2, \ldots, v_{d-1})$. In this case, let w be the unique vector for which $\|w\| = 1$, $w \dashv f$ and the orientation of the set $(v_1, v_2, \ldots, v_{d-1}, w)$ is the same as that of X. Now define ϑ by

$$\vartheta_x(v_1, v_2, \ldots, v_{d-1}) = \begin{cases} \langle \omega, w \rangle(x) \mu_B^{d-1}(P_{d-1}) & \text{if } \dim P_{d-1} = (d-1), \\ 0 & \text{otherwise.} \end{cases}$$

It follows that ϑ is a smooth $d-1$ form on X such that

$$\int_{\partial \mathcal{M}} \vartheta = \int_{\partial \mathcal{M}} \langle \omega, u \rangle(y)\, d\mu_B^{d-1}(y).$$

Next, Stokes' theorem for general differential forms implies that

$$\int_{\partial \mathcal{M}} \vartheta = \int_{\mathcal{M}} d\vartheta.$$

As before, let P_d denote the parallelotope spanned by an ordered set of vectors $(v_1, v_2, \ldots, v_d)$. Then both $d\vartheta$ and $\pm \mu_B^d(P_d)$ (where the sign depends on the orientation of the ordered set of vectors) are d-forms on X. Since the space of d-forms at a point is one-dimensional, at each point $x \in X$ there is a scalar $\varphi(x)$ such that $d\vartheta = \varphi(x) \mu_B^d(P_d)$ and φ is a smooth function on X. Thus

$$\int_{\mathcal{M}} d\vartheta = \int_{\mathcal{M}} \varphi(x)\, d\mu_B^d(x).$$

Combining the last three equations gives

$$\int_{\partial \mathcal{M}} \langle \omega, u \rangle(x)\, d\mu_B^{d-1}(x) = \int_{\mathcal{M}} \varphi(x)\, d\mu_B^d(x). \tag{5.32}$$

5.5 Further properties of **I**

Observe next that ϑ, $d\vartheta$ and φ are all defined on the whole of X and are independent of the set $\mathcal{M}$. Therefore, in Equation (5.32) take $\mathcal{M}$ to be $B[\epsilon, x]$. Then

$$\int_{\partial B[\epsilon,x]} \langle \omega, u \rangle(y) \, d\mu_B^{d-1}(y) = \int_{B[\epsilon,x]} \varphi(y) \, d\mu_B^d(y).$$

Dividing both sides by $V_\epsilon = \mu_B^d(B[\epsilon, x])$ and taking limits as $\epsilon \to 0$ yield

$$\text{div } \omega(x) = \varphi(x).$$

Finally, substituting this expression for φ in Equation (5.32) produces the required result. ∎

The next goal is to define a Laplacian on X and then to investigate its properties. As on any differentiable manifold, an intrinsic notion of differential exists in a Minkowski space. If $g : X \mapsto \mathbb{R}$ is a differentiable function then the *differential* dg_x is a linear map from X to $\mathbb{R}$ (*i.e.* an element of X^*) and is defined in the usual way by the requirement that

$$\lim_{h \to 0} \frac{|g(x+h) - g(x) - dg_x(h)|}{\|h\|} = 0.$$

Since all norms on X are equivalent, both the definition of $\lim_{h \to 0}$ and whether or not the above limit is zero are independent of which norm is used. Therefore, dg_x is independent of the norm. At each point x the functional dg_x is an element of X_x^* and if $g \in C^\infty(X)$ this assignment is smooth. Therefore, $dg \in \Omega^1(X)$, which implies that the Laplacian can be defined as Minkowski divergence composed with the differential. Since the divergence depends on the norm so does the Laplacian.

Definition 5.5.5 *Let (X, B) be a Minkowski space. Let $g(x) \in C^\infty(X)$. Define $(\Delta_B g)(x) \in C^\infty(X)$, the **Minkowski Laplacian** of g, by*

$$(\Delta_B g)(x) := [\text{div}_B(dg)](x).$$

Proposition 5.5.6 *The operator Δ acting on $C^\infty(X)$ is a differential operator which is homogeneous of order 2 and translation invariant.*

Proof. Since both div_B and d are linear, it is clear that Δ_B is a linear operator from $C^\infty(X) \to C^\infty(X)$. To prove that Δ_B is a differential operator we use a theorem of Peetre [405] (see also Helgason [253]) which states that a linear operator from $C^\infty(X) \to C^\infty(X)$ that shrinks supports is a differential operator. If $x \notin \text{supp } g$ then there is an open neighbourhood U_x of x such that $U_x \cap \text{supp } g = \emptyset$. Thus dg vanishes on U_x and, therefore, so does $\Delta_B = \text{div } dg$. Hence $\text{supp } \Delta_B g \subseteq \text{supp } g$ and Δ_B is a differential operator.

For $g(x) \in C^\infty(X)$ and for $\xi \in \mathbb{R}^+$, let $g_\xi(x) := g(\xi x)$. To establish the homogeneity we must prove that $\Delta_B g_\xi(x) = \xi^2 \Delta_B g(x)$. For this, we have

$$\Delta_B g_\xi(x) = \lim_{\epsilon \to 0} V_\epsilon^{-1} \int_{\partial B[x,\epsilon]} \langle dg_\xi, u \rangle \, d\mu_B^{d-1}$$

$$= \lim_{\epsilon \to 0} V_\epsilon^{-1} \int_{\partial B[x,\epsilon]} \langle \xi dg, u \rangle \, d\mu_B^{d-1}$$

$$= \lim_{\epsilon \to 0} \xi^d V_{\xi\epsilon}^{-1} \xi^{1-d} \int_{\partial B[x,\xi\epsilon]} \langle \xi dg, u \rangle \, d\mu_B^{d-1}$$

$$= \xi^2 \lim_{\xi\epsilon \to 0} V_{\xi\epsilon}^{-1} \int_{\partial B[x,\xi\epsilon]} \langle dg, u \rangle \, d\mu_B^{d-1}$$

$$= \xi^2 \Delta_B g(x).$$

The first and last equations use the definitions of Δ_B and div; the second equation uses the fact that $dg_\xi = \xi dg$; the third equation uses the homogeneity of volume and surface area; and the fourth equation is just a change of variables in the limit.

Similarly, for $y \in X$, let $g_y(x) := g(x+y)$. Translation invariance means that

$$\Delta_B(g_y(x)) = (\Delta_B g)(x+y),$$

which follows immediately from the translation invariance of both d and div_B. ∎

The final step in this discussion of differential operators is to show that Δ_B is elliptic and hence can be used to introduce an intrinsic ellipsoid into the Minkowski space X.

Choose any basis for X and let $(\zeta_1, \zeta_2, \ldots, \zeta_d)$ be the coordinate functions associated with that basis.

Proposition 5.5.6 implies that $(\Delta_B g)(x)$ can be expressed in the form

$$(\Delta_B g)(x) = \sum_{1 \le i,j \le d} \tau_{ij} \frac{\partial^2 g}{\partial \zeta_i \partial \zeta_j}(x), \tag{5.33}$$

where $\tau_{ij} \in \mathbb{R}$. Furthermore, without loss of generality, we may assume that the matrix (τ_{ij}) is symmetric.

Theorem 5.5.7 *The matrix (τ_{ij}) is positive definite.*

Proof. Since the matrix (τ_{ij}) is real and symmetric, there is a basis of eigenvectors of (τ_{ij}) for which the corresponding matrix is diagonal. Let $\eta_1, \eta_2, \ldots, \eta_d$ denote the diagonal entries. Then $\eta_j = 2^{-1} \Delta_B(\zeta_j^2)$, where $\zeta_1, \zeta_2, \ldots, \zeta_d$ now denote the coordinate functions associated with the new basis. We must show that $\eta_j > 0$ for $1 \le j \le d$.

Let $C_\epsilon(0)$ be the cube defined by

$$C_\epsilon(0) := \{x : -\epsilon/2 \le \zeta_i \le \epsilon/2, \ 1 \le i \le d\}.$$

5.5 Further properties of I

Using both the divergence theorem and the fact that $d(\zeta_j^2) = 2\zeta_j d\zeta_j$, we have

$$\eta_j = 2^{-1}\Delta_B(\zeta_j^2)\big|_{x=0}$$

$$= \lim_{\epsilon \to 0}\left[\int_{\partial C_\epsilon(0)} \langle \zeta_j \, d\zeta_j, u(x)\rangle \, d\mu_B^{d-1}\right] \bigg/ \mu_B^d(C_\epsilon(0)).$$

Let F_k^+ denote the facet of $C_\epsilon(0)$ containing the point ϵe_k where e_k is the kth eigenvector of (τ_{ij}) and let F_k^- denote the opposite facet. Thus

$$\eta_j = \lim_{\epsilon \to 0}[\mu_B^d(C_\epsilon(0))]^{-1} \sum_{k=1}^{d} \left[\int_{F_k^+} \langle \zeta_j d\zeta_j, u(x)\rangle \, d\mu_B^{d-1}\right.$$

$$\left.+ \int_{F_k^-} \langle \zeta_j d\zeta_j, u(x)\rangle \, d\mu_B^{d-1}\right].$$

When $j \neq k$, the integral over F_k^+ is the negative of the integral over F_k^-, but when $j = k$, the two integrals are the same. Hence

$$\eta_j = 2\lim_{\epsilon \to 0}[\mu_B^d(C_\epsilon(0))]^{-1} \int_{F_j^+} \langle \zeta_j d\zeta_j, u(x)\rangle \, d\mu_B^{d-1}.$$

Now $[\mu_B^d(C_\epsilon(0))]^{-1} = C\epsilon^d$, where $C := \mu_B^d(C_1(0)) > 0$. Thus

$$\eta_j = \lim_{\epsilon \to 0} 2C^{-1}\epsilon^{-d}\mu_B^{d-1}(F_j^+)\epsilon u(d\zeta_j)$$

$$= 2C^{-1}\alpha_j u(d\zeta_j),$$

where $\alpha_j := \epsilon^{1-d}\mu_B^{d-1}(F_j^+)$ is the Minkowski area of the jth facet of $C_1(0)$, which is independent of ϵ and is strictly positive. In the first equation above we used the fact that the value of ζ_j on F_j^+ is ϵ.

Finally, $u(d\zeta_j)$ is the absolute value of the jth component of the normal to the jth coordinate plane, which is also strictly positive (otherwise B has empty interior). Hence $\eta_j > 0$ as needed. ∎

Corollary 5.5.8 *The operator Δ_B is elliptic.*

With respect to any basis for X, we have shown that Δ_B is of the form

$$(\Delta_B g)(x) = \sum_{1 \leq i,j \leq d} \tau_{ij} \frac{\partial^2 g}{\partial \zeta_i \partial \zeta_j}(x).$$

Definition 5.5.9 *The symbol of Δ_B is the following polynomial in $\phi_1, \ldots, \phi_d$,*

$$\varsigma(\Delta_B) := \sum_{1 \leq i,j \leq d} \tau_{ij} \phi_i \phi_j.$$

Since (τ_{ij}) is positive definite, $\varsigma(\Delta_B) > 0$ for all vectors $(\phi_1, \phi_2, \ldots, \phi_d) \in X^* \setminus \{0\}$.

It is also possible to define $\varsigma(\Delta_B)$ in a coordinate-free way. For each linear functional f in X^*, let g_f be the function on X defined by

$$g_f(x) := \exp(f(x)).$$

Then we have

$$\varsigma(\Delta_B)(f) = (\Delta_B g_f)(x)/g_f(x).$$

From this formula it is clear that $\varsigma(\Delta_B)$ is a function on X^*. To see that this expression agrees with the earlier one, choose a basis for X. If, with respect to this basis and its dual basis, $x = (\zeta_1, \zeta_2, \ldots, \zeta_n)$ and $f = (\phi_1, \phi_2, \ldots, \phi_n)$ then $f(x) = \sum_i \phi_i \zeta_i$ and $g_f(x) = \exp(\sum_i \phi_i \zeta_i)$. A simple calculation now shows that the two definitions of $\varsigma(\Delta_B)$ coincide.

Definition 5.5.10 *The **associated ellipsoid** $E^\circ \subseteq X^*$ is defined by*

$$E^\circ := \{(\phi_1, \ldots, \phi_n) \in X^* : \varsigma(\Delta_B) \leq 1\}.$$

The associated ellipsoid $E \subseteq X$ is defined as $E^{\circ\circ}$.

The second definition of $\varsigma(\Delta_B)$ (or the fact that the definition of the Laplacian is independent of basis) implies that E and E° are intrinsic in the sense of being independent of basis. Therefore, if B is transformed by an invertible linear transformation T then the ellipsoid associated with $T(B)$ is $T(E)$. Evidently, the ellipsoid E induces a Euclidean norm on X.

It would be interesting to know if this ellipsoid has a connection with any of the ellipsoids considered by Milman and Pajor [382] or by Lewis [321]. It would also be interesting to know whether, if the ball B is replaced by the isoperimetrix $\tilde{I}$, the associated ellipsoid E changes. Another way to ask this last question is as follows. The concept of *unit normal* used in the definition of divergence is based on *normality* (Definition 3.2.2). One might use *transversality* (Definition 5.5.2) instead. If this change is made one will still obtain a Laplacian as an elliptic differential operator and hence an ellipsoid. How is this ellipsoid related to the original one?

5.5.3 Minimal positions

The final item in this section is the Minkowski analogue of a theorem of Petty [414]. Recall from Definition 2.5.18 that the image of K under a volume-preserving affine transformation is called a position of K.

5.5 Further properties of **I**

Definition 5.5.11 *The body K is said to be in a **minimal position** if the (Euclidean) surface area of K is minimal among all positions of K.*

Theorem 5.5.12 (Petty) *A convex body K in $\mathbb{R}^d$ is in minimal position if and only if*

$$\int_{\partial K} \hat{u}^t \hat{u} \, d\lambda(\hat{u}) = d^{-1}\lambda(\partial K)\mathbf{1}.$$

Here $\hat{u}$ is a Euclidean unit vector normal to ∂K (and hence because it is in X^* is a row vector) at a point where the surface element is $d\lambda(\hat{u})$ and $\hat{u}^t \hat{u}$ denotes the matrix product of $\hat{u}^t$ with $\hat{u}$ (yielding a rank 1 symmetric matrix). A fuller explanation of the notation is given below.

The Minkowski analogue of this theorem is due to Clack [113]. Although it is not a full generalization in that it gives only a necessary condition, we shall give the proof of only the Minkowski version. It is closely patterned on Petty's original proof of Theorem 5.5.12.

The notation for this theorem requires rather extensive explanation. First, the integrals follow the notation established prior to Theorem 5.2.2. Next, $\vec{\nabla}\sigma$ is the gradient of the support function σ of $\mathbf{I}_B$. If $\mathbf{I}_B$ is strictly convex then $\vec{\nabla}\sigma$ is single valued and, for $f \in X^*$, $\vec{\nabla}\sigma(f)$ is that point of $\mathbf{I}_B$ (*i.e.* a point in X) at which f is a supporting function (see *e.g.*, [51], p. 28). Given a vector $y \in X$ and $f \in X^*$ the *tensor product* $y \otimes f$ is the rank 1 linear operator on X defined by

$$y \otimes f(x) := f(x)y.$$

However, in the following proof it is convenient to use coordinates relative to some basis. If y has coordinates $(\eta_1, \eta_2, \ldots, \eta_d)^t$ and f has coordinates $(\phi_1, \phi_2, \ldots, \phi_d)$ then the matrix for $y \otimes f$ is the rank 1, $d \times d$ matrix

$$(\eta_1, \eta_2, \ldots, \eta_d)^t (\phi_1, \phi_2, \ldots, \phi_d).$$

To avoid extensive displays we shall allow the notation for a vector to also stand for its coordinates and the notation for a linear transformation to also stand for its matrix (all relative to some fixed basis).

Theorem 5.5.13 (Clack) *Let (X, B) be a Minkowski space such that $\mathbf{I}_B$ is strictly convex. If a convex body K is in (Minkowski) minimal position then*

$$\int_{\partial K} \vec{\nabla}\sigma(\tilde{f}) \otimes \tilde{f} \, d\mu_B(\tilde{f}) = d^{-1}\mu_B(\partial K)\mathbf{1}. \tag{5.34}$$

Here $\tilde{f}$ represents a normal to K that is of unit length with respect to $\mathbf{I}^\circ$.

Proof. Since Minkowski volume is preserved if and only if Euclidean volume is preserved, we wish to minimize $\mu_B(\partial T(K))$ under all linear maps T of determinant

1. By Theorem 5.3.2, we have

$$\mu_B(\partial T(K)) = \mu_{T^{-1}(B)}(\partial K)$$

and so we may, equivalently, minimize the area of K under all Minkowski measures deriving from balls of the form $T(B)$. But from Theorem 5.3.3 with $\det T = 1$, we have

$$\mu_{T(B)}(\partial K) = \int_{\partial K} \sigma(T^*\hat{f})\, d\lambda(\hat{f}).$$

To find necessary conditions for a minimum of this latter expression we consider the partial derivatives of

$$\int_{\partial K} \sigma(T^*\hat{f})\, d\lambda(\hat{f}) + \alpha(1 - \det T)$$

with respect to the d^2 variables τ_{ij}, the entries in the matrix for T, and set all the partial derivatives equal to 0. Thus

$$\frac{\partial}{\partial \tau_{ij}} \int_{\partial K} \sigma(T^*\hat{f})\, d\lambda(\hat{f}) = \alpha \frac{\partial (\det T)}{\partial \tau_{ij}}.$$

The coefficient of τ_{ij} in $\det T$ is its cofactor, which, since $\det T = 1$, is the jith entry of T^{-1}. Also

$$\partial(\sigma(T^*\hat{f})/\partial \tau_{ij} = \sigma_{B,j}(T^*\hat{f})\hat{f}_i,$$

where $\sigma_{B,j}$ is the partial derivative of σ with respect to the jth coordinate. Therefore, we obtain

$$\int_{\partial K} \sigma_{B,j}(T^*\hat{f})\hat{f}_i\, d\lambda(\hat{f}) = \alpha[T^{-1}]_{ji}.$$

Combining the above d^2 equations into a single matrix equation yields

$$\int_{\partial K} \vec{\nabla}\sigma(T^*\hat{f}) \otimes \hat{f}\, d\lambda(\hat{f}) = \alpha T^{-1}. \tag{5.35}$$

Thus if T is a linear transformation that places K in a minimal position then it must satisfy (5.35). If K is already in a minimal position then **1** is such a matrix and hence

$$\int_{\partial K} \vec{\nabla}\sigma(\hat{f}) \otimes \hat{f}\, d\lambda(\hat{f}) = \alpha \mathbf{1}. \tag{5.36}$$

To determine α we evaluate the trace of both sides of this expression:

$$\alpha d = \int_{\partial K} \hat{f} \vec{\nabla}\sigma(\hat{f})\, d\lambda(\hat{f}) = \int_{\partial K} \sigma(\hat{f})\, d\lambda(\hat{f}) = \mu_B(\partial K). \tag{5.37}$$

For the central equality we have used Euler's formula for the positive homogeneous function σ.

5.5 Further properties of **I**

Finally we normalize the various factors appropriately:

$$\int_{\partial K} \vec{\nabla}\sigma(T^*\hat{f}) \otimes \hat{f} \, d\lambda(\hat{f}) = \int_{\partial K} \vec{\nabla}\sigma(T^*\hat{f}/\sigma(\hat{f})) \otimes \hat{f}\sigma(\hat{f})^{-1}\sigma(\hat{f}) \, d\lambda(\hat{f})$$
$$= \int_{\partial K} \vec{\nabla}\sigma(T^*\tilde{f}) \otimes \tilde{f} \, d\mu(\tilde{f}). \tag{5.38}$$

(Recall that $\vec{\nabla}\sigma$ is homogeneous of degree 0.)

Combining (5.36), (5.37) and (5.38) we get

$$\int_{\partial K} \vec{\nabla}\sigma(\tilde{f}) \otimes \tilde{f} \, d\mu(\tilde{f}) = d^{-1}\mu_B(\partial K)\mathbf{1},$$

where $\tilde{f}$ is the unit vector with respect to the ball $\mathbf{I}^\circ$ in X^*. If we prefer to use $\tilde{\mathbf{I}}^\circ$ then we need an extra factor of σ_B on the right. ∎

This result raises a number of questions. First, do such minimal positions exist? Can there be more than one? And is (5.34) sufficient? Again using Petty's proof, Clack has shown that a convex body K always has a minimal position. However, he has also shown that it may be not unique and not all positions satisfying (5.34) are minimal. These latter questions of uniqueness and sufficiency appear to be complicated. We conclude with the existence of minimal positions.

Theorem 5.5.14 *A convex body K in a Minkowski space (X, B) has a minimal position.*

Proof. First, as we remarked at the beginning of the proof of Theorem 5.5.13,

$$\mu_B(TK) = \mu_{T^{-1}(B)}(\partial K) = dV(K[d-1], \mathbf{I}_{T^{-1}(B)}).$$

Thus we need to show the existence of a position $\mathbf{I}_B^-$ of $\mathbf{I}_B$ such that $V(K[d-1], \mathbf{I}_B^-) = V^-$, where $V^- := \inf\{V(K[d-1], T\mathbf{I}_B) : \det T = 1\}$. Let $\mathbf{I}_j$ be a sequence of positions of $\mathbf{I}_B$ such that $V(K[d-1], \mathbf{I}_j) \to V^-$. Choose $\beta > 0$ so that $\beta E \subseteq K$ and let L_j be a line segment of maximal length $2\alpha_j$ contained in $\mathbf{I}_j$. Since $V(K[d-1], \mathbf{I}_j)$ converges, it is a bounded sequence. Let c be an upper bound. Then, by standard properties of mixed volumes, we have

$$c \geq V(K[d-1], \mathbf{I}_j) \geq V(\beta E[d-1], L_j) = 2\alpha_j \beta^{d-1}\epsilon_{d-1}/d.$$

This shows that the sequence $\{\alpha_j\}$ is bounded above by c' (say) and hence each $\mathbf{I}_j \subseteq c'E$. The Blaschke selection theorem guarantees a subsequence of $\{\mathbf{I}_j\}$ that converges to a body $\mathbf{I}^-$. Corollary 2.5.19 asserts that this limit is a position of $\mathbf{I}_B$ which must be a minimizing position. ∎

5.6 Notes

The idea of considering surface area relative to some general convex set B via mixed volumes was introduced and first investigated by Minkowski [390], §27. The idea was extensively developed by Busemann in a series of papers on Minkowski geometry; see especially [65], [66], [68] and [71]. In [65] (which deals with situations more general than Minkowski geometry) Busemann gives a list of requirements for "intrinsic" area which, because of the more general context, differ somewhat from ours. Essential to both, however, are (a) the "intrinsicness" (Requirements 5.1.1(a)) that area in a hyperplane should depend only on the metric in that hyperplane and not on the space in which it is embedded; and (b) that it should coincide with the usual notion if the metric is Euclidean. Barthel [24] also gives a list of similar axioms for Minkowski measures of arbitrary dimension and develops some of the properties of any measure which satisfies his axioms. In [22] Barthel discusses the Busemann definition of measure in order to create a formalism based on the volumes of parallelotopes. A recent paper that also uses the Busemann definition is that of Boju and Funar [46].

Two monotonicity requirements might be imposed in addition to those listed in Requirements 5.1.1. The first was introduced by Kolmogoroff [297] and is stressed by Busemann in [65]. It is that if a space has two metrics one of which dominates the other then the smaller metric should induce smaller areas. In Minkowski geometry this means that if $B_1 \subseteq B_2$ (equivalently $\|.\|_1 \geq \|.\|_2$) then $\sigma(B_1) \geq \sigma(B_2)$. Since the last inequality is equivalent to $\mathbf{I}(B_1) \supseteq \mathbf{I}(B_2)$ the requirement asserts that $\mathbf{I}$ should be inclusion reversing. It will be shown in Chapters 6 and 7 that the Holmes–Thompson and Busemann definitions both have this property. The other possible monotonicity condition asserts that if K_1 and K_2 are two convex bodies in a Minkowski space (X, B) with $K_1 \subseteq K_2$ then $\mu_B(\partial K_1) \leq \mu_B(\partial K_2)$. Minkowski [390] emphasized this requirement and showed that it is equivalent to being able to represent μ_B as a mixed volume; in other words it is equivalent to the convexity of σ.

In addition to Requirement 5.1.1(b) one might also ask about the continuity of $\mu_B(\partial K)$ with respect to K. If σ is convex so that μ_B is represented as a mixed volume, then the continuity is automatic (see, *e.g.*, Schneider [479], p. 275, for the continuity of mixed volumes). The continuity in K can also be seen as an instance of a more general theorem about valuations. A *valuation* is a mapping υ from $\mathcal{C}_b$ (or some larger class of sets) into $\mathbb{R}^+$ (or some more general abelian semi-group) such that

$$\upsilon(K \cap L) + \upsilon(K \cup L) = \upsilon(K) + \upsilon(L)$$

whenever $\upsilon(K \cup L)$ is in $\mathcal{C}_b$. It is an important theorem of McMullen [367] that every monotonic valuation is continuous. For more information about valuations see either the survey article by McMullen and Schneider [371] or the more recent one by McMullen [370].

Busemann (and his co-investigators – Ewald, Shephard and Straus) have written a great deal about what it means to require that σ^k be convex for dimensions k other than 1 and $d-1$. The investigation of this question began in [68] but is the central theme of [76]–[79], [81]–[84], [86] and [87]. The article [76] is less technical and is a good introduction to Busemann's ideas on the subject. We shall comment on these ideas at greater length in the next chapter after the necessary concepts and notation have been introduced.

One of the open problems in the subject is that alluded to in §5.1: to what extent do Requirements 5.1.1 (or these requirements plus additional conditions) characterize the area function? Busemann [65] shows that a stronger version of Kolmogoroff's monotonicity condition is sufficient to characterize intrinsic area. The extensive calculations that led to [262] convinced us that $\mu_B(\partial B) = \mu_{B^\circ}(\partial B^\circ)$ is sufficient to characterize μ_B, at least for $d=3$. In Chapter 7 we shall see that Example 5.1.4(i) is the only one that coincides with Hausdorff measure (see Rogers [445]), while results in Chapter 6 (see also Schneider and Wieacker [483]) show that Example 5.1.4(ii) is the only one that yields certain integro-geometric formulas. There are uniqueness results for valuations (see [370] or [371]) but they are not relevant here, where we are concerned with a much finer classification.

The results given in §5.2 are almost entirely due to Busemann. He first considered the isoperimetric problem in the Minkowski plane [66] and then for arbitrary dimension [68]. He built on the isoperimetric results to introduce other fundamental geometric concepts for Minkowski geometry in [71]. His motivation (see Busemann [63, 64]) was the eventual study of Finsler spaces from a more synthetic point of view than was customary – to try and do something about "the impenetrable forest whose entire vegetation consists of tensors". For this "the study of Minkowskian geometry ought to be the first and main step, the passage from there to general Finsler spaces the second and simpler step". Both of these quotations are from his highly readable article [70].

In both [66] and [68] Busemann considers extensions of the theory to spaces in which the ball B need not be convex. For this he needs generalizations of the Brunn–Minkowski theory due to Lusternik [332] (see also Dinghas and Schmidt [129]). In [68] Busemann also obtains the solution of the isoperimetric problem when the function σ is not necessarily convex. Therefore, as he observes in [71], the convexity of σ is needed not to construct the isoperimetrix but rather for a variety of other desirable geometric properties of area.

The ideas that go into the construction of the isoperimetrix were developed well before Busemann in the theory of crystallography. An account of that development is given, *e.g.*, by Taylor [504]. In crystallography, as presented by Taylor, the function $\sigma(f)$ is the "surface tension function", which is defined on the boundary of the Euclidean ball in the dual space and represents the energy density at the point of a surface where the outward unit normal is f. The integral $\int \sigma(f_x)\, d\lambda(x)$ of Equation (5.13) represents the total energy of the surface over which the integral is evaluated. For a physically stable surface, this integral is to be minimized. Therefore, when σ is convex the construction of $\mathbf{I}$ coincides with the Wulff construction [538] of the "crystal" associated with the integrand σ (see Taylor [504], p. 573). The integrands in crystallography need not be convex but the Wulff construction is an intersection of half-spaces and, as in Busemann [68], is the polar of the body $\{f \in X^* : \sigma(f) \leq 1\}$. Consequently, even if σ is not convex there is a convex function which yields the same crystal as the original function and, among all functions which have the same crystal, the convex one is the smallest. For many purposes ([504], Theorem 3.3) the original function can be replaced by the convex one.

Wulff did not prove that his construction yields a minimal surface. Taylor [504] refers to Liebmann [325], von Laue [523], Dinghas [128] and Herring [258] (in chronological order) for "successively more complete proofs". All but the last of these predate Busemann's proof. A recent proof that the construction yields a minimal surface under very general conditions has been given by Brothers and Morgan [59]. Their proof is a generalization of Gromov's [207] proof of the Brunn–Minkowski inequality in that it is based on the divergence theorem.

Taylor's work ([503]; see also Morgan [393, 394] and the references given there) deals largely with minimal surfaces with prescribed boundary, a topic which is well beyond the scope of this book. In that work it is frequently assumed that the Wulff crystal is a zonohedron. This assumption appears to be partly because many crystals are, in fact, zonohedra and partly because the calculations are simpler in that case.

The behaviour of σ and μ_B under invertible linear maps is implicit in the work of Busemann but Clack [112] made the situation clear. It was R. D. Holmes who pointed out that $\mathbf{I}_B$ behaves like a representative of the dual ball. He was thinking of the examples in [262] but the calculations in Examples 5.3.5 indicate that Requirement 5.1.1(a) forces a relationship of this sort. The normalization $\tilde{\mathbf{I}}$ to satisfy (5.19) is due to Busemann [68, 71].

There will be considerably more information about the isoperimetrices arising from Examples 5.1.4(i) and (ii) in the next two chapters. For his definition, Busemann provided no information about what $\mathbf{I}(B)$ looks like for particular values of B. Computer-drawn pictures of intersection bodies were made by P. Fred Pickel (personal communication via Erwin Lutwak). R. Gardner has beautiful computer-drawn pictures in [171] and in his recent book [172]. The illustrations of projection bodies were made by Brian Ingalls, an honours student at Dalhousie University, to whom I am much indebted. Brian Ingalls's work was based on earlier programs made by Barbara Taylor. One of the advantages of the fact that a projection body of a polytope is a zonotope is that projection bodies are easier to visualize and construct than are intersection bodies. The projection body of a double cone over a regular n-gon was considered by Chilton and Coxeter [110], and they also showed that as n increases the projection body converges to the solid of revolution of a cosine curve which is, in our terminology, $\mathbf{I}(cylinder)$.

5.6 Notes

The idea of transversality was introduced by Busemann in [71] and explored in much greater depth (and in arbitrary intermediate dimensions k, $1 \leq k \leq d-1$) in [87] (joint work with Straus). Divergence, the divergence theorem and the Laplacian were developed in [506]. Clack's result generalizing Petty's theorem [414] appeared first in [112] and then in [113]. Green [199] had earlier considered the problem of minimizing the length of a convex curve under area-preserving affine maps and gave a criterion for that minimum in terms of the Fourier coefficients of the support function. He also conjectured that the maximum (over all convex curves) of the above minimum is attained by an equilateral triangle. This conjecture was proved by Gustin [233]. This last result was generalized to d dimensions by Ball [16], who showed that a convex body has an affine image for which the isoperimetric ratio is no bigger than that of a regular simplex. Other extremum problems for convex bodies in the plane under affine transformations were considered by Behrend [28, 29] and by John [270]. Ader [1] proved the three-dimensional analogues of some of the results of Behrend and John. A quite different type of minimal problem was considered by Petty in his important paper [419]. He considered the mixed volumes $V(K[d-1], B)$ and found the minimum over convex bodies B (normalized to avoid the minimum being 0). The general setting of this paper provides a broad context for many of the ideas in this chapter.

6
Special properties of the Holmes–Thompson definition

Chapters 6 and 7 present some of the more pertinent facts relating to the first two definitions of Minkowski area from Example 5.1.4. In some cases (*e.g.*, the convexity of both definitions) the property is common to both but the proofs are different. In other cases the properties differ but exhibit a certain dual character. For this reason the material in the two chapters is presented, as far as possible, in parallel sections following an idea of Lutwak that he used effectively in [337].

The definition of area that is explored in this chapter was given in [262] and elaborated in [273] and [509]. As shown in Equation (5.25), the solution to the isoperimetric problem that this definition yields is a scalar multiple of the projection body of $B°$. Sometimes the exact multiple is irrelevant. At other times we shall need to be more precise. There is a large literature on projection bodies from which some of the main theorems come. We refer particularly to Bolker [47], Schneider and Weil [481], Bourgain and Lindenstrauss [54], Lutwak [334, 337] and Petty [418]. An account of all the material on projection bodies that will be used in this chapter can be found in Gardner [172].

6.1 The convexity of the area function σ

We begin by recalling the definition of Example 5.1.4(ii). For this definition of Minkowski measures $\gamma(C)$ is the normalized volume product of C (see Equation (5.9)):

$$\gamma(C) := \lambda(C)\lambda^*(C°)/\epsilon_d.$$

Therefore, Haar measure on a hyperplane $f^\perp$ is normalized by prescribing that the $(d-1)$-dimensional measure μ_B of the cross-section $(B \cap f^\perp)$ shall be

$$\mu_B(B \cap f^\perp) := \lambda(B \cap f^\perp)\lambda^*((B \cap f^\perp)°)/\epsilon_{d-1},$$

where λ denotes some Haar measure on $f^\perp$ and λ^* its dual on $(f^\perp)^*$. Calculations are more easily effected if one introduces an auxiliary Euclidean structure and takes both λ and λ^* to be Lebesgue measure. The function $\sigma_B(f)$, which is defined first for unit vectors $\hat{f}$ as the ratio of Minkowski measure to Lebesgue measure in hyperplanes parallel to $\hat{f}^\perp$ and then extended to be positively homogeneous in f, is given by (5.10):

$$\sigma_B(f) = \lambda((B \cap f^\perp)^\circ)|f|/\epsilon_{d-1}. \tag{6.1}$$

The translation of (6.1) using the fact that the dual of a cross-section is the projection of the dual (Corollary 2.2.10) is more useful (see (5.24)):

$$\sigma_B(f) = (2\epsilon_{d-1})^{-1} \int_{\partial B^\circ} |f(\hat{x})| \, d\lambda(\hat{x}) := (2\epsilon_{d-1})^{-1} \int_{S^{d-1}} |f(\hat{x})| \, d\nu_{B^\circ}(\hat{x}), \tag{6.2}$$

where $\hat{x}$ denotes a unit normal to ∂B° and ν_{B° is the surface area measure induced by B° (see Definition 2.3.12). Unless specific reference to B is needed, we omit the subscript.

From the properties of absolute value and the integral, it follows that σ defined by Equation (6.2) is subadditive and therefore is the support function of a convex set in X. However, (6.2) has connections with other important topics in convexity theory. We present three of these connections. Each offers information about the body $\mathbf{I}$ whose support function σ is.

(1) Approximation by polytopes. This was outlined in the discussion following Equation (5.24). If B is a polytope then so is B° and, in this case, the integral in (6.2) reduces to a sum

$$\sigma(f) = (2\epsilon_{d-1})^{-1} \sum_{i=1}^{k} \alpha_i |f(\hat{x}_i)|. \tag{6.3}$$

In this equation the summation is over the facets $\{F_1, \ldots, F_k\}$ of B° which have outward unit Euclidean normals $\hat{x}_i$ and Euclidean areas α_i. Thus $\sigma = \epsilon_{d-1}^{-1} \sum_{i=1}^{k} \sigma_i$, where $\sigma_i(f) := |f(\alpha_i \hat{x}_i)/2|$; i.e. each σ_i is the support function of the line segment $[\alpha_i \hat{x}_i]$. Hence σ is the support function of the Minkowski sum of these line segments. This shows that σ is the support function not just of a convex set, but of a zonotope whose generating line segments are in the direction of the normals to the facets of B° (the vertices of B) and whose lengths are the Euclidean area of the corresponding facets of B°. For terminology on zonotopes see, e.g., Coxeter [117] or the survey article of Schneider and Weil [481].

This representation of $\mathbf{I}_B$ allows a quite detailed analysis of its facial structure. First consider a linear functional $f \in X^*$ such that $f^\perp$ contains no vertices of B. For each normalized vertex $\hat{x}_i$ we can choose a sign $\kappa_i = \pm 1$ such that $f(\kappa_i \hat{x}_i) > 0$. Then $\sigma_i(f) = f(\kappa_i \alpha_i \hat{x}_i/2)$ and $\sigma(f) = (2\epsilon_{d-1})^{-1} \sum_{i=1}^{k} f(\kappa_i \alpha_i \hat{x}_i)$. Thus f supports $\mathbf{I}_B$ at $(2\epsilon_{d-1})^{-1} \sum_{i=1}^{k} \kappa_i \alpha_i \hat{x}_i$. Since this is true for all functionals sufficiently close to f, this point is a vertex of $\mathbf{I}_B$. All vertices of $\mathbf{I}_B$ are of this type and all sums of this form represent a vertex provided only that the vectors

$\kappa_i \hat{x}_i$ all lie on one side of some hyperplane $f^\perp$ in X. Next, if $f \in X^*$ is such that $f^\perp$ contains a single pair of opposite vertices $\pm x_i$ of B, then f supports $\mathbf{I}_B$ along an edge (one-dimensional face) that is parallel to x_i. Conversely, all edges of $\mathbf{I}_B$ are parallel to some vertex of B. The lengths of these edges are proportional to the areas of the corresponding facets of B°.

In general, if f in X^* is such that $f^\perp$ contains k linearly independent vertices of B, then f and $-f$ support $\mathbf{I}_B$ in k-dimensional faces whose edges are parallel to the vertices in $f^\perp$. If these k linearly independent vertices and their negatives are the only vertices in $f^\perp$ then the two faces parallel to $f^\perp$ consist of k-dimensional parallelotopes. In particular, the facets of $\mathbf{I}_B$ correspond to those f in X^* whose null spaces $f^\perp$ contain $(d-1)$ linearly independent vertices of B. These facets are parallelotopes if $f^\perp$ contains precisely $(d-1)$ vertices and their negatives. If F is a facet of $\mathbf{I}_B$ then among the vertices of B that generate F there are $d-2$ linearly independent ones $x_1, x_2, \ldots, x_{d-2}$ that generate two opposite facets of F along which F is joined to two neighbouring facets F' and F'' of $\mathbf{I}_B$. These same vectors also generate two facets of F' and F'' and so on. In this way we obtain a zone of facets of $\mathbf{I}_B$ with each member of the zone joined to the next by a face generated by the vectors $x_1, x_2, \ldots, x_{d-2}$. In general the number of facets in a zone of $\mathbf{I}$ depends on the zone. However, it is possible to regard each face of $\mathbf{I}$ as a union of parallelotopes and, when this is done, the number of parallelotopes in a zone is constant. Finally, we observe that each facet of $\mathbf{I}$ is centrally symmetric and that this property characterizes zonotopes among polytopes. Figures 5.5, 5.6, 5.7 and 5.8 give a number of examples of this construction of $\mathbf{I}$.

In the case when B is not a polytope one may approximate B by a sequence $\{B_n\}$ of polytopes converging in the Hausdorff metric to B. It then follows (see Theorem 6.2.1) that the corresponding zonotopes $\mathbf{I}(B_n)$ converge to $\mathbf{I}(B)$. This means that $\mathbf{I}$ is a *zonoid* (a limit of zonotopes) and, *a fortiori*, is convex.

(2) Mixed volumes. Equation (6.2) can also be translated into the language of mixed volumes. First we look at the heuristic geometric idea based on a three-dimensional picture. If $\hat{u}$ is a Euclidean unit vector then $[\hat{u}]$ (see Example 0.2.3(iv)) is a line segment of length 1. If K is a convex body then $K + [\hat{u}]$ looks like K "pulled apart" in the direction $\hat{u}$. Alternatively, $K + \xi[\hat{u}]$ can be viewed as K together with a "cylinder" of height ξ, whose generators are parallel to $\hat{u}$, that is inserted into K around the shadow boundary (see Figure 6.1). The volume of this cylinder is $\xi \lambda(\mathrm{Proj}_{\hat{u}} K)$. Therefore,

$$\lambda(K + \xi[\hat{u}]) = \lambda(K) + \xi \lambda(\mathrm{Proj}_{\hat{u}} K),$$

i.e. $V(K[d-1], [\hat{u}]) = \lambda(\mathrm{Proj}_{\hat{u}} K)$. Thus, in our case,

$$\begin{aligned}\sigma(f) &= \lambda(\mathrm{Proj}_f B^\circ)|f|/\epsilon_{d-1} \\ &= V(B^\circ[d-1], [f])/\epsilon_{d-1} \\ &= h_{\Pi B^\circ}(f)/\epsilon_{d-1}.\end{aligned} \qquad (6.4)$$

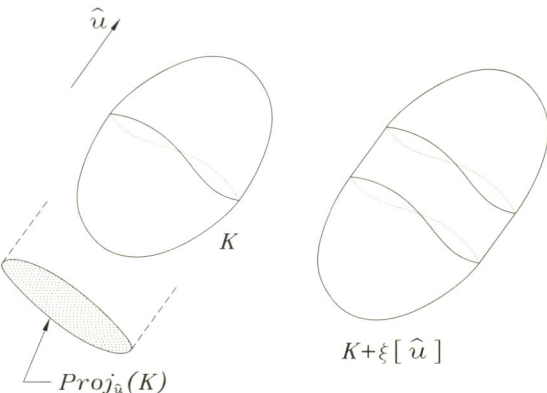

Figure 6.1

We remark that each term in (6.4) is in the right space; we do not need to mix X and X^*.

A more formal proof of the correctness of this formula is obtained by comparing (6.2) with the integral representation for a mixed volume (Equation (2.14)), recalling that the support function of the line segment $[f]$ is just $|f(x)|/2$ and that in (6.2) the usual roles of X and X^* are reversed.

The convexity of σ now follows from the properties of mixed volumes (Proposition 2.3.7) and is a restatement of Proposition 2.3.10.

There are a variety of consequences, connected with Cauchy's formula for the surface area of a convex body and Kubota's formula for projection measure integrals, that can be most easily derived from this view of σ. These ideas will be discussed separately in §6.4 below.

(3) Vector measures. This connection uses ideas that are well presented by Bolker [47] (except that in his terminology the projection body is the dual of the usual definition).

Here we emphasize the role of the surface area measure ν_{B° in Equation (6.2). Abbreviating ν_{B° to ν, (6.2) becomes

$$\sigma(f) = (2\epsilon_{d-1})^{-1} \int_{S^{d-1}} |f(\hat{x})| \, d\nu(\hat{x}). \qquad (6.5)$$

Since $f(\hat{x}) \geq 0$ on precisely one hemisphere S_f^+ of S^{d-1} and since B° is symmetric, we can rewrite (6.5) as

$$\sigma(f) = \epsilon_{d-1}^{-1} \int_{S_f^+} f(\hat{x}) \, d\nu(\hat{x}).$$

This equation suggests a new measure ν_f on S^{d-1} defined by

$$\nu_f(U) := \epsilon_{d-1}^{-1} \int_U f(\hat{x}) \, dv(\hat{x}) \tag{6.6}$$

for each Borel set $U \subseteq S^{d-1}$. We observe that these measures are linear in f and hence can be combined into a single vector measure $\boldsymbol{\nu}$,

$$\boldsymbol{\nu}(U) := \epsilon_{d-1}^{-1} \int_U \hat{x} \, dv(\hat{x}). \tag{6.7}$$

It is clear that $f(\boldsymbol{\nu}(U)) \leq f(\boldsymbol{\nu}(S_f^+)) = \sigma(f)$ and hence

$$\sigma(f) = \sup\{f(y) : y \in \boldsymbol{\nu}(U), \ U \subseteq S^{d-1}\}.$$

Thus σ is the support function of the convex hull of the range of the vector measure $\boldsymbol{\nu}$. If B is smooth so that B° is strictly convex then the vector measure $\boldsymbol{\nu}$ is atomless. In this case, the theorem of Liapunov [324] (see also, e.g., Halmos [238] or Lindenstrauss [326]) shows that the range of $\boldsymbol{\nu}$ is convex and hence that σ is the support function of the range of $\boldsymbol{\nu}$, i.e. $\mathbf{I} = \{\boldsymbol{\nu}(U) : U \subseteq S^{d-1}\}$. In general, when $\boldsymbol{\nu}$ is not atomless (e.g., when B is a polytope, $\boldsymbol{\nu}$ consists only of atoms located at the normalized vertices of B with weights proportional to the areas of the faces of B°) then $\mathbf{I} = \mathrm{conv}\{\boldsymbol{\nu}(U) : U \subseteq S^{d-1}\}$.

In the final part of this first section we consider the extent to which these ideas apply in other dimensions. If M is a $(d-k)$-dimensional subspace of X, then from (5.9)

$$\gamma(B \cap M) = \lambda^{d-k}(B \cap M)\lambda^{*(d-k)}((B \cap M)^\circ)/\epsilon_{d-k}$$

(here the polar is taken with respect to M and so $(B \cap M)^\circ \subseteq M^*$) and hence

$$\sigma(B \cap M) = \lambda^{*(d-k)}(B \cap M)^\circ/\epsilon_{d-k} = \lambda^{*(d-k)}(\mathrm{Proj}_M B^\circ)/\epsilon_{d-k}. \tag{6.8}$$

Before proceeding with this argument we need to present some multilinear algebra. For the details of the ideas briefly outlined here we refer to Greub [200].

Let $\mathcal{A}$ denote the vector space of alternating k-linear forms on $(X^*)^k$; i.e. $\omega \in \mathcal{A}$ whenever $\omega(f_1, f_2, \ldots, f_k)$ is a real-valued function that is linear in each variable separately and for which

$$\omega(f_1, f_2, \ldots, f_i, f_{i+1}, \ldots, f_k) = -\omega(f_1, f_2, \ldots, f_{i+1}, f_i, \ldots, f_k).$$

The vector operations in $\mathcal{A}$ are the usual pointwise ones for functions. The *wedge product* $\Lambda^k = \bigwedge_{i=1}^k X^*$ is defined to be the dual space of $\mathcal{A}$. If $(e_1, e_2, \ldots, e_d)$ is a basis for X^* then there are $\binom{d}{k}$ different ways of choosing k of these basis elements. An element ω of $\mathcal{A}$ may be assigned values at each of these choices arbitrarily but then, by the alternating property and the k-linearity, its values on the whole of $(X^*)^k$ are determined. Therefore, the dimensions of both $\mathcal{A}$ and Λ^k are $\binom{d}{k}$.

Among the elements of Λ^k are the *simple* elements that are the evaluation maps

$$(f_1 \wedge f_2 \wedge \cdots \wedge f_k)(\omega) := \omega(f_1, f_2, \ldots, f_k).$$

The simple elements consisting of evaluation at a choice of k basis vectors $e_{i_1} \wedge e_{i_2} \wedge \cdots \wedge e_{i_k}$ ($1 \le i_1 < i_2 < \cdots < i_k \le d$) form a basis for Λ^k and hence each element of Λ^k is a linear combination of simple elements.

When X^* has a Euclidean structure determined by an inner product $\langle .,. \rangle$ then Λ^k can also be given an inner product (for which we use the same notation $\langle .,. \rangle$) defined for simple elements $\varphi = f_1 \wedge f_2 \wedge \cdots \wedge f_k$ and $\psi = g_1 \wedge g_2 \wedge \cdots \wedge g_k$ by

$$\langle \varphi, \psi \rangle := \det \langle f_i, g_j \rangle$$

and extended bilinearly to the whole of Λ^k. The Euclidean norm of $\varphi = f_1 \wedge f_2 \wedge \cdots \wedge f_k$ is $\langle \varphi, \varphi \rangle^{1/2} = \det \langle f_i, f_j \rangle^{1/2}$. If volume in X^* is measured using Lebesgue measure from the given Euclidean structure then it can be verified that

$$|f_1 \wedge f_2 \wedge \cdots \wedge f_k| = \langle \varphi, \varphi \rangle^{1/2} = \lambda^k \left(\sum_{i=1}^{k} [f_i] \right).$$

The last expression is the k-dimensional volume of the parallelotope spanned by the vectors $f_1, f_2, \ldots, f_k$.

We now return to Equation (6.8). If $f_1, f_2, \ldots, f_k$ are k linearly independent vectors in X^* such that $M = \bigcap_{i=1}^{k} f_i^{\perp}$ then it can be shown that

$$\lambda^{*(d-k)}(\mathrm{Proj}_M B^{\circ}) = V(B^{\circ}[d-k], [f_1], [f_2], \ldots, [f_k])/\lambda^{*k}\left(\sum_{i=1}^{k}[f_i]\right).$$

Thus, analogously to the derivation of (6.4), we may define σ on the cone of simple elements in Λ^k (which Busemann, Ewald and Shephard [77–84] call the *Grassmann cone*) by the following equation.

$$\sigma(f_1 \wedge f_2 \wedge \cdots \wedge f_k) := \epsilon_{d-k}^{-1} V(B^{\circ}[d-k], [f_1], [f_2], \ldots, [f_k]). \quad (6.9)$$

This definition ensures that σ is positively homogeneous. Returning to Requirement 5.1.1(d) one can ask whether σ defined in this way is subadditive. Using the properties of mixed volumes from Proposition 2.3.7, Equation (6.9) implies that $\sigma(f_1 \wedge f_2 \wedge \cdots \wedge f_k)$ is subadditive in each variable separately, and this implies that whenever φ_1 and φ_2 are simple elements whose sum $\varphi_1 + \varphi_2$ is also a simple element then

$$\sigma(\varphi_1 + \varphi_2) \le \sigma(\varphi_1) + \sigma(\varphi_2).$$

Unfortunately one cannot argue inductively that if $\sum \alpha_i \varphi_i$ is a simple element then

$$\sigma\left(\sum \alpha_i \varphi_i\right) \le \sum \alpha_i \sigma(\varphi_i). \quad (6.10)$$

6.1 The convexity of the area function σ

For this it would be sufficient if σ could be extended to a convex function on the whole of Λ^k. Busemann, Ewald and Shephard [77] have shown that, for a general convex body K, such an extension is not possible. It is, however, the intermediate concept expressed by (6.10) that is of geometric significance, *i.e.* whether k-dimensional Minkowski content is minimized by a flat k-dimensional surface (see Busemann and Shephard [86] and Busemann [76]). This remains an open problem.

Partly because of the light it sheds on the nature of σ and partly because of the interesting questions it raises, we give a brief outline of the example of Busemann, Ewald and Shephard which shows that in general σ cannot be extended to a convex function on Λ^k.

Let K be a convex polytope in $\mathbb{R}^d$ and suppose, for convenience, that 0 is an interior point of K. Then $K + [x]$ is a new polytope whose boundary contains a subset of the form $S + [x]$, where S is a subset of the shadow boundary of K and is homeomorphic to S^{d-2}. The body K is said to have a *sharp* shadow boundary in the direction x if S is unique. Likewise, if $x_1, x_2, \ldots, x_k$ are linearly independent vectors in $\mathbb{R}^d$ which span a k-dimensional subspace M then consider $K + \sum_{i=1}^{k}[x_i]$. The boundary of this polytope contains a subset of the form $S + \sum_{i=1}^{k}[x_i]$, where $S \subseteq \partial K$ and S is homeomorphic to S^{d-k-1}. If S is unique then K is said to have *sharp* shadow boundary in the direction M. Moreover, we think of M as represented by the simple element $x_1 \wedge x_2 \wedge \cdots \wedge x_k \in \bigwedge_{i=1}^{k} X$.

Since K is a polytope, for a fixed dimension k the set of directions for which the shadow boundary is sharp is a relatively open subset of the set of simple unit elements in $\bigwedge_{i=1}^{k} X$ and this open subset consists of a finite number n of pairs of components $\pm\mathcal{G}_1, \pm\mathcal{G}_2, \ldots, \pm\mathcal{G}_n$. For directions in one of these components, the shadow boundary is constant; *i.e.* K has n different sharp shadow boundaries, $S_1, S_2, \ldots, S_n$. Moreover, if each shadow boundary is given an orientation, then there is a one–one correspondence between the components $=\mathcal{G}_1, \ldots, \pm\mathcal{G}_n$ and the *oriented* sharp shadow boundaries. Corresponding to each oriented sharp shadow boundary S_j we construct a *vector area* $\nu(S_j)$ as follows.

Join each vertex of S_j to the origin to form a surface S'_j homeomorphic to S^{d-k}. Each $(d-k)$-dimensional face F_i of S'_j has λ^{d-k} content α_i and a unit normal (compatible with the orientation of S_j) $\varphi_i \in \Lambda^k$. Let

$$\nu(S_j) := \sum_i \alpha_i \varphi_i.$$

Note that in the case $k = 1$ and $K = B^\circ$ these vectors are just the vertices of $\mathbf{I}$. Finally, define the linear functional Φ_j on Λ^k by

$$\Phi_j(\psi) := \langle \nu(S_j), \psi \rangle = \sum_i \alpha_i \langle \varphi_i, \psi \rangle. \tag{6.11}$$

If ψ is a simple element and is in the component $\mathcal{G}_j$ then it is not difficult to show that

$$\sigma(\psi) = \Phi_j(\psi). \tag{6.12}$$

Here the orientation of S_j is such that $\Phi_j(\psi) > 0$; if $\psi \in -\mathcal{G}_j$ then $\nu(-S_j) = -\nu(S_j)$ and, with the obvious notation, $\Phi_{-j}(\psi) = \Phi_j(-\psi) = \sigma(\psi)$. Thus, if we knew that $\sigma(\psi) \geq |\Phi_{j'}(\psi)|$ for all $j' = 1, 2, \ldots, n$, (6.12) would immediately imply that

$$\sigma(\psi) = \sup_j |\Phi_j(\psi)|. \tag{6.13}$$

Since the right-hand side represents the restriction to the simple elements of a convex function on Λ^k, Equation (6.13) implies that σ is extensible to a convex function on Λ^k. Conversely, Busemann, Ewald and Shephard show that if σ is extensible to a convex function on Λ^k then (6.12) must hold not only on the component $\mathcal{G}_j$ but on its convex hull, which, they show, has non-empty interior in Λ^k; i.e. the linear functional Φ_j coincides with σ on a set with interior. It is then clear that $\sigma(\psi) \geq \Phi_j(\psi)$ for all simple elements ψ and that σ is given by (6.13). Thus σ is extensible to a convex function on Λ^k if and only if

$$\sigma(\psi) \geq |\Phi_j(\psi)|, \qquad j = 1, 2, \ldots, n, \tag{6.14}$$

which is so if and only if σ is the support function of

$$\mathrm{conv}\{\pm \nu(S_j) : j = 1, 2, \ldots, n\}.$$

Again we remark that this is the case when $k = 1$. In that case the above convex hull is (up to a scalar multiple) the isoperimetrix $\mathbf{I}$ with vertices given by the vectors $\pm \nu(S_j)$.

Example 6.1.1 In $\mathbb{R}^4$ with the usual basis $(e_1, \ldots, e_4)$, consider the set $K = \mathrm{conv}\{x_1, x_2, x_3, x_4, x_5, x_6\}$, where

$$x_i := (\cos 2i\pi/3, \sin 2i\pi/3, \cos i\pi/3, \sin i\pi/3)^t.$$

Let $M_1 := \mathrm{span}\{e_1, e_2\}$ and $M_2 := \mathrm{span}\{e_3, e_4\}$. Then the shadow boundary S_1 of K in the direction M_1 is the hexagonal curve $[x_1, x_2, x_3, x_4, x_5, x_6, x_1]$. The vector area $\nu(S_1)$ spanned by this curve is $2^{-1} \sum_{i=1}^{6} x_i \wedge x_{i+1}$, where $x_7 := x_1$. Routine calculations show that

$$\nu(S_1) = 3\sqrt{3}/2 (e_1 \wedge e_2 + e_3 \wedge e_4).$$

Thus, with the notation above, $\Phi_1(e_1 \wedge e_2) = \Phi_1(e_3 \wedge e_4) = 3\sqrt{3}/2$. Now $\sigma(e_1 \wedge e_2) = V(K[d-2], [e_1], [e_2]) = 3\sqrt{3}/2$; this is the area of the regular hexagon with vertices $(0, 0, \cos i\pi/3, \sin i\pi/3)^t$ $(i = 1, 2, \ldots, 6)$ in M_2. On the other hand, $\sigma(e_3 \wedge e_4) = V(K[d-2], [e_3], [e_4]) = 3\sqrt{3}/4$; this is the area of the equilateral triangle with vertices $(\cos 2i\pi/3, \sin 2i\pi/3, 0, 0)^t$ $(i = 1, 2, 3)$ in M_1. So, for this example, we have $\sigma(e_3 \wedge e_4) < \Phi_1(e_3 \wedge e_4)$ and $\sigma(\psi) \neq \sup\{|\Phi_j(\psi)|\}$. Hence σ cannot be extended to a convex function on Λ^2. The reason for this discrepancy is that the projection of K onto M_1 covers the triangle twice. The vector calculations involved in the linear functionals Φ_j keep track of multiplicities; the scalar calculations for σ do not.

Remark. Busemann, Ewald and Shephard show that this phenomenon cannot arise with zonotopes or zonoids; *i.e.* for zonoids all projections do extend to convex functions on Λ^k.

6.2 Properties of the mapping I

Now that we know that $\mathbf{I}(B) = \Pi(B^\circ)$ is a convex set we can study the nature of the mapping $\mathbf{I}$ as a function on the collection of symmetric convex sets. Note that, like $(.)^\circ$, $\Pi(.)$ is a map from convex sets in X to convex sets in X^* (and *vice versa*). Hence $\mathbf{I}$ maps the set of symmetric convex bodies in X into itself.

Theorem 6.2.1

(i) *The mapping $\mathbf{I}$ is order reversing with respect to inclusion.*

(ii) *If $\xi > 0$ then $\mathbf{I}(\xi B) = \xi^{1-d}\mathbf{I}(B)$.*

(iii) *With respect to the metric Δ_2, the mapping $\mathbf{I}$ is a Lipschitz mapping with Lipschitz constant $(d-1)$.*

Proof

(i) If B_1 and B_2 are two unit balls in a d-dimensional space X and if $B_1 \supseteq B_2$ then $B_1^\circ \subseteq B_2^\circ$ and hence $V(B_1^\circ[d-1], [f]) \leq V(B_2^\circ[d-1], [f])$. Therefore, $\sigma_{B_1}(f) \leq \sigma_{B_2}(f)$ and $\mathbf{I}(B_1) \subseteq \mathbf{I}(B_2)$.

(ii) This proof is a similar straightforward calculation.

(iii) If B_1 and B_2 are two unit balls with $\Delta_2(B_1, B_2) = \delta$ then there exist positive scalars ξ, η with $B_1 \subseteq \xi B_2$, $B_2 \subseteq \eta B_1$ and $\log \max\{\xi, \eta\} = \delta$. From (i) and (ii) we get $\mathbf{I}(B_1) \supseteq \mathbf{I}(\xi B_2) = \xi^{1-d}\mathbf{I}(B_2)$ and $\mathbf{I}(B_2) \supseteq \eta^{1-d}\mathbf{I}(B_1)$, from which the result follows. ∎

Remarks

(i) Part (i) shows that this definition satisfies Kolmogoroff's monotonicity requirement [297] – namely, that if X has two norms $\|.\|_1$ and $\|.\|_2$ with $\|.\|_1 \leq \|.\|_2$ then the smaller distances generate smaller areas.

(ii) One can equally well use the function Δ_1, which is invariant under scalar multiples. For this function it is immaterial whether one uses $\mathbf{I}$ or the normalization $\tilde{\mathbf{I}}$ of Definition 5.3.6 and also whether one includes the factor ϵ_{d-1}^{-1} or not. The same is true throughout most of this section.

(iii) Part (iii) shows that **I** is contractive in $\mathbb{R}^2$. In fact in this case **I** is an involution and is an isometry.

(iv) For $d > 2$, the Lipschitz constant $(d - 1)$ seems far from the best possible. An earlier conjecture that, perhaps, **I** is contractive with respect to Δ_1 in all dimensions has been shown by Clack [112] to be false. It is possible that $\mathbf{I}^2$ is contractive. In this context $\mathbf{I}^2$ is the more natural map to consider because it preserves the general shape of the convex body (see Figures 5.5, 5.6, 5.7 and 5.8).

(v) In view of Equation (5.21), one might also consider the Banach–Mazur distance Δ. The same calculations show that $\tilde{\mathbf{I}}$ is Lipschitz with constant $(d-1)$ with respect to Δ. The few calculations we have made do not contradict the statement that $\tilde{\mathbf{I}}$ is contractive with respect to Δ.

Theorem 6.2.2 *The range of **I** is the class of zonoids with centre at 0.*

Proof. In §6.1 we saw that if B is a polytope then $\mathbf{I}(B)$ is a zonotope. By Theorem 2.5.1, we may approximate B by a sequence of polytopes P_n converging in the Hausdorff metric to B. It follows from Theorem 6.2.1 that $\mathbf{I}(P_n)$ is a sequence of zonotopes converging to $\mathbf{I}(B)$. Hence, by definition, $\mathbf{I}(B)$ is a zonoid.

Conversely, if Z is a zonoid, then (see, *e.g.*, [47] or [481]) there is a symmetric measure ν on the unit sphere S^{d-1} such that

$$h_Z(f) = \int_{S^{d-1}} |f(u)| \, d\nu(u).$$

We now appeal to Minkowski's theorem (see Schneider [479], p. 392) to conclude that there is a centrally symmetric convex body B° whose surface measure (Definition 2.3.12) is given by $\nu(u)$, *i.e.* $\Pi(B^\circ) = Z$ and hence, up to a scalar multiple, $\mathbf{I}(B) = Z$. ∎

The next step is to show that **I** is injective. This theorem is due to Aleksandrov [3] and, as Schneider and Weil [481] comment, has been re-proved many times. We begin with Proposition 2.3.14, which, together with Equation (2.14) and the definition of the surface measure ν (Definition 2.3.12), shows that in Minkowski's theorem the correspondence between convex body and surface measure is injective.

The outline of the rest of the proof of Aleksandrov's theorem is as follows. If $\Pi B_1^\circ = \Pi B_2^\circ$ then $V(B_1^\circ[d-1], [f]) = V(B_2^\circ[d-1], [f])$ for all vectors f in X^*. Hence, by linearity and continuity, $V(B_1^\circ[d-1], Z) = V(B_1^\circ[d-1], Z)$ for all zonoids Z. We need to show that the zonoids are "weakly dense" among the symmetric convex bodies in the sense that the preceding equation implies $V(B_1^\circ[d-1], K) = V(B_2^\circ[d-1], K)$ for all symmetric K and then apply Proposition 2.3.14. Petty [415] uses an approximation argument to show this step directly. An alternative, more analytic approach is to use support functions rather than convex sets. For then one can allow subtraction and consider the linear space generated by the support functions of line segments. Bourgain and Lindenstrauss [54] give a proof, due to Choquet, that this linear space is uniformly dense in the linear space of all continuous (even) functions on the unit ball $C_e(S^{d-1})$. This approach is somewhat comparable to the proof of the corresponding fact in Chapter 7 and is the one we outline here.

Theorem 6.2.3 *If $\Pi B_1^\circ = \Pi B_2^\circ$ then $B_1 = B_2$.*

Proof. Let ν_i be the surface area measure on S^{d-1} induced by B_i° ($i = 1, 2$) (Definition 2.3.12). If $\Pi B_1^\circ = \Pi B_2^\circ$ then their support functions are equal and hence

$$\int_{S^{d-1}} |\langle f, \hat{v}\rangle| \, d\nu_1(\hat{v}) = \int_{S^{d-1}} |\langle f, \hat{v}\rangle| \, d\nu_2(\hat{v}) \tag{6.15}$$

for all $f \in S^{d-1}$. Instead of thinking of $|\langle f, \hat{v}\rangle|$ as the support function of the line segment $[2\hat{v}]$ we regard it as an even function $\hat{v} \mapsto |\langle f, \hat{v}\rangle|$ on S^{d-1}, i.e. as the support function of $[2f]$. Let Y denote the closed linear span of these support functions of line segments. By the linearity and continuity of the integral, it is clear that Equation (6.15) extends to functions in Y. We next show that Y is the whole of $C_e(S^{d-1})$.

Since every ellipsoid is a zonoid, the support function of an ellipsoid is in Y. Thus if $T = (\tau_{ij})$ is a positive definite symmetric matrix then

$$\phi_T(\hat{v}) := \left(\sum_1^d \tau_{ij} v_i v_j\right)^{1/2}$$

is the support function of an ellipse and hence is in Y. Next differentiate $\phi_T(\hat{v})$ with respect to τ_{ij} (this is a limit of differences of elements of Y and so is in Y) and evaluate the derivative at the identity matrix (recalling that $\hat{v} \in S^{d-1}$ so that $\sum_1^d v_i^2 = 1$). It follows that the map $\hat{v} \mapsto v_i v_j \in Y$. Repeated differentiation with respect to the various entries in T shows that every even polynomial in the v_i's is in Y and hence, by the Stone–Weierstrass theorem, $Y = C_e(S^{d-1})$.

Hence Equation (6.15) implies that

$$\int_{S^{d-1}} \psi(\hat{v}) \, d\nu_1(\hat{v}) = \int_{S^{d-1}} \psi(\hat{v}) \, d\nu_2(\hat{v})$$

for every $\psi \in C_e(S^{d-1})$. In particular, with $\psi(\hat{v}) = h_K(\hat{v})$ (where we now need to use the Euclidean structure to identify X and X^*) we get

$$V(B_1^\circ[d-1], K) = \int_{S^{d-1}} h_K(\hat{v}) \, d\nu_1(\hat{v})$$
$$= \int_{S^{d-1}} h_K(\hat{v}) \, d\nu_2(\hat{v}) = V(B_2^\circ[d-1], K)$$

for all symmetric bodies K, which, by Proposition 2.3.14, yields $B_1^\circ = B_2^\circ$ and hence the result. ∎

If the domain of **I** is extended to include non-symmetric bodies then **I** is not one–one. As an example of this consider a cube C with vertices v_1, v_2, v_3, v_4 on one facet and their negatives $-v_3, -v_4, -v_1, -v_2$ on the opposite facet. Then $T := \text{conv}\{v_1, v_3, -v_2, -v_4\}$ is a regular tetrahedron. The normals to the faces of T are $\{v_2, v_4, -v_1, -v_3\}$. It follows that, up to a scalar multiple,

$$\Pi(T) = \Pi(T^\circ) = \sum_{i=1}^{4} [v_i] = \Pi(C^\circ).$$

This body is a rhombic dodecahedron (see Figure 5.4(b)) and is homothetic to $(T - T)^\circ$ (see Example 1.1.17).

Given a convex body K_0, let $[K_0]$ be the equivalence class consisting of all convex bodies K such that $\mathbf{I}(K) = \Pi(K^\circ) = \mathbf{I}(K_0)$. Then, by Theorem 6.2.3, $[K_0]$ contains precisely one centrally symmetric body. We shall be able to say more about this after we have defined Blaschke addition and scalar multiplication. These ideas are also needed to show how **I** behaves in relation to the algebraic structure on convex sets.

Consider the pairing

$$V(B[d-1], K) = \int_{S^{d-1}} h_K(f)\, dv_B(f), \qquad (6.16)$$

where B and K are symmetric convex bodies and v_B is the surface measure of B. The mapping $K \mapsto h_K$ is linear when Minkowski addition and scalar multiplication are used in $\mathcal{C}_b$ (Definition 2.1.2) and pointwise operations are used for h_K. The measures on S^{d-1} also have a natural semi-linear structure defined pointwise $(\alpha v + \beta v')(U) := \alpha(v(U)) + \beta(v'(U))$ for all Borel sets $U \subseteq S^{d-1}$ and all $\alpha, \beta \geq 0$.

The Blaschke semi-linear structure on $\mathcal{C}_b$ is defined so that the mapping $B \mapsto v_B$ is semi-linear (see, *e.g.*, Firey [159, 160]).

Definition 6.2.4 *If K_1 and K_2 are convex bodies in $\mathbb{R}^d$ with 0 as an interior point define the **Blaschke scalar product** $\xi \cdot K_1$ by*

$$\xi \cdot K_1 := \xi^{1/(d-1)} K_1 \qquad (\xi \geq 0)$$

*and the **Blaschke sum** $K_1 \dotplus K_2$ by the equation*

$$v_{K_1 \dotplus K_2}(U) := v_{K_1}(U) + v_{K_2}(U)$$

for all Borel subsets U of S^{d-1}.

In this definition, essential use is made of Minkowski's theorem that there is, indeed, a unique (up to translation) convex body with the property that its surface area measure is given by $v_{K_1} + v_{K_2}$. It is straightforward to conclude from (6.16) that

$$V((\xi \cdot K_1 + \xi_2 \cdot K_2)[d-1], K) = \xi_1 V(K_1[d-1], K) + \xi_2 V(K_2[d-1], K) \quad (6.17)$$

for all convex sets K.

If now we put $K = [u]$ in (6.17) we see, via the support functions, that

$$\Pi(\xi_1 \cdot K_1 + \xi_2 \cdot K_2) = \xi_1 \Pi K_1 + \xi_2 \Pi K_2, \quad (6.18)$$

where, on the right, we have the Minkowski sum. For the mapping **I** the situation is slightly complicated by the polar reciprocal.

Proposition 6.2.5 *If B_1 and B_2 are centrally symmetric convex bodies then*

$$\mathbf{I}(\xi \cdot B_1) = \xi^{-1} \mathbf{I}(B_1), \quad (6.19)$$

$$\mathbf{I}((B_1^\circ + B_2^\circ)^\circ) = \mathbf{I}(B_1) + \mathbf{I}(B_2). \quad (6.20)$$

Proof. This follows directly from (6.18) and the definition of **I**. ∎

One might call the operation $(B_1^\circ + B_2^\circ)^\circ$ the *harmonic* Blaschke sum of B_1 and B_2. Note that one of the virtues of the normalization $\tilde{\mathbf{I}}$ is that instead of (6.19) we have $\tilde{\mathbf{I}}(\xi B_1) = \xi \tilde{\mathbf{I}}(B_1)$ (see (5.21)). On the other hand, this normalization appears to complicate the additive structure.

We can now return, briefly, to the question of extending the domain of **I** to non-centrally symmetric convex bodies. It is evident that for every convex body K, $\Pi(-K) = \Pi(K)$. Hence if we set

$$\mathbf{D}(K) := 2^{-1} \cdot K + 2^{-1} \cdot (-K)$$

(and call it the *Blaschke difference body* of K) then from (6.18) we have

$$\Pi(\mathbf{D}K) = \Pi K,$$

i.e.

$$\mathbf{I}((\mathbf{D}K)^\circ) = \mathbf{I}(K^\circ).$$

Thus the equivalence class $[K]$ contains the symmetric body $(\mathbf{D}K^\circ)^\circ$. Since there is precisely one symmetric body in $[K]$, if $\mathbf{I}(K_1) = \mathbf{I}(K_2)$ then $(\mathbf{D}K_1^\circ)^\circ = (\mathbf{D}K_2^\circ)^\circ$. Note that if for K we take the regular tetrahedron T as in the example above then $\mathbf{D}T = C^\circ$. Thus the Blaschke difference body of a regular tetrahedron is an octahedron.

There is another characterization of the symmetric member of $[K]$. From the Kneser–Süss inequality (which is the counterpart of the Brunn–Minkowski inequality for Blaschke addition; see [479], p. 394) it follows that

$$\lambda(K) \leq \lambda(\mathbf{D}K)$$

and hence
$$\lambda(K) \geq \lambda((\mathbf{D}K^\circ)^\circ).$$

Thus $(\mathbf{D}K^\circ)^\circ$ has the smallest volume among all the elements of $[K]$.

The last result of this section is another algebraic equation satisfied by the mapping $\mathbf{I}$. We give a direct approach and point out first that here it is preferable to use the normalized isoperimetrix $\tilde{\mathbf{I}}$ (Definition 5.3.6). First some more notation.

Let (X, A) be a Minkowski space of dimension d_1 and (Y, B) a Minkowski space of dimension d_2. Then we may consider the space $X \times Y$ of dimension $d_1 + d_2$ whose dual space $(X \times Y)^*$ is isomorphic to $X^* \times Y^*$. We think of $X \times Y$ as the direct sum $X \times \{0\} + \{0\} \times Y$ and write $(x, 0)$, $(0, y)$ for elements of these summands and $(A, 0)$ for $A \times \{0\}$, etc. In $X \times Y$ we consider the ball $A * B = \text{conv}((A, 0) \cup (0, B))$. Then one can show that $(A * B)^\circ = A^\circ \times B^\circ \subseteq X^* \times Y^*$.

Theorem 6.2.6 *With the notation just given, there exist scalars κ_1 and κ_2 depending only on the dimensions d_1 and d_2 such that*
$$\tilde{\mathbf{I}}(A * B) = \kappa_1 \tilde{\mathbf{I}}(A) \times \kappa_2 \tilde{\mathbf{I}}(B).$$

Proof. We prove the result for polytopes; the general result follows by means of an approximation and continuity argument.

Suppose A has vertices $a_1, a_2, \ldots, a_k$ and the areas of the corresponding facets of A° are $\alpha_1, \alpha_2, \ldots, \alpha_k$. Define $b_1, b_2, \ldots, b_\ell$, $\beta_1, \beta_2, \ldots, \beta_\ell$ likewise for B and B°. Then
$$h_{\tilde{\mathbf{I}}(A)}(f) = \eta_A^{-1} \sum_{i=1}^k \alpha_i |f(a_i)|$$
for all $f \in X^*$, where $\eta_A := \lambda^{d_1}(A^\circ) \epsilon_{d_1-1}/\epsilon_{d_1}$, and
$$h_{\tilde{\mathbf{I}}(B)}(g) = \eta_B^{-1} \sum_{j=1}^\ell \beta_j |g(b_j)|$$
for all $g \in Y^*$, where $\eta_B := \lambda^{d_2}(B^\circ) \epsilon_{d_2-1}/\epsilon_{d_2}$. Now $(A * B)^\circ = A^\circ \times B^\circ$. This set has facets of two kinds: (a facet of A°) $\times B^\circ$ and $A^\circ \times$ (a facet of B°) with normals $(a_i, 0)$ and $(0, b_j)$ and areas $\alpha_i \lambda^{d_2}(B^\circ)$ and $\beta_j \lambda^{d_1}(A^\circ)$ respectively. Hence
$$h_{\tilde{\mathbf{I}}(A*B)}(f, g) = \eta_{A*B}^{-1} \left[\sum_{i=1}^k \alpha_i \lambda^{d_2}(B^\circ)|f(a_i)| + \sum_{j=1}^\ell \beta_j \lambda^{d_1}(A^\circ)|g(b_j)| \right];$$
here $(f, g)(a, 0) = f(a)$, $(f, g)(0, b) = g(b)$ and
$$\eta_{A*B} := \lambda^{d_1+d_2}(A^\circ \times B^\circ) \epsilon_{d_1+d_2-1}/\epsilon_{d_1+d_2}.$$

But now $\lambda^{d_1+d_2}(A° \times B°) = \lambda^{d_1}(A°)\lambda^{d_2}(B°)$ and so

$$h_{\tilde{\mathbf{I}}(A*B)}(f, g) = \kappa_1 h_{\tilde{\mathbf{I}}(A)}(f) + \kappa_2 h_{\tilde{\mathbf{I}}(B)}(g),$$

which implies that

$$\tilde{\mathbf{I}}(A * B) = \kappa_1 \tilde{\mathbf{I}}(A) \times \kappa_2 \tilde{\mathbf{I}}(B),$$

where $\kappa_i := \epsilon_{d_1+d_2}\epsilon_{d_i-1}/\epsilon_{d_1+d_2-1}\epsilon_{d_i}$. ∎

Remark. To see the connection with (6.20), note that on the right $\tilde{\mathbf{I}}(A) \times \tilde{\mathbf{I}}(B)$ can be thought of as the Minkowski sum $(\tilde{\mathbf{I}}(A), 0) + (0, \tilde{\mathbf{I}}(B))$. However, as we see from the expressions for the normals and areas of the faces of $A° \times B°$ we have

$$A° \times B° = \lambda^{d_2}(B°) \cdot (A°, 0) + \lambda^{d_1}(A°) \cdot (B°, 0).$$

Thus, $A * B$ is closely related to the harmonic Blaschke sum of A and B mentioned above.

6.3 Cauchy's formula for surface areas

In this section we arrive at the results that motivated the work of [262]. After my re-proving of Schäffer's result (Theorem 4.3.8; see [508]), it was suggested by R. D. Holmes that this might be made the basis of a definition of surface area in higher dimensions. It was with some surprise that, after a great many calculations, a few of which appear in [262], we realized that not only was this possible but that in $\mathbb{R}^3$ it appeared to define the areas *uniquely*.

Theorem 6.3.1 *If A and B are two unit balls in a Minkowski space X and if $A°$ and $B°$ are their dual balls in X^* then*

$$\mu_B(\partial A) = \mu_{A°}(\partial B°).$$

Proof. Suppose, first, that A and B are smooth and strictly convex. Then from (5.13) we have

$$\mu_B(\partial A) = \int_{\partial A} \sigma_B(\hat{f}) \, d\lambda(\hat{f}) = \int_{S^{d-1}} \sigma_B(\hat{f}) \, d\nu_A(\hat{f}),$$

where $\hat{f}$ denotes a Euclidean unit normal to ∂A and ν_A is the surface area measure on S^{d-1} induced by A (Definition 2.3.12).

Substituting for σ_B from (6.6) we have

$$\mu_B(\partial A) = (2\epsilon_{d-1})^{-1} \int_{S^{d-1}} \left(\int_{S^{d-1}} |\hat{f}(\hat{x})| \, d\nu_{B°}(\hat{x}) \right) d\nu_A(\hat{f}),$$

where dv_{B° is the surface area measure on S^{d-1} induced by ∂B°. (Note that the two copies of S^{d-1} are in X^* and X respectively.) Likewise,

$$\mu_{A^\circ}(\partial B^\circ) = (2\epsilon_{d-1})^{-1} \int_{S^{d-1}} \left(\int_{S^{d-1}} |\hat{f}(\hat{x})| dv_A(\hat{f}) \right) dv_{B^\circ}(\hat{x}),$$

where we have used the fact that $(A^\circ)^\circ = A$. An application of Fubini's theorem completes the proof in this case. The general case follows by an approximation argument. ∎

Corollary 6.3.2 *If (X, B) is a Minkowski space and (X^*, B°) its dual space then*

$$\mu_B(\partial B) = \mu_{B^\circ}(\partial B^\circ).$$

Proof. Take $A = B$ in Theorem 6.3.1. ∎

Remarks

(i) It is interesting to rephrase Theorem 6.3.1 in the language of mixed volumes. From (5.15) we have

$$\mu_B(\partial A) = d\epsilon_{d-1}^{-1} V(A[d-1], \Pi B^\circ)$$

and

$$\mu_{A^\circ}(\partial B^\circ) = d\epsilon_{d-1}^{-1} V(B^\circ[d-1], \Pi A).$$

Thus 6.3.1 is the statement that

$$V(K[d-1], \Pi K') = V(K'[d-1], \Pi K)$$

for all (centrally symmetric) convex bodies K in X and K' in X^*. For this form of the equation and its use in a variety of geometric inequalities, see Lutwak [336, 337].

(ii) At first sight, the proof of 6.3.1 (specialized to $\mathbb{R}^2$) seems different from that of Theorem 4.3.8. But the proof of 4.3.8 is, essentially, an integration by parts formula which can be viewed as an application of Fubini's theorem.

The same interchange in the order of integration yields a Minkowski version of Cauchy's theorem and its extension to Kubota's formula (Bonnesen and Fenchel [51], §32).

Theorem 6.3.3 *If K is a convex body in a Minkowski space (X, B) then*

$$\mu_B(\partial K) = d\epsilon_{d-1}^{-1} \int_{\partial B^\circ} V(K[d-1], [\hat{x}]) \, d\lambda(\hat{x}).$$

6.3 Cauchy's formula for surface areas

Proof. We have, exactly as in the proof of Theorem 6.3.1,

$$\mu_B(\partial K) = \int_{S^{d-1}} \sigma_B(\hat{f}) \, dv_K(\hat{f})$$

$$= (2\epsilon_{d-1})^{-1} \int_{S^{d-1}} \int_{S^{d-1}} |\hat{f}(\hat{x})| \, dv_{B^\circ}(\hat{x}) \, dv_K(\hat{f})$$

$$= (2\epsilon_{d-1})^{-1} \int_{S^{d-1}} \int_{S^{d-1}} |\hat{f}(\hat{x})| \, dv_K(\hat{f}) \, dv_{B^\circ}(\hat{x}).$$

We interpret $|\hat{f}(\hat{x})|/2$ as the support function of the line segment $[\hat{x}]$ and use Equation (2.14) to interpret the inner integral as a mixed volume. Thus

$$\mu_B(\partial K) = d\epsilon_{d-1}^{-1} \int_{S^{d-1}} V(K[d-1], [\hat{x}]) \, dv_{B^\circ}(\hat{x})$$

$$= d\epsilon_{d-1}^{-1} \int_{\partial B^\circ} V(K[d-1], [\hat{x}]) \, d\lambda(\hat{x}). \quad \blacksquare$$

This formula expresses the Minkowski surface area of K as an average of the projections of K in the directions $\hat{x}$. It looks, however, rather mixed because of the various Euclidean measures in the integral on the right. It is not clear what are the best normalizations to make but the following ones eliminate most of the Euclidean measures. Firstly, multiply and divide by $\sigma_B^d = \lambda^*(B^\circ)/\epsilon_d$. This gives, using the definition of V_B just before Equation (5.17),

$$\mu_B(\partial K) = \frac{\epsilon_d d}{\epsilon_{d-1} \lambda^*(B^\circ)} \int_{\partial B^\circ} V_B(K[d-1], [\hat{x}]) \, d\lambda_{B^\circ}(\hat{x}).$$

Secondly, inside the integral, divide and multiply by $\sigma_{B^\circ}(\hat{x})$. This normalizes $\hat{x}$ to lie on the surface of $(\mathbf{I}(B^\circ))^\circ$ (in X) and changes the Euclidean surface area measure of B° to the Minkowski surface area measure (relative to B° as unit ball in X^*). Thirdly, multiply and divide by $\sigma_{B^\circ}^d$ to change $\mathbf{I}(B^\circ)$ to $\tilde{\mathbf{I}}(B^\circ)$. Together these give

$$\mu_B(\partial K) = \kappa_B \int_{\partial B^\circ} V_B(K[d-1], [\tilde{x}]) \, d\mu_{B^\circ}^{d-1}(\tilde{x}), \quad (6.21)$$

where $d\mu_{B^\circ}^{d-1}(\tilde{x})$ is the Minkowski *area element* of B° at the point whose outward normal is $\tilde{x}$ and $\tilde{x}$ is normalized to lie on $\partial(\tilde{\mathbf{I}}(B^\circ))^\circ$. The constant

$$\kappa_B := \frac{d\epsilon_d^2}{\epsilon_{d-1} \lambda(B) \lambda^*(B^\circ)} = \frac{d\epsilon_d}{\epsilon_{d-1} \mu_B^d(B)}$$

depends on the dimension and on the ball. The last equation comes from replacing the normalized volume product $\lambda(B)\lambda^*(B^\circ)/\epsilon_d$ with $\mu_B^d(B)$ (from (5.9) and (5.1)). Note, also, that $V_B(K[d-1], [\tilde{x}])$ is a function on $\partial(\tilde{\mathbf{I}}(B^\circ))^\circ$. The same construction that was used in §2.3 to produce the surface area measure ν_K on S^{d-1} could be used to construct a surface area measure carried on $\partial(\tilde{\mathbf{I}}(B^\circ))^\circ$ instead, in which case the integral in (6.21) should be interpreted as an integral over this surface (and similarly for the integrals below).

Another interpretation of (6.21) is to use (5.18) to write $\mu_B(\partial K)$ as a mixed volume:

$$V_B(K[d-1], \tilde{\mathbf{I}}(B)) = d^{-1}\kappa_B \int_{\partial B^\circ} V_B(K[d-1], [\tilde{x}])\, d\mu_{B^\circ}(\tilde{x}). \qquad (6.22)$$

This suggests writing $V_B(K[d-1], \tilde{\mathbf{I}}(B))$ as $W_{B,1}(K)$ and calling it the *first projection measure integral* of K. Likewise, we set $V_B(K[d-1], [\tilde{x}]) = W'_{B,0}(K)$. More generally we make the following definitions.

Definition 6.3.4 *For each n ($0 \le n \le d$) the nth **projection measure integral** of K (with respect to B) is written $W_{B,n}(K)$ and is defined by the equation*

$$W_{B,n}(K) := V_B(K[d-n], \tilde{\mathbf{I}}(B)[n]).$$

The numbers $W'_{B,n-1}(K, \tilde{x}) := V_B(K[d-n], \tilde{\mathbf{I}}(B)[n-1], [\tilde{x}])$ are the $(n-1)$st projection measure integrals of the projections of K onto the hyperplane perpendicular to $\tilde{x}$.

With this notation, $W_{B,0}(K)$ is the Minkowski volume of K and $W_{B,1}(K)$ is the Minkowski surface area of K. Cauchy's formula is now written as

$$W_{B,1}(K) = \kappa_B \int_{\partial B^\circ} W'_{B,0}(K, \tilde{x})\, d\mu_{B^\circ}(\tilde{x}). \qquad (6.23)$$

This can be extended to a version of Kubota's formula ([51], §32; [479], p. 295) in exactly the same way as in the Euclidean case. As usual, we do the calculations in the Euclidean framework and then renormalize the resulting formulas. The particular approach we take here comes from Lutwak [335] where more details can be found.

If $K_1, K_2, \ldots, K_{d-1}$ are convex bodies in $\mathbb{R}^d$ then there is a mixed area measure $\nu^{d-1}(K_1, K_2, \ldots, K_{d-1}; \cdot)$ on S^{d-1} such that for all convex bodies K

$$V(K_1, K_2, \ldots, K_{d-1}, K) = d^{-1} \int_{S^{d-1}} h(K, f)\, d\nu^{d-1}(K_1, \ldots, K_{d-1}; f). \qquad (6.24)$$

Moreover, since this area measure satisfies the conditions of Minkowski's theorem, there is a (unique) convex body $[K_1, K_2, \ldots, K_{d-1}]$ whose surface area measure ν is $\nu^{d-1}(K_1, K_2, \ldots, K_{d-1}; \cdot)$. Repeating the same steps as in the proof of Theorem 6.3.3 we get

$$V(K_1, K_2, \ldots, K_{d-1}, \mathbf{I}(B)) = (d\epsilon_{d-1})^{-1} \int_{S^{d-1}} \sigma_B(f)\, d\nu([K_1, \ldots, K_{d-1}]; f)$$

$$= \epsilon_{d-1}^{-1} \int_{\partial B^\circ} V([K_1, \ldots, K_{d-1}][d-1], [x])\, d\lambda(x).$$

But it is clear from (6.24) that $V([K_1, \ldots, K_{d-1}][d-1], [x]) = V(K_1, \ldots, K_{d-1}, [x])$ and so

$$V(K_1, \ldots, K_{d-1}, \mathbf{I}(B)) = \epsilon_{d-1}^{-1} \int_{\partial B^\circ} V(K_1, \ldots, K_{d-1}, [x]) \, d\lambda(x).$$

Finally, if we normalize as before, we have

$$V_B(K_1, \ldots, K_{d-1}, \tilde{\mathbf{I}}(B)) = \kappa_B \int_{\partial B^\circ} V_B(K_1, \ldots, K_{d-1}, [\tilde{x}]) \, d\mu_{B^\circ}^{d-1}(\tilde{x}). \quad (6.25)$$

Theorem 6.3.5 (Kubota's recursion formula) *If (X, B) is a Minkowski space then*

$$W_{B,n}(K) = \kappa_B \int_{\partial B^\circ} W'_{B,n-1}(K, \tilde{x}) \, d\mu_{B^\circ}^{d-1}(\tilde{x}),$$

where the various terms are as defined in the preceding discussion.

Proof. In Equation (6.25) substitute K for the first $(d-n)$ of the K_i's and $\tilde{\mathbf{I}}(B)$ for the remaining $(n-1)$ of them. ∎

6.4 Integral geometry in Minkowski spaces

This section is concerned with some analogues of various classical formulas from integral geometry. Extensions of such formulas to an affine setting were first considered by Busemann [75]. We consider them in a general Minkowski space and place them here because the ones dealing with area are intimately connected with the particular definition of surface area considered in this chapter. For the details of this aspect we refer to Wieacker [533]. The two sources for this section are El-Ekhtiar [142] and Wieacker [533], although there are a variety of other papers dealing with integral geometry in the Minkowski plane, *e.g.* Chakerian [97], Constantin [116] and Peri [408, 409]. A more comprehensive treatment of the results of this section can now be found in Schneider and Wieacker [483].

The idea is that the metric geometry of a Minkowski space $(X, \|.\|)$ imposes a translation-invariant measure on sets of flats in X. Here we will restrict attention to lines and hyperplanes. One can then use these measures to define the areas of rectifiable surfaces. The surface area of a convex body can be specified, for example, as the measure of the set of lines meeting the convex body. Alternatively, the surface area can be obtained as the measure of the collection of those sets consisting of $(d-1)$ hyperplanes whose intersection meets the convex body. In both cases this leads to a surface area consistent with that of Example 5.1.4(ii). There is, however, a complication in the case of measures on hyperplanes. One would like the length of a curve to be proportional to the measure of the set of those hyperplanes that meet it. In particular, one would like to recover the metric as the length of a line segment in this way. However, this is not possible in general. In order that the metric be generated by a measure on hyperplanes it is necessary and sufficient that the metric be a *hypermetric* (see Kelly [285]). This condition can be characterized by saying that the unit ball B must be the dual of a zonoid.

As in much of the preceding work, it is convenient if the space $(X, \|.\|)$ has an auxiliary Euclidean structure with respect to which one can make various calculations. In fact, we shall define the measures on lines and hyperplanes in terms of this structure.

A line L is determined by a direction $\hat{u}_L$ and a point x_L in the hyperplane H_L through 0 and perpendicular to $\hat{u}_L$. Thus

$$L = \{y : y = x_L + \alpha \hat{u}_L, \ \alpha \in \mathbb{R}, \ x_L \in H_L\}.$$

The measure on sets of points x in H_L is $(d-1)$-dimensional Lebesgue measure coming from the Euclidean structure. The measure on a set of directions $\hat{u}$ is the surface area measure induced by B°. The measure of a set $\mathcal{L}$ of lines is the product of these measures. More precisely:

Definition 6.4.1 *If $\mathcal{L}$ is a set of lines in $(X, \|.\|)$ then, for each direction $\hat{u}$, set $x_\mathcal{L}(\hat{u}) := \{x_L : L \in \mathcal{L}, \ L \text{ has direction } \hat{u}\}$. The **measure** ν **of** $\mathcal{L}$ is then*

$$\nu(\mathcal{L}) := \int_{\partial B^\circ} \lambda^{d-1}(x_\mathcal{L}(\hat{u})) \, d\lambda^{d-1}(\hat{u}) = \int_{S^{d-1}} \lambda^{d-1}(x_\mathcal{L}(\hat{u})) \, d\nu_{B^\circ}(\hat{u}). \quad (6.26)$$

Similarly, a hyperplane H is determined by a Euclidean unit vector $\hat{f}_H$ in X^* and a real number α_H. Thus

$$H = \{x \in X : \hat{f}_H(x) = \alpha_H\}.$$

The measure on a set of real numbers is Lebesgue measure on the line and the measure on a set of directions $\hat{f}$ is the surface area measure induced by B. Then the measure ν of a set of hyperplanes $\mathcal{H}$ is the product of these measures.

Definition 6.4.2 *If $\mathcal{H}$ is a set of hyperplanes in $(X, \|.\|)$ then, for each direction $\hat{f}$, set $\alpha_\mathcal{H}(\hat{f}) := \{\alpha_H : H \in \mathcal{H}, \ H \text{ has direction } \hat{f}\}$. The **measure** ν **of** $\mathcal{H}$ is then*

$$\nu(\mathcal{H}) := \int_{\partial B} \lambda^1(\alpha_\mathcal{H}(\hat{f})) \, d\lambda^{d-1}(\hat{f}) = \int_{S^{d-1}} \lambda^1(\alpha_\mathcal{H}(\hat{f})) \, d\nu_B(\hat{f}). \quad (6.27)$$

These measures are less arbitrary than they appear. Indeed, it can be shown, in the same way that Schneider [476] did for measures on hyperplanes, that each translation-invariant measure ν' on lines must be of the form

$$\nu'(\mathcal{L}) = \int_{S^{d-1}} \lambda(x_\mathcal{L}(\hat{u})) d\theta(\hat{u}),$$

where θ is some even measure on the Euclidean sphere S^{d-1}. By Minkowski's theorem this can be interpreted as the surface measure of a symmetric convex body B'. The same is true (by Schneider's result [476]) for translation-invariant measures on hyperplanes. Therefore, in Definitions 6.4.1 and 6.4.2 the only arbitrariness is the choice of convex body for the surface area measure.

Theorem 6.4.3 (El-Ekhtiar) *If K is a convex body in a Minkowski space $(X, \|.\|)$ then*

$$\int_{L \cap K \neq \emptyset} dv(L) = \epsilon_{d-1} \mu_B(\partial K).$$

Proof. From the definition of v we have

$$\int_{L \cap K \neq \emptyset} dv(L) = \int_{\partial B^\circ} \int_{L \cap K \neq \emptyset} d\lambda(x(\hat{u})) \, d\lambda(\hat{u}).$$

But the region in the hyperplane H perpendicular to $\hat{u}$ that is determined by those lines L in the direction $\hat{u}$ that intersect K is precisely the orthogonal projection of K on H. Therefore,

$$\int_{L \cap K \neq \emptyset} d\lambda(x(\hat{u})) = V(K[d-1], [\hat{u}]) = 2^{-1} \int_{\partial K} |\hat{f}(\hat{u})| \, d\lambda(\hat{f}).$$

Hence,

$$\int_{L \cap K \neq \emptyset} dv(L) = 2^{-1} \int_{\partial B^\circ} \int_{\partial K} |\hat{f}(\hat{u})| \, d\lambda(\hat{f}) \, d\lambda(\hat{u})$$

$$= 2^{-1} \int_{\partial K} \int_{\partial B^\circ} |\hat{f}(\hat{u})| \, d\lambda(\hat{u}) \, d\lambda(\hat{f})$$

$$= \epsilon_{d-1} \int_{\partial K} \sigma_B(\hat{f}) \, d\lambda(\hat{f}) = \epsilon_{d-1} \mu_B(\partial K). \blacksquare$$

The preceding argument can be reversed, as the next result shows.

Theorem 6.4.4 *If v is a translation-invariant measure on lines and if the surface area of a convex body is defined by $v(\partial K) := \int_{L \cap K \neq \emptyset} dv(L)$ then $v(\partial K) = \epsilon_{d-1} \mu_B(\partial K)$ for some symmetric convex body B.*

Proof. Since v is translation invariant, Schneider's argument in [476] shows that

$$v(\mathcal{L}) = \int_{S^{d-1}} \lambda(x_\mathcal{L}(\hat{u})) \, d\theta(\hat{u})$$

for some even finite measure θ on the boundary of the Euclidean ball. But then Minkowski's theorem shows that there is a symmetric body B° such that θ is the surface area measure of B°; hence,

$$v(\mathcal{L}) = \int_{\partial B^\circ} \lambda(x_\mathcal{L}(\hat{u})) \, d\lambda(\hat{u}).$$

Then, as before,

$$v(\partial K) = \int_{L\cap K\neq \emptyset} dv(L) = \int_{\partial B^\circ} V(K[d-1],[\hat{u}])\, d\lambda(\hat{u})$$

$$= \int_{\partial B^\circ} \int_{\partial K} |\hat{f}(\hat{u})|\, d\lambda(\hat{f})\, d\lambda(\hat{u})$$

$$= \epsilon_{d-1} \int_{\partial K} \sigma_B(\hat{f})\, d\lambda(\hat{f}) = \epsilon_{d-1}\mu_B(\partial K). \qquad \blacksquare$$

Theorem 6.4.5 (El-Ekhtiar) *If K is a convex body in a d-dimensional Minkowski space $(X, \|.\|)$ then*

$$\int \mu(L \cap K)\, dv(L) = d\epsilon_d\, \mu(K).$$

On the left, $\mu(L \cap K)$ denotes the Minkowski length of the line segment $L \cap K$ and, on the right, $\mu(K)$ denotes the Minkowski volume of K.

Proof. The Minkowski length $\mu(L \cap K)$ of the line segment $L \cap K$ can be expressed in Euclidean terms by

$$\mu(L \cap K) = \lambda(L \cap K)/\rho_B(\hat{u}) = h_{B^\circ}(\hat{u})\lambda(L \cap K),$$

where $\hat{u}$ is the direction of L. Hence, using Definition 6.4.1,

$$\int \mu(L \cap K)\, dv(L) = \int_{\partial B^\circ} \left(\int \lambda(L \cap K)\, d\lambda(x(\hat{u})) \right) h_{B^\circ}(\hat{u})\, d\lambda(\hat{u}).$$

As in the previous theorems, the effective region of integration of the first integral is the region where $L \cap K \neq \emptyset$ so that $\lambda(L \cap K) \geq 0$ and this is the orthogonal projection of K on the hyperplane perpendicular to $\hat{u}$. Integrating $\lambda(L \cap K)$ over this region is precisely the evaluation of the (Euclidean) volume of K. Therefore,

$$\int \mu(L \cap K)\, dv(L) = \int_{\partial B^\circ} \lambda(K) h_{B^\circ}(\hat{u})\, d\lambda(\hat{u}) = \lambda(K)\lambda(B^\circ) d.$$

However, $\mu(K) = \sigma_B^d \lambda(K) = \epsilon_d^{-1} \lambda(B^\circ)\lambda(K)$ and so

$$\int \mu(L \cap K)\, dv(L) = d\epsilon_d \mu(K). \qquad \blacksquare$$

Remark. Since Theorem 6.4.5 deals with volumes rather than areas, the essential feature that $\int \mu(L \cap K)\, dv(L)$ is proportional to the volume of K is independent of which area function is used (the area function does not enter the statement or the proof). However, the result that the constant of proportionality is precisely $d\epsilon_d$ (the surface area of a Euclidean unit ball) depends on which normalization of d-dimensional Haar measure is used.

6.4 Integral geometry in Minkowski spaces

We now turn our attention to measures on hyperplanes. The first result is independent of which particular surface measure is used.

Theorem 6.4.6 *If K is a convex body in a d-dimensional Minkowski space $(X, \|.\|)$ then*

$$\int \mu(H \cap K) \, dv(H) = \mu_B(\partial B)\lambda(K).$$

Proof. From the definition of v we have

$$\int \mu(H \cap K) \, dv(H) = \int_{\partial B} \int_{\mathbb{R}} \mu(H \cap K) \, d\alpha \, d\lambda(\hat{f})$$

$$= \int_{\partial B} \int_{\mathbb{R}} \lambda(H \cap K) \, d\alpha \, \sigma(\hat{f}) \, d\lambda(\hat{f})$$

$$= \int_{\partial B} \lambda(K)\sigma(\hat{f}) \, d\lambda(\hat{f}) = \lambda(K)\mu_B(\partial B). \blacksquare$$

Next we consider the possibility of defining length by means of a measure on hyperplanes.

Theorem 6.4.7 *If $(X, \|.\|)$ is a d-dimensional Minkowski space with unit ball B then there exists a translation-invariant measure v_B on hyperplanes such that*

$$\|x\| = 2^{-1} \int_{H \cap [0,x] \neq \emptyset} dv_B(H)$$

if and only if B° is a zonoid.

Proof. First, suppose that B° is a zonoid so that B° is the projection body of some symmetric convex body K in X. To prove that there is a suitable measure, consider the measure v defined on hyperplanes by Definition 6.4.2 but with K replacing B in (6.27). Then, for a fixed direction $\hat{f}$, we have

$$\lambda\{\alpha_H : H \cap [0, x] \neq \emptyset, \hat{f}_H = \hat{f}\} = |\hat{f}(x)|$$

and hence

$$2^{-1} \int_{H \cap [0,x] \neq \emptyset} dv(H) = 2^{-1} \int_{\partial K} |\hat{f}(x)| \, d\lambda(\hat{f}) = V(K[d-1], [x])$$

$$= h_{\Pi(K)}(x) = h_{B^\circ}(x) = \|x\|_B.$$

Conversely, suppose a functional $\|.\|$ is defined by

$$\|x\| := 2^{-1} \int_{H \cap [0,x] \neq \emptyset} dv(H)$$

for some translation-invariant measure ν on hyperplanes. Then, using Schneider's result [476], $\nu(\mathcal{H}) = \int_{S^{d-1}} \lambda(\alpha_\mathcal{H}(\hat{f})) \, d\theta(\hat{f})$ for some even measure θ on the surface of the Euclidean ball. Therefore,

$$\|x\| = 2^{-1} \int_{H \cap [0,x] \neq \emptyset} d\nu(H) = 2^{-1} \int_{\partial E} |\hat{f}(x)| \, d\theta(\hat{f}).$$

As before, we apply Minkowski's theorem to assert that θ arises as the surface measure of some convex body K and hence

$$\|x\| = 2^{-1} \int_{\partial K} |\hat{f}(x)| \, d\lambda(\hat{f}) = h_{\Pi K}(x).$$

This equation shows that $\|.\|$ is a norm and that it arises as the support function of the zonoid ΠK, which implies that the unit ball corresponding to $\|.\|$ is the dual of that zonoid. ∎

Remark. Note that the factor $\frac{1}{2}$ is natural here because the measure ν counts each hyperplane twice, once with direction $\hat{f}$ and once with direction $-\hat{f}$.

El-Ekhtiar [142] has shown that a similar result holds if one considers, instead of a hyperplane, translates of the boundary of K where $\Pi(K) = B^\circ$. The integration, however, is with respect to the Euclidean measure on X.

Theorem 6.4.8 *If $(X, \|.\|)$ is a Minkowski space with unit ball B and if $B^\circ = \Pi K$ then, for all x in X, we have*

$$\|x\|_B = 2^{-1} \int_X \#\{(y + \partial K) \cap [0, x]\} \, d\lambda(y)$$

where $\#\{A\}$ denotes the cardinality of A and $\lambda(y)$ denotes d-dimensional Lebesgue measure at y in X.

Proof. Note, first, that $(y + \partial K) \cap [0, x] \neq \emptyset$ if and only if $\partial K \cap [-y, x - y] \neq \emptyset$ so that we shall think of translating $[0, x]$ rather than translating ∂K. Next, to avoid complications due to the segment $[-y, x - y]$ meeting ∂K in more than one point, divide ∂K into three pieces along the shadow boundary S_x determined by x, i.e. $\partial K = K_x^+ \cup S_x \cup K_x^-$, where

$$K_x^+ := \{y \in \partial K : y + \alpha x \in \operatorname{int} K \text{ for small } \alpha > 0\},$$
$$K_x^- := \{y \in \partial K : y + \alpha x \in \operatorname{int} K \text{ for small } \alpha < 0\},$$
$$S_x := \{y \in \partial K : y + \alpha x \notin \operatorname{int} K \ \forall \alpha\}.$$

Since all measurements are translation invariant, translate these pieces to new positions $K_x'^+ := K_x^+ - 2x$, $S_x' := S_x$ and $K_x'^- := K_x^- + 2x$ so that a line segment

$[-y, x - y]$ meets at most one piece. Then, as in the proof of Theorem 6.4.2, we have

$$\lambda \left(\{y : [-y, x - y] \cap K_x'^+ \neq \emptyset\}\right) = \lambda \left(\{y : [-y, x - y] \cap K_x'^- \neq \emptyset\}\right)$$
$$= V(K[d - 1], [x])$$

and $\lambda(\{y : [-y, x - y] \cap S_x' \neq \emptyset\}) = 0$. Hence

$$\int_X \#\{(y + \partial K) \cap [0, x]\} d\lambda(y)$$

$$= \int_X \#\{\partial K \cap [-y, x - y]\} d\lambda(y)$$

$$= \int_X \#\{K_x^+ \cap [-y, x - y]\} d\lambda(y) + \int_X \#\{S_x \cap [-y, x - y]\} d\lambda(y)$$

$$+ \int_X \#\{K_x^- \cap [-y, x - y]\} d\lambda(y)$$

$$= \int_X \#\{K_x'^+ \cap [-y, x - y]\} d\lambda(y) + 0 + \int_X \#\{K_x'^- \cap [-y, x - y]\} d\lambda(y)$$

$$= 2V(K[d - 1], [x])$$

$$= 2h_{\Pi K}(x) = 2h_{B^\circ}(x) = 2\|x\|. \qquad \blacksquare$$

The two preceding theorems can both be extended to polygonal arcs composed of finitely many line segments.

Theorem 6.4.9 *If $B^\circ = \Pi(K)$ and if $(X, \|.\|)$ is the Minkowski space with unit ball B then, for every polygonal arc $\wp$,*

(i) $\mu_B(\wp) = 2^{-1} \int \#\{H \cap \wp\} d\nu_K(H)$,
(ii) $\mu_B(\wp) = 2^{-1} \int \#\{(y + \partial K) \cap \wp\} d\lambda(y)$.

In (i) ν_K denotes the measure defined by (6.27) with K replacing B and in (ii) λ is Lebesgue measure in X.

With some care about the way in which a rectifiable curve is approximated by a sequence of polygonal arcs one can use a limiting argument to extend this last theorem to rectifiable arcs in X [142, 533]. Wieacker [533] (see also [532]) has extended the ideas to a general setting. The length of a rectifiable curve $\mathcal{M}_1$ is the integral $2^{-1} \int \#\{\mathcal{M}_1 \cap H\} d\nu(H)$. If $\mathcal{M}_2$ is a two-dimensional rectifiable surface then $c_2 \int \#\{\mathcal{M}_2 \cap H_1 \cap H_2\} d\nu(H_1) d\nu(H_2)$ is the two-dimensional surface area of $\mathcal{M}_2$. In general, if $\mathcal{M}_k$ is a k-dimensional rectifiable Borel subset of X (*i.e.* the image in X of a bounded set in $\mathbb{R}^k$ under a Lipschitz map) then its k-dimensional surface area is

$$\nu_k(\mathcal{M}_k) := c_k \int \#\{\mathcal{M}_k \cap H_1 \cap H_2 \cap \cdots \cap H_k\}\, dv(H_1)\ldots dv(H_k),$$

where c_k is some constant depending only on k. In particular, when $k = (d-1)$ we can regard $H_1 \cap H_2 \cap \cdots \cap H_{d-1}$ as defining a line and the product measure $\nu \times \nu \times \cdots \times \nu$ ($d-1$ factors) as a translation-invariant measure on lines. In this case we recover the situation of Theorem 6.4.3. Wieacker proves that if $(X, \|.\|)$ is a Minkowski space whose unit ball B is the dual of a zonoid then

$$\nu_k(\mathcal{M}_k) = \mu_B^{(k)}(\mathcal{M}_k);$$

i.e. the surface measure ν_k coincides with that of Example 5.1.4(ii) in all dimensions.

6.5 Bounds for the surface area of B

One of the central theorems in Chapter 4 is Golab's theorem that the length of the boundary of the unit ball lies in the interval [6, 8]:

$$6 \leq \mu_B(\partial B) \leq 8$$

with equality on the left if and only if B is an affine regular hexagon and equality on the right if and only if B is a parallelogram. Since $\mu_B(\partial B) = \mu_{B°}(\partial B°)$, in each case the respective bound is also attained by $B°$. In both cases, however, $B°$ is an affine image of B. In this section and the corresponding one in the next chapter we are concerned with the analogues of these inequalities in higher dimensions and ask what numbers replace 6 and 8 for $d \geq 3$.

It is an immediate consequence of the Blaschke–Santaló inequality (Theorem 2.3.3) that every upper bound for Busemann's definition of surface area is also an upper bound in the present case and every lower bound here is a lower bound there. We shall see in the next chapter that for the Busemann definition there is a precise upper bound which is attained only by parallelotopes. That bound is, therefore, an upper bound here but no longer a precise one; it is, however, the only result in this direction. In the other direction an inequality of Petty, the Petty projection inequality, combined with other inequalities gives a reasonable (but not exact) lower bound if B is either a zonoid or the dual of a zonoid. For other balls one needs Mahler's conjecture (see §2.6) to obtain the same result.

The following proposition is a simple consequence of the Blaschke–Santaló inequality.

Proposition 6.5.1 *If K is a convex body in a d-dimensional Minkowski space and if $\mu_{(1)}$ and $\mu_{(2)}$ refer to the Minkowski surface measures from Examples 5.1.4(i)* *and (ii) respectively then*

$$\mu_{(2)}(\partial K) \leq \mu_{(1)}(\partial K).$$

Proof. We have

$$\mu_{(2)}(\partial K) = \int_{\partial K} \epsilon_{d-1}^{-1} \lambda^*((B \cap f^\perp)^\circ) \, d\lambda(\hat{f})$$

and

$$\mu_{(1)}(\partial K) = \int_{\partial K} \epsilon_{d-1} \lambda(B \cap f^\perp)^{-1} \, d\lambda(\hat{f}).$$

However, since $\lambda^*((B \cap f^\perp)^\circ)\lambda(B \cap f^\perp) \leq \epsilon_{d-1}^2$ (Theorem 2.3.3), the first integrand is less than or equal to the second, from which the result follows. ∎

In Theorem 7.4.1 we shall show that $\mu_{(1)}(\partial B) \leq 2d\epsilon_{d-1}$ with equality if and only if B is a parallelotope. It follows that this is also a bound for $\mu_{(2)}(\partial B)$. For the rest of the chapter, μ_B (or μ) refers to $\mu_{(2)}$.

Theorem 6.5.2 *If B is the unit ball in a d-dimensional Minkowski space then $\mu_B(\partial B) \leq 2d\epsilon_{d-1}$. The only case when equality occurs here is if $d = 2$ and B is a parallelogram.*

Proof. The first statement is a direct consequence of Proposition 6.5.1 and Theorem 7.4.1 below.

In the case when $d = 2$, we showed in Theorem 4.3.7 that equality occurs if and only if B is a parallelogram. If $d \geq 3$ and if B is not a parallelotope then $\mu_B(\partial B) \leq \mu_{(1)}(\partial B) < 2d\epsilon_{d-1}$ (see Theorem 7.4.1); if B is a parallelotope then $\mu_B(\partial B) = \mu_{B^\circ}(\partial B^\circ) \leq \mu_{(1)}(\partial B^\circ) < 2d\epsilon_{d-1}$ (from Corollary 6.3.2 and Theorem 7.4.1). ∎

Next we turn to the question of possible lower bounds for $\mu_B(\partial B)$. For this we need the Mahler–Reisner inequality (Theorem 2.3.4), which states that

$$\lambda(K)\lambda^*(K^\circ) \geq 4^d/d!$$

if K is a zonoid or the dual of a zonoid and that equality holds if and only if K is a parallelotope or cross-polytope.

In what follows and in §7.4 we shall let

$$m_d := \min\{\lambda(B)\lambda^*(B^\circ) : B \text{ a symmetric convex body in } \mathbb{R}^d\}. \qquad (6.28)$$

Bourgain and Milman [56] have proved that $m_d \geq c^d \epsilon_d^2$ for some positive constant c. Some of the history of the information that is known about m_d was given in §2.6.

Theorem 6.5.3 *If B is the unit ball of a d-dimensional Minkowski space then* $\mu_B(\partial B) \geq 2m_{d-1}/\epsilon_{d-1}$.

Proof. Since $\mu_B(\partial K) = dV(K[d-1], \mathbf{I})$ it follows from Proposition 2.3.7(ii) that if $K \subseteq B$ then $\mu_B(\partial K) \leq \mu_B(\partial B)$.

Consider a cross-section of B by an arbitrary hyperplane $f^\perp$ and let $K_\eta := \{x \in B : -\eta \leq f(x) \leq \eta\}$ so that K_η is a thin "disc" about the cross-section $B \cap f^\perp$. Since $K_\eta \subseteq B$ we have $\mu_B(\partial K_\eta) \leq \mu_B(\partial B)$. But as $\eta \to 0$,

$$\lim_{\eta \to 0} \mu_B(\partial K_\eta) = 2\epsilon_{d-1}^{-1} \lambda(B \cap f^\perp) \lambda^*((B \cap f^\perp)^\circ) \geq 2m_{d-1}/\epsilon_{d-1},$$

from which the result follows. ∎

Corollary 6.5.4 *If B has a cross-section that is either a zonoid or the dual of a zonoid then* $\mu_B(\partial B) \geq 2^{2d-1}/\epsilon_{d-1}(d-1)!$.

Remarks

(i) Corollary 6.5.4 should be compared with the value for a cube C (and cross-polytope) for which $\mu_C(\partial C) = d2^{2d-1}/\epsilon_{d-1}(d-1)!$.

(ii) Corollary 6.5.4 raises the question of what kind of cross-sections an arbitrary symmetric convex body can have. For example, is it possible for a symmetric polytope to have all its cross-sections neither zonotopes nor duals of zonotopes? It is not possible for a polytope to have all its cross-sections either parallelotopes or cross-polytopes, so the use of $4^{d-1}/(d-1)!$ is clearly not the best possible here. This leads to the following minimax problem,

$$\min_B \max_f \{\lambda(B \cap f^\perp) \lambda^*((B \cap f^\perp)^\circ)\},$$

where the maximum is over all linear functionals f and the minimum is over all symmetric convex bodies or over all symmetric polytopes.

For the next theorem we follow the argument given in [509]. In addition to the inequalities already given in Chapter 2 we need the Petty projection inequality [418]. Lutwak [342] (see also [334]) gives a full account of this inequality and its relationship to the Busemann intersection inequality. That inequality is needed in the next chapter. Proofs of both inequalities are contained in Gardner [172].

Theorem 6.5.5 (Petty projection inequality) *If K is a convex body in $\mathbb{R}^d$ then* $\lambda(K)^{d-1}\lambda((\Pi K)^\circ) \leq (\epsilon_d/\epsilon_{d-1})^d$ *with equality if and only if K is an ellipsoid.*

Theorem 6.5.6 *If B is the unit ball in a d-dimensional Minkowski space then*

$$\mu_B(\partial B) \geq (d/\epsilon_d)(4^d/d!)^{1/d}(m_d)^{(d-1)/d}.$$

Proof. Since $\mu_B(\partial K) = dV(K[d-1], \mathbf{I})$ if we use the Minkowski inequality (Theorem 2.3.8) and the identification of $\mathbf{I}$ (Equation (5.25)) then we get

$$\mu_B(\partial B)^d = d^d V(B[d-1], \mathbf{I})^d \geq d^d \lambda(B)^{d-1} \lambda(\mathbf{I})$$
$$= (d/\epsilon_{d-1})^d \lambda(B)^{d-1} \lambda(\Pi B^\circ).$$

Since ΠB° is always a zonoid, we may use the Mahler–Reisner inequality (Theorem 2.3.4) to obtain

$$\mu_B(\partial B)^d \geq \left(\frac{4d}{\epsilon_{d-1}}\right)^d \frac{\lambda(B)^{d-1}}{d!\lambda^*((\Pi B^\circ)^\circ)}.$$

The next step is to apply the Petty projection inequality to B° in X^*. This gives

$$\mu_B(\partial B)^d \geq \left(\frac{4d}{\epsilon_d}\right)^d \frac{\lambda(B)^{d-1} \lambda^*(B^\circ)^{d-1}}{d!}.$$

Finally, the definition of m_d gives the stated result. ∎

Corollary 6.5.7 *If B is the unit ball in a d-dimensional Minkowski space and if B is either a zonoid or the dual of a zonoid then*

$$\mu_B(\partial B) \geq 4^d/\epsilon_d(d-1)!.$$

Proof. If B is either a zonoid or the dual of a zonoid then in the last step of the proof of the theorem we may use the Mahler–Reisner inequality once more and replace m_d by $4^d/d!$. ∎

Remarks

(i) In the case when $d = 2$, Corollary 6.5.7 gives $\mu_B(\partial B) \geq 16/\pi$. Considering the variety of inequalities that have been used, this is surprisingly close to the actual bound of 6.

(ii) When $d = 3$, Corollary 6.5.7 gives $\mu_B(\partial B) \geq 24/\pi$ (when B is a zonoid). The smallest value of $\mu_B(\partial B)$ that we have found (see [262]) is $36/\pi$ in the case when B is either the rhombic dodecahedron (which is a zonoid) or its dual (the cubo-octahedron). These are Z_3 and $S_3 - S_3$ (see Example 1.1.17) respectively.

6.6 Miscellaneous properties

The main unresolved problem concerning the isoperimetrix is whether or not, in dimensions $d \geq 3$, there are unit balls B (other than ellipsoids) for which B and $\mathbf{I}(B)$ are homothetic. This question can be restated by asking if there are analogues in higher dimensions of the Radon curves discussed in Chapter 4. The conjecture is that there are not. This view is based on the fact that other characterizations of Radon curves either by symmetry of normality or by projections of norm 1 do characterize ellipsoids for $d > 2$. In this section we give some results that are partial answers to this question. These results are followed by a final theorem (6.6.11) that, while unrelated to its predecessors, gives some further algebraic properties of the Minkowski volume and area functions considered in this chapter.

First we modify the problem stated above and ask instead (see Busemann and Petty [85]) for what bodies B is the normalized isoperimetrix $\tilde{\mathbf{I}}(B)$ actually equal to B. The Petty projection inequality gives a precise answer to this question.

Theorem 6.6.1 *If B is the unit ball in a Minkowski space $(X, \|.\|)$ then $\lambda^*(\tilde{\mathbf{I}}(B)^\circ) \leq \lambda^*(B^\circ)$ with equality if and only if B is an ellipsoid.*

Proof. From the facts that $\mathbf{I}(B) = \epsilon_{d-1}^{-1}(\Pi(B^\circ))$ and that $\tilde{\mathbf{I}}(B) = (\sigma_B^d)^{-1}\mathbf{I}(B)$ we have

$$\tilde{\mathbf{I}}(B) = \epsilon_d \lambda^*(B^\circ)^{-1} \epsilon_{d-1}^{-1}(\Pi(B^\circ))$$

and therefore

$$\tilde{\mathbf{I}}(B)^\circ = (\epsilon_{d-1}/\epsilon_d)\lambda^*(B^\circ)(\Pi B^\circ)^\circ.$$

Hence the statement of the theorem is exactly a reformulation of the Petty projection inequality (Theorem 6.5.5). ∎

Corollary 6.6.2 *If $\tilde{\mathbf{I}}(B) \subseteq B$ then B is an ellipsoid and $\tilde{\mathbf{I}}(B) = B$.*

Proof. If $\tilde{\mathbf{I}}(B) \subseteq B$ then $\tilde{\mathbf{I}}(B)^\circ \supseteq B^\circ$ and the result follows from the theorem. ∎

Remark. It follows that if $\tilde{\mathbf{I}}$ is homothetic to B then $\tilde{\mathbf{I}} = \rho B$ for some scalar $\rho \geq 1$ with equality if and only if B is an ellipsoid. This is so even when $d = 2$. In this case $\tilde{\mathbf{I}}$ is homothetic to B if and only if ∂B is a Radon curve. For such curves other than ellipses, $\rho > 1$; e.g. if B is an affine regular hexagon then $\rho = \pi/3$.

Corollary 6.6.3 *If the space X is considered with the two balls B and $\tilde{\mathbf{I}}$ and if the Minkowski volumes with respect to these balls are denoted by μ_B and $\mu_{\tilde{\mathbf{I}}}$ respectively then $\mu_{\tilde{\mathbf{I}}}(A) \leq \mu_B(A)$ for every measurable set A in X.*

Proof. From the normalization of Haar measure λ in X we have

$$\mu_{\tilde{\mathbf{I}}}(A) = \sigma_{\tilde{\mathbf{I}}}^d \lambda(A) \quad \text{and} \quad \mu_B(A) = \sigma_B^d \lambda(A).$$

Because $\sigma_B^d = \epsilon_d^{-1}\lambda^*(B^\circ)$ and $\sigma_{\tilde{\mathbf{I}}}^d = \epsilon_d^{-1}\lambda^*(\tilde{\mathbf{I}}^\circ)$ the result is a direct consequence of the theorem. ∎

Remark. In particular, in Corollary 6.6.3 we may take A to be either $\tilde{\mathbf{I}}$ or B, which yield $\mu_{\tilde{\mathbf{I}}}(\tilde{\mathbf{I}}) \leq \mu_B(\tilde{\mathbf{I}})$ and $\mu_{\tilde{\mathbf{I}}}(B) \leq \mu_B(B)$ respectively. However,

$$\mu_B(B) = \gamma(B) = \epsilon_d^{-1}\lambda(B)\lambda^*(B^\circ)$$

(Equation (5.9)) and similarly for $\mu_{\tilde{\mathbf{I}}}(\tilde{\mathbf{I}})$. Therefore, the Blaschke–Santaló inequality gives $\mu_B(B) \leq \epsilon_d$ and $\mu_{\tilde{\mathbf{I}}}(\tilde{\mathbf{I}}) \leq \epsilon_d$. It would be highly satisfactory if these five quantities were linearly ordered in the following chain:

$$\mu_{\tilde{\mathbf{I}}}(B) \leq \mu_B(B) \underset{(*)}{\leq} \mu_{\tilde{\mathbf{I}}}(\tilde{\mathbf{I}}) \leq \epsilon_d \underset{(**)}{\leq} \mu_B(\tilde{\mathbf{I}}).$$

Inequality ($*$), which is true with equality when $d = 2$, means that the volume product does not decrease under the mapping $\tilde{\mathbf{I}}$ and hence that $\tilde{\mathbf{I}}$ is, in general, "more ellipsoidal" than B. Inequality ($**$), which by Theorem 4.4.2 is also true with *equality* when $d = 2$, is equivalent to a strengthening of the Petty projection inequality by a factor of $\lambda^*(\Pi K)\lambda((\Pi K)^\circ)/\epsilon_d^2$. This is Petty's conjectured projection inequality, which Lutwak [342] describes as "possibly the major open problem in the area of affine isoperimetric inequalities" (see also Lutwak [338]).

These inequalities are also connected to the isoperimetric ratio in the space $(X, \|.\|)$.

Theorem 6.6.4 *If K is a convex body in a Minkowski space $(X, \|.\|)$ with unit ball B then*

$$\frac{\mu_B(\partial K)^d}{\mu_B(K)^{d-1}} \geq \frac{(4d)^d}{d!\epsilon_d}.$$

Proof. Since $\mathbf{I}$ is the solution to the isoperimetric problem in X we have

$$\frac{\mu_B(\partial K)^d}{\mu_B(\partial K)^{d-1}} \geq \frac{\mu_B(\partial \mathbf{I})^d}{\mu_B(\mathbf{I})^{d-1}} = \frac{\mu_B(\partial \tilde{\mathbf{I}})^d}{\mu_B(\tilde{\mathbf{I}})^{d-1}} = d^d \mu_B(\tilde{\mathbf{I}})$$

(the last equation is (5.19)). Hence, using Corollary 6.6.3 with $A = \tilde{\mathbf{I}}$, we get

$$\frac{\mu_B(\partial K)^d}{\mu_B(K)^{d-1}} \geq d^d \mu_{\tilde{\mathbf{I}}}(\tilde{\mathbf{I}}) = \frac{d^d \lambda(\tilde{\mathbf{I}})\lambda^*(\tilde{\mathbf{I}}^\circ)}{\epsilon_d}.$$

Since $\tilde{\mathbf{I}}$ is a zonoid we may now use the Mahler–Reisner inequality to complete the proof. ∎

Remark. Inequality ($**$) (Petty's conjectured projection inequality) would imply that $\mu_B(\partial \mathbf{I})^d/\mu_B(\mathbf{I})^{d-1} \geq d^d \epsilon_d$, i.e. that the isoperimetric ratio is not less than that in Euclidean space with equality if and only if the space is Euclidean.

Further evidence that for $d \geq 3$ it is unlikely that $\mathbf{I}(B)$ is homothetic to B is given by the next theorem. If $d = 2$ and B is a polygon with n vertices (and edges) then $\mathbf{I}(B)$ is also an n-gon. This does not happen in higher dimensions, as the following combinatorial argument shows. We observe that the statement of the theorem is also true for $d = 1, 2$.

Theorem 6.6.5 *If $d \geq 3$ and if B is a polytope in $\mathbb{R}^d$ with $2n$ vertices then $\mathbf{I}(B)$ has at least $2^{d-1}(n - (d - 2))$ vertices.*

Proof. The proof is by induction on the dimension d. In order to clarify the proof of the inductive step we begin with $d = 3$.

Let $v_1, -v_1$ be two opposite vertices of B and consider a plane H through the origin that contains these vertices. If H is rotated about the line $[v_1, -v_1]$ it passes, in turn, through all the remaining vertices of B. There are a finite number, k say, of positions of H in which it contains other vertices. Let us suppose that in the ith position it contains $2n_i$ vertices in total (including $\pm v_1$). Then, on the one hand, we obtain

$$2n = \sum_{i=1}^{k}(2n_i - 2) + 2,$$

i.e. since $d = 3$,

$$\sum_{i=1}^{k}(n_i - (d - 2)) = n - (d - 2).$$

On the other hand, to the ith position of H there correspond two facets of $\mathbf{I}(B)$ parallel to that position, each facet containing $2n_i$ vertices. The set of $2k$ such facets forms a zone of $\mathbf{I}(B)$ with each facet joined to the next by a line segment parallel to $[-v_1, v_1]$ containing two vertices. Consequently, the total number of vertices in this zone is $2\sum_{i=1}^{k}(2n_i - 2)$. But $\mathbf{I}(B)$ has at least as many vertices as this zone does and therefore $\mathbf{I}(B)$ has at least $4\sum_{i=1}^{k}(n_i - 1) = 4(n - 1)$ vertices. However, since $d = 3$, we may write $4(n - 1)$ as $2^{d-1}(n - (d - 2))$.

Now suppose the proposition is valid for dimension $(d - 1)$. Consider $(d - 2)$ linearly independent vertices $v_1, v_2, \ldots, v_{d-2}$ (and their negatives). Consider a hyperplane H through the origin containing these vertices. If H is rotated about the subspace $L = \text{span}\{v_1, v_2, \ldots, v_{d-2}\}$ it passes, in turn, through the remaining vertices of B. We may suppose L contains no other vertices. Again there are a finite number, k, of positions in which H contains other vertices and we suppose that the total number in position i is $2n_i$. Hence

$$2n = \sum_{i=1}^{k} 2(n_i - (d - 2)) + 2(d - 2),$$

$$\sum_{i=1}^{k}(n_i - (d - 2)) = n - (d - 2). \tag{6.29}$$

Corresponding to the ith position, $\mathbf{I}(B)$ has a facet which is a $(d-1)$-dimensional zonotope. By the induction hypothesis, this zonotope contains at least $2^{d-2}(n_i - (d-3))$ vertices. The set of $2k$ such facets forms a zone of $\mathbf{I}(B)$ with each facet joined to the next by a $(d-2)$-dimensional cube parallel to L. Since such a cube has 2^{d-2} vertices, the total number of vertices in this zone is

$$2\sum_{i=1}^{k}[2^{d-2}(n_i - (d-3)) - 2^{d-2}] = 2^{d-1}\sum_{i=1}^{k}(n_i - (d-2)). \tag{6.30}$$

Hence $\mathbf{I}(B)$ has at least this many vertices. Combining Equations (6.29) and (6.30) we get that $\mathbf{I}(B)$ has at least $2^{d-1}(n - (d-2))$ vertices and hence the result. ∎

Corollary 6.6.6 *If $d \geq 3$ and if B is a polytope in $\mathbb{R}^d$ then $\mathbf{I}(B)$ is not homothetic to B.*

Proof. Suppose that B has $2n$ vertices. Since these form n pairs of opposite vertices we have $n \geq d$. Hence $n - (d-2) \geq 2$ and $\mathbf{I}(B)$ has at least 2^d vertices (exactly 2^d vertices when B is a cross-polytope with $2d$ vertices). There are two cases:

(i) If $n > 2(d-2)$ then we have $n - (d-2) > n/2$ and hence

$$2^{d-1}(n - (d-2)) > 2^{d-2}n \geq 2n,$$

the last inequality since $d \geq 3$.

(ii) If $d \leq n \leq 2(d-2)$ (for this case to arise we have $d \geq 4$), then $2^{d-1} > 2(d-2)$ and hence $2^d > 2n$.

Thus, in either case, the number of vertices in $\mathbf{I}(B)$ is strictly greater than the number in B and therefore $\mathbf{I}(B)$ cannot be homothetic to B. ∎

We have seen that if B is a polytope then so is $\mathbf{I}(B)$ and that the general shape of $\mathbf{I}(B)$ resembles that of B°. Since in Minkowski spaces B is smooth if and only if B° is strictly convex, the following results can be viewed as further instances of the preceding idea. These results, however, are also important in their own right as answers to the question, When are transversals unique?

Theorem 6.6.7 *If B is smooth then $\mathbf{I}(B)$ is strictly convex.*

Proof. We consider the gradient $\vec{\nabla}\sigma_f$ of the support function σ of $\mathbf{I}(B)$ in the direction f. Recall from Theorem 5.5.13 that

$$\vec{\nabla}\sigma_f = \{x \in \partial\mathbf{I}(B) : f \text{ supports } \mathbf{I}(B) \text{ at } x\}.$$

Therefore, $\mathbf{I}(B)$ is strictly convex if and only if $\vec{\nabla}\sigma_f$ is single-valued. However,

$$\sigma(f) = (2\epsilon_{d-1})^{-1}\int_{\partial B^\circ}|f(x)|\,d\lambda(x) = \epsilon_{d-1}^{-1}\int_{U}f(x)\,d\lambda(x),$$

where U is an arbitrary measurable subset of $\partial B°$ such that

$$H_f^{++} := \{g \in \partial B° : f(\hat{x}_g) > 0\} \subseteq U \subseteq H_f^+ := \{g \in \partial B° : f(\hat{x}_g) \geq 0\}$$

and where $\hat{x}_g$ is the unit normal to $B°$ at g. It follows that $\vec{\nabla}\sigma_f = \{y : y = \epsilon_{d-1}^{-1} \int_U x \, d\lambda(x)\}$. Thus $\mathbf{I}(B)$ is strictly convex if and only if $H_f^+ \setminus H_f^{++}$ is of measure zero for all f in X^*. For this it is sufficient if $B°$ is strictly convex and it is well known that if B is smooth then $B°$ is strictly convex. ∎

Theorem 6.6.8 *If B is strictly convex then $\mathbf{I}(B)$ is smooth.*

Proof. The argument is similar to the preceding one. If B is strictly convex it follows that $B°$ is smooth. However, if $f \neq g$ and if $B°$ is smooth, then the symmetric difference $(H_f^+ \setminus H_g^+) \cup (H_g^+ \setminus H_f^+)$ is of positive measure and hence $\vec{\nabla}\sigma_f \cap \vec{\nabla}\sigma_g = \emptyset$. This implies that at each point x of $\partial \mathbf{I}(B)$ there is a unique supporting hyperplane and hence that $\mathbf{I}(B)$ is smooth. ∎

Corollary 6.6.9 *If B is both smooth and strictly convex then so is $\mathbf{I}(B)$.*

Remark. Examples (due to Johnson) to show that if B is smooth then $\mathbf{I}(B)$ need not be smooth and if B is strictly convex $\mathbf{I}(B)$ need not be so are given in [273]. Those examples correct oversights in [262].

The final theorem of this section (first proved by R. D. Holmes [262]) is of a different nature. It consists of a number of formulas that are very striking and deserve greater prominence. On the other hand, they are disconnected from the rest of the material. They do, however, concern the two affine invariants $\mu_B(B)$ and $\mu_B(\partial B)$ that have been the subject of this section.

The first of these invariants is (apart from a constant depending on the dimension) the volume product and can be expressed as an integral:

$$\mu_B(B) = \epsilon_d^{-1} \int_B \int_{B°} d\lambda^d(f) \, d\lambda^d(x).$$

The second is the self-surface-area and can be expressed either as a mixed volume or as another double integral:

$$\mu_B(\partial B) = \epsilon_{d-1}^{-1} V(B[d-1], \Pi B°)$$

$$= (2\epsilon_{d-1})^{-1} \int_{\partial B} \int_{\partial B°} |\hat{f}(\hat{x})| \, d\lambda^{d-1}(\hat{x}) \, d\lambda^{d-1}(\hat{f}).$$

The formulas relate the behaviour of these affine invariants to the operations of $\times$ and $*$ considered in §6.2.

If (X, A) and (Y, B) are two Minkowski spaces with unit balls A and B respectively then the Cartesian product $X \times Y$ can be given either of the following two unit balls: (i) the Cartesian product $A \times B$ or (ii) the suspension $A * B = \mathrm{conv}\{(A \times \{0\}) \cup (\{0\} \times B)\}$. In terms of the norms these are the ℓ_∞ and ℓ_1 constructions respectively, *i.e.*

$$\|(x, y)\|_{A \times B} = \max\{\|x\|_A, \|y\|_B\},$$

and

$$\|(x, y)\|_{A*B} = \|x\|_A + \|y\|_B.$$

In Theorem 6.2.7 and Equation (6.21) certain factors depending on the dimension arose naturally. In order that the formulas in the next theorem appear more striking we exclude similar factors by means of the following normalization. (Note that d is a superscript, not an exponent.)

Definition 6.6.10 *If (X, B) is a d-dimensional Minkowski space then set*

$$\rho^d(B) := \epsilon_d d! \mu_B^d(B),$$
$$\rho^{d-1}(\partial B) := \epsilon_{d-1}(d-1)! \mu_B^{d-1}(\partial B).$$

Theorem 6.6.11 *If (X, A) and (Y, B) are two Minkowski spaces of dimensions m and n respectively and if $X \times Y$ is equipped with either of the unit balls $A \times B$ or $A * B$ then*

(i) $\rho^m(A) = \rho^m(A^\circ)$ *and* $\rho^{m-1}(\partial A) = \rho^{m-1}(\partial A^\circ)$,
(ii) $\rho^{m+n}(A \times B) = \rho^m(A)\rho^n(B)$ *and* $\rho^{m+n}(A * B) = \rho^m(A)\rho^n(B)$,
(iii) $\rho^{m+n-1}(\partial(A \times B)) = \rho^{m-1}(\partial A)\rho^n(B) + \rho^m(A)\rho^{n-1}(\partial B)$
$= \rho^{m+n-1}(\partial(A * B))$.

Proof. The first statement in (i) is clear from the definition of $\mu_B(B)$ and the second is Corollary 6.3.2. The first part of (ii) is a consequence of the following formulas for Euclidean volume:

$$\lambda(A \times B) = \lambda(A)\lambda(B),$$
$$\lambda(A * B) = \lambda(A)\lambda(B) \, n!\, m!/(n+m)!$$

The second part of (ii) now follows from (i) and the fact that $(A * B)^\circ = A^\circ \times B^\circ$. The proof of the first part of (iii) can be made first for polytopes and then followed by a continuity argument. The proof for polytopes is not difficult and the details are given in [262]. The second equation follows from the first by the same argument as in (ii). ∎

Remark. It follows from Theorem 6.6.11 that if $\mathbb{R}^d$ is equipped with a unit ball B that is constructed from d copies of the interval $[-1, 1]$ by means of an arbitrary sequence of either $\times$ or $*$ operations, then $\rho(B)$ and $\rho(\partial B)$ and hence $\mu_B(B)$ and $\mu_B(\partial B)$ are independent of the particular choice of sequence. For every such ball (which includes the cube and cross-polytope as the extreme cases of a single operation repeated d times) we have

$$\rho^d(B) = \rho^d([-1,1]^d) = \rho^d([-1,1])^d = 4^d$$

and hence $\mu_B(B) = 4^d/d!\epsilon_d$. Also

$$\rho^{d-1}(\partial B) = \rho^{d-1}(\partial[-1,1]^d) = d\,\rho^{d-1}([-1,1]^{d-1})\rho^0(\partial[-1,1]) = d\,2^{2d-1}$$

so that

$$\mu_B(\partial B) = d\,2^{2d-1}/\epsilon_{d-1}(d-1)!.$$

6.7 Notes

A number of historical and bibliographical references have already been given in this chapter. These will be supplemented here with additional references on some of the topics.

The study of zonotopes began, perhaps, with crystallography. Fedorov [149] investigated their combinatorial properties. Taylor [505] recently pointed out, however, that Fedorov's definition is a little more general in that it allows the opposite edges of a facet to be of different lengths (but parallel). In the mathematical literature zonotopes have been studied from various points of view. Kelly and Moser [286] and Coxeter [118] (see also [117]) considered combinatorial and classification questions (initially related to problems in projective geometry). In $\mathbb{R}^3$ every space-filling polytope is a zonotope. This aspect has been studied by (among others) Shephard [487], McMullen [366] and Groemer [204]. The characterization of those convex bodies which tile d-dimensional space was given by Venkov [519] and generalized by Aleksandrov [6]. The same result was proved independently by McMullen [368] (see [369] for the history). McMullen [365] was also interested in other aspects – in particular, a duality relation between zonotopes. Also, in $\mathbb{R}^3$, the zonohedra are the polyhedra that are equidecomposable by translations with a cube (Hadwiger [236]). For the corresponding problem for $d > 3$ see Mürner [395]. Zonotopes are characterized as those convex polytopes all of whose two-dimensional faces are symmetric and also as those polytopes that are projections of some higher-dimensional cube. A proof of the first characterization is in Schneider [479], p. 182. For the latter see, *e.g.*, Hadwiger [235], Naumann [398] and Chakerian and Filliman [103]. For a number of interesting properties of zonotopes but especially in connection with Hadwiger's covering problem and related illumination problems see Martini [354].

There are, fortunately, several good articles dealing with zonoids. In addition to those of Bolker [47], Lutwak [334], Schneider and Weil [481] and the book by Gardner [172] mentioned in the introduction there are more recent surveys by Goodey and Weil [191] and by Martini [359]. The second of these deals with intersection bodies as well as projection bodies and the bibliography is particularly extensive. The parallelism between projection bodies and intersection bodies was first explored by Lutwak [337]. A number of recent articles explore the connection more fully and will be commented on in §7.5.

In this chapter we have used a definition of Λ^k that is different from that in Chapter 5. The "algebraic" view used in §6.1 is well explained by Greub [200]. In this view Λ^k is the dual space of the space of all alternating k-linear forms on $(X^*)^k$. The "analytic" view used in Chapter 5 is that Λ^k is the space of all alternating k-linear forms on X^k; *i.e.* the two dual operations are missing. One way to connect the two is via the Alt mapping (see Spivak [494], p. 78). The reason for using the two definitions was to avoid having to introduce Alt in either place.

The concept of what is meant by *convexity* on the Grassmann manifold of intermediate dimension ($1 < k < d - 1$) is perhaps not yet adequately elucidated despite the extensive work of Busemann, Ewald and Shephard [77, 78]; see also their separate articles [79]–[84]. That work is surveyed in both Busemann's article [76] and that of Busemann and Shephard [86]. A much more recent attack on this problem, although from a somewhat different perspective, is that of Goodman and Pollack [193].

There are two or three further comments regarding the work of Busemann, Ewald and Shephard. Firstly, the examples they give are intrinsically non-symmetric. Although it seems likely that there are symmetric counter-examples of a similar type, their examples are not immediately "symmetrizable". Secondly, it is of interest that in addition to the negative result established by Example 6.1.1, Busemann, Ewald and Shephard [77] show that such an example cannot occur for zonoids. Shephard [79] also shows that the projection functions arising from a simplex are convex (but not "totally" convex in the terminology of Busemann and Straus [87]). One might ask whether the same is true either for intersection bodies or for the duals of either class. The reason for the discrepancy that produces the negative result of Example 6.1.1 is that in one calculation the projected area is counted according to multiplicity, *i.e.* by the number of times the projection covers the projected region, while in the other calculation it is not. This raises the question of whether, in the definition of σ as a projected area, the multiplicity should perhaps be taken into account. As Busemann and Shephard [86] point out, the important question is whether, although σ is not convex in the strong sense of being extensible to a convex function on the whole of Λ^k, it is convex in the weaker sense that flat regions minimize surface content.

Twice in §6.2 (first when it is shown that the range of **I** is the class of zonoids and second when Blaschke sums are introduced) and several times in §6.4, reference has been made to Minkowski's theorem. Formally, the theorem asserts that if ν is a measure on the unit sphere S^{d-1} in a d-dimensional Euclidean space such that

$$\int_{S^{d-1}} \hat{u}\, d\nu(\hat{u}) = 0$$

and such that ν is not concentrated on a great circle then there is a convex body

K (unique up to translation) for which ν is the area measure. Minkowski [385] proved the theorem for $d = 3$, first for polyhedra (*i.e.* when ν is an atomic measure concentrated on a finite set of points which span the space) and then in greater generality by an approximation argument. The general version was proved by Fenchel and Jessen [151] and by Aleksandrov [4]. Minkowski conjectured but was unable to prove that if κ is a smooth, strictly positive function from S^{d-1} into $\mathbb{R}$ for which

$$\int_{S^{d-1}} \hat{u}/\kappa(\hat{u}) \, d\lambda = 0,$$

then there is a smooth closed convex hypersurface $\mathcal{M}$ such that the Gaussian curvature of $\mathcal{M}$ at the point whose outward normal is $\hat{u}$ is given by $\kappa(\hat{u})$. Here "smooth" may mean either k times differentiable for some k or infinitely differentiable. Gluck [179] gives a very good survey of the history of the solution of this problem; see also Busemann [74]. Gluck refers, in particular, to the work of Lewy [322], Nirenberg [399], Aleksandrov [5] and Pogorelov [431]. Cheng and Yau [108] suggest that there may be gaps in the work of Pogorelov and give an independent proof of the regularity of solutions to Minkowski's problem. The solution of what Gluck calls the "generalized Minkowski problem" is due to Gluck himself [178].

Theorem 6.2.1 is a convenient way to show that $\mathbf{I}$ and $\tilde{\mathbf{I}}$ are continuous functions on the collection of symmetric convex bodies. It is unsatisfactory because the factor $d - 1$ has nothing to do with the shape of the sets and therefore appears extraneous to the problem. For this reason the normalization $\tilde{\mathbf{I}}$ is more appropriate but then one loses the inclusion-reversing property. It would be satisfying to show that either $\tilde{\mathbf{I}}$ or $\tilde{\mathbf{I}}^2$ is contractive with respect to Δ.

Theorem 6.2.3 asserts that a symmetric convex body is determined by the areas of all of its $(d - 1)$-dimensional projections. Schneider and Weil [480] show that this is no longer true if a symmetric open set of directions is omitted. There is also the question of whether any convex body is completely determined by the areas of its projections; *i.e.* are there convex bodies K for which the equivalence class $[K]$ is a singleton? Martini [353] showed that the only polytopes with this property are parallelotopes. Gardner and Volčič [173] showed that there are no other convex bodies with the property. When projections onto k-dimensional subspaces for $1 < k < (d - 1)$ are considered, Goodey, Schneider and Weil [189] have shown that the situation is different. In that case there are other convex bodies uniquely determined by their k-dimensional projections.

Theorem 6.2.3 also shows that $\mathbf{I}$ is injective on the collection of symmetric convex bodies. It is natural to ask about the continuity of the inverse map. Instead of the body $\mathbf{I}(B)$ one can consider its support function σ_B as a function on the Euclidean sphere S^{d-1}. There are then a variety of norms to measure the distance between two support functions (the uniform norm corresponds to the Hausdorff metric). Goodey [187] has shown that $\mathbf{I}^{-1}$ is continuous but not uniformly continuous with respect to the Hausdorff metric. For the case $d = 3$, Campi [89] considers the L_2 norm on the support functions and obtains a rather strong type of continuity. Bourgain and Lindenstrauss [54] have stronger results of this type.

It is too much to expect that $\mathbf{I}^{-1}$ is monotonic. An important question of Shephard's [486] asks for the weaker property: does the inequality $\sigma(B_1) \leq \sigma(B_2)$ (between the functions $\sigma(B_i)$ on S^{d-1}) imply that $\lambda(B_1^\circ) \leq \lambda(B_2^\circ)$? In other words, since $\sigma(B)$ represents the projected area of B°, if K_1 and K_2 are two (symmetric) convex bodies such that each projection of K_1 onto a $(d-1)$-dimensional subspace has smaller area than the corresponding projection of K_2, does it follow that K_1 also has the smaller volume; or, in terms of mixed volumes, does $V(K_1[d-1],[u]) \leq V(K_2[d-1],[u])$ for all u imply that $\lambda(K_1) \leq \lambda(K_2)$? For Minkowski geometry this question is similar in style to Kolmogoroff's monotonicity requirement. It asks: if (X, B_1) and (X, B_2) are two Minkowski spaces (with the same underlying linear space X) and if $\mu_{B_1}^{d-1}(\mathcal{M}) \leq \mu_{B_2}^{d-1}(\mathcal{M})$ for every rectifiable surface $\mathcal{M}$, does it follow that $\mu_{B_1}^d(U) \leq \mu_{B_2}^d(U)$ for every measurable region U in X?

Shephard showed that the answer is no if non-symmetric bodies are allowed. Petty [416] and Schneider [471] independently showed that the answer is still no in general for symmetric bodies but yes if the larger set K_2 (B_2° for our application to Minkowski geometry) is a zonoid. Weil [529] has used the idea of the Schneider–Petty result to characterize projection bodies as follows. A symmetric convex body K is a zonoid if and only if $V(K_1[d-1], K) \leq V(K_2[d-1], K)$ for all K_1 and K_2 with the property that $V(K_1[d-1],[u]) \leq V(K_2[d-1],[u])$ for all u in X. Shephard [486] pointed out that this problem is dual to one of Busemann and Petty [85] and this aspect has been fully explored by Lutwak [334, 336, 342]. Further results about how large projected areas can be compared with those of a Euclidean ball of the same volume are given by Ball [17]. A complete account of Shephard's problem is given in Chapter 4 of Gardner's book [172].

Another question dealing with projections of convex bodies was settled by Rogers [444]. He showed that if every projection of K_1 onto a two-dimensional subspace is homothetic to the corresponding projection of K_2 then K_1 is homothetic to K_2. The same problem with homothetic replaced by direct congruence is more difficult. Partial answers have been given, *e.g.*, by Golubyatnikov [185, 186]. For a fuller discussion of these problems see Gardner [172].

The Kneser–Süss inequality mentioned in §6.2 is the analogue for Blaschke addition and multiplication of the Brunn–Minkowski inequality for Minkowski addition and multiplication. It was proved in [294].

For the background to §6.4 in classical integral geometry, one should read the book by Santaló [457]. An up-to-date survey of the more recent literature is that of Schneider and Wieacker [482]. Conversations with John Wieacker in Freiburg, his manuscript [533] and the thesis of El-Ekhtiar [142] led to §6.4. The results of the section have now been subsumed in the comprehensive treatment by Schneider and Wieacker [483]. For an integral geometry definition of arc length in Finsler geometry, see Owens [403]. Alexander [7] gives information on hypermetric spaces and their relation to Hilbert's Fifth Problem. The geometry of Minkowski spaces whose ball is the dual of a zonoid is discussed by Witsenhausen [535, 536]. In [536] he shows that the unit ball is the dual of a zonotope if and only if the norm is piecewise linear and satisfies the following inequality:

$$\|x\| + \|y\| + \|z\| + \|x + y + z\| \geq \|x + y\| + \|y + z\| + \|z + x\|.$$

One of the items in [535] is the question of the smallest p such that the ℓ_p ball in $\mathbb{R}^d$ is a zonoid.

In [418] Petty proved the inequality 6.5.5 and conjectured the stronger statement:

$$\epsilon_d^2 \lambda(\Pi B) \lambda(B)^{1-d} \geq (\epsilon_{d-1}/\epsilon_d)^d.$$

For discussion of Petty's inequality and the conjecture see Lutwak [334, 338, 342]. In [339] Lutwak shows that if the conjecture is true then it is the first in a series of inequalities involving mixed projection bodies. Lutwak proves the last in the series. Recent new proofs of Petty's projection inequality have been given by Schmuckenschlaeger [470] and Makai and Martini [350]. The complementary inequality $\binom{2d}{d} d^{-d} \leq \lambda(K)^{d-1} \lambda((\Pi K)^\circ)$ with equality if and only if K is a simplex was established by Zhang [540].

As stated in the text, the outstanding open problem in this chapter is whether there exist bodies B other than ellipsoids for which $\mathbf{I}(B) := \Pi(B^\circ)$ is homothetic to B. My guess is that such bodies do not exist. However, Schneider's [473] construction of zonoids (not ellipsoids) whose polars are zonoids shows that the unexpected can occur. This question is equivalent to finding those bodies B for which $\Delta_1(B, \mathbf{I}(B)) = 0$. If one could show that $\mathbf{I}$ is contractive for Δ_1 then one might use Edelstein's contractive mapping theorem [136] to solve this problem. Examples of a number of iterates of the mapping $\mathbf{I}$ were given in Chapter 5 (see Figures 5.5, 5.6, 5.7 and 5.8). The sets appear to get rounder under iteration by $\mathbf{I}$ but nevertheless keep rather large facets. Iterates of the mapping Π were considered by Weil [528] and he was able to characterize $\Pi^n(K)$ for polytopes K.

Another approach is to look for numerical affine invariants that increase (or decrease) under the action of $\mathbf{I}$. Natural candidates are the volume product $\lambda(B)\lambda(B^\circ)$, the Minkowski isoperimetric ratio $\mu_B(\partial B)^d/\mu_B(B)^{d-1}$ and the one occurring in the Petty projection inequality $\pi(B) := \lambda(B)^{d-1}\lambda(\Pi(B)^\circ)$. Since for some B, $\mathbf{I}(B)$ is a multiple of B° (e.g., the cross-polytope and the double cone) the first two cannot be strictly monotonic except possibly on the set of zonoids. A number of calculations (for those few polyhedra for which it is possible) suggest that π may be strictly increasing under the mapping $B \mapsto \Pi(B)^\circ$.

A closely related problem is to consider non-symmetric sets and ask if $D(K) := K - K$ is homothetic to $(\Pi K)^\circ$. Martini [355, 357, 358] (see also Heil and Martini [252]) has shown that among polytopes and if $d > 2$ this is true if and only if K is a simplex. This important result (only one of a number of characterizations of simplices that Martini gives) provides an alternative proof of Corollary 6.6.6. Perhaps it also gives added interest to the balls considered in Example 1.1.17. For $d = 2$ Martini makes the observation that any set whose difference body is bounded by a Radon curve has this property.

Filliman [156] considered the following maximum problem for zonotopes. Which zonotopes, among those that are either projections of a unit cube or that have the sum of the lengths of their generators uniformly bounded, have the largest volume?

Another open problem is the one discussed in §6.5: which balls in $\mathbb{R}^d$ maximize (minimize) the self-surface-area? In [509] the following values for the self-surface-areas of the Euclidean ball, the cube (and cross-polytope) and the balls Z_d and $S_d - S_d$ (Example 1.1.17) were given. For an ellipsoid E, $\mu_E(\partial E) = d\epsilon_d$. For the cube C and its dual the cross-polytope,

$$\mu_C(\partial C) = \frac{2^{2d-1}}{(d-1)!\epsilon_{d-1}},$$

while for the balls Z_d and $S_d - S_d$ (for which we will use the abbreviation Z),

$$\mu_Z(\partial Z) = \frac{d(d+1)(2d-2)!}{(d-1)!^3}.$$

For $d = 3$ the smallest known value of $\mu_B(\partial B)$ (viz. $36/\pi$ [262]) occurs when B is either Z_3 or $S_3 - S_3$, the rhombic dodecahedron and the cubo-octahedron respectively. In the calculation of $\mu_Z(\partial Z)$ use is made of the value of $\lambda(S_d - S_d)$ which comes from the case of equality in the Rogers and Shephard inequality [446],

$$\lambda(K - K) \leq \binom{2n}{n}\lambda(K).$$

Rogers and Shephard [447] obtain this inequality and a number of related ones from a general inequality about the m-dimensional sections and n-dimensional projections of an $(m+n)$-dimensional convex body which is very like the equation for $\lambda(A * B)$ used in the proof of Theorem 6.6.11.

As d increases, the self-surface-area of the cube (and cross-polytope) decreases more rapidly than those of either E or Z. In particular, routine estimates show that, for $d \geq 5$, $\mu_C(\partial C) \leq \mu_E(\partial E)$. Furthermore, for $d \geq 10$, $\mu_C(\partial C) \leq \mu_Z(\partial Z)$. These inequalities suggest that the situation here is dual to that in the next chapter and that, for large d, $\mu_B(\partial B)$ is minimal for the cube and cross-polytope. For large d, it may also be the case that $\mu_B(\partial B)$ is maximal for ellipsoids.

7
Special properties of the Busemann definition

The pattern of this chapter closely follows that of the preceding one. The definition of area that is explored in this chapter was investigated intensively by Busemann in a series of important papers [66, 68, 71]. The idea, however, goes back to Bouligand and Choquet [53]. Going further back, we will also show in §7.3 that the definition agrees with the $(d-1)$-dimensional Hausdorff measure arising from the metric on (X, B). Since the solution to the isoperimetric problem $\mathbf{I}_B$ that this definition yields is the dual of the intersection body of B, recent work on intersection bodies is relevant. We refer in particular to Lutwak [337, 340], Ball [14], Milman and Pajor [382], Gardner [168, 169] and Zhang [541, 542, 544]. All the material on intersection bodies used in this chapter can be found in Gardner [172].

7.1 The convexity of the area function σ

We begin by recalling Busemann's definition; see Example 5.1.4(i) and Equations (5.7) and (5.8). Given a Minkowski space (X, B) Haar measure in each hyperplane in X is normalized by prescribing the $(d-1)$-dimensional measure μ_B of the cross-section $(B \cap f^\perp)$ of the unit ball by a hyperplane $f^\perp$ to be

$$\mu_B(B \cap f^\perp) := \epsilon_{d-1}.$$

Furthermore, if λ is an auxiliary Lebesgue measure then the function $\sigma_B(f)$, defined first for unit vectors $\hat{f}$ as the ratio of Minkowski measure to Lebesgue measure in hyperplanes parallel to $\hat{f}^\perp$ and then made positively homogeneous in f, is given by

$$\sigma_B(f) = \frac{|f|\epsilon_{d-1}}{\lambda^{d-1}(B \cap f^\perp)}. \tag{7.1}$$

Unless specific reference to B is needed the subscript will be omitted.

The proof that σ is convex is due to Busemann [67] and is modeled on the proof of the Brunn–Minkowski theorem given in Bonnesen and Fenchel [51]. The proof given by Milman and Pajor [382] (Theorem 3.9), like that of Barthel given in [24], uses Busemann's method but the notation and calculations are considerably simplified. They also achieve a little greater generality. It is their version which we present here.

Theorem 7.1.1 *If (X, B) is a Minkowski space with unit ball B then σ defined by Equation (7.1) is a convex function on X^*.*

Proof. Since σ is positively homogeneous we need only show that it is subadditive. For this, let $f_1, f_2 \in X^*$ and let $f_3 := f_1 + f_2$. Let L be the plane spanned by f_1 and f_2 and let M be the orthogonal complement of L. Let $M_i := \mathrm{span}(M, f_i)$, $i = 1, 2, 3$.

If φ is a positive real number, define $\rho_i(\varphi)$ by

$$\rho_i(\varphi) := \lambda^{d-2}[B \cap (M + \varphi f_i/|f_i|)], \qquad i = 1, 2, 3$$

so that

$$V_i := \int_0^\infty \rho_i(\varphi)\, d\varphi = 2^{-1}\lambda^{d-1}(B \cap M_i), \qquad i = 1, 2, 3.$$

See Figure 7.1, in which only the parts of B and $B \cap M_i$ on one side of the plane L are shown.

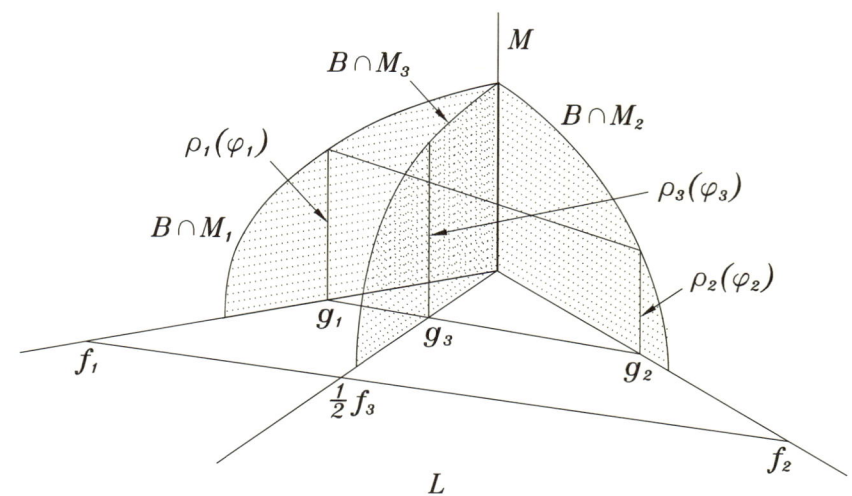

Figure 7.1

7.1 The convexity of the area function σ

To prove that σ is convex we need to prove that

$$\frac{V_3}{|f_3|} \geq \left(\frac{|f_1|}{V_1} + \frac{|f_2|}{V_2}\right)^{-1}. \qquad (7.2)$$

Let φ_1, φ_2 be positive real numbers and let $g_i := \varphi_i f_i/|f_i|$, $i = 1, 2$. Define g_3 as the point at which the interval $[g_1, g_2]$ meets the ray from 0 through f_3. Then $g_3 = \varphi_3 f_3/|f_3|$ for some positive number φ_3. For convenience in the following calculations, we also set $\xi_i := |f_i|\varphi_i^{-1}$ so that $f_i = \xi_i g_i$.

We can calculate φ_3 in two different ways. Firstly, from the areas of triangles $\triangle 0g_1g_2$, $\triangle 0g_1g_3$, $\triangle 0g_3g_2$, we have

$$\varphi_1\varphi_2 \sin(g_1 0 g_2) = \varphi_1\varphi_3 \sin(g_1 0 g_3) + \varphi_2\varphi_3 \sin(g_3 0 g_2),$$

i.e.

$$\sin(g_1 0 g_2)/\varphi_3 = \sin(g_1 0 g_3)/\varphi_2 + \sin(g_3 0 g_2)/\varphi_1.$$

Likewise, using $\triangle 0 f_1 f_2$, etc. and recalling that $f_3 = f_1 + f_2$, we get

$$\sin(f_1 0 f_2)/|f_3| = \sin(f_1 0 f_3)/|f_2| = \sin(f_3 0 f_1)/|f_1|$$

and hence

$$|f_3|/\varphi_3 = |f_2|/\varphi_2 + |f_1|/\varphi_1, \qquad \text{i.e. } \xi_3 = \xi_1 + \xi_2. \qquad (7.3)$$

Secondly, expressing g_3 as a convex combination of g_1 and g_2, we have $g_3 = \alpha g_1 + (1-\alpha)g_2$ for some α with $0 < \alpha < 1$, i.e.

$$(\xi_1 + \xi_2)^{-1}(f_1 + f_2) = \xi_3^{-1} f_3 = \alpha \xi_1^{-1} f_1 + (1-\alpha)\xi_2^{-1} f_2. \qquad (7.4)$$

Hence $\alpha = \xi_1/(\xi_1 + \xi_2)$ and $(1-\alpha) = \xi_2/(\xi_1 + \xi_2)$.

In the calculation of V_i, we need only consider the bounded interval over which $\rho_i(\varphi) \neq 0$, i.e. those φ's for which $B \cap (M + \varphi f_i/|f_i|) \neq \emptyset$. For $i = 1, 2$ we reparametrize these intervals so that the range of integration is $[0, 1]$ in both cases. Thus

$$t = V_1^{-1} \int_0^{\varphi_1} \rho_1(u)\,du = V_2^{-1} \int_0^{\varphi_2} \rho_2(u)\,du, \qquad 0 \leq t \leq 1.$$

Hence

$$\frac{d\varphi_1}{dt} = \frac{V_1}{\rho_1(\varphi_1)} \quad \text{and} \quad \frac{d\varphi_2}{dt} = \frac{V_2}{\rho_2(\varphi_2)}.$$

Also, by differentiating (7.3), we have

$$\frac{|f_3|}{\varphi_3^2}\frac{d\varphi_3}{dt} = \frac{|f_1|V_1}{\varphi_1^2 \rho_1(\varphi_1)} + \frac{|f_2|V_2}{\varphi_2^2 \rho_2(\varphi_2)},$$

i.e.

$$\varphi_3^{-1}\xi_3\, d\varphi_3/dt = \xi_1 V_1/\varphi_1 \rho_1(\varphi_1) + \xi_2 V_2/\varphi_2 \rho_2(\varphi_2).$$

Since B is convex,

$$\rho_3(\varphi_3) \geq \lambda^{d-2} \{\alpha [B \cap (M + g_1)] + (1 - \alpha) [B \cap (M + g_2)]\}$$

and hence, using the Brunn–Minkowski inequality,

$$\rho_3(\varphi_3)^{1/(d-2)} \geq \alpha(\rho_1(\varphi_1))^{1/(d-2)} + (1-\alpha)(\rho_2(\varphi_2))^{1/(d-2)},$$

which gives

$$\rho_3(\varphi_3) \geq \rho_1(\varphi_1)^\alpha \rho_2(\varphi_2)^{1-\alpha}.$$

We now have the information necessary to estimate the left-hand side of (7.2) as follows. First,

$$\frac{V_3}{|f_3|} = \frac{1}{|f_3|} \int_0^1 \rho_3(\varphi_3(t)) \frac{d\varphi_3}{dt} dt. \tag{7.5}$$

Furthermore,

$$\frac{\rho_3(\varphi_3(t))}{|f_3|} \frac{d\varphi_3}{dt}$$

$$\geq \rho_1(\varphi_1)^\alpha \rho_2(\varphi_2)^{1-\alpha} \xi_3^{-1} \left(\frac{\xi_1 V_1}{\varphi_1 \rho_1(\varphi_1)} + \frac{\xi_2 V_2}{\varphi_2 \rho_2(\varphi_2)} \right)$$

$$= (\xi_1 + \xi_2)^{-2} \left(\frac{\xi_1^2 V_1}{|f_1|\rho_1(\varphi_1)} + \frac{\xi_2^2 V_2}{|f_2|\rho_2(\varphi_2)} \right) \rho_1(\varphi_1)^{\xi_1/(\xi_1+\xi_2)} \rho_2(\varphi_2)^{\xi_2/(\xi_1+\xi_2)},$$

where we have used Equations (7.3) and (7.4).

The arithmetic–geometric mean inequality gives

$$\xi_1 \left(\frac{\xi_1 V_1}{|f_1|\rho_1(\varphi_1)} \right) + \xi_2 \left(\frac{\xi_2 V_2}{|f_2|\rho_2(\varphi_2)} \right)$$

$$\geq (\xi_1 + \xi_2) \left(\frac{\xi_1 V_1}{|f_1|\rho_1(\varphi_1)} \right)^{\xi_1/\xi_1+\xi_2} \left(\frac{\xi_2 V_2}{|f_2|\rho_2(\varphi_2)} \right)^{\xi_2/\xi_2+\xi_2}$$

so that the integrand in (7.5) is at least

$$(\xi_1 + \xi_2)^{-1} \left(\frac{\xi_1 V_1}{|f_1|} \right)^{\xi_1/(\xi_1+\xi_2)} \left(\frac{\xi_2 V_2}{|f_2|} \right)^{\xi_2/(\xi_1+\xi_2)}.$$

A final use of the arithmetic–geometric mean inequality shows that

$$\left(\frac{\xi_1 V_1}{|f_1|} \right)^{\xi_1/(\xi_1+\xi_2)} \left(\frac{\xi_2 V_2}{|f_2|} \right)^{\xi_2/(\xi_1+\xi_2)} \geq \left(\frac{\xi_1}{\xi_1+\xi_2} \frac{|f_1|}{\xi_1 V_1} + \frac{\xi_2}{\xi_1+\xi_2} \frac{|f_2|}{\xi_2 V_2} \right)^{-1}$$

$$= (\xi_1 + \xi_2) \left(\frac{|f_1|}{V_1} + \frac{|f_2|}{V_2} \right)^{-1}$$

and the proof of (7.2) is complete. ∎

As for the convexity of σ on Λ^k (we use the same notation as in §6.1) for values of k other than 1 and $d-1$, nothing seems to be known. The preceding proof does, of course, extend to the case of k-dimensional subspaces lying in a common $(k+1)$-dimensional subspace Y simply by considering the Minkowski space $(Y, B \cap Y)$.

7.2 Properties of the mapping I

In §5.4 the solution to the isoperimetric problem for the Busemann definition was shown to be $\mathbf{I}(B) = \epsilon_{d-1}(IB)^\circ$ (Equation (5.22)). Figure 5.3(a)–(d) gave some idea of the shape of $\mathbf{I}(B)$ in some simple cases. As in §6.2 we now study the nature of the mapping $\mathbf{I}$ (given by $\mathbf{I}(B) = (IB)^\circ$ – the factor ϵ_{d-1} is not important) as a function on symmetric convex sets. The material in this section is drawn almost exclusively from the work of Lutwak [337] (but see also [340]). Note that like $(.)^\circ$ and $\Pi(.)$ the mapping $I(.)$ takes convex sets in X to convex sets in X^* (and *vice versa*).

Theorem 7.2.1

(i) *The mapping $\mathbf{I}$ is order-reversing with respect to inclusion.*
(ii) *If $\xi > 0$ then $\mathbf{I}(\xi B) = \xi^{1-d}\mathbf{I}(B)$.*
(iii) *With respect to the metric Δ_2, the mapping $\mathbf{I}$ is a Lipschitz mapping with Lipschitz constant $(d-1)$.*

Proof

(i) If B_1 and B_2 are two unit balls in a d-dimensional space X and if $B_1 \supseteq B_2$ then $B_1 \cap f^\perp \supseteq B_2 \cap f^\perp$ and hence $\lambda(B_1 \cap f^\perp) \geq \lambda(B_2 \cap f^\perp)$. Therefore, $\sigma_{B_1}(f) \leq \sigma_{B_2}(f)$ and $\mathbf{I}(B_1) \subseteq \mathbf{I}(B_2)$.
(ii) This proof is a similar straightforward calculation.
(iii) If B_1 and B_2 are two unit balls with $\Delta_2(B_1, B_2) = \delta$ then there exist positive scalars ξ, η with $B_1 \subseteq \xi B_2$, $B_2 \subseteq \eta B_1$ and $\log \max\{\xi, \eta\} = \delta$. From (i) and (ii) we get $\mathbf{I}(B_1) \supseteq \mathbf{I}(\xi B_2) = \xi^{1-d}\mathbf{I}(B_2)$ and $\mathbf{I}(B_2) \supseteq \eta^{1-d}\mathbf{I}(B_1)$ from which the result follows. ∎

Remarks

(i) Part (i) shows that this definition satisfies Kolmogoroff's monotonicity requirement [297] – namely, that if X has two norms $\|.\|_1$ and $\|.\|_2$ with $\|.\|_1 \leq \|.\|_2$ then the smaller distances generate smaller areas.
(ii) As in Chapter 6, one can equally well use the function Δ_1 in this theorem. The advantage of doing so is that then it makes no difference whether we use $\mathbf{I}$ or $\tilde{\mathbf{I}}$ (and whether we include the factor ϵ_{d-1} or not). The same is true throughout most of this section.

(iii) In $\mathbb{R}^2$, the mapping **I** coincides with that of Chapter 6 and is an involutory isometry. For $d > 2$, the Lipschitz constant $(d-1)$ seems far from the best possible.

(iv) In view of Equation (5.21) following Definition 5.3.6, one might also consider the Banach–Mazur distance Δ. The same calculations show that $\tilde{\mathbf{I}}$ is Lipschitz with constant $(d-1)$ with respect to Δ. As in the remarks following Theorem 6.2.1, it is conceivable that $\tilde{\mathbf{I}}$ is contractive with respect to Δ.

The purpose of the next part of this section is to show that **I** is injective. Prior to presenting this result and to put it in context, it is worthwhile to digress in order to explain Lutwak's ideas about dual mixed volumes [333, 337, 340] (see also Zhang [542]) and the linear structure that accompanies them. The setting for the theory of dual mixed volumes is the collection of star-shaped sets (not convex bodies). However, to emphasize the similarity with the Brunn–Minkowski theory, the same letters will be used for star-shaped sets as for convex sets.

We reinterpret (7.1) in terms of $\mathbf{I}^\circ$. Since the radial function of $\mathbf{I}^\circ$ is the reciprocal of the support function of **I** we have

$$\rho(IB, f) = \epsilon_{d-1}^{-1}\, \lambda^{d-1}(B \cap f^\perp) \qquad (|f| = 1).$$

The polar-coordinate formula for volume (2.13) can be used to represent this as an integral,

$$\rho(IB, f) = (\epsilon_{d-1}(d-1))^{-1} \int_{S^{d-1} \cap f^\perp} \rho^{d-1}(B, \hat{u})\, d\lambda(\hat{u}). \qquad (7.6)$$

Hence, in the study of intersection bodies, radial functions play a role comparable to that of support functions in the study of projection bodies.

In the same way that the support function is linear with respect to Minkowski addition and scalar multiplication, one can define another *addition* of sets so that the radial function is linear. Since we already have $\rho_{\alpha A}(.) = \alpha \rho_A(.)$ for $\alpha \geq 0$, it remains to define $\tilde{+}$ so that $\rho_{A\tilde{+}B}(.) = \rho_A(.) + \rho_B(.)$. This is done for sets that are star-shaped about 0 as follows.

Definition 7.2.2 (Lutwak [337]) *If* $x_1, x_2, \ldots, x_k \in \mathbb{R}^d$ *and if each* $x_i = \alpha_i u$ *for some* $u \in \mathbb{R}^d$ *then*

$$x_1 \tilde{+} x_2 \tilde{+} x_3 \tilde{+} \cdots \tilde{+} x_k := \left(\sum_{i=1}^{k} \alpha_i\right) u.$$

Otherwise, $x_1 \tilde{+} x_2 \tilde{+} \cdots \tilde{+} x_k := 0$.

If $A_1, A_2, \ldots, A_k$ *are star-shaped subsets of* $\mathbb{R}^d$ *then*

$$A_1 \tilde{+} A_2 \tilde{+} \cdots \tilde{+} A_k := \{y \in \mathbb{R}^d : y = x_1 \tilde{+} \cdots \tilde{+} x_k,\; x_i \in A_i\}.$$

This addition is associative on star-shaped sets (not on the vectors) and $(\alpha + \beta)A = \alpha A \tilde{+} \beta A$; $\alpha(A \tilde{+} B) = \alpha A \tilde{+} \alpha B$. The reason for using star-shaped sets here is that if A and B are convex then the set whose radial function is $\rho_A + \rho_B$ need not be convex. We say that a set A is a *star body* if its radial function $\rho_A(.)$ is a continuous, positive real, valued function on S^{d-1}. If it is symmetric then ρ_A is even. Recall (Definition 0.2.2) that throughout the book *symmetric* means symmetric with respect to 0. In what follows we shall deal mainly with the collection of symmetric star bodies, which we denote by $\mathcal{S}$.

We summarize the preceding discussion by saying that, restricted to non-negative scalars, ρ is a linear map from the set $(\mathcal{S}, \tilde{+}, \cdot)$ to the vector space $C_e^+(S^{d-1})$:

$$\rho(\alpha A \tilde{+} \beta B, \hat{u}) = \alpha \rho(A, \hat{u}) + \beta \rho(B, \hat{u}), \qquad \forall \hat{u} \in S^{d-1}. \tag{7.7}$$

The next step is to consider the volume of a radial sum

$$\lambda(\alpha_1 K_1 \tilde{+} \alpha_2 K_2 \tilde{+} \cdots \tilde{+} \alpha_k K_k)$$

for star bodies K_i. Since $\lambda(A) = d^{-1} \int_{S^{d-1}} \rho_A(\hat{u})^d \, d\lambda(\hat{u})$ we can use (7.7) to show that $\lambda(\alpha_1 K_1 \tilde{+} \cdots \tilde{+} \alpha_k K_k)$ is a homogeneous polynomial

$$\lambda(\alpha_1 K_1 \tilde{+} \alpha_2 K_2 \tilde{+} \cdots \tilde{+} \alpha_k K_k) = \sum \tilde{V}(K_{i_1}, K_{i_2}, \ldots, K_{i_d}) \alpha_{i_1} \alpha_{i_2} \ldots \alpha_{i_d}$$

in the α_i's whose coefficients $\tilde{V}(K_{i_1}, K_{i_2}, \ldots, K_{i_d})$ are called the *dual mixed volumes* of the sets $K_1, K_2, \ldots, K_k$ and which, from the above derivation, can be expressed as integrals:

$$\tilde{V}(K_{i_1}, K_{i_2}, \ldots, K_{i_d}) = d^{-1} \int_{S^{d-1}} \rho(K_{i_1}, \hat{u}) \rho(K_{i_2}, \hat{u}) \cdots \rho(K_{i_d}, \hat{u}) \, d\lambda(\hat{u}). \tag{7.8}$$

If we apply Hölder's inequality to the integral in (7.8) we can obtain (see Lutwak [337] for the details) the following dual form of Minkowski's inequality:

$$\tilde{V}(K[d-1], K')^d \leq \lambda(K)^{d-1} \lambda(K'), \tag{7.9}$$

with equality if and only if K and K' are homothetic.

There is also the following analogue of Proposition 2.2.13.

Lemma 7.2.3 *If $B_1, B_2 \in \mathcal{S}$ and if $\tilde{V}(B_1[d-1], K) = \tilde{V}(B_2[d-1], K)$ for all symmetric star bodies K then $B_1 = B_2$.*

Proof. First we let $K = B_2$ and get

$$\tilde{V}(B_1[d-1], B_2) = \lambda(B_2). \tag{7.10}$$

This, together with the dual Minkowski inequality (7.9), gives us

$$\lambda(B_2) \leq \lambda(B_1).$$

Interchanging B_1 and B_2 implies that $\lambda(B_1) = \lambda(B_2)$. Using this in (7.10) shows that

$$\tilde{V}(B_1[d-1], B_2)^d = \lambda(B_1)^{d-1}\lambda(B_2)$$

and hence that B_1 and B_2 are homothetic. But since B_1 and B_2 have the same volume and are symmetric, $B_1 = B_2$. ∎

There is another important piece of machinery that we need here. For $\varphi \in C_e(S^{d-1})$ and $f \in X^*$, $\|f\| = 1$ we define the *Radon transform* $R\varphi$ by

$$(R\varphi)(f) := (d-1)^{-1} \int_{S^{d-1} \cap f^\perp} \varphi(\hat{u}) \, d\lambda(\hat{u}), \tag{7.11}$$

where $S^{d-1} \cap f^\perp$ is oriented so that if φ is in C_e^+ then so is $R\varphi$. The space $C_e(S^{d-1})$ has an inner product,

$$\langle \varphi_1, \varphi_2 \rangle := \int_{S^{d-1}} \varphi_1(\hat{u}) \varphi_2(\hat{u}) \, d\lambda(\hat{u}), \tag{7.12}$$

and it is well known (see, *e.g.*, [253]) that R is self-adjoint and has dense range in $C_e(S^{d-1})$. For more complete discussions of the properties of the Radon transform needed here, see Groemer [206] and Garder [172].

Theorem 7.2.4 *If* $(IB_1)^\circ = (IB_2)^\circ$ *then* $B_1 = B_2$.

Proof. Since the polar reciprocal is injective for convex bodies, we may suppose $IB_1 = IB_2$. By Lemma 7.2.3 it is sufficient to show that this equation implies that

$$\tilde{V}(B_1[d-1], K) = \tilde{V}(B_2[d-1], K)$$

for all symmetric star bodies K. Since $IB_1 = IB_2$ we certainly have

$$\langle \rho(IB_1), \rho(K) \rangle = \langle \rho(IB_2), \rho(K) \rangle$$

for all such K. From (7.6) and (7.11) we see that

$$\rho(IB) = \epsilon_{d-1}^{-1} R\rho^{d-1}(B)$$

and so

$$\langle R\rho^{d-1}(B_1), \rho(K) \rangle = \langle R\rho^{d-1}(B_2), \rho(K) \rangle.$$

Next we use the self-adjoint property of R to conclude that

$$\langle \rho^{d-1}(B_1), R\rho(K) \rangle = \langle \rho^{d-1}(B_2), R\rho(K) \rangle,$$

i.e. $\tilde{V}(B_1[d-1], K') = \tilde{V}(B_2[d-1], K')$ for all symmetric star bodies K' whose radial function is in the range of R. Finally, we need the facts that the range of R is dense and $\tilde{V}$ is continuous to get the desired conclusion. ∎

The dual mixed volume $\tilde{V}(K[d-1], L)$ is linear in the second variable L (with respect to radial addition). One can define a *Blaschke radial addition*, denoted by $\tilde{+}$, to make $\tilde{V}(K[d-1], L)$ linear in the first variable in the same way in which the ordinary mixed volume $V(K[d-1], L)$ can be made linear in the first variable by defining Blaschke addition to make it so.

If the domain of **I** is extended to include non-symmetric bodies then it is no longer one–one. Given a convex body K, let $[K]$ be the equivalence class consisting of all convex bodies K' such that $\mathbf{I}(K') = (IK')^\circ = \mathbf{I}(K)$. Then, by Theorem 7.2.4, $[K]$ contains precisely one symmetric body. Lutwak [337] shows that this symmetric body is $\tilde{D}(K)$, where $\tilde{D}$ is defined by

$$\tilde{D}(K) := \tfrac{1}{2} \cdot K \,\tilde{+}\, \tfrac{1}{2} \cdot K.$$

He also shows that this body is characterized as the body with least volume in $[K]$.

For the two radial sums it is essential that one work in the larger class of star bodies. There is no analogue of Minkowski's theorem to make sure that the radial functions that are constructed are the radial functions of convex sets. Lutwak [337] shows that within this larger framework algebraic results for $\mathbf{I}(B) = (IB^\circ)$ analogous to those of the preceding chapter are possible.

7.3 Area and Hausdorff measures

In a metric space there is available another notion of measure that is due to Hausdorff. A full account of this theory can be found in the book by Rogers [445]; for a briefer introduction see Morgan [392]. The idea is to cover an arbitrary set A with countably many sets A_i of small diameter (diameter less than η) and give each of these covering sets a "measure" that is a function of the diameter. An approximation to the measure of A is the sum of the measures of the A_i's. Then take the infimum over all such covers. As η decreases there are fewer available covers and hence the value of this infimum increases. The value for the outer measure of A is the limit (equivalently, the supremum) as $\eta \to 0$. One must then construct the measurable sets, *e.g.* by using the Carathéodory condition. It turns out that the measure constructed in this way is a regular measure and all Borel sets are measurable.

In the case of Minkowski spaces we will first show that the largest set with a fixed diameter is a ball. The precise statement is an inequality known as the isodiametric inequality, which states that, among all closed sets with diameter 2, the ones with maximal volume are translates of the unit ball. Thus the most economical covers by sets of prescribed diameter are covers by balls. This inequality enables us to prove the main result of this section, which is that, in Minkowski spaces, Hausdorff measure (which relies only on the metric structure) coincides with Busemann's definition of Minkowski measure.

Definition 7.3.1 *In a metric space* (X, δ) *the **diameter** of a set A,* diam A, *is defined by*

$$\operatorname{diam} A := \sup\{\delta(a_1, a_2) : a_i \in A\}.$$

Theorem 7.3.2 (isodiametric inequality) *If λ is a Haar measure on a Minkowski space $(X, \|.\|)$ and if A is a measurable subset of X then*

$$2^d \lambda(A) \leq (\operatorname{diam} A)^d \lambda(B). \tag{7.13}$$

Proof (Barthel [25]). Firstly, the inequality is obvious if either A is a null set ($\lambda(A) = 0$) or A is unbounded (diam $A = +\infty$). Hence we may consider bounded sets of positive measure. Secondly, since diam $A = \operatorname{diam}(c\ell A)$ and $\lambda(c\ell A) \geq \lambda(A)$ it is sufficient to prove the inequality for closed and bounded (*i.e.* compact) sets. Thirdly, for a compact set A, the diameter ρ is attained: $\rho = \operatorname{diam} A = \|a_1 - a_2\|$ for some $a_1, a_2 \in A$. If a_1 is such a point then $B[a_1, \rho]$ contains A. Since $B[a_1, \rho]$ is convex it also contains conv A. Similarly for every other pair of points at which ρ is attained. Hence diam(conv A) = diam A. Since $\lambda(\operatorname{conv} A) \geq \lambda(A)$ it is sufficient to prove (7.13) for convex bodies.

Let K be a convex body with diameter ρ. We have $\|x - y\| \leq \rho$ for all x, y in K. Therefore, $K + (-K) \subseteq \rho B$. Hence, using the Brunn–Minkowski inequality, we have

$$2^d \lambda(K) = [\lambda(K)^{1/d} + \lambda(-K)^{1/d}]^d \leq \lambda(K + (-K)) \leq \rho^d \lambda(B),$$

which establishes (7.13). Equality occurs here if and only if K is homothetic to $-K$ and then equality occurs in the second inequality if and only if both K and $-K$ are homothetic to B. ∎

Remarks

(i) The second step, passing from A to $c\ell A$, shows that (7.13) is true more generally for arbitrary sets A if we interpret λ as outer Haar measure.

(ii) The conditions for equality show that in order to have equality in (7.13) the ball B must be symmetric.

Now we turn to the definition of Hausdorff measures on a metric space (X, δ). The books by Rogers [445] and Morgan [392] are good references, although the definition Rogers gives is more general than the one we use.

Definition 7.3.3 *Let A be an arbitrary subset of X and let m be a non-negative real number. The **m-dimensional Hausdorff outer measure** of A is $\nu_m(A)$, where*

$$\nu_m(A) := \lim_{\eta \to 0} \nu_{m,\eta}(A) = \sup_{\eta > 0} \nu_{m,\eta}(A);$$

7.3 Area and Hausdorff measures

here

$$v_{m,\eta}(A) := \epsilon_m \inf\left\{\sum_{i=1}^{\infty}(\operatorname{diam} A_i/2)^m : \operatorname{diam} A_i < \eta,\ A \subseteq \bigcup A_i\right\}.$$

Recall that $\epsilon_m := \pi^{m/2}/\Gamma(m/2+1)$ is well defined for $m \geq 0$ and that for integral values of m it represents the Euclidean volume of a Euclidean unit ball. The outer measure v_m generates a measure which is defined on all Borel subsets of X. The pertinent fact is that, for a fixed set A and varying values of m, there is a particular value m_0 such that $v_m(A) = +\infty$ for $m < m_0$ and $v_m(A) = 0$ for $m > m_0$. The number m_0 is called the *Hausdorff dimension* of the set A. The greater generality in Rogers [445] is to consider a wider class of functions $\psi(t)$ in place of the particular ones $\varphi_m(t) = \epsilon_m(t/2)^m$ that are used here to measure the size of the sets A_i.

Next we consider the effect of restricting the types of covering sets A_i.

Definition 7.3.4 *With the same notation as before let*

$$v'_{m,\eta}(A) := \epsilon_m \inf\left\{\sum_{i=1}^{\infty}\rho_i^m : 2\rho_i < \eta\ \text{and}\ A \subseteq \bigcup B[x_i, \rho_i]\right\};$$

and

$$v''_{m,\eta}(A) := \epsilon_m \inf\left\{\sum_{i=1}^{\infty}\rho_i^m : 2\rho_i < \eta\ \text{and}\ A \subseteq \bigcup B[x_i, \rho_i]\ \text{and}\ x_i \in A\right\}.$$

Then set

$$v'_m(A) := \sup_{\eta>0} v'_{m,\eta}(A) \quad \text{and} \quad v''_m(A) := \sup_{\eta>0} v''_{m,\eta}(A).$$

Since the sets over which the infima are being taken get progressively smaller we have

$$v_{m,\eta}(A) \leq v'_{m,\eta}(A) \leq v''_{m,\eta}(A) \quad \text{for all } \eta > 0$$

and hence

$$v_m(A) \leq v'_m(A) \leq v''_m(A). \tag{7.14}$$

We now return to the situation in which $(X, \|.\|)$ is a Minkowski space. It is clear that v_m, v'_m and v''_m are all invariant under isometries of the space and hence are all translation invariant. Thus the content of the next theorem is not so much that they are all multiples of Haar measure as that they are all the same multiple of Haar measure. The development here is due to Busemann [65].

Theorem 7.3.5 (Busemann) *Let $(X, \|.\|)$ be a Minkowski space of dimension d and let λ be a Haar measure on X. For each measurable subset U of X we have*

$$\nu_d(U) = \nu'_d(U) = \nu''_d(U) = \epsilon_d \lambda(U)/\lambda(B).$$

Proof. Let $\xi := \epsilon_d/\lambda(B)$. For an arbitrary ball $B[x, \rho]$ we have

$$\lambda(B[x, \rho]) = \xi^{-1} \epsilon_d \rho^d. \tag{7.15}$$

Let $\eta, \zeta > 0$ be chosen. Then there is an open set G such that $U \subseteq G$ and $\lambda(G \setminus U) < \zeta/2$. Moreover, the collection of all balls $B[x, \rho]$ with radius $\rho < \eta/2$ satisfies the conditions of Vitali's covering theorem. Hence by Vitali's theorem (the particular form used here can be found in McShane [373], p. 366) there exist countably many points x'_i in U and $\rho'_i < \eta/2$ such that the balls $B[x'_i, \rho'_i]$ are pairwise disjoint and $C := \bigcup_{i=1}^{\infty} B[x'_i, \rho'_i] \subseteq G$ and $\lambda(U \setminus C) = 0$. Furthermore, there exist points x''_i and $\rho''_i < \eta/2$ such that $U \setminus C \subseteq \bigcup_{i=1}^{\infty} B[x''_i, \rho''_i]$ and $\sum \lambda(B[x''_i, \rho''_i]) < \zeta/2$. Combining these two collections of balls into a single family $\{B[x_i, \rho_i] : i = 1, 2, \ldots\}$ we have

$$U \subseteq \bigcup_{i=1}^{\infty} B[x_i, \rho_i] \quad \text{and} \quad \sum_{i=1}^{\infty} \lambda(B[x_i, \rho_i]) < \lambda(U) + \zeta.$$

Equation (7.15) then gives

$$\xi^{-1} \epsilon_d \sum_{i=1}^{\infty} \rho_i^d < \lambda(U) + \zeta.$$

Since ζ is arbitrary, we get

$$\nu''_d(U) \leq \lambda(U)\xi. \tag{7.16}$$

On the other hand, if A_i is an arbitrary cover of U by sets with diameters less than η then

$$\lambda(E) \leq \lambda\left(\bigcup_{i=1}^{\infty} A_i\right) \leq \sum_{i=1}^{\infty} \lambda(A_i) \leq \sum_{i=1}^{\infty} (\text{diam } A_i/2)^d \lambda(B),$$

where the last inequality comes from Theorem 7.3.2. Using the definition of ξ we get $\xi \lambda(U) \leq \epsilon_d \sum_{i=1}^{\infty} (\text{diam } A_i/2)^d$, and hence

$$\xi \lambda(U) \leq \nu_d(U). \tag{7.17}$$

Combining inequalities (7.14), (7.16) and (7.17) we get

$$\nu_d(U) = \nu'_d(U) = \nu''_d(U) = \xi \lambda(U),$$

as required. ∎

Corollary 7.3.6 *Let $(X, \|.\|)$ be a d-dimensional Minkowski space; then the d-dimensional Hausdorff measure of the unit ball B, $\nu_d(B)$, is ϵ_d.*

Proof. Take $U = B$ in Theorem 7.3.5. ∎

In contrast to the Minkowski measures (where we explicitly assume that there is an appropriate measure to apply to a particular set), the strength of the Hausdorff measures lies in the fact that the definition is independent of dimension (indeed, their values can be used to define dimension). Suppose now that Y is an m-dimensional subspace of the d-dimensional Minkowski space $(X, \|.\|)$. Then Y is a Minkowski space in its own right with unit ball $Y \cap B$. Thus, if $A \subseteq Y$ we can consider its r-dimensional Hausdorff measure either as a subset of Y or as a subset of X. However, if $\{B[x_i, \rho_i] : i = 1, 2, \ldots\}$ is a cover of A by balls in X with centres x_i in A then $\{B[x_i, \rho_i] \cap Y : i = 1, 2, \ldots\}$ is a cover of A by balls (of the same radii ρ_i) in Y. Conversely, balls in Y with centres in A are the intersection with Y of balls in X. Hence, with an obvious notation, $\nu''_{r,X}(A) = \nu''_{r,Y}(A)$ for all $A \subseteq Y$ and all $r \geq 0$. This is the essential part of the proof of the main theorem.

Theorem 7.3.7 *Let $(X, \|.\|)$ be a Minkowski space. If Y is an m-dimensional subspace of X then the Hausdorff measure ν_m on X coincides with the Busemann m-dimensional Minkowski measure μ on all Borel subsets of Y.*

Proof. Since both μ and ν_m are Haar measures on Y it is sufficient to prove that $\mu(Y \cap B) = \nu_m(Y \cap B)$ for the unit ball $Y \cap B$ in Y. By Theorem 7.3.5 we have $\nu_m(Y \cap B) = \nu''_m(Y \cap B)$ and the preceding discussion shows that $\nu''_{m,X}(Y \cap B) = \nu''_{m,Y}(Y \cap B)$. Finally, Corollary 7.3.6 shows that, in the m-dimensional Minkowski space Y, $\nu''_{m,Y}(Y \cap B) = \nu_{m,Y}(Y \cap B) = \epsilon_m$. Since Busemann defines $\mu(Y \cap B) = \epsilon_m$ the proof is complete. ∎

Remarks

(i) There are three main ingredients to the foregoing theorems. Firstly, covers by arbitrary sets are equivalent to covers by balls with centres in the given set. Secondly, in a subspace Y, covering by balls in X of radius ρ and with centres in Y is equivalent to covering by balls in Y of the same radius ρ. Finally, the Hausdorff measure ν_m is determined by a function, $\varphi_m(t) = \epsilon_m(t/2)^m$, of the diameter of the covering sets that depends only on m. These three things together mean that the measure to be assigned to a unit ball in a subspace of dimension m depends only on m and not on the shape of the ball. The particular number ϵ_m is to ensure consistency with the Euclidean case. Therefore, it coincides with the Busemann definition of m-dimensional content.

(ii) Wieacker [533] (and see Schneider and Wieacker [482]) has extended the application of Theorem 7.3.7 to quite general "m-dimensional surfaces" in

X. He shows that if $\mathcal{M}$ is the image $f(A)$ of some bounded subset A of m-dimensional Euclidean space $(\mathbb{R}^m, |.|)$ by a Lipschitz mapping f of A into X then the m-dimensional Hausdorff measure $\nu_m(\mathcal{M})$ coincides with the Busemann m-dimensional Minkowski "surface area" of $\mathcal{M}$.

7.4 Bounds for the surface area of B

For Busemann's definition of surface area we ask the same questions as in Chapter 6: How large and how small can the self-surface-area of the unit ball be? In other words, in dimension d what numbers correspond to 6 and 8 in dimension 2?

In this case we can give a precise upper bound of $2d\epsilon_{d-1}$, which is attained if and only if the ball is a parallelotope. We used this fact in the proof of Theorem 6.5.2. The precise lower bound presents a more difficult problem. We give two inexact lower bounds one of which is very elementary while the other involves the considerable machinery of geometric inequalities. Moreover, this second result involves the best lower bound for the volume product (Definition 2.3.2) and hence is only replacing one unsolved problem by another. If we use Mahler's conjecture, the numbers we obtain for $d = 2, 3$ are reasonable and suggest that for $d \geq 3$ the lower bound may be attained by the Euclidean ball.

Theorem 7.4.1 *If B is the unit ball in a d-dimensional Minkowski space $(X, \|.\|)$ then $\mu_B(\partial B) \leq 2d\epsilon_{d-1}$ with equality if and only if B is a parallelotope.*

Proof (Busemann and Petty [85]). From the definition of $\mu_B(\partial B)$ we have

$$\mu_B(\partial B) = \epsilon_{d-1} \int_{\partial B} \lambda(B \cap \hat{f}^\perp)^{-1} d\lambda(\hat{f})$$

$$= \epsilon_{d-1} \int_{\partial B} h_B(\hat{f})[h_B(\hat{f})\lambda(B \cap \hat{f}^\perp)]^{-1} d\lambda(\hat{f}).$$

Let $H(\alpha) = \{x \in X : \hat{f}(x) = \alpha\}$ and let $\varphi(\alpha) = \lambda(B \cap H(\alpha))$ so that $\varphi(0) = \lambda(B \cap \hat{f}^\perp)$. Then we have $0 \leq \varphi(\alpha) \leq \varphi(0)$ and

$$\lambda^d(B) = \int_{-h_B(\hat{f})}^{h_B(\hat{f})} \varphi(\alpha) d\alpha \leq 2h_B(\hat{f})\varphi(0).$$

Using this inequality in the earlier equation, we get

$$\mu_B(\partial B) \leq 2\epsilon_{d-1} \lambda^d(B)^{-1} \int_{\partial B} h_B(\hat{f}) d\lambda(\hat{f})$$

$$= 2\epsilon_{d-1} \lambda^d(B)^{-1} d\lambda^d(B),$$

which concludes the proof of the inequality.

Certainly equality holds when B is a parallelotope. Conversely, suppose that equality holds in the above inequalities. Then, for almost all directions $\hat{f}$ for the tangent hyperplanes, we have $\varphi(\alpha)$ constant so that $\lambda^d(B) = 2h_B(\hat{f})\lambda^{d-1}(B \cap \hat{f}^\perp)$. But there can be at most $2d$ distinct tangent hyperplanes to B for which this is so. Therefore, ∂B is contained in at most $2d$ hyperplanes, which implies that B is a polytope. A symmetric polytope has, however, at least $2d$ facets. Therefore, B has exactly $2d$ facets and is a parallelotope. ∎

Chakerian and Talley [106] give a different proof of this theorem. It is included here because it accomplishes a little more than the preceding one. For each direction $\hat{f}$ and each z in B consider $\varphi(\hat{f}(z))$, where φ is the same function as before; i.e. $\varphi(\hat{f}(z))$ is the (Euclidean) area of the cross-section of B by a hyperplane through z parallel to $\hat{f}^\perp$. Then let

$$\tau(B, z) := \epsilon_{d-1} \int_{\partial B} \varphi(\hat{f}(z))^{-1} \, d\lambda(\hat{f}).$$

One can think of $\tau(B, z)$ as the surface area of B relative to a point z different from 0. It is evident that $\tau(B, 0) = \mu_B(\partial B)$. We consider the average value of $\tau(B, z)$.

Theorem 7.4.2 *If B is the unit ball (not necessarily symmetric) of a d-dimensional Minkowski space $(X, \|.\|)$ then*

$$\int_B \tau(B, z) \, d\lambda^d(z) = d\epsilon_{d-1} V(B[d-1], B - B).$$

Proof. Using the definition of τ and then a change of order of integration, we get

$$\int_B \tau(B, z) \, d\lambda^d(z) = \epsilon_{d-1} \int_B \int_{\partial B} \varphi(\hat{f}(z))^{-1} \, d\lambda^{d-1}(\hat{f}) \, d\lambda^d(z)$$

$$= \epsilon_{d-1} \int_{\partial B} \int_B \varphi(\hat{f}(z))^{-1} \, d\lambda^d(z) \, d\lambda^{d-1}(\hat{f}).$$

Now think of $\int_B \varphi(\hat{f}(z))^{-1} \, d\lambda^d(z)$ as a double integral, first over $B \cap H(\hat{f}(z))$ (defined in the proof of Theorem 7.4.1), which yields a value of 1, and then over the interval $[-h_B(-\hat{f}), h_B(\hat{f})]$. Thus

$$\int_B \tau(B, z) \, d\lambda^d(z) = \epsilon_{d-1} \int_{\partial B} (h_B(\hat{f}) + h_B(-\hat{f})) \, d\lambda^{d-1}(\hat{f})$$

$$= \epsilon_{d-1} dV(B[d-1], B - B). \quad ■$$

Corollary 7.4.3 *If B is symmetric then the average value of $\tau(B, z)$ is given by*

$$\lambda(B)^{-1} \int_B \tau(B, z) \, d\lambda(z) = 2\epsilon_{d-1} d.$$

Proof. In this case $B - B = 2B$ and $V(B[d-1], B - B) = 2\lambda(B)$. ∎

Theorem 7.4.1 is contained in this corollary since, when B is symmetric, $\varphi(\hat{f}(z)) \leq \varphi(\hat{f}(0))$ and so $\tau(B,z)$ attains its minimum at 0. Consequently

$$\lambda(B)^{-1} \int_B \tau(B,z) d\lambda(z) \geq \tau(B,0) = \mu_B(\partial B).$$

Next we turn to lower bounds for $\mu_B(\partial B)$. First, Chakerian and Talley give a lower bound in the non-symmetric case, which, for symmetric B, is easy to establish. Recall that since $\mu_B(\partial K) = dV(K[d-1],\mathbf{I})$ if $K_1 \subseteq K_2$ then $\mu_B(\partial K_1) \leq \mu_B(\partial K_2)$.

Theorem 7.4.4 *If B is the unit ball in a d-dimensional Minkowski space $(X, \|.\|)$ then $\mu_B(\partial B) \geq 2\epsilon_{d-1}$.*

Proof. Consider an arbitrary cross-section of B by a hyperplane $f^\perp$. Then, by definition, $\mu_B(B \cap f^\perp) = \epsilon_{d-1}$. However, instead of the cross-section, consider the convex body

$$K_\eta := B \cap \{x : -\eta \leq f(x) \leq \eta\}.$$

Then K_η is (for small η) a thin "disc" surrounding the cross-section. We have $K_\eta \subseteq B$ and $\lim_{\eta \to 0} \mu_B(\partial K_\eta) = 2\epsilon_{d-1}$, from which the result follows. ∎

For the next theorem we shall need a variety of inequalities from Chapter 2 and a further one due to Busemann [72]. The latter one is called the Busemann intersection inequality. As well as Busemann's original paper, proofs can be found in Lutwak [342] and Gardner's book [172].

Theorem 7.4.5 (Busemann's intersection inequality) *If K is a convex body in a d-dimensional space then*

$$\lambda^*(IK) \leq (\epsilon_{d-1}/\epsilon_d)^d \epsilon_d^2 \lambda(K)^{d-1}.$$

We present it in this form (rather than cancelling the ϵ_d^2) to emphasize the similarity to the Petty projection inequality used in the preceding chapter. The dual nature of the two inequalities has been well pointed out by Lutwak [337, 340].

Since the precise lower bound for the volume product is known only for zonoids we require the definition of m_d from Equation (6.28).

Theorem 7.4.6 *If B is the unit ball in a d-dimensional Minkowski space then*

$$\mu_B(\partial B) \geq (d\epsilon_d)(m_d/\epsilon_d^2)^{1/d}.$$

Proof. We have

$$\mu_B(\partial B)^d = d^d V(B[d-1],\mathbf{I})^d \geq d^d \lambda(B)^{d-1} \lambda(\mathbf{I})$$
$$= d^d \lambda(B)^{d-1} \epsilon_{d-1}^d \lambda((IB)^\circ).$$

Here the inequality comes from the Minkowski inequality 2.3.8 and the last equation comes from the definition of **I**, Equation (5.22). Now use the definition of m_d (6.28) and the Busemann intersection inequality 7.4.5 to give

$$\mu_B(\partial B)^d \geq (d\epsilon_{d-1})^d \lambda(B)^{d-1} m_d / \lambda^*(IB)$$
$$\geq (d\epsilon_d)^d m_d / \epsilon_d^2,$$

and the proof is complete. ∎

Remarks

(i) The right-hand side is arranged in this way to emphasize that $d\epsilon_d$ is the self-surface-area of the Euclidean ball and m_d/ϵ_d^2 is the ratio of the minimal to the maximal volume product.

(ii) For $d = 2$, where $m_d = 4^2/2!$, we get $\mu_B(\partial B) \geq 4\sqrt{2}$. This is very close to 6 considering that three different inequalities with different conditions for equality have been used.

(iii) For $d = 3$ and assuming the validity of Mahler's conjecture we get $\mu_B(\partial B) \geq 4(6\pi)^{1/3} = 10.6443\ldots$. In this case both the rhombic dodecahedron Z_3 (see Example 1.1.17) and the Euclidean ball have self-surface-area 4π. The zonotope Z_3 can be viewed as a three-dimensional analogue of the affine regular hexagon in $\mathbb{R}^2$. For $d \geq 4$ the self-surface-area of a Euclidean ball is smaller than that of Z_d.

Regardless of the actual value of m_d, it is not unreasonable to suppose that $(m_d/\epsilon_d^2)^{1/d}$ is not too far from 1. If we assume Mahler's conjecture we get an asymptotic value $\geq 2/\pi e$.

Thus there are a variety of reasons to conjecture that the minimum value of $\mu_B(\partial B)$ for $d \geq 3$ is the Euclidean value $d\epsilon_d$. We note also that the validity of this conjecture would be an easy consequence of the strengthening of Busemann's intersection inequality by a factor of

$$\lambda^*(IK)\lambda((IK)^\circ)/\epsilon_d^2;$$

i.e. if it were true that, for $d \geq 3$,

$$\lambda(K)^{d-1}\lambda((IK)^\circ) \geq (\epsilon_d/\epsilon_{d-1})^d \tag{7.18}$$

then $\mu_B(\partial B) \geq d\epsilon_d$. The inequality (7.18) bears a striking resemblance to the Petty projection inequality 6.5.5. Such a strengthened form of 7.4.5 is also what is needed to prove the analogue of Theorem 6.6.1.

7.5 Notes

As stated in the introduction to this chapter, Bouligand and Choquet [53] gave a definition of area equivalent to that of Busemann. Their motivation was for applications to Finsler spaces. The large literature dealing with Finsler geometry is beyond our scope here. For references to the early work in the subject see the book by Rund [451].

Busemann's theorem (7.1.1) has been generalized by Barthel [23] and further generalized by Barthel and Franz [26]. These results are interesting because they show that the convexity of σ is a special case of a general affine inequality relating the $(d-1)$-measure of two compact sets C_1 and C_2 (lying respectively in two half-hyperplanes) to the $(d-1)$-measure of a certain *harmonic linear combination* of C_1 and C_2. The properties of this harmonic linear combination are studied in [23]. The generality in [26] is that no convexity assumption is needed on C_1 and C_2. For a generalization in a different direction, see Ewald [145].

In contrast to Busemann's positive result on the convexity of $I(B)$ for symmetric convex bodies B, Croft [119] has shown that this result is tight in the following sense. Every convex body contains a point x such that the intersection body relative to that point as origin is not convex. Moreover, there are convex bodies such that every interior point behaves in this way. Finally there are symmetric convex bodies for which the centre is the only point with respect to which the intersection body is convex.

These results of Croft suggest that a general study of the convexity of σ in intermediate dimensions is fraught with more difficulties than was the case in Chapter 6. Indeed, the work of Busemann, Ewald and Shephard [77–84] deals largely with the convexity of projection functions and not with cross-section functions. On the other hand, work of Lutwak [337], the recent thesis of Klain [289] and results leading (among other things) to the solution of the Busemann–Petty problem (see below) indicate that the results of Croft were misleading in the sense that they focused attention on the lack of convexity as a negative feature when the exact opposite is true. The proper setting for the theory of intersection bodies is the class of star bodies.

Filliman [157] is concerned with critical values for the area of k-dimensional cross-sections of polytopes. He shows that there may be saddle points as well as extreme points, which again demonstrates that these functions are not convex. The work of Grinberg [201] also deals with general k-sections but otherwise, as Martini [359] observes, until recently little work had been done.

Closely related to the intersection body is the *cross-section body* $C(K)$ one definition of which is $C(K) := \bigcup_{x \in K} I_x(K)$, where I_x refers to the intersection body relative to an arbitrary point x. The term *cross-section body* was introduced by Makai and Martini [350] but the concept is older; see, *e.g.*, Petty [412]. Petty [412] and, independently, Martini [356] have shown that, for $d \geq 3$, $C(K) \subseteq \Pi(K)$ with equality if and only if K is an ellipsoid and hence that (for $d \geq 3$)

$$I(K) \subseteq \Pi(K)$$

with equality if and only if K is an ellipsoid with centre at 0. This is one of the few results which relate the mappings Π and I. The pictures of isoperimetrices in Chapter 5 suggest a closer connection between the mappings $\Pi(.)°$ and $(I(.))°$. The survey by Martini [359] is wide-ranging and (as noted in Chapter 6) particularly useful for its extensive bibliography. In [359] Martini asks whether $C(K)$ is always convex for convex K; this is Problem 8.11 in [172]. It is of some interest that in [350] the authors show that the cross-section body of a tetrahedron is a cube. The above results and further discussion of cross-section bodies are contained in Gardner [172].

Although the construction of the isoperimetrix as the dual of the intersection body of B is due to Busemann [68], the term *intersection body* was coined by Lutwak [337]. In [337] Lutwak showed that results dealing with projection bodies, mixed volumes, Minkowski addition and the cosine transform are mirrored by results dealing with intersection bodies, dual mixed volumes, radial addition and the Radon transform. Projection bodies have been recognized for a long time as an important subclass of convex bodies. Lutwak's paper [337] pointed to the corresponding importance of intersection bodies in the class of star-shaped bodies. This observation has led not only to the solution of the Busemann–Petty problem but also to other interesting results. Goodey and Weil [192] have proved a conjecture of Lutwak's (Problem 8.1 in [172]) that a star-shaped body is an intersection body if and only if it is the "radial limit" of a sequence of radial sums of ellipsoids with centre at 0. This result is dual to the characterization of zonoids as Hausdorff limits of Minkowski sums of ellipsoids (Gardner [172], Corollary 4.1.12). This result can also be compared with one of Grinberg and Zhang's [203], that every symmetric convex body is the uniform limit of Blaschke sums of ellipsoids. Grinberg and Zhang [203] also show that every intersection body can be approximated by intersection bodies with analytic radial functions. Fallert, Goodey and Weil [148] have proved Lutwak's conjecture that intersection bodies behave with respect to cross-sections as projection bodies do with respect to projections – in particular, that a (central) cross-section of an intersection body is again an intersection body. All of the above results are based on a study of the connection between intersection bodies and the spherical Radon transform.

In Theorem 7.2.4 we briefly touched on the connection with the Radon transform by giving the equation

$$\rho_{IB} = R\rho_B^{d-1}. \tag{7.19}$$

Here, the factor ϵ_{d-1}^{-1} has been built into the definition of R. For a more extensive discussion of the role of R in the study of convex sets, see Groemer [206], Gardner [172], Goodey and Weil [190] and Lutwak [340]. Equation (7.19) shows that intersection bodies are those whose radial functions are images under R of certain even functions on S^{d-1}. Recently, Lutwak [340] has suggested that the definition be enlarged so that a star body is an intersection body if and only if its radial function is the Radon transform of an even *measure* on S^{d-1}. One immediate advantage of this definition is to make the class of intersection bodies closed in the *radial metric topology*. This topology is the one generated by the uniform norm on the space of radial functions. Measures can also be thought of as linear functionals on certain function spaces and functional analytic techniques can then be brought to bear on these problems; see Goodey, Lutwak and Weil [188].

Proofs of the injectivity of the mapping I (equivalent to that of $\mathbf{I}$) were first given by Funk [165] for $d = 3$ and by Petty [415] for general d. A more general theorem was proved by Schneider [472]. All of these proofs use spherical harmonics and the Funk–Hecke theorem (see, *e.g.*, Groemer [206] or Schneider [479], p. 431). A well-written account of this and other theorems proved in the same way is given by Falconer [147]. The development given in §7.2 is from Lutwak [337].

Corresponding to the Shephard [486] problem on projections and volumes is the older Busemann–Petty problem [85] on cross-sections and volumes. Stated precisely, this asks: if K_1 and K_2 are symmetric convex sets and if

$$\lambda^{d-1}(K_1 \cap f^\perp) \leq \lambda^{d-1}(K_2 \cap f^\perp) \tag{7.20}$$

for each linear functional f, does it follow that

$$\lambda^d(K_1) \leq \lambda^d(K_2)?$$

If the answer is yes then we say that K_1 and K_2 have the B–P property. Note that (7.20) can be rewritten as $I(K_1) \subseteq I(K_2)$. This problem originally arose in Busemann's work on Minkowski geometry. There it is reformulated in the following way. If $(X, \|.\|_1)$ and $(X, \|.\|_2)$ are two Minkowski spaces and if $\mu_1^{d-1}(\mathcal{M}) \geq \mu_2^{d-1}(\mathcal{M})$ for all rectifiable hypersurfaces $\mathcal{M}$, does it follow that $\mu_1^d(U) \geq \mu_2^d(U)$ for measurable regions U in X?

It is known that there exist non-symmetric convex bodies K with constant cross-sectional area (through some interior point), *i.e.* $I(K) = E_d$. It is also known that such a body has volume greater than that of the Euclidean ball of the same cross-sectional area. Hence, for some α, $K_1 := \alpha K$ and $K_2 := E_d$ do not have the B–P property (which is the reason for the restriction to symmetric sets). Lutwak [337] proved the theorem corresponding to the Petty–Schneider result (see §6.7) that if K_1 (this time the *smaller* set) is an intersection body then for each K_2 satisfying (7.20), K_1 and K_2 have the B–P property. Subsequently, [340] he extended this result to the larger class of intersection bodies defined above. This key result has generated much recent research in the area. Lutwak [337] also showed that if K_1 is a symmetric star body whose radial function is smooth and positive and if K_1 is *not* an intersection body then there is a symmetric body K_2 such that K_1 and K_2 do not have the B–P property.

Building on Lutwak's results, Zhang [542] was able to characterize intersection bodies by means of dual mixed volumes. He proved that a symmetric star body K

is an intersection body if and only if

$$\tilde{V}(K_1[d-1], K) \leq \tilde{V}(K_2[d-1], K)$$

for all symmetric star bodies K_1 and K_2 such that (a) $I(K_1) \subseteq I(K_2)$ and (b) both K_1 and K_2 have positive Gauss curvature and twice-differentiable boundary. With his extended definition Lutwak [340] improved this result by deleting condition (b). This makes the result exactly analogous to Weil's characterization of projection bodies [529].

In the meantime, much effort went into showing that in general the answer to the Busemann–Petty problem is no. Here we summarize that work by mentioning the earliest and the most recent results. Comprehensive surveys of the intermediate results are given by both Gardner [169, 172] and Martini [359]. The first proof of the existence of counter-examples was by Larman and Rogers [309]. For $d \geq 12$ a probabilistic argument showed the existence without constructing concrete examples. This was followed by Ball [13], who showed that for $d \geq 10$ the cube (as K_1) and a suitable multiple of the Euclidean ball (as K_2) do not have the B–P property. From Lutwak's result this means that for $d \geq 10$ the cube is not an intersection body. The most recent results are those of Zhang [541, 544] and Gardner [170]. In his remarkable papers, Zhang showed first [544] that the cube is not an intersection body for $d \geq 4$ (and hence the Busemann–Petty answer is no for these d), and then [541] that for $d \geq 4$ *no* polytope is an intersection body. On the affirmative side Gardner [169, 170] has shown that every symmetric convex body in $\mathbb{R}^3$ with sufficiently smooth radial function is an intersection body. Since this collection is dense it follows that the B–P property holds for all pairs of symmetric convex bodies in $\mathbb{R}^3$ satisfying (7.20). Thus the Busemann–Petty answer is yes for $d = 2, 3$ and the problem is solved. Note that with Lutwak's extended definition of intersection body the class is closed and the smoothness condition can be removed from Gardner's statement.

Milman and Pajor [382] point out that a modification of the Busemann–Petty problem is equivalent to what they describe as "the major open problem of this theory". The modification asks: is there a constant c (independent of dimension) such that if K_1 and K_2 satisfy (7.20) then $\lambda(K_1) \leq c\lambda(K_2)$?

Related to the Busemann–Petty problem are the work of Grinberg and Rivin [202] on the stability of the inequality under perturbations and results which give bounds for the Lebesgue measure of cross-sections of convex sets. Since zonotopes are projections of cubes it is not surprising that sections of cubes have received particular attention. For the unit cube, Hensley [256] gave an upper bound of 5 for $(d-1)$-dimensional cross-sections, which was improved by Ball [12] to $\sqrt{2}$ (the best possible). Vaaler [515] established the best lower bound of 1 for k-dimensional cross-sections and Ball [15] gave an upper bound for all k which is best possible for certain k. The corresponding problem for general convex bodies K has been considered also by Hensley [257] and better bounds have been given by Ball [14].

In [444] Rogers showed that if K_1 and K_2 are convex bodies with the property that every two-dimensional cross-section of K_1 through some point x_1 is homothetic to the parallel cross-section of K_2 through a point x_2 then K_1 and K_2 are homothetic. The corresponding result for cross-sections of higher dimension (and convex sets replaced by star bodies) is Problem 7.2 in Gardner [172]. Problem 7.3 in the same work asks the same question with congruence replacing homothety.

Rogers [445] observes that the first Hausdorff measure was introduced by Carathéodory [94] in 1914. The paper of Hausdorff [249] appeared in 1919. The concept has been studied extensively, particularly by Besicovitch. The bibliography in [445] lists 31 papers by Besicovitch over the 40 years from 1928 to 1967.

The isodiametric inequality 7.3.2 was extended by Barthel and Pabel [27] to include star bodies as "unit ball". For extensions and a fuller account of the results of §7.3 we refer to the recent paper of Schneider and Wieacker [483].

Busemann's intersection inequality was proved in [72]. It was extended to starshaped sets by Petty [415] and to cross-sections of intermediate dimension by Busemann and Straus [87] and, independently, by Grinberg [201]. The case of equality for this extended result is a formula for the volume of an ellipsoid in terms of the k-volumes of its k-dimensional cross-sections due to Furstenberg and Tzkoni [166]. A short proof of this latter result was given by Miles [379]. A very similar formula but involving both the volume and surface area of an ellipsoid is due to Guggenheimer [229]. Lutwak [341] proved generalizations of these results. As was pointed out, Busemann's inequality is not sufficient to prove the analogue of Theorem 6.6.1. Thus there are several questions still open in this area. Are there bodies other than ellipsoids for which $B = \tilde{\mathbf{I}}(B)$? Are there bodies other than ellipsoids for which B and $\mathbf{I}(B)$ are homothetic? Are there bodies other than ellipsoids for which B and $I^2(B)$ are homothetic? (The second is an old question of Busemann and Petty [85] and has often been raised since.)

Finally, there is the problem of either establishing or disproving the conjecture that for large d ($d \geq 3$?) ellipsoids E have the smallest self-surface-area. In this connection, note that we observed in Chapter 6 that lower bounds established there serve also as lower bounds here. We have not mentioned them explicitly, however, because the lower bounds given by Corollary 6.5.4 and Theorem 6.5.6 are not as good as those given by Theorems 7.4.4 and 7.4.6 respectively.

8
Trigonometry

The main concerns of this chapter are to define and then to investigate the notions of sine and cosine in Minkowski space. These functions are connected to the two concepts of perpendicularity that we have called *normality* (Definition 3.2.2) and *transversality* (Definitions 4.6.2 and 5.5.2). We then go on to show that these trigonometric functions retain some features of the Euclidean case. In particular, there is a sine formula for triangles and a variety of trigonometric identities. Both functions are defined in a natural way by evaluating a linear functional at a vector. The formula for cosine is the more self-evident; because it is desirable for the sine function to be related to area and volume by the usual formulas for the volume of a parallelotope, its definition depends on the choice of area function. As far as possible we shall leave this undetermined, and speak of the function σ, the isoperimetrix $\mathbf{I}$, its polar $\mathbf{I}^\circ$ and the normalization $\tilde{\mathbf{I}}$ of $\mathbf{I}$, introduced in Definition 5.3.6, that has the property that $d\mu(\tilde{\mathbf{I}}) = \mu(\partial\tilde{\mathbf{I}})$. In some cases we shall need to be specific about the choice of area function. The last two sections deal largely with two-dimensional subspaces of a Minkowski space X and so form a sequel to Chapter 4.

8.1 The functions cm and sm

The trigonometric functions cm and sm are functions of two variables since we must consider the position of the angle in space and not just the size of the angle. They are dual to each other in the sense that cm is defined when the first variable is a point in the Minkowski space X and sm is defined when the first variable is a hyperplane, *i.e.* a point in the Minkowski space X^*. Another way of viewing this duality is that sm is defined in terms of the cm between a vector and the normal to the hyperplane.

The cosine function is straightforward except that it does require a smoothness condition on the ball. Let $(X, \|.\|)$ be a Minkowski space with unit ball B. Let $x_1 \in X$, $x_1 \neq 0$, and suppose that B is smooth at $x_1/\|x_1\|$. Then, up to a positive scalar factor, there is a unique linear functional f_1 that attains its norm at x_1, i.e. such that $f_1(x_1) = \|x_1\| \, \|f_1\|$.

Definition 8.1.1 *With the above notation, we define the **Minkowski cosine**, denoted by* cm, *as follows: if* $x_2 \in X$, $x_2 \neq 0$, *then*

$$\mathrm{cm}(x_1, x_2) := f_1(x_2)/\|x_2\| \, \|f_1\|. \tag{8.1}$$

More generally, if Y is a subspace of X then we define $\mathrm{cm}(x_1, Y)$ *by*

$$\mathrm{cm}(x_1, Y) := \max\{\mathrm{cm}(x_1, x_2) \, : \, x_2 \in Y \setminus \{0\}\}. \tag{8.2}$$

Remarks

(i) The smoothness of B at x_1 is sufficient to guarantee the existence of $\mathrm{cm}(x_1, x_2)$ for all $x_2 \in X$. If x_1 and x_2 are specified then we need only consider the subspace M spanned by x_1 and x_2. If $B_M := B \cap M$ is smooth at x_1 then $\mathrm{cm}(x_1, x_2)$ is defined as above, either by considering f_1 to be an element of M^* or by using the Hahn–Banach theorem to extend such an $f_1 \in M^*$ to the whole of X with preservation of norm. If B itself is not smooth at x_1 then this extension may not be unique but all extensions will have the same values on M.

(ii) It is clear from the definition that $\mathrm{cm}(\xi x_1, \eta x_2) = \mathrm{cm}(x_1, x_2)$ for all $\xi, \eta > 0$ and also that $\mathrm{cm}(x_1, -x_2) = -\mathrm{cm}(x_1, x_2)$. Moreover, if f_1 supports B at x_1 then $-f_1$ supports B at $-x_1$ and hence $\mathrm{cm}(-x_1, x_2) = -\mathrm{cm}(x_1, x_2)$.

(iii) We have $\mathrm{cm}(x_1, x_1) = 1$. Also, for $x_2 \neq x_1$, $|\mathrm{cm}(x_1, x_2)| \leq 1$ with $\mathrm{cm}(x_1, x_2) = 1$ if and only if the line segment $[x_1/\|x_1\|, x_2/\|x_2\|] \subset \partial B$.

Having defined a cosine function, it is natural to consider those vectors x_2 such that $\mathrm{cm}(x_1, x_2) = 0$ to be "perpendicular" to x_1.

Proposition 8.1.2 *We have* $\mathrm{cm}(x_1, x_2) = 0$ *if and only if* $x_1 \dashv x_2$ *in the sense of Definition 3.2.2. More generally, if Y is a subspace of X*, $\mathrm{cm}(x_1, Y) = 0$ *if and only if* $Y \subseteq f_1^\perp$, *i.e. if and only if* $x_1 \dashv x_2$ *for all* $x_2 \in Y$.

Proof. The statements in the proposition follow from the fact that $\mathrm{cm}(x_1, x_2) = 0$ if and only if $x_2 \in f_1^\perp$ where f_1 supports B at x_1. ∎

The sine function is more complicated. Firstly, because we need an odd function, we need to be concerned about orientation and, secondly, in order that it be related to area and volume, the normalizing factors need to be different from (8.1). The basic idea in the definition of sm comes from Proposition 5.5.1, which we will first use to construct the absolute value of the sm function. We will decide about the sign afterwards.

8.1 The functions cm and sm

Suppose that $(X, \|.\|)$ is a Minkowski space. Let H be a hyperplane in X (with $0 \in H$) and let f be a (non-zero) linear functional on X such that $f(H) = 0$. If P is a parallelotope in H and if $x \in X$ with $f(x) \neq 0$ then the Euclidean volume of the parallelotope spanned by x and P, which we denote by $[x] + P$, is given by

$$\lambda^d([x] + P) = |x|\lambda^{d-1}(P)\sin(H, x)$$
$$= \lambda^{d-1}(P)|f(x)|/|f|.$$

The ratio of Minkowski volume to Euclidean volume in X is σ_B and the ratio of Minkowski area to Euclidean area in H is $\sigma(f)/|f|$. It follows that the Minkowski volume of $[x] + P$ is given by

$$\mu^d([x] + P) = \sigma_B^d \lambda^{d-1}(P)|f(x)|/|f|$$
$$= \sigma_B^d \mu^{d-1}(P)|f(x)|/\sigma(f)$$
$$= \|x\|\mu^{d-1}(P)[\sigma_B^d|f(x)|/\|x\|\sigma(f)].$$

Thus, in order to keep the desirable formula

$$\mu^d([x] + P) = \|x\|\mu^{d-1}(P)\mathrm{sm}(H, x),$$

we require that

$$|\mathrm{sm}(H, x)| = \sigma_B^d|f(x)|/\|x\|\sigma(f) = |f(x)|/\|x\|\tilde{\sigma}(f),$$

where $\tilde{\sigma}$ is the support function of $\tilde{\mathbf{I}}(B) = (\sigma_B^d)^{-1}\mathbf{I}_B$ (Definition 5.3.6).

We now need to work out the sign of $\mathrm{sm}(H, x)$, which is equivalent to deciding which of the two possible normals to H, f or $-f$, to choose. To make this choice requires both an orientation for X and a basis for H. Suppose, therefore, that X has been given an orientation. Then each basis $(b_1, b_2, \ldots, b_d)$ for X has a sign ($+$ or $-$) depending on whether the orientation of the basis (equivalently, the orientation of the parallelotope spanned by the basis vectors in the same order) agrees with that of X or not. Let $(x_1, x_2, \ldots, x_{d-1})$ be a basis for H. Since $f(x) \neq 0$, the ordered set $(x_1, x_2, \ldots, x_{d-1}, x)$ is a basis for X. Choose the sign of f so that the sign of $f(x)$ is the same as that of the basis $(x_1, x_2, \ldots, x_{d-1}, x)$. In other words, we choose the sign of f and hence of $\mathrm{sm}(H, x)$ so that $f(x) > 0$ if the orientation of the basis $(x_1, x_2, \ldots, x_{d-1}, x)$ agrees with the orientation of X and $f(x) < 0$ otherwise.

Definition 8.1.3 *Let H and x be, respectively, a hyperplane through the origin and a non-zero vector in an oriented Minkowski space $(X, \|.\|)$. Suppose also that H has a basis $(x_1, x_2, \ldots, x_{d-1})$. Then the **Minkowski sine** $\mathrm{sm}(H, x)$ is defined by*

$$\mathrm{sm}(H, x) := f(x)/\|x\|\tilde{\sigma}(f), \tag{8.3}$$

where f is a linear function in X^* such that $f^\perp = H$ and whose sign is such that

254   Trigonometry

$f(x)$ has the same sign as the basis $(x_1, x_2, \ldots, x_{d-1}, x)$ for X, and where $\tilde{\sigma}$ is the norm in X^* induced by the isoperimetrix $\tilde{\mathbf{I}}_B$ in X (see Definition 5.3.6).

Thus sm is a function defined on ordered sets of d linearly independent vectors (bases). We think of it, however, as the sine of the angle between the hyperplane spanned by the first $(d-1)$ of them and the line spanned by the last one. In fact, we can allow x to lie in the hyperplane H because then $f(x) = 0 = \text{sm}(H, x)$. The situation is most easily handled when $\dim X = 2$ for then sm is a function of ordered pairs of vectors.

More generally, if H is a hyperplane and if Y is a subspace of X then we can define $\text{sm}(H, Y)$ by

$$\text{sm}(H, Y) := \max\{\text{sm}(H, x) \,:\, x \in Y \setminus \{0\}\}. \tag{8.4}$$

In [71] Busemann defined the Minkowski sine function in terms of area and volume but it is equivalent to the above definition. He also extended the definition to a slightly more general situation, but that is now easily accomplished as follows.

If L and M are subspaces of X with the property that their span $L + M$ is of dimension 1 greater than $\max\{\dim M, \dim L\}$ then we may define $\text{sm}(L, M)$. Suppose, for definiteness, that $\dim L \geq \dim M$, then $\dim(L + M) = \dim L + 1$ and we regard L as a hyperplane in $L + M$. Then $\text{sm}(L, M)$ is defined by (8.4), where everything is considered relative to $L + M$, i.e.

$$\text{sm}(L, M) := \max\{f(x)/\tilde{\sigma}_{L+M}(f) \,:\, x \in M \cap B\},$$

where $f \in (L + M)^*$ and $f^\perp = L$.

Remark. It is important to observe that $\tilde{\sigma}_{L+M}(f)$ is the norm of f in $(L + M)^*$ induced by the isoperimetrix of the space $(L+M, B\cap(L+M))$. There is no obvious relationship between this isoperimetrix in the subspace and the isoperimetrix in the whole space X.

It is worthwhile to consider the special case when L and M are both one-dimensional, spanned by vectors x_1 and x_2. Let (Y, B_Y) be the two-dimensional space spanned by x_1 and x_2, and denote by P the parallelogram spanned by x_1 and x_2 (see Figure 8.1).

In Euclidean terms we have

$$\lambda(P) = |x_1|\,|x_2|\sin(x_1, x_2) = |x_1||f_1(x_2)|/|f_1|,$$

where $f_1 \in X^*$ is such that $f_1(x_1) = 0$ and signs are kept consistent with orientation. Hence

$$\mu(P) = \sigma_{B_Y}|x_1||f_1(x_2)|/|f_1|$$
$$= \|x_1\|\,\|x_2\|\frac{\sigma_{B_Y}|x_1||f_1(x_2)|}{\|x_1\|\,\|x_2\|\,|f_1|}.$$

8.1 The functions cm and sm

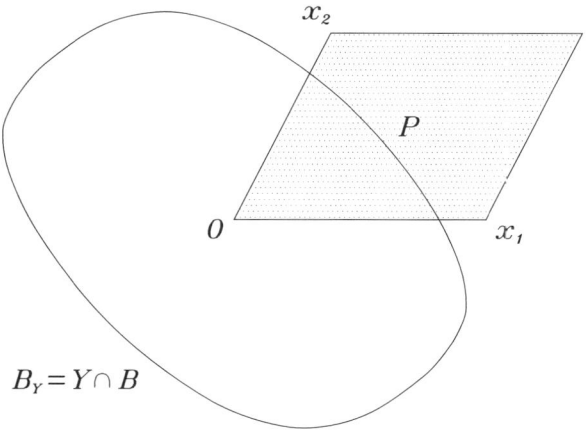

Figure 8.1

Note that in two-dimensional spaces the superscript on σ to denote the dimension is omitted (Equation (4.11)) to avoid confusion with an exponent.

Therefore, if we want to be consistent with our previous formulas, *i.e.*

$$\mu(P) = \|x_1\| \, \|x_2\| \text{sm}(x_1, x_2)$$

and

$$\text{sm}(x_1, x_2) = \frac{f_1(x_2)}{\|x_2\| \tilde{\sigma}(f_1)} = \frac{\sigma_{B_Y} f_1(x_2)}{\|x_2\| \sigma(f_1)},$$

then we require $\sigma(f_1) = \|x_1\| \, |f_1|/|x_1|$ for that f_1 for which $f_1(x_1) = 0$. In particular, if we normalize both x_1 and f_1 to be Euclidean unit vectors then $\sigma(f_1) = \|x_1\|$. In other words, the support function of $\mathbf{I}$ is the same as the support function of B° except (since f_1 and x_1 are orthogonal in the Euclidean sense) for a 90° rotation. Thus, as we saw in §4.4, $\mathbf{I} = B^\circ$ rotated through 90°.

Returning to the general case, we can summarize the discussion preceding Definition 8.1.3 in the following proposition.

Proposition 8.1.4 *If $x_1, x_2, \ldots, x_k$ span a parallelotope $P_{1,2,\ldots,k}$ in $(X, \|.\|)$ then $\mu(P_{1,2,\ldots,k}) = \|x_k\| \mu(P_{1,2,\ldots,k-1}) \, |\text{sm}(L, x_k)|$, where L is the subspace spanned by $x_1, \ldots, x_{k-1}$. In particular, with $k = 2$, $\mu(P_{1,2}) = \|x_1\| \, \|x_2\| \, |\text{sm}(x_1, x_2)|$.*

Moreover, if the space spanned by L and x_k is oriented then the signed volume of $P_{1,2,\ldots,k}$ is given by the same formula without the absolute value signs, the sign of μ depending on the orientation of the parallelotope.

Remark. Note that in the case when $k = 2$, we have adopted the preferable view of sm as a function on ordered sets of vectors and replaced the variable L by x_1, where

L is the one-dimensional subspace spanned by x_1. We shall do this consistently in two-dimensional spaces in the subsequent sections. This convention makes sm positively homogeneous of degree 0 in both variables.

Corollary 8.1.5 *In the case when $k = 2$, $\text{sm}(x_1, x_2) = -\text{sm}(x_2, x_1)$.*

Proof. This is a direct consequence of the last statement of the proposition, the change in sign coming from the change in orientation. ∎

These ideas may also be expressed in other notation. If $(x_1, x_2, \ldots, x_d)$ is an ordered set of d vectors in an oriented d-dimensional Minkowski space and if P is the parallelotope spanned by this set, then we may define the Minkowski determinant of $(x_1, x_2, \ldots, x_d)$ by

$$\text{mdet}(x_1, x_2, \ldots, x_d) := \pm \mu(P),$$

where the sign depends on whether the orientation of $(x_1, x_2, \ldots, x_d)$ agrees with that of X or not. Then we have

$$\text{mdet}(x_1, x_2, \ldots, x_d) = \|x_i\| \mu(P_i) \text{sm}(L_i, x_i),$$

where P_i and L_i are the parallelotope and linear space respectively, spanned by all the vectors except x_i. Further, given an ordered set of $(d-1)$ linearly independent vectors $(x_1, x_2, \ldots, x_{d-1})$ in X we may define the Minkowski wedge product $\text{m} \bigwedge_{i=1}^{d-1} x_i$ to be that linear functional f such that $f(x_i) = 0$ $(i = 1, 2, \ldots, d-1)$, $\tilde{\sigma}(f) = \mu(P_{1,2,\ldots,d-1})$ and the sign of f is chosen in accordance with the orientation of the x_i's. Then

$$\text{mdet}(x_1, x_2, \ldots, x_d) = f(x_d) = \left(\text{m} \bigwedge_{i=1}^{d-1} x_i \right)(x_d).$$

In the same way that the cosine function leads to a notion of perpendicularity, so does the sine function. This is a central idea in the work of Busemann. It cannot be, however, that x is perpendicular to H if $\text{sm}(H, x) = 1$ since Definition 8.1.3 implies that $\sup\{\text{sm}(H, x)\}$ is not, in general, equal to 1 (and this is so whether the supremum is taken over x or over H). Following Busemann [71] we make the following definitions.

Definition 8.1.6 *For each hyperplane H in X let*

$$\alpha(H) := \sup\{\text{sm}(H, x) : x \in X\}$$

and for each vector x in X let

$$\alpha(x) := \sup\{\text{sm}(H, x) : H \text{ a hyperplane in } X\}.$$

It follows from Definitions 8.1.3 and 8.1.6 and the definition of dual norms that $\alpha(H) = \|f\|/\tilde{\sigma}(f) = \|f\|/\|f\|_{\tilde{1}^\circ}$, where $f^\perp = H$, and $\alpha(x) = \|x\|_{\tilde{1}}/\|x\|$, and both suprema are attained. Since B and $\tilde{\mathbf{I}}$ play major roles in Minkowski geometry it is natural that these ratios of norms might occur in places where, in Euclidean space, there is an unobtrusive 1.

Recall from Definition 3.2.2 that a vector x is normal to the hyperplane H if and only if a translate of H supports B at $x/\|x\|$. Likewise, from Definition 5.5.2, a vector x is transversal to the hyperplane H if and only if a translate of H supports $\tilde{\mathbf{I}}$ at $x/\|x\|_{\tilde{1}}$.

Remark. Busemann doubles the usage of these words by also writing: the hyperplane H is transversal (resp. normal) to the vector x if x is normal (resp. transversal) to H. Since it is imperative to keep the concepts separate, even in two-dimensional spaces where hyperplanes and vectors may get identified, we shall try to avoid this confusion.

Proposition 8.1.7 *The vector x is normal to the hyperplane H if and only if $|\mathrm{sm}(H, x)| = \alpha(H)$. The vector x is transversal to the hyperplane H if and only if $|\mathrm{sm}(H, x)| = \alpha(x)$.*

Proof. We have that $|\mathrm{sm}(H, x)| = \alpha(H)$ if and only if $|f(x)| = \|f\|\,\|x\|$ (where $f^\perp = H$), i.e. if and only if f attains its B° norm at x, which means that f supports B as required. Likewise $|\mathrm{sm}(H, x)| = \alpha(x)$ if and only if $|f(x)| = \|x\|_{\tilde{1}}\tilde{\sigma}(f)$. Since $\tilde{\sigma}$ is the norm dual to $\|x\|_{\tilde{1}}$, this last equation means that f supports $\tilde{\mathbf{I}}$ at $x/\|x\|_{\tilde{1}}$. ■

Corollary 8.1.8

(i) *Given H, there is a vector x normal to H; x is unique if B is strictly convex.*

(ii) *Given H, there is a vector x transversal to H; x is unique if $\tilde{\mathbf{I}}$ is strictly convex.*

(iii) *Given x, there is a hyperplane H such that x is normal to H; H is unique if B is smooth at $x/\|x\|$.*

(iv) *Given x, there is a hyperplane H such that x is transversal to H; H is unique if $\tilde{\mathbf{I}}$ is smooth at $x/\|x\|_{\tilde{1}}$.*

Proof. (i) follows directly from Proposition 8.1.7 and the convexity of B; (iii) also follows from these statements and an application of the Hahn–Banach theorem. In the same way, (ii) follows from 8.1.7 and the Requirement 5.1.1(d) that σ be convex so that $\tilde{\mathbf{I}}$ is convex; and (iv) is another application of the Hahn–Banach theorem. ■

258   Trigonometry

Remarks

(i) This corollary means that there is some interest in theorems that give conditions on B to guarantee that $\tilde{I}$ is either smooth or strictly convex (or both). Such theorems depend on the definition of σ. Theorems 6.6.7 and 6.6.8 are examples for Example 5.1.4(ii). Examples of similar theorems for the Busemann area can be found in Busemann [71] and also in Busemann and Straus [87].

(ii) Much of the motivation for the work of Busemann [71, 76] and Busemann and Straus [87] came from the connections among transversality, the convexity of σ and the requirement that plane surfaces minimize area.

(iii) From the definitions of normality and transversality it follows that B is homothetic to $\tilde{I}$ if and only if for all x in X "x is normal to H" is equivalent to "x is transversal to H". One would like to be able to deduce that then normality between vectors (Definition 3.2.2) is symmetric and hence (if $d \geq 3$) B is an ellipsoid, but the gap seems large.

8.2 The function α

In the preceding section (Definition 8.1.6) we defined the function α on the Minkowski space X by

$$\alpha(x) = \sup\{\operatorname{sm}(H, x) \ : \ H \text{ a hyperplane in } X\} = \|x\|_{\tilde{I}}/\|x\|. \tag{8.5}$$

Therefore, like sm and cm, α is homogeneous of degree 0 and all three should, properly, be regarded as functions on rays rather than on vectors. On ∂B, α coincides with $\|.\|_{\tilde{I}}$ and on $\partial \tilde{I}$ it coincides with $\|.\|^{-1}$. It is clear that if X is Euclidean then α is identically 1. Thus α is a measure of how far X is from being Euclidean. We also have that

$$(\min \alpha)\tilde{I} \subseteq B \subseteq (\max \alpha)\tilde{I}$$

so that these extreme values of α also measure the distance between B and $\tilde{I}$. Finally, as we mentioned above, this ratio of norms occurs in a number of trigonometric formulas in §8.3. For these reasons it seems appropriate to gather the basic facts about this function into a single section.

By the *Minkowski radial function* of $\partial \tilde{I}$ we mean the function ρ_B defined on ∂B such that if $\|x\| = 1$ then $\rho_B(x)x \in \partial \tilde{I}$.

Proposition 8.2.1 *The Minkowski radial function of $\partial \tilde{I}$ is $1/\alpha(x)$.*

Proof. Since $\rho_B(x)x \in \partial \tilde{I}$ if and only if $\rho_B(x)\|x\|_{\tilde{I}} = \|\rho_B(x)x\|_{\tilde{I}} = 1$, it follows that $\rho_B(x) = 1/\|x\|_{\tilde{I}} = 1/\alpha(x)$ from (8.5) because $\|x\| = 1$. ∎

Corollary 8.2.2 *The isoperimetrix $\tilde{I}$ coincides with B if and only if $\alpha \equiv 1$.*

The proof of the next theorem depends on using Equation (5.25) to define $\mathbf{I}_B$ and may not be true for other definitions.

Theorem 8.2.3 *If X is a Minkowski space and if σ and $\mathbf{I}$ come from Example 5.1.4(ii) then* $\min\{\alpha(x) : x \in X\} \leq 1$ *with equality if and only if X is Euclidean.*

Proof. If $\min \alpha(x) \geq 1$ then $\|x\|_{\tilde{\mathbf{I}}} \geq \|x\|$ and hence $\tilde{\mathbf{I}}^\circ \subseteq B^\circ$. In this case, Corollary 6.6.2 implies that B is an ellipsoid and $\tilde{\mathbf{I}} = B$. ∎

For the remainder of this section we assume that X is two-dimensional. The function $\alpha(x)$ is constant if and only if $\tilde{\mathbf{I}}$ is a multiple of B, which, in the two-dimensional case, occurs if and only if ∂B is a Radon curve. As we have just seen, this multiple is ≥ 1 with equality only if the Radon curve is an ellipse. However, in dimension 2, we always have $\mathbf{I}^2(B) = B$ so that $\tilde{\mathbf{I}}^2(B) := \tilde{\mathbf{I}}(\tilde{\mathbf{I}}(B))$ is a multiple of B. It is interesting to see what that multiple is.

Proposition 8.2.4 *If (X, B) is a two-dimensional Minkowski space and if area is normalized using Equation (4.14) then*

$$\tilde{\mathbf{I}}^2(B) = \pi^2[\lambda(B)\lambda^*(B^\circ)]^{-1} B \supseteq B.$$

If area is normalized using Equation (4.13) then

$$\tilde{\mathbf{I}}^2(B) = \lambda(B)\lambda^*(B^\circ)\pi^{-2} B \subseteq B.$$

In either case $\tilde{\mathbf{I}}^2(B) = B$ if and only if B is an ellipse.

Proof. From Definition 5.3.6 we have that $\tilde{\mathbf{I}}(B) = \sigma_B^{-1} \mathbf{I}(B)$ and hence

$$\mathbf{I}(\tilde{\mathbf{I}}(B)) = \sigma_B \mathbf{I}^2(B) = \sigma_B B.$$

(See the comment about notation just after Figure 8.1.) The first equation comes from either Theorem 6.2.1(ii) or Theorem 7.2.1(ii) and the second from the statement following (4.7) that $\mathbf{I}(B) = \Lambda(B^\circ)$ so that $\mathbf{I}^2(B) = B$. Therefore, using Definition 5.3.6 again, $\tilde{\mathbf{I}}(\tilde{\mathbf{I}}(B)) = \sigma_{\tilde{\mathbf{I}}}^{-1} \sigma_B B$ and we need to compute $\sigma_{\tilde{\mathbf{I}}}^{-1} \sigma_B$.

If we use Equation (4.14) then $\sigma_B = \lambda^*(B^\circ)/\pi$ and hence $\sigma_{\tilde{\mathbf{I}}} = \lambda^*(\tilde{\mathbf{I}}^\circ)/\pi = \sigma_B^2 \lambda^*(\mathbf{I}^\circ(B))/\pi = \sigma_B^2 \lambda(B)/\pi$. Therefore, in this case, $\sigma_{\tilde{\mathbf{I}}}^{-1} \sigma_B = \pi/\lambda(B)\sigma_B = \pi^2/\lambda(B)\lambda^*(B^\circ)$.

In the other case a similar calculation using $\sigma_B = \pi/\lambda(B)$ yields the corresponding result.

The final statement follows from the case of equality in the Blaschke–Santaló inequality (Theorem 2.3.3). ∎

We now return to the function α and establish the best possible lower bound for max α in the case when $\sigma_B = \lambda^*(B^\circ)/\pi$. We follow Petty's argument [413] in a dual form.

Theorem 8.2.5 *If X is a two-dimensional Minkowski space and if area is normalized using Equation (4.14) then*

$$\max\{\alpha(x) : x \in X\} \geq 3/\pi$$

with equality if and only if B is an affine regular hexagon.

Proof. The maximum inscribed and minimum circumscribed isoperimetrices to B are $(\min \alpha)\tilde{\mathbf{I}}$ and $(\max \alpha)\tilde{\mathbf{I}}$ respectively. Therefore, $(\max \alpha)^{-1}\|f\| \leq \tilde{\sigma}(f) \leq (\min \alpha)^{-1}\|f\|$ for all f in X^*.

We consider $\hat{f}$ on the Euclidean circle S^1 and integrate $\|\hat{f}\|$ around S^1 with respect to *arc length measure* $\nu_{\tilde{\mathbf{I}}}$ (the one-dimensional version of the surface area measure) induced by $\tilde{\mathbf{I}}$. From the first inequality above we get

$$\int_{S^1} \|\hat{f}\| \, d\nu_{\tilde{\mathbf{I}}}(\hat{f}) \leq \max(\alpha) \int_{S^1} \tilde{\sigma}(\hat{f}) \, d\nu_{\tilde{\mathbf{I}}}(\hat{f}).$$

Now multiply both sides of this inequality by σ_B. Then, on the left, we have

$$\sigma_B \int_{S^1} \|\hat{f}\| \, d\nu_{\tilde{\mathbf{I}}}(\hat{f}) = \int_{S^1} h_B(\hat{f}) \, d\nu_{\mathbf{I}}(\hat{f}) = V(B, \mathbf{I}_B) = \mu_B(\partial B).$$

From (2.12), the right-hand integral is $\sigma_B 2\lambda(\tilde{\mathbf{I}}) = 2\mu_B(\tilde{\mathbf{I}})$. Hence

$$\mu_B(\partial B) \leq 2 \max(\alpha)\mu_B(\tilde{\mathbf{I}}) = 2\pi \max(\alpha).$$

The last equality is from Theorem 4.4.2(ii). But since $\mu_B(\partial B) \geq 6$ (Theorem 4.3.6) we have $\max \alpha \geq 3/\pi$.

Conversely, if $\max \alpha = 3/\pi$ then the previous inequality gives $\mu_B(\partial B) \leq 6$ which is possible only if $\mu_B(\partial B) = 6$ and B is an affine regular hexagon. ∎

8.3 Trigonometric formulas

In this section we are concerned entirely with two-dimensional subspaces of a normed space X. Since the trigonometric functions in such a subspace are defined in terms of the geometry of that space we may as well assume that X is two-dimensional. Since we shall need the Minkowski cosine function we shall also assume that the unit ball B is smooth. Our aim is to give Minkowski analogues of various familiar trigonometric formulas. The addition formulas are due to Busemann [71] and Petty [413]. However, Guggenheimer's idea [226, 227] of expressing a change-of-basis matrix in terms of these functions and then obtaining the addition formulas from the product of two such matrices (as with the usual Euclidean rotations) seems preferable to the original proofs as a means of deriving them. In the next section we shall use these addition formulas (in the way familiar from elementary calculus) to calculate the derivatives of Minkowski sine and cosine.

First, we give the sine formula. Recall from the remarks following Proposition 8.1.4 that in two-dimensional spaces both arguments of sm are taken to be vectors.

Theorem 8.3.1 *Let x, y, z be vectors in a Minkowski space X such that $x + y + z = 0$. Then we have*

$$\|z\|^{-1}|\text{sm}(x, y)| = \|x\|^{-1}|\text{sm}(y, z)| = \|y\|^{-1}|\text{sm}(z, x)|.$$

Proof. As in the familiar Euclidean proof we consider the triangle spanned by x and $-y$ whose third side is z. If we express the Minkowski area τ of this triangle in three ways using Proposition 8.1.4 we get

$$2\tau = \|x\| \, \|y\| \, |\text{sm}(x, y)| = \|y\| \, \|z\| \, |\text{sm}(y, z)| = \|z\| \, \|x\| \, |\text{sm}(z, x)|$$

and hence the result. ∎

In addition to our assumption that X is two-dimensional with smooth ball B, we now suppose that X is oriented. In this case, if each of the bases (x, y), (y, z) and (z, x) in Theorem 8.3.1 is positively oriented then the absolute value signs may be deleted.

Next we construct, for each $x \in \partial B$, an "orthogonal complement" $x^{\dashv}$ so that the pair $(x, x^{\dashv})$ form a "normal" basis.

Definition 8.3.2 *Let $x \in X$ with $\|x\| = 1$. Define $x^{\dashv}$, **the normal** to x, as that unique vector for which*

 (i) $\tilde{\sigma}(x^{\dashv}) = 1$,
 (ii) *the pair $(x, x^{\dashv})$ is positively oriented, and*
 (iii) $f(x^{\dashv}) = 0$ *for each f such that $f(x) = \|f\| \, \|x\|$, i.e. $x \dashv x^{\dashv}$.*

*For each x in X with $\|x\| = 1$, we call the pair $(x, x^{\dashv})$ a **normal basis** for X.*

Proposition 8.3.3 *If $(x_1, x_1^{\dashv})$ is a normal basis then there is a unique linear functional g such that $g(x_1) = 0$ and $g(x_1^{\dashv}) = \tilde{\sigma}(g) = \|g\|_{\tilde{\mathbf{I}}^{\circ}} = 1$.*

Moreover, if $x_2 \in \partial B$ then $\text{sm}(x_1, x_2) = \pm g(x_2)$ the sign depending on the orientation of the pair (x_1, x_2).

Proof. Apart from the sign, there is a unique linear functional g such that $g(x_1) = 0$ and $\tilde{\sigma}(g) := \|g\|_{\tilde{\mathbf{I}}^{\circ}} = 1$. We need to show that this g supports $\tilde{\mathbf{I}}$ at $x_1^{\dashv}$. Let f be such that $f(x_1) = \|f\| = 1$. Then x_1 supports B° at f. Consequently, from the definition of $x_1^{\dashv}$ and because $\mathbf{I} = \Lambda(B^{\circ})$ (see Figure 8.2), g supports $\tilde{\mathbf{I}}$ at $x_1^{\dashv}$, as required. Choose the sign of g so that $g(x_1^{\dashv}) = 1$.

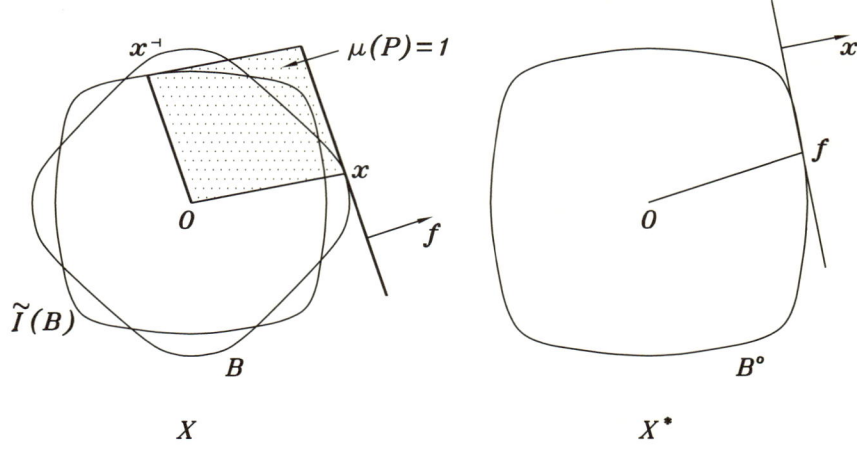

Figure 8.2

Finally, if $\|x_2\| = 1$ it follows from Definition 8.1.3 that $\text{sm}(x_1, x_2) = \pm g(x_2)$. ∎

Remark. We do not need $\tilde{\mathbf{I}}$ to be smooth at $x_1^{\dashv}$. Among the functionals that support $\tilde{\mathbf{I}}$ at $x_1^{\dashv}$ we choose the one for which $g(x_1) = 0$. However, if $\tilde{\mathbf{I}}$ is smooth at $x_1^{\dashv}$ then the proposition says that the unit tangent to $\tilde{\mathbf{I}}$ at $x_1^{\dashv}$ is x_1.

Proposition 8.3.4 *If $(x, x^{\dashv})$ is a normal basis then the Minkowski area of the parallelogram spanned by x and $x^{\dashv}$ is 1.*

Proof. If $Q(x)$ denotes the given parallelogram then we have

$$\mu(Q(x)) = \|x\| \, \|x^{\dashv}\| \text{sm}(x, x^{\dashv}) = \|x^{\dashv}\| \text{sm}(x, x^{\dashv}).$$

With g as in Proposition 8.3.3, we have $\text{sm}(x, x^{\dashv}) = g(x^{\dashv})/\|x^{\dashv}\| = 1/\|x^{\dashv}\|$. ∎

Proposition 8.3.5 *If $(x_1, x_1^{\dashv})$ and $(x_2, x_2^{\dashv})$ are two normal bases for X then $(x_2, x_2^{\dashv})^t = T(x_1, x_1^{\dashv})^t$, where*

$$T = \begin{pmatrix} \text{cm}(x_1, x_2) & \text{sm}(x_1, x_2) \\ -\xi \text{sm}(x_1 x_2) & \text{cm}(x_2 x_1) \end{pmatrix},$$

$\det T = 1$ *and* $\xi = \alpha(x_1^{\dashv})^{-1} \alpha(x_2^{\dashv})^{-1}$.

Proof. Define the matrix $T := (\tau_{ij})$ by the equations

$$x_2 = \tau_{11} x_1 + \tau_{12} x_1^{\dashv}, \tag{8.6}$$

$$x_2^{\dashv} = \tau_{21} x_1 + \tau_{22} x_1^{\dashv}. \tag{8.7}$$

Since both $(x_1, x_1^{\to})$ and $(x_2, x_2^{\to})$ are normal bases it follows from Proposition 8.3.4 that T preserves area and therefore $\det T = 1$. If f, with $\|f\| = 1$, supports B at x_1 then applying f to (8.6) and using Definitions 8.3.2 and 8.1.1 yields $\tau_{11} = \text{cm}(x_1, x_2)$. Likewise, choose g as in Proposition 8.3.3 and apply it to (8.6) to give $\tau_{12} = \text{sm}(x_1, x_2)$.

Since $\det T = 1$,

$$T^{-1} = \begin{pmatrix} \tau_{22} & -\tau_{12} \\ -\tau_{21} & \tau_{11} \end{pmatrix}$$

and therefore

$$x_1 = \tau_{22} x_2 - \tau_{12} x_2^{\to}. \tag{8.8}$$

If f' is the functional of norm 1 which supports B at x_2 and if we apply f' to (8.8) then we get $\tau_{22} = \text{cm}(x_2, x_1)$. Unfortunately, the calculation which obtained τ_{12} from (8.6) does not give τ_{21} from (8.8). Instead, we apply the same f as before to Equation (8.7) to give $\tau_{21} = f(x_2^{\to})$. We interpret this equation as follows. Since $f(x_1^{\to}) = 0$, it follows from Definition 8.1.3 that

$$\text{sm}(x_1^{\to}, x_2^{\to}) = \frac{-f(x_2^{\to})}{\|x_2^{\to}\|\tilde{\sigma}(f)}$$

(the negative sign coming from the orientation of $x_1^{\to}$ and $x_2^{\to}$). Hence

$$\tau_{21} = f(x_2^{\to}) = -\|x_2^{\to}\|\tilde{\sigma}(f)\text{sm}(x_1^{\to}, x_2^{\to}).$$

Now, from Definition 8.1.6 and the formulas that immediately follow it we get

$$\tau_{21} = -\alpha(x_2^{\to})^{-1}\alpha(x_1^{\to})^{-1}\text{sm}(x_1^{\to}, x_2^{\to}). \qquad \blacksquare$$

Remark. Another interpretation of τ_{21} is possible. One can consider the space X with ball $\tilde{I}(B) = \tilde{I}$, which is the *first derived geometry* of Guggenheimer [226]. The isoperimetrix for $(X, \tilde{I})$ is $\tilde{I}^2(B) = \zeta B$, where ζ is given by Proposition 8.2.4. Calculations show that $\text{sm}_{\tilde{I}}(x_1^{\to}, x_2^{\to})$, the Minkowski sine of the pair $(x_1^{\to}, x_2^{\to})$ in $(X, \tilde{I})$, is given by

$$\text{sm}_{\tilde{I}}(x_1^{\to}, x_2^{\to}) = -\zeta^{-1} f(x_2^{\to})$$

and that $\tau_{21} = -\zeta \text{sm}_{\tilde{I}}(x_1^{\to}, x_2^{\to})$. The advantage of this form of the expression is that the factor ζ is independent of $x_1^{\to}$ and $x_2^{\to}$. The function $\text{sm}_{\tilde{I}}$ is the second sine function that Guggenheimer uses.

Corollary 8.3.6 *For all non-zero* $x_1, x_2 \in X$ *we have*

$$\text{cm}(x_1, x_2)\text{cm}(x_2, x_1) + \alpha(x_1^{\to})^{-1}\alpha(x_2^{\to})^{-1}\text{sm}(x_1, x_2)\text{sm}(x_1^{\to}, x_2^{\to}) = 1.$$

Proof. Since sm and cm are independent of scalar multiples we can suppose that $\|x_1\| = \|x_2\| = 1$. Then the given formula is a restatement of the equation $\det T = 1$. ∎

Remark. In Definition 8.3.2 $x^{\dashv}$ was defined only for x with $\|x\| = 1$. In Corollary 8.3.6 all the functions are homogeneous of degree 0 and we can suppose that $x_i^{\dashv} := (x_i/\|x_i\|)^{\dashv}$ for $i = 1, 2$.

If, in 8.3.6, we restrict x_i to belong to $\partial \tilde{\mathbf{I}}$ then from (8.5) we can replace $\alpha(x_i^{\dashv})^{-1}$ by $\|x_i^{\dashv}\|$. The same applies to statement (i) in the next theorem.

Theorem 8.3.7 *If $x_1, x_2, x_3 \in X \setminus \{0\}$ we have*

(i) $\mathrm{cm}(x_1, x_3) = \mathrm{cm}(x_2, x_3)\mathrm{cm}(x_1, x_2) - \alpha(x_1^{\dashv})^{-1}\alpha(x_2^{\dashv})^{-1}\mathrm{sm}(x_2, x_3)\mathrm{sm}(x_1^{\dashv}, x_2^{\dashv})$,

(ii) $\mathrm{sm}(x_1, x_3) = \mathrm{cm}(x_2, x_3)\mathrm{sm}(x_1, x_2) + \mathrm{sm}(x_2, x_3)\mathrm{cm}(x_2, x_1)$,

(iii) $\alpha(x_2^{\dashv})\mathrm{sm}(x_1^{\dashv}, x_3^{\dashv}) = \alpha(x_1^{\dashv})\mathrm{sm}(x_2^{\dashv}, x_3^{\dashv})\mathrm{cm}(x_1, x_2) + \alpha(x_3^{\dashv})\mathrm{sm}(x_1^{\dashv}, x_2^{\dashv})$
$\times \mathrm{cm}(x_3, x_2)$.

Proof. If T maps $(x_1, x_1^{\dashv})$ to $(x_2, x_2^{\dashv})$ and S maps $(x_2, x_2^{\dashv})$ to $(x_3, x_3^{\dashv})$ then ST maps $(x_1, x_1^{\dashv})$ to $(x_3, x_3^{\dashv})$ and the equations follow from Proposition 8.3.5 and the usual product formula for matrices. ∎

A final trigonometric formula comes directly from the definitions.

Proposition 8.3.8 *If $x_1, x_2 \in X$ then $\mathrm{sm}(x_1^{\dashv}, x_2) = \alpha(x_1^{\dashv})\mathrm{cm}(x_1, x_2)$.*

Proof. Choose f that supports B at x_1. Then

$$\mathrm{cm}(x_1, x_2) = \frac{f(x_2)}{\|f\|\,\|x_2\|} \quad \text{and} \quad \mathrm{sm}(x_1^{\dashv}, x_2) = \frac{f(x_2)}{\|x_2\|\tilde{\sigma}(f)}$$

and the result now follows from the definition of $\alpha(x_1^{\dashv}) = \alpha(f^{\perp})$. ∎

8.4 Differentiation of the trigonometric functions

The purpose of this section is to establish Minkowski analogues of the familiar formulas from the calculus of the trigonometric functions. Moreover, the pattern of proof is familiar. We begin with the two fundamental limits, $\sin t/t$ and $(1 - \cos t)/t$, and use them together with the addition formulas to establish the differentiation results. Throughout the section we deal with an oriented two-dimensional normed space with smooth ball B.

The main difficulty is to decide upon the numerical variable to use as the independent variable for differentiation. The functions sm and cm are each functions of two rays and what we require is a measure of the *angle* between these rays. The

8.4 Differentiation of the trigonometric functions

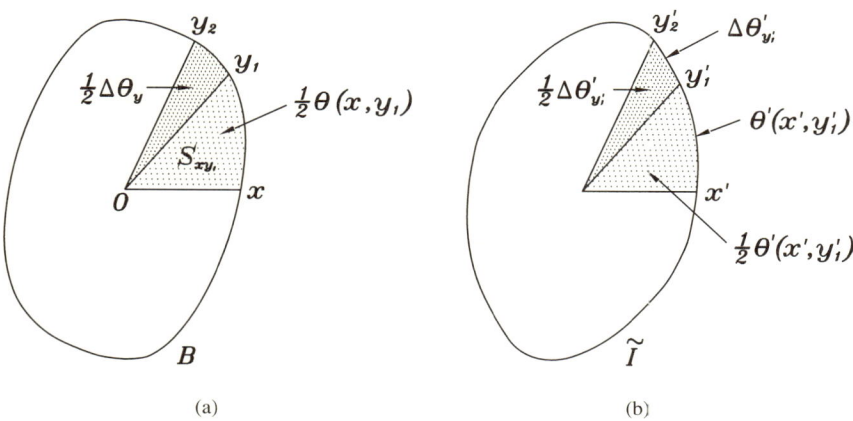

Figure 8.3

measurement of angles raises exactly the same sort of questions as the measurement of area did in Chapter 5: what properties should angle measure have? Without attempting an answer to this question we proceed pragmatically and suggest two possible measures which satisfy any reasonable list of criteria.

The most obvious method is to consider either that sector of B or that sector of $\tilde{I}$ between the two rays and then measure the angle by either (twice) the area or the arc length of that sector (see Figure 8.3). There appear to be four choices altogether but Equation (5.20) shows that the arc length of $\partial \tilde{I}$ between two rays coincides with twice the area of $\tilde{I}$ between the same two rays.

In the case of the Busemann normalization of area in which $\mu(B) = \pi$ (Equation (4.13)) the natural choice among the three given in the preceding paragraph is to define the angle measure by the area of the sector of B as in Figure 8.3(a). It is this definition which is used by Busemann [71], Guggenheimer [226, 227], Petty [413] and Biberstein [34].

With the dual normalization of area (Equation (4.14)) in which $\mu(\tilde{I}) = \pi$ and $\mu(\partial \tilde{I}) = 2\pi$ (Theorem 4.4.2) it is more natural to define the angle measure either by the area or, equivalently, by the arc length of $\partial \tilde{I}$ as in Figure 8.3(b).

In both cases the measure of the angle between opposite rays is constant (and equal to π). Also, if we append a sign to the measure to take account of the orientation of the angle then the measure is additive. However, the second definition has a disadvantage for the purposes of this section. Because $\mathrm{sm}(x_1, x_2)$ is twice the area of the triangle spanned by the unit vectors x_1 and x_2 whereas the angle between these vectors is twice the area of the sector of $\tilde{I}$ between them, $\lim_{x_2 \to x_1} \mathrm{sm}(x_1, x_2)/\mathrm{angle}(x_1, x_2) \neq 1$. The definition of angle measure can be modified to avoid this discrepancy but only with the loss of the properties just listed. In particular, the angle "measure" is no longer additive. Since the geometry is not Euclidean and there are few rotational isometries it is a genuine question as to whether this is a disadvantage or an advantage in that it more accurately reflects the geometry. These questions have not been adequately explored and for that reason we shall adopt the Busemann definition. Where appropriate, however, we shall indicate what happens with the alternative definition.

The second difficulty is to develop a satisfactory notation. All the functions we consider are defined on rays but it will be convenient if we think of a ray as represented by the point in which it intersects ∂B (resp., for the alternative definition, $\partial \tilde{\mathbf{I}}$). Hence for the remainder of this section, vectors x, y (with or without subscripts) will denote points on ∂B.

Definition 8.4.1 (and notation) *If x and y are two vectors on ∂B with y following x with respect to the orientation then*

  (i) *we write $x \leq x_1 \leq y$ to mean that $x_1 \in \partial B$ between x and y with respect to the orientation;*
  (ii) *the **measure of the angle** between x and y is $\theta(x, y)$ defined by*

$$\theta(x, y) := 2\mu_B(\mathcal{S}_{xy}),$$

where $\mathcal{S}_{xy}$, the sector between x and y, is given by $\mathcal{S}_{xy} := \{z : z = \eta x_1, \ 0 \leq \eta \leq 1, \ x \leq x_1 \leq y\}$;
  (iii) $\theta(y, x) := -\theta(x, y)$;
  (iv) *if $x \leq y_1 \leq y_2$ then $\Delta\theta_y := \theta(y_1, y_2)$, $\Delta\theta_x$ is defined similarly if $x_1 \leq x_2 \leq y$.*

If f is a real-valued function of y we shall write

$$\frac{df}{d\theta_y}(y_1) := \lim_{y_2 \to y_1} \frac{f(y_2) - f(y_1)}{\Delta\theta_y}. \tag{8.9}$$

For comparison with the alternative definition, vectors x', y' will denote points on $\partial \tilde{\mathbf{I}}$, $\theta'(x', y')$ and $\Delta\theta'_{y'}$ will denote twice the area of the sector of $\tilde{\mathbf{I}}$ between x' and y' and between y'_1 to y'_2 respectively (Figure 8.3(b)). We shall write

$$\frac{df}{d\theta'_{y'}}(y'_1) := \lim_{y'_2 \to y'_1} \frac{f(y'_2) - f(y'_1)}{\Delta\theta'_{y'}}. \tag{8.10}$$

The expressions $df/d\theta_x$ and $df/d\theta'_{x'}$ are defined similarly.

We can now state and prove the two basic limits.

Theorem 8.4.2 *With the notation explained above we have*

$$(i) \ \lim_{y_2 \to y_1} \frac{\text{sm}(y_1, y_2)}{\Delta\theta_y} = 1; \qquad (i)' \ \lim_{y'_2 \to y'_1} \frac{\text{sm}(y'_1, y'_2)}{\Delta\theta'_{y'}} = \alpha(y_1)^2.$$

8.4 Differentiation of the trigonometric functions

$$(ii) \lim_{y_2 \to y_1} \frac{1 - \text{cm}(y_1, y_2)}{\Delta\theta_y} = 0; \qquad (ii)' \lim_{y'_2 \to y'_1} \frac{1 - \text{cm}(y'_1, y'_2)}{\Delta\theta'_{y'}} = 0.$$

Proof

(i) Let τ denote the Minkowski area of the triangle spanned by y_1 and y_2 and let τ_B denote the area of the sector of B between y_1 and y_2. Then

$$\tau = 2^{-1} \|y_1\| \|y_2\| \text{sm}(y_1, y_2) = \text{sm}(y_1, y_2)/2$$

and, from Definition 8.4.1, $\Delta\theta_y = 2\tau_B$. Hence $\text{sm}(y_1, y_2)/\Delta\theta_y = \tau/\tau_B$. However, the ratio of areas τ/τ_B is the same if measured with an auxiliary Lebesgue measure and elementary calculus shows that $\lim_{y_2 \to y_1} \tau/\tau_B = 1$, which yields the result.

(i)' With an obvious change of notation, we have

$$\tau' = 2^{-1} \|y'_1\| \|y'_2\| \text{sm}(y'_1, y'_2),$$

and $\Delta\theta'_{y'} = 2\tau'_{\mathbf{I}}$. As before, we have $\lim_{y'_2 \to y'_1} \tau'/\tau'_{\mathbf{I}} = 1$. Therefore,

$$\lim_{y'_2 \to y'_1} \frac{\text{sm}(y'_1, y'_2)}{\Delta\theta'_{y'}} = \|y'_1\|^{-2} = \alpha(y'_1)^2.$$

The last equation comes from (8.5).

(ii) and (ii)'. We will show that

$$\lim_{y_2 \to y_1} \frac{1 - \text{cm}(y_1, y_2)}{\text{sm}(y_1, y_2)} = 0.$$

This result together with (i) establishes (ii). Furthermore, since cm and sm are homogeneous of degree 0, this limit and (i)' will also establish (ii)'.

We suppose that X has an auxiliary Euclidean metric $|.|$ and that ∂B has equation $r = r(\phi)$ in polar coordinates. Let $y_1 = (r_1, 0)$, and $y_2 = (r_2, \phi)$. Let P be the parallelogram spanned by y_1 and y_2. (See Figure 8.4.) Then, from Proposition 8.1.4, we get

$$\text{sm}(y_1, y_2) = \mu(P).$$

Let f be the linear functional of norm 1 which supports B at y_1. Then the equation of the tangent to B at y_1 is $f(x) = 1$ and, from Definition 8.1.1, $\text{cm}(y_1, y_2) = f(y_2)$. Therefore, if L is the line through y_2 parallel to the tangent at y_1, then the equation of L is $f(x) = \text{cm}(y_1, y_2)$. Let z be the point where L meets the line spanned by y_1, let $(y_1)^{\dashv}$ be defined as in Definition 8.3.2 and let Q be the parallelogram spanned by y_1 and $(y_1)^{\dashv}$. Proposition 8.3.4 implies that $\mu(Q) = 1$. Moreover, the line L divides Q into two subparallelograms R (spanned by z and $(y_1)^{\dashv}$) and $Q \setminus R$ whose Minkowski areas are $\text{cm}(y_1, y_2)$ and

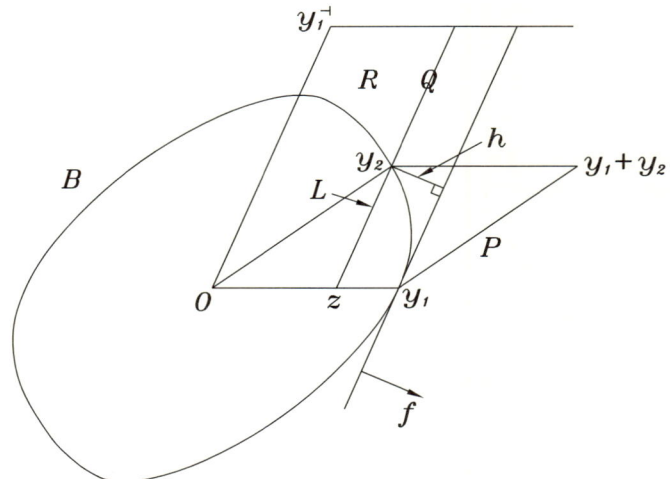

Figure 8.4

$1 - \operatorname{cm}(y_1, y_2)$ respectively. Hence

$$\frac{1 - \operatorname{cm}(y_1, y_2)}{\operatorname{sm}(y_1, y_2)} = \frac{\mu(Q \setminus R)}{\mu(P)} = \frac{\lambda(Q \setminus R)}{\lambda(P)} = \frac{|(y_1)^{-1}|h}{r_1 r_2 \sin \phi},$$

where h is the height of $Q \setminus R$.

From the equations for the tangent and L it follows that

$$h = |f|^{-1}(1 - f(y_2))$$

and elementary calculations show that

$$h = |f|^{-1} r_1^{-2}(r_1^2 - r_1 r_2 \cos \phi + r_1' r_2 \sin \phi),$$

where $r_1' = dr/d\phi$ evaluated at $\phi = 0$. The required result now follows from an application of L'Hôpital's rule. ∎

Equation 8.4.2(i)′ complicates the differential formulas for sm and cm in the alternate situation. For this reason it may be advisable to redefine the measurement of angles in this case. A possibility is to set

$$\vartheta(x', y') := \alpha(x')\alpha(y')\theta'(x', y'). \tag{8.11}$$

If, in the derivation of 8.4.2(i)′, we replace $\Delta\theta'_{y'}$ by $\Delta\vartheta_{y'}$ and use (8.5) then the factors $\|y_1'\|$ and $\|y_2'\|$ cancel with $\alpha(y_1')$ and $\alpha(y_2')$ so that

$$\lim_{y_2' \to y_1'} \frac{\operatorname{sm}(x', y')}{\Delta\vartheta_{y'}} = 1. \tag{8.12}$$

8.4 Differentiation of the trigonometric functions

To offset the advantage of having this limit, the angle between x' and $-x'$ is now not constant but $\alpha^2(x')\pi$. Even worse, if $x' \leq y' \leq z'$ on $\partial \tilde{I}$ then the equation

$$\vartheta(x', y') + \vartheta(y', z') = \vartheta(x', z')$$

may no longer hold. This equation will be valid, however, if $\alpha(x') = \alpha(y') = \alpha(z')$ and that will certainly be true (Corollary 5.3.4) if there are isometries T_1 and T_2 for which $T_1(x') = y'$ and $T_2(y') = z'$. Thus, although this definition of angle measure is more complicated it does reflect the geometric interplay between B and $\tilde{I}$. For the rest of the chapter, when we make reference to the alternative definition of angle measure we shall be referring to Equations (8.11) and (8.12).

Using Theorems 8.3.7 and 8.4.2 and Equation (8.12) we can derive the differentiation formulas.

Theorem 8.4.3 *With the notation established above we have*

(i) $\quad \dfrac{d}{d\theta_y} \operatorname{sm}(x, y) = \operatorname{cm}(y, x);$

(i)′ $\quad \dfrac{d}{d\vartheta_{y'}} \operatorname{sm}(x', y') = \operatorname{cm}(y', x');$

(ii) $\quad \dfrac{d}{d\theta_x} \operatorname{sm}(x, y) = -\operatorname{cm}(x, y);$

(ii)′ $\quad \dfrac{d}{d\vartheta_{x'}} \operatorname{sm}(x', y') = -\operatorname{cm}(x', y');$

(iii) $\quad \dfrac{d}{d\theta_y} \operatorname{cm}(x, y) = -\alpha(x^{\dashv})^{-1}\alpha(y^{\dashv})^{-1}\operatorname{sm}(x^{\dashv}, y^{\dashv});$

(iii)′ $\quad \dfrac{d}{d\vartheta_{y'}} \operatorname{cm}(x', y') = -\alpha(x'^{\dashv})^{-1}\alpha(y'^{\dashv})^{-1}\operatorname{sm}(x'^{\dashv}, y'^{\dashv}).$

Proof

(i) From Theorem 8.3.7(ii) we have

$$\operatorname{sm}(x, y_2) - \operatorname{sm}(x, y_1) = \operatorname{sm}(x, y_1)[\operatorname{cm}(y_1, y_2) - 1] + \operatorname{sm}(y_1, y_2)\operatorname{cm}(y_1, x);$$

hence,

$$\frac{\operatorname{sm}(x, y_2) - \operatorname{sm}(x, y_1)}{\Delta\theta_y} = \operatorname{sm}(x, y_1)\left[\frac{\operatorname{cm}(y_1, y_2) - 1}{\Delta\theta_y}\right] + \frac{\operatorname{sm}(y_1, y_2)}{\Delta\theta_y}\operatorname{cm}(y_1, x).$$

Now take limits as $y_2 \to y_1$ and use Theorem 8.4.2 (and set $y_1 = y$). The proof of (i)′ is identical except that it uses Equation (8.12).

(ii) and (ii)′. Since $\operatorname{sm}(x, y) = -\operatorname{sm}(y, x)$ these are a direct consequence of (i) and (i)′ respectively.

(iii) and (iii)'. These follow from Theorem 8.3.7(i) in the same way that (i) and (i)' follow from 8.3.7(ii). ∎

Remark. Note the change of order of the variables in (i) and (i)'. Note also that the negative signs in (ii) and (ii)' are because, as x increases, the angle between x and y decreases.

The last two formulas (for the derivative of $cm(x, y)$ with respect to the first variable, and its partner) are treated separately because they require further explanation. We need to introduce the notion of *curvature*. In a two-dimensional normed space X suppose c is a rectifiable curve. If x_1 and x_2 are two points of c with unit tangents $t(x_1)$ and $t(x_2)$ respectively then

$$\Delta\psi := \theta(t(x_1), t(x_2))$$

and if $t'(x_i) := t(x_i)/\|t(x_i)\|_{\tilde{I}}$

$$\Delta'\psi := \vartheta(t'(x_1), t'(x_2)).$$

If Δs is the (Minkowski) arc length from x_1 to x_2 we define two *Minkowski curvatures* for c at x_1

$$\kappa(c, x_1) := \lim_{x_2 \to x_1} \frac{\Delta\psi}{\Delta s} \quad \text{and} \quad \kappa'(c, x_1) := \lim_{x_2 \to x_1} \frac{\Delta'\psi}{\Delta s}.$$

The reciprocal of κ (respectively κ') is the *radius of curvature* ρ (respectively ρ') of c at x_1.

In the proof of the next theorem we shall consider the following situation. For the curve c we shall take $\partial\tilde{I}$ and we shall think of it as being parametrized by ∂B; i.e. to each $x \in \partial B$ we consider $x^{\dashv} \in c$. Moreover, since B is assumed to be smooth, this parametrization is continuous. The final observation that we need is the one following Proposition 8.3.3 – namely, that x is the unit tangent to $c = \partial\tilde{I}$ at $x^{\dashv}$. The radius of curvature of c at $x^{\dashv}$ will be denoted by $\rho(\tilde{I}, x^{\dashv})$ (resp. $\rho'(\tilde{I}, x^{\dashv})$).

Theorem 8.4.4 *With the above notation we have*

(i) $\quad \dfrac{dcm(x, y)}{d\theta_x} := \lim\limits_{x_1, x_2 \to x} \dfrac{cm(x_2, y) - cm(x_1, y)}{\Delta\theta_x} = \rho(\tilde{I}, x^{\dashv}) sm(x, y);$

(i)' $\quad \dfrac{dcm(x', y')}{d\vartheta_{x'}} := \lim\limits_{x'_1, x'_2 \to x'} \dfrac{cm(x'_2, y') - cm(x'_1, y')}{\Delta\vartheta_{x'}} = \rho'(\tilde{I}, x'^{\dashv}) sm(x', y').$

Proof. In the case of (i), from Theorem 8.3.7(i) we have

$cm(x_2, y) - cm(x_1, y)$
$= cm(x_1, y)[cm(x_2, x_1) - 1] - \alpha(x_2^{\dashv})^{-1}\alpha(x_1^{\dashv})^{-1} sm(x, y) sm(x_2^{\dashv}, x_1^{\dashv}).$

Therefore, if we divide by $\Delta\theta_x$ and take limits as $x_1, x_2 \to x$ (i.e. as $\Delta\theta_x \to 0$) then Theorem 8.4.2 shows that the first term on the right tends to 0.

For the second term, we have (from (8.5))

$$-\frac{\alpha(x_1^{\to})^{-1}\alpha(x_2^{\to})^{-1}\operatorname{sm}(x_2^{\to}x_1^{\to})}{\Delta\theta_x} = \frac{\|x_1^{\to}\|\,\|x_2^{\to}\|\operatorname{sm}(x_1^{\to}x_2^{\to})}{\Delta\theta_x}$$

$$= \frac{2\tau'}{\Delta\theta_x} = \frac{2\tau'}{\Delta s}\frac{\Delta s}{\Delta\theta_x},$$

where τ' is the area of the triangle spanned by $x_1^{\to}$ and $x_2^{\to}$ and Δs is the length of c between these two points. The initial minus sign is absorbed by the change in order of the sm variables.

Since $\Delta s = 2\theta'(x_1^{\to}, x_2^{\to})$, we have that $2\tau'/\Delta s \to 1$ and also, since x_i is the unit tangent to $\partial\tilde{\mathbf{I}} = c$ at $x_i^{\to}$, we have by definition that $\Delta\theta_x = \Delta\psi$. Thus

$$\lim_{x_1,x_2\to x}\frac{-\alpha(x_1^{\to})^{-1}\alpha(x_2^{\to})^{-1}\operatorname{sm}(x_2^{\to}x_1^{\to})}{\Delta\theta_x} = \lim_{x_1,x_2\to x}\frac{2\tau'}{\Delta s}\frac{\Delta s}{\Delta\theta_x} = \rho(\tilde{\mathbf{I}}, x^{\to}).$$

Thus we get the desired equation.

The proof of (i)' is the same except that now in the denominator we have $\Delta\vartheta_{x'}$ but, again by definition, we have $\Delta\vartheta_{x'} = \Delta'\psi$ and there is no change to the other factors. ∎

These formulas show that sm and cm satisfy differential equations analogous to that of classical *simple harmonic motion*.

Corollary 8.4.5 *With the above notation we have*

(i) $\dfrac{d^2}{d\theta_x^2}\operatorname{sm}(x, y) = -\rho(\tilde{\mathbf{I}}, x^{\to})\operatorname{sm}(x, y);$

(ii) $\dfrac{d^2}{d\theta_y^2}\operatorname{sm}(x, y) = \rho(\tilde{\mathbf{I}}, y^{\to})\operatorname{sm}(y, x) = -\rho(\tilde{\mathbf{I}}, y^{\to})\operatorname{sm}(x, y);$

(iii) $\dfrac{d}{d\theta_x}\left(\rho(\tilde{\mathbf{I}}, x^{\to})^{-1}\dfrac{d}{d\theta_x}\operatorname{cm}(x, y)\right) = -\operatorname{cm}(x, y).$

Remark. With the alternative definition of angle we get exactly the same set of equations but with primes on all the variables.

8.5 Notes

The definition of the cosine function has been given by many people. Busemann [71] and Petty [413] attribute it to Finsler [158], although his definition looks considerably different from ours. The definition of the sine function in terms of area and volume is due to Busemann [71] and also has an appearance different from ours. Barthel [22] develops a formalism for the definition and a satisfactory treatment of both the sine function and the notion of curvature. The particular form of the definition of sine given in Definition 8.1.3 can be found in [262].

The definition of the function α also comes from Busemann [71] but most of the properties discussed in §8.2, especially those dealing with the extreme values of α, are due to Petty [413]. However, as stated above, Petty's values for these extrema are "dual" to those presented here because he uses the dual normalization of volume (area in two-dimensional spaces).

The formula (8.5) for α as a ratio of norms implies that the extreme values of α measure the Δ_1 (or the Δ_2) distance between B and $\tilde{\mathbf{I}}$:

$$\Delta_1(\tilde{\mathbf{I}}, B) = \log(\max \alpha) - \log(\min \alpha).$$

Furthermore, as Petty [413] pointed out, these values can be substituted in the two-dimensional Bonnesen isoperimetric inequality (Theorem 4.5.5) as follows. If (X, B) is a two-dimensional Minkowski space then

$$\mu(\partial B)^2 - 4\pi \mu(B) \geq \pi^2[(\max \alpha)^2 - (\min \alpha)^2].$$

Proposition 8.2.1 and its simple corollary are due to Busemann [71], who then raises the natural question of what can be said if α is constant (*i.e.* if $\tilde{\mathbf{I}}$ and B are homothetic). In particular, if dim $X \geq 3$, does max α = min α imply that the space is Euclidean? This question has been restated many times in various contexts since then (including several in this book!) for both definitions of $\mathbf{I}$. In the case of the Holmes–Thompson definition a partial answer is a Corollary to Theorem 8.2.3, that if the constant is ≥ 1 then indeed X is Euclidean.

The observation that the parallelogram spanned by a normal basis has area 1 (Proposition 8.3.4) was made by Barthel [24].

Both Busemann [71] and Petty [413] give the addition formulas for sm and cm. In fact, Busemann used the addition formula to define the Minkowski cosine function. I am grateful to Professor Guggenheimer for making available to me a copy of his lecture notes [227]. The proofs in §8.3 and §8.4 rely heavily on his work [226, 227]. The word *pseudo* in the title of [226] refers to the fact that Guggenheimer does not assume that the metric is symmetric. This generality complicates some of the formulas but, because the sine function requires attention to orientation in any case, the complication is not as large as one might expect. See also Zaustinsky [539], who makes a similar comment.

The question of how to measure angles in situations more general than that of §8.4 was discussed by Bliss [43]. Busemann [69] discusses axioms for angle measure. He posits additivity and the measure of a straight angle being π as necessary and shows that this is sufficient to obtain many of the usual relationships between angle measure and curvature. Lippmann [330] considers a wide variety of angle measures which are in the literature and investigates the metric properties that can be recovered from each of them. For the situation in the Minkowski plane see, in addition to the papers already mentioned, Lippmann [329], Golab [183] and Graham, Witsenhausen and Zassenhaus[198].

One of the purposes of Proposition 8.3.4 and the differentiation formulas in §8.4 is to prepare the ground for the differential geometry of Minkowski spaces. A treatment of this topic has not been included partly for lack of space and partly because the ambiguity about the definition of angle made it a more extensive task than I had anticipated. Readers are referred to the work of Petty [413] and Guggenheimer [226, 227] for two-dimensional differential geometry and to the ground-breaking thesis of Biberstein [34] for an excellent introduction to both the two- and three-dimensional cases. A number of applications of Biberstein's work have been given by Constantin [115, 116]. The paper of Rund [450] has applications to Finsler geometry in mind and is much more analytic in outlook. The work of Laugwitz [314] is more general in two ways. It discusses affine differential geometry and it deals with both n-dimensional and infinite dimensional spaces.

An interesting point of view for the Minkowski trigonometric functions has been investigated by Petty and Barry [422] and by Guggenheimer [228] and concerns the differential equations which these functions satisfy (Theorem 8.4.4). One may begin with the ordinary differential equation

$$d^2y/dt^2 + p(t)y = 0,$$

where $p(t)$ is a function on $\mathbb{R}$. This equation is known as Hill's equation. Of course, if p is a positive constant then this equation represents simple harmonic motion and the solutions are the usual trigonometric functions ($\sin t$, $\cos t$). In general, if this equation has two independent solutions $u_1(t)$ and $u_2(t)$ whose Wronskian $u_1'u_2 - u_1u_2'$ is identically 1 then one can plot the curve in $\mathbb{R}^2$ given parametrically by $(x, y) = (u_1(t), u_2(t))$. Petty and Barry give conditions (Theorem 3.1 in [422]) for this curve to be closed and, if $p(t)$ is positive, it then represents a closed convex curve (the boundary of a non-symmetric Minkowski unit ball). Heil [251] relates the volume product for this convex set to the eigenvalues of the equation.

From the solutions to Hill's equation, one can recover the Minkowski sine and cosine functions by

$$\mathrm{sm}(t_1, t_2) = u_1(t_2)u_2(t_1) - u_1(t_1)u_2(t_2)$$

and

$$\mathrm{cm}(t_1, t_2) = u_1'(t_1)u_2(t_2) - u_1(t_2)u_2'(t_1).$$

Moreover, the *hodograph* given parametrically by $(x, y) = (u_2'(t), u_1'(t))$ is the equation of the isoperimetrix **I**. A similar interpretation can be found in the beautiful paper by Wallen [526] except that there the roles of **B** and **I** are interchanged.

Wallen points out that one can view the boundary of the isoperimetrix as the curve for which equal areas are swept out in equal times (by a particle travelling with constant Minkowski speed) and then the boundary of the unit ball is the hodograph curve. In particular, Wallen observes that if B is a circle with origin *not* at the centre then **I** is an ellipse with one focus at the origin.

9

Various numerical parameters

As is abundantly clear from the previous chapters there is a huge variety of non-isometric Minkowski spaces in each dimension. It is not feasible to seek numerical invariants that will completely classify all these spaces. However, there are a variety of properties that stand out as being significant in the theory of infinite dimensional Banach spaces. It frequently happens that these properties can be framed in terms of numerical parameters that depend on the "shape" of the space and are invariant under isometries. The projection constant in §9.1 below is a good example of this situation. In finite dimensional spaces such numerical parameters lead to several interesting questions. Firstly, one can try to calculate the exact value of the parameter for particular spaces. This is often a difficult problem. Secondly, one can ask for the bounds, as precisely as possible, of the parameter over all Minkowski spaces of a fixed dimension. Instances of this problem were considered in §6.5 and §7.4, where bounds for the parameter $\mu_B(\partial B)$ were given for each of the two definitions of μ_B. Thirdly, one can investigate the asymptotic behaviour of the parameter as the dimension gets large. The asymptotic behaviour is often significant for the infinite dimensional theory. Finally, one can use differences in the value of a particular parameter for two distinct spaces to measure how far apart (in the Banach–Mazur or some other metric) the two spaces are. Various examples of these four problems will be considered in this last chapter.

The first section is concerned with two parameters known as *projection constants*. The first of these relates to a characterization of Euclidean spaces given in Chapter 3. If there is a projection of norm 1 onto every subspace of a Minkowski space X (dim $X \geq 3$) then X is Euclidean. Here we consider non-Euclidean spaces and ask how large the norms of projections onto subspaces must be. The second parameter is similar but, instead of considering projections onto subspaces Y of X, we ask the same question about the norms of projections from a containing space Y onto X. The second section is a brief one dealing with a parameter known as Macphail's constant. The bulk of the chapter consists of the last four sections, which are concerned with the work of J. J. Schäffer.

In his long and interesting paper [459] Schäffer introduced four ways of measuring the size of the unit sphere in a normed space. He called these parameters the *girth, perimeter, inner radius* and *inner diameter*. The purpose of §§9.3–9.6 is to present that part of his work that deals with finite dimensional spaces.

With the exception of the theorem of Witsenhausen (9.4.5) and a few calculations in §9.5, the material presented here is all contained in Schäffer's book [466]. Moreover, the more interesting results of Schäffer [461, 462], Schäffer and Sundaresan [467], Karlovitz [282] and Harrell and Karlovitz [246, 247] deal with infinite dimensional spaces. In particular, the startling facts (a) that a complete normed space X is superreflexive (James [268]) if and only if the girth of X is strictly larger than 4, and (b) that there exist *flat* Banach spaces in which not only is the girth 4 but there is a centrally symmetric, closed curve in ∂B with length 4, are both infinite dimensional results.

Consequently, in presenting only that part dealing with finite dimensional spaces, the material appears somewhat eviscerated. However, as with the parameters in §9.1 and §9.2 there is considerable challenge in the evaluation of Schäffer's parameters in various spaces. For these calculations we concentrate on the simplest spaces whose unit balls are zonotopes and their duals. We make this restriction for a variety of reasons. Firstly, these are spaces for which exact evaluations are possible. Note that even in the plane, the perimeter of an ℓ_p ball for $p \neq 1, 2, \infty$ involves an elliptic integral. Secondly, the results give some indication of what to expect in analogous infinite dimensional spaces. Thirdly, zonotopes and their duals play such a prominent role in the theory developed in Chapter 6 that further investigation of their properties seems interesting. Finally, these spaces afford the extreme values needed in the last section.

In §9.3 we define the notion of curve and use this to define the inner metric. We also present without proof three of the basic facts needed subsequently. Schäffer's four parameters are defined in §9.4 and some of the immediate facts about them presented. This section also contains the inequality of Witsenhausen. The fifth section contains the detailed calculations alluded to above. These calculations are for the balls C_d and CP_d from Example 1.1.13 and for Z_d and $S_d - S_d$ from Example 1.1.17. The final section discusses the relationship between the Banach–Mazur distance between two spaces and the parameters for those spaces. In this section we also list the extreme values for the parameters in each dimension. Throughout this chapter projections will be denoted by P.

9.1 Projection constants

In the introduction we referred to the fact that if there exists a projection of norm 1 onto each subspace of X then X must be Euclidean. This result can be quantified

9.1 Projection constants

as follows. If X is a Minkowski space with subspace Y let

$$\varpi(X; Y) := \min\{\|P\| : P \text{ is a projection of } X \text{ onto } Y\}$$

and then set

$$\varpi_s(X) := \max\{\varpi(X; Y) : Y \subseteq X\}. \tag{9.1}$$

Theorem 3.4.6 can now be restated by saying that a Minkowski space X with $\dim X \geq 3$ is Euclidean if and only if $\varpi_s(X) = 1$. A thorough investigation of the parameter ϖ_s for other spaces (especially the ℓ_p^d spaces) was made by Sobczyk [491]. The following results are from his paper. The first states that ϖ_s is bounded by a function of the Banach–Mazur distance from Euclidean space.

Recall, from the notation following Examples 1.1.12–1.1.14, that ℓ_p^d denotes the space $(\mathbb{R}^d, \|.\|_p)$ for $1 \leq p \leq \infty$. As a convenient abbreviation let $\eta := \exp \Delta(X, \ell_2^d)$ in the following theorem.

Theorem 9.1.1 *If $(X, \|.\|)$ is a d-dimensional Minkowski space then $\varpi_s(X) \leq (\eta + 1)/2$.*

Proof. If $\Delta(X, \ell_2^d) = \log \eta$ then there is an ellipsoid E in X (the image of the unit ball in ℓ_2^d under some linear transformation T) such that $E \subseteq B \subseteq \eta E$. Then for all $x \in X$ we have

$$\eta^{-1}|x| \leq \|x\| \leq |x|,$$

where $|.|$ is the norm induced by E. Let Y be an arbitrary subspace of X. Then, with respect to the Euclidean structure induced by E, there is an orthogonal projection P of X onto Y. Let $T := 2P - 1$. Straightforward calculations show that T is also orthogonal with respect to E, leaves each point y of Y fixed and is an involution: $T^2 = \mathbf{1}$. Hence

$$\|Tx\|^2 \leq |Tx|^2 = \langle Tx, Tx \rangle = \langle T^2 x, x \rangle = \langle x, x \rangle = |x|^2 \leq \eta^2 \|x\|^2.$$

Consequently, $\|T\| \leq \eta$ and therefore

$$\|P\| = \|(T + \mathbf{1})/2\| \leq (\eta + 1)/2.$$

It follows that $\varpi(X; Y) \leq (\eta + 1)/2$ and, since Y is arbitrary, that $\varpi_s(X) \leq (\eta + 1)/2$, as required. ∎

Corollary 9.1.2 *If $(X, \|.\|)$ is a d-dimensional Minkowski space then*

$$\varpi_s(X) \leq (\sqrt{d} + 1)/2.$$

Proof. This statement follows at once from Theorems 9.1.1 and 3.3.6. ∎

By calculating the exact distance from ℓ_p^d to ℓ_2^d and by giving a lower estimate for $\varpi(\ell_p^d)$ Sobczyk also showed that the preceding estimate is very close to the best possible. The details of these calculations are not complicated but are a little too extensive to include here. We therefore state only the results.

Theorem 9.1.3 (Sobczyk) *If $d = 2^k$ for some positive integer k and if $X = \ell_p^d$ and if η is defined as above then*

  (i) $\eta = d^{|p^{-1} - 1/2|}$;
  (ii) $(\eta - 1)/2 \leq \varpi_s(X)$.

Remarks

(i) Observe that if q is the conjugate of p (i.e. $1/p + 1/q = 1$) then $|p^{-1} - \frac{1}{2}| = |q^{-1} - \frac{1}{2}|$. When $p = 1$ inequality 9.1.3(ii) implies that $\varpi_s(\ell_1^d) \geq (\sqrt{d} - 1)/2$ (at least when $d = 2^k$).

(ii) We make the restriction on d because the method of proof consists of constructing a certain $2^k \times 2^k$ matrix as the tensor product of a 2×2 matrix with itself k times.

(iii) The subspaces Y for which $\varpi(X; Y)$ is large are those with both large dimension and large codimension. Indeed, if either $\dim Y = 1$ or $\operatorname{codim} Y = 1$ then we have the following results, the first of which is yet another application of the Hahn–Banach theorem.

Proposition 9.1.4 *If $(X, \|.\|)$ is a normed linear space and if Y is a one-dimensional subspace of X then there is a projection P of X onto Y with $\|P\| = 1$.*

Proof. Let $Y := \operatorname{span}\{x_1\}$ with $\|x_1\| = 1$. By the Hahn–Banach theorem (1.3.4), there is a linear functional f such that $f(x_1) = \|f\| = 1$. Thus $X = Y \oplus f^\perp$ and each x in X has a unique representation in the form $x = \alpha x_1 + z$ with $f(z) = 0$. Note that $|\alpha| = |f(x)| \leq \|x\|$. Define P by the equation $Px := \alpha x_1$. Then $\|Px\| = |\alpha| \leq \|x\|$. Therefore, $\|P\| \leq 1$. ∎

Theorem 9.1.5 (Bohnenblust) *If $(X, \|.\|)$ is a d-dimensional Minkowski space and if Y is a hyperplane in X then there is a projection P of X onto Y with $\|P\| \leq 2(1 - d^{-1})$.*

To save space we omit the proof of this result. The proof is interesting because it works by induction on the dimension and the inductive step makes clever use of Helly's intersection theorem for convex sets (see §0.3).

The next parameter comes from turning the question of projections around. Instead of looking at subspaces we consider those spaces Y in which a particular Minkowski space X can be embedded and then ask about projections of Y onto

9.1 Projection constants

X. With $\varpi(Y; X)$ defined above (but note the change of order) we let

$$\varpi^s(X) := \sup\{\varpi(Y; X) : X \subseteq Y\}.$$

Definition 9.1.6 *The number $\varpi^s(X)$ is called the **projection constant** of X.*

Frequently the notation $\lambda(X)$ is used for this parameter. The choice of sub- and superscript s is meant to indicate projections onto subspaces and from superspaces respectively.

As was the case for subspaces, the first question asks which spaces X (if any) have $\varpi^s(X) = 1$. This question is closely related (in fact it turns out to be equivalent; see Proposition 9.1.12 below) to the following one. If X, Y and Z are normed linear spaces with $X \subseteq Y$ and if T is a bounded linear transformation from X into Z, is there an extension $\bar{T}$ from Y into Z with $\|\bar{T}\| = \|T\|$? In other words, are there circumstances under which the Hahn–Banach extension theorem holds for normed linear spaces Z other than the scalar field? One way to answer this question is to use the Hahn–Banach theorem in each coordinate to prove the following proposition.

Proposition 9.1.7 *If X is a subspace of a normed linear space Y and if T is a bounded linear mapping from X into ℓ_∞^d then there is an extension $\bar{T}$ from Y into ℓ_∞^d with $\|\bar{T}\| = \|T\|$.*

Proof. Let P_i denote the projection of ℓ_∞^d onto the ith coordinate subspace. Then $f_i := P_i T$ is a bounded linear functional on X with $\|f_i\| = \|P_i T\| \le \|T\|$. By the Hahn–Banach theorem (1.3.3) f_i has an extension $\bar{f}_i$ from Y into $\mathbb{R}$ with $\|\bar{f}_i\| = \|f_i\|$. Define $\bar{T}$ by

$$\bar{T}(y) := (\bar{f}_1(y), \bar{f}_2(y), \ldots, \bar{f}_d(y)).$$

It follows that $\bar{T}$ is an extension of T and that

$$\|\bar{T}\| = \max_i\{\|\bar{f}_i\|\} = \max_i\{\|f_i\|\} \le \|T\|$$

and hence that $\|\bar{T}\| = \|T\|$. ∎

Remarks

(i) The same proof will work for the infinite dimensional space ℓ_∞.

(ii) Since there is an isomorphism between ℓ_∞^d and an arbitrary Minkowski space Z which is continuous in both directions, a bounded linear map from X into an arbitrary Minkowski space Z has a continuous extension from Y to Z.

Corollary 9.1.8 *If $X = \ell_\infty^d$ then $\varpi^s(X) = 1$.*

Proof. Suppose $X \subseteq Y$. The identity $\mathbf{1}_X$ on X is a bounded linear mapping from X into ℓ_∞^d and hence, by the proposition, has an extension P from Y into $\ell_\infty^d = X$ with $\|P\| = \|\mathbf{1}_X\| = 1$. Such an extension is necessarily a projection. ∎

The converse is also true – namely, that if $(X, \|.\|)$ is a d-dimensional Minkowski space and if $\varpi^s(X) = 1$ then X is isometrically isomorphic to ℓ_∞^d. The infinite dimensional version of this result is that a Banach space X with $\varpi^s(X) = 1$ is isometrically isomorphic to the space of all continuous functions on some extremally disconnected compact Hausdorff space (with the supremum norm). Here we will prove a limited version of the finite dimensional theorem using a result of Zippin [545]. It requires two preliminary lemmas.

Lemma 9.1.9 *If $(X, \|.\|)$ is a Minkowski space whose unit ball B is a polytope with $2k$ facets then $(X, \|.\|)$ is isometrically isomorphic to a subspace of ℓ_∞^k.*

Proof. If B has $2k$ facets then $B^\circ \subseteq X^*$ is a polytope with $2k$ vertices: $\{\pm v_1, \pm v_2, \ldots, \pm v_k\}$. If n of these vertices are linearly independent (which we can suppose are the first n with a consistent choice of sign) then consider the vector space $Y := X^* \oplus \mathbb{R}^{k-n}$ with basis $\{v_1, v_2, \ldots, v_n, e_1, e_2, \ldots, e_{k-n}\}$ and unit ball $C_p := \mathrm{conv}\{\pm v_1, \pm v_2, \ldots, \pm v_n, \pm e_1, \pm e_2, \ldots, \pm e_{k-n}\}$. The Minkowski space (Y, C_p) is isometrically isomorphic to ℓ_1^k and the linear map P defined by $P(v_j) := v_j$ for $j = 1, 2, \ldots, n$ and $P(e_i) := v_{n+i}$ for $i = 1, 2, \ldots, (n-k)$ is a projection (with norm 1) of Y onto X^* and B° is the image under P of the cross-polytope C_p. Hence, by Theorems 2.2.9 and 0.1.1, $B = C_p^\circ \cap X$; i.e. B is the cross-section determined by X of the unit ball C_p° in $Y^* = X \oplus \mathbb{R}^{k-n}$. ∎

Remark. It follows by an approximation argument that every Minkowski space can be isometrically embedded in ℓ_∞. In fact the Hahn–Banach theorem can be used to show that a separable Banach space can be embedded in ℓ_∞.

Lemma 9.1.10 *If X is a subspace of a normed linear space Y and if P is a projection of Y onto X with $\|P\| = 1$ then there is a subspace M of Y^* such that*

   (i) *M is isometrically isomorphic to X^*, and*
   (ii) *there is a projection P' of Y^* onto M with $\|P'\| = 1$.*

Proof. If P from Y to X is a linear mapping then P^* from X^* to Y^* defined by Equation (0.5) is a linear map with $\|P^*\| = \|P\|$ (Theorem 1.3.9) and the range of P^* is a subspace M of Y^*. If P is a projection then this range is given by $M = \{h \in Y^* : h(y) = h(Py), \, \forall y \in Y\}$.

Define P' from Y^* to M by the same equation as P^*: $P'h := hP$, i.e. $P'h(y) := h(Py)$. Routine calculations show that since P is a projection the range of P' is

M, that P' is a projection and that $\|P'\| = \|P^*\| = \|P\|$. Finally, P^* is a one–one linear map of X^* onto M with $\|P^*\| = 1$ so that X^* is isometrically isomorphic to M. ∎

Theorem 9.1.11 *If $(X, \|.\|)$ is a d-dimensional Minkowski space whose unit ball is a polytope and if $\varpi^s(X) = 1$ then there is an isometry of X onto ℓ_∞^d.*

Proof. Suppose B has $2k$ facets. By Lemma 9.1.9 $(X, \|.\|)$ can be represented as a subspace of ℓ_∞^k. Since $\varpi^s(X) = 1$ there is a projection of norm 1 from ℓ_∞^k onto X. Hence, by Lemma 9.1.10, there is a projection Q of norm 1 from ℓ_1^k onto a subspace Z isometric to X^*. It is sufficient to show that Z is isometric to ℓ_1^d and, for this, we use an argument of Zippin [545].

We think of Q as a linear operator on ℓ_1^k and represent it, with respect to the usual basis, as a $k \times k$ matrix (τ_{ij}). Then

$$Qe_i = \sum_{j=1}^{k} \tau_{ji} e_j.$$

If B_Z denotes the unit ball in Z then B_Z is a polytope with a number, n say, of pairs of opposite vertices. Choose one vertex v_m ($m = 1, 2, \ldots, n$) from each pair. Each such vertex v_m is the image under Q of a vertex of the ball in ℓ_1^k, i.e.

$$v_m = \pm Q e_{i(m)}.$$

Thus we have

$$\pm v_m = Q e_{i(m)} = \sum_{1}^{k} \tau_{j\,i(m)} e_j = \sum_{1}^{k} \tau_{j\,i(m)} Q e_j. \tag{9.2}$$

The last equality comes from the fact that Q is a projection and so $Q^2 e_{i(m)} = Q e_{i(m)}$.

Now $\|\pm v_m\| = 1 = \|Q e_{i(m)}\| = \sum |\tau_{j\,i(m)}|$ and therefore, if we choose the signs according to the sign of $\tau_{j\,i(m)}$, $\sum_{1}^{k} \tau_{j\,i(m)} Q e_j = \sum_{1}^{k} \pm \tau_{j\,i(m)}(\pm Q e_j)$ is a convex linear combination of the points $\pm Q e_j$ in B_Z. Since v_m is a vertex, and hence an extreme point, of B_Z it follows from (9.2) that $\pm v_m = Q e_{i(m)} = \pm Q e_j$ for all those j such that $\tau_{j\,i(m)} \neq 0$. In other words, for each m there is a subset $J_m \subseteq \{1, 2, \ldots, k\}$ such that $j \in J_m$ if and only if $\tau_{j\,i(m)} \neq 0$; and if $j \in J_m$ then $Q e_j = \pm v_m$. It is clear from this last equation that if $m \neq m'$ then $J_m \cap J_{m'} = \emptyset$. Moreover, from (9.2), we have $v_m = \pm \sum_{j \in J_m} \tau_{j\,i(m)} e_j$.

Since the vertices of B_Z span Z there are d linearly independent vertices $\{v_1, v_2, \ldots, v_d\}$ in $\{v_1, v_2, \ldots, v_n\}$. If $n > d$ then

$$v_{d+1} = \sum_{m=1}^{d} \beta_m v_m = \sum_{m=1}^{d} \sum_{j \in J_m} \beta_m \tau_{j\,i(m)} e_j$$

for some scalars β_m. Therefore, $J_{d+1} \subseteq \bigcup_{m=1}^{d} J_m$ but this contradicts the fact that these sets are disjoint. Hence $n = d$ and B_Z is a cross-polytope. ∎

Remark. Goodner [194] shows in general that if $\varpi^s(X) = 1$ then the distance between any two extreme points of the unit ball of X is 2. Goodner's result gives an alternative proof of Theorem 9.1.11 and shows that the restriction that the unit ball be a polytope is unnecessary.

With regard to the exact value of ϖ^s for particular spaces not a great deal is known. Rutovitz [452] has shown that $\varpi^s(\ell_2^d) = 2\epsilon_{d-1}/\epsilon_d$. Earlier Grünbaum [223] had proved that this was an upper bound and, in the same paper, that $\varpi^s(\ell_1^d) = 2\epsilon_{d-1}/\epsilon_d$ when d is odd and is $2\epsilon_{d-2}/\epsilon_{d-1}$ when d is even. Using the John result, Kadeč and Snobar [278] showed that for every d-dimensional Minkowski space $\varpi^s(X) \leq \sqrt{d}$.

Finally in this section we consider the relationship between ϖ^s and the Banach–Mazur distance. The result is most easily established by first returning to the question of extending bounded linear maps.

Proposition 9.1.12

  (i) If X and Y are normed linear spaces and if $X \subseteq Y$ then every bounded linear map from X into a third space Z has an extension $\bar{T}$ from Y into Z with $\|\bar{T}\| \leq \varpi^s(X)\|T\|$. Conversely, if every bounded linear map T from X into an arbitrary space Z has an extension $\bar{T}$ from Y into Z with $\|\bar{T}\| \leq \beta\|T\|$ then $\varpi^s(X) \leq \beta$.

  (ii) If X is a separable normed linear space and if Y and Z are two normed linear spaces with $Z \subseteq Y$ then every bounded linear map T from Z into X has an extension $\bar{T}$ from Y into X with $\|\bar{T}\| \leq \varpi^s(X)\|T\|$. Conversely, if every bounded linear map T from Z into X has an extension $\bar{T}$ from Y into X with $\|\bar{T}\| \leq \beta\|T\|$ (Y and Z are arbitrary except that $Z \subseteq Y$) then $\varpi^s(X) \leq \beta$.

Proof

  (i) Since $X \subseteq Y$ there is a projection P from Y onto X with $\|P\| \leq \varpi^s(X)$. If we set $\bar{T} := TP$ then the map $\bar{T}$ has the desired properties. Conversely, if every such map has an extension then consider, in particular, the identity $\mathbf{1}_X$ on X. It has an extension P from Y onto X with $\|P\| \leq \beta \|\mathbf{1}_X\| = \beta$ and so $\varpi^s(X) \leq \beta$.

  (ii) The second proof is only slightly more complicated. First (see the remark following Lemma 9.1.9) X can be embedded in ℓ_∞ and hence T can be regarded as a mapping from Z into ℓ_∞. By the remark following Theorem 9.1.7, T has an extension $\tilde{T}$ from Y into ℓ_∞. Moreover, there is a projection P of ℓ_∞ onto X with $\|P\| \leq \varpi^s(X)$. Hence $\bar{T} := P\tilde{T}$ has the desired properties. The proof of the converse is exactly as before. ∎

Theorem 9.1.13 *If X and Y are isomorphic normed linear spaces then $\varpi^s(X)/\varpi^s(Y) \leq \exp(\Delta(X, Y))$.*

Proof. Let T be an isomorphism of X onto Y and suppose that Z is some space with $X \subseteq Z$. By Proposition 9.1.12(ii), there is an extension $\bar{T}$ from Z onto Y with $\|\bar{T}\| \leq \varpi^s(Y)\|T\|$. It is easily checked that $P := T^{-1}\bar{T}$ is a projection of Y onto X with $\|P\| \leq \varpi^s(Y)\|T^{-1}\|\|T\|$. Since Z and P are arbitrary, $\varpi^s(X) \leq \varpi^s(Y)\|T^{-1}\|\|T\|$, from which inequality the result follows. ∎

The following corollary is immediate.

Corollary 9.1.14 *The function ϖ^s is continuous with respect to Δ.*

9.2 Macphail's constant

As with real numbers, if (x_i) is a sequence of elements in a Banach space then the series $\sum x_i$ is said to *converge absolutely* if $\sum \|x_i\|$ converges. Likewise, the series is said to *converge unconditionally* if every rearrangement of the series converges. An absolutely convergent series converges unconditionally and, in every Minkowski space, the converse is also true. For ℓ_∞^d this fact is established by considering each coordinate separately and the general case follows from Corollary 1.2.5. Banach [21] asked whether there are any infinite dimensional spaces in which the two concepts are equivalent. The problem was quantified by Macphail [345] in the following way.

Let $\mathcal{S} = (x_1, x_2, \ldots, x_m)$ be a finite sequence (sequence rather than subset because repetitions are allowed) in a normed linear space $(X, \|.\|)$. Construct "norms" for $\mathcal{S}$ with the following equations:

$$|\mathcal{S}| := \sum_{i=1}^{m} \|x_i\|,$$

$$|\mathcal{S}|_* := \sup\left\{\|\sum_{j \in J} x_j\| : J \subseteq \{1, 2, \ldots, m\}\right\}.$$

Now we can define Macphail's constant $\varrho(X)$ for X.

Definition 9.2.1 *With the previous notation, **Macphail's constant** for X is*

$$\varrho(X) := \inf\{|\mathcal{S}|_*/|\mathcal{S}| : \mathcal{S} \text{ a finite sequence in } X\}.$$

Macphail [345] showed that $\varrho(X) > 0$ if and only if every unconditionally convergent series in X is absolutely convergent. He also showed that $\varrho(X) = 0$ for the two cases $X = \ell_1$ and $X = L_1[0, 1]$. Dvoretzky and Rogers [134] showed that if X is a d-dimensional Minkowski space then $\varrho(X) \leq 2/d^{1/4}$ and hence that $\varrho(X) = 0$ for all infinite dimensional Banach spaces. Therefore, unconditional convergence implies absolute convergence if and only if the space is finite dimensional.

In view of the Dvoretzky–Rogers result the interest in calculating $\varrho(X)$ lies not so much in its inherent significance as in its relationship with other parameters. We illustrate this relationship with a few isolated results of Rutovitz [452] and Gordon [195].

Theorem 9.2.2 *If X and Y are d-dimensional Minkowski spaces then*

$$\varrho(X)/\varrho(Y) \leq \exp(\Delta(X, Y)).$$

Proof. Let $\eta = \exp(\Delta(X, Y))$; then there is a linear mapping T of Y onto X such that $\|T\| = 1$ and $\|T^{-1}\| = \eta$. If $\mathcal{S} = (y_1, y_2, \ldots, y_m)$ is an arbitrary finite sequence in Y and if $T(\mathcal{S}) := (Ty_1, Ty_2, \ldots, Ty_m)$ then

$$\varrho(X) \leq |T(\mathcal{S})|_*/|T(\mathcal{S})| \leq |\mathcal{S}|_*/\|T^{-1}\|^{-1}|\mathcal{S}| = \eta|\mathcal{S}|_*/|\mathcal{S}|.$$

Taking the infimum over all finite sequences $\mathcal{S}$ in Y now gives the result. ∎

As with ϖ^s it is immediate from Theorem 9.2.2 that ϱ is continuous with respect to Δ.

Theorem 9.2.3 *If $(X, \|.\|)$ is a d-dimensional Minkowski space then $\varrho(X) \leq \varpi^s(X)/2d$.*

Proof. Since both ϱ and ϖ^s are continuous it is sufficient to consider those X's whose unit ball is polyhedral. Moreover, by Lemma 9.1.9, we can regard X as a subspace of ℓ_∞^k for some k. Also $\varpi^s(X) = \varpi(\ell_\infty^k, X)$.

Let P be a projection of ℓ_∞^k onto X with $\|P\| = \varpi^s(X)$. Consider the finite set $\mathcal{S} := (y_1, y_2, \ldots, y_{2k}) \subseteq \ell_\infty^k$, where $y_i := e_i$ for $i = 1, 2, \ldots, k$ and $y_i := -e_{i-k}$ for $i = k+1, k+2, \ldots, 2k$. We calculate $|P(\mathcal{S})|_*/|P(\mathcal{S})|$. First, if $J \subseteq 1, 2, \ldots, 2k$ then $\|\sum_{j \in J} y_j\|_\infty \leq 1$ and hence

$$\left\|\sum_{j \in J} Py_j\right\| = \left\|P(\sum_{j \in J} y_j)\right\| \leq \|P\| = \varpi^s(X).$$

Therefore, $|P(\mathcal{S})|_* \leq \varpi^s(X)$.

Representing P as a matrix (τ_{ij}) we have $Pe_i = \sum_{j=1}^k \tau_{ji} e_j$ and therefore

$$\sum_{i=1}^{2k} \|Py_i\| = 2\sum_{i=1}^k \left\|\sum_{j=1}^k \tau_{ji} e_j\right\| = 2\sum_{i=1}^k \max_j |\tau_{ji}| \geq 2\sum_{i=1}^k |\tau_{ii}| = 2\operatorname{tr} P = 2d.$$

The last equality comes from the fact that trace is independent of basis and there is a basis for ℓ_∞^k with respect to which the matrix for P has d 1's on the diagonal and 0's elsewhere. Therefore, $|P(\mathcal{S})| \geq 2d$.

Combining these inequalities yields $\varrho(X) \leq |P(\mathcal{S})|_*/|P(\mathcal{S})| \leq \varpi^s/2d$. ∎

Remark. If $X = \ell_\infty^d$ then $\varpi^s(X) = 1$ and $\varrho(X) = 1/2d$ and hence equality occurs in Theorem 9.2.3.

The next theorem gives a lower bound for $\varrho(X)$.

Theorem 9.2.4 *If $(X, \|.\|)$ is a d-dimensional Minkowski space then $\varrho(X) \geq d^{-3/2}\epsilon_{d-1}/\epsilon_d$.*

Proof. Denote the unit ball in X by B. By Theorem 3.3.6, there is a Euclidean unit ball E with

$$E \subseteq B \subseteq \sqrt{d}E. \tag{9.3}$$

Let $\mathcal{S} := (x_1, x_2, \ldots, x_m)$ be an arbitrary finite sequence in X. Using the notation of Example 0.2.3(iv), let $Z := \sum_{i=1}^m 2[x_i]$. Then Z is a zonotope with vertices of the form $\sum_{i=1}^m \pm x_i$. Hence

$$\zeta := \max\{\|x\| : x \in Z\} \leq 2\max\left\{\left\|\sum_{j \in J} x_j\right\| : J \subseteq \mathcal{S}\right\} = 2|\mathcal{S}|_*. \tag{9.4}$$

On the other hand,

$$\zeta = \max\{|f(x)| : x \in Z, f \in B^\circ\}$$
$$= \max\{|f(x)|/\|f\| : x \in Z, f \in X^* \setminus \{0\}\}.$$

Hence $\|f\|\zeta \geq \max\{|f(x)| : x \in Z\} = h_Z(f)$ (see Definition 2.2.1).

Now integrate the last inequality over the surface of the Euclidean unit ball E° in X^*. Thus

$$\zeta \int_{S^{d-1}} \|f\| \, d\lambda \geq \int_{S^{d-1}} h_Z(f) \, d\lambda$$
$$= dV(E^\circ[d-1], Z)$$
$$= d\sum_{i=1}^m V(E^\circ[d-1], 2[x_i])$$
$$= 2\epsilon_{d-1} \sum_{i=1}^m \|x_i\|_E$$
$$\geq 2\epsilon_{d-1} \sum_{i=1}^m \|x_i\|$$
$$= 2\epsilon_{d-1}|\mathcal{S}|.$$

Note that in this Euclidean calculation we must identify the linear spaces X and X^* in order that the mixed volumes make sense. Since, from (9.3), $B \subseteq \sqrt{d}E$ we have $d^{-1/2}E^\circ \subseteq B^\circ$ and hence if $f \in \partial E^\circ$ then $\|f\| \leq \sqrt{d}$. Therefore,

$$\zeta d^{3/2}\epsilon_d \geq 2\epsilon_{d-1}|\mathcal{S}|.$$

If we now substitute for ζ from (9.4) we get

$$2|\mathcal{S}|_* d^{3/2}\epsilon_d \geq 2\epsilon_{d-1}|\mathcal{S}|.$$

Rearranging the terms and noting that $\mathcal{S}$ is arbitrary yields the required result. ∎

The argument in the preceding proof was originally used by Mayer [360] to prove that $\varrho(\ell_2^d) = \epsilon_{d-1}/d\epsilon_d$. For the case when $X = \ell_p^d$ first Rutovitz [452] and then Gordon [195] used the same argument with more careful estimates of the integrals involved. Rutovitz considered $p \geq 2$ and showed that $\varrho(\ell_p^d)$ is asymptotic to $d^{p^{-1}-1}$; Gordon's result is that if $1 \leq p \leq 2$ then $\varrho(\ell_p^d)$ is asymptotic to $d^{-1/2}$. The last result we mention is that of Kadeč and Snobar [278]: for every d-dimensional Minkowski space

$$d^{-1} \leq 2\varrho(X) \leq d^{-1/2}.$$

(Note that their definition of Macphail's constant is equal to 2ϱ.)

9.3 The inner metric

In this section we use the notion of a curve in a metric space to define the inner metric for a subset of the metric space. Although the concepts are quite general, our context is always that of a Minkowski space. For a more complete discussion see Busemann [74].

We begin by recalling the definitions of curve and the length of a curve given in §4.3 for two-dimensional spaces, which are equally valid in general. By a *curve* c in X we mean the *range* of a continuous function φ that maps a closed bounded interval $[\alpha, \beta]$ into X. The curve c is said to *lie* in a subset A, or to be *in* A, if $c \subset A$.

The curve c defined by a function $\varphi : [\alpha, \beta] \mapsto X$ is said to be *rectifiable* if the set of all Riemann sums

$$\left\{ \sum_{i=1}^n \|\varphi(t_i) - \varphi(t_{i-1})\| : (t_0, t_1, \ldots, t_n) \text{ is a partition of } [\alpha, \beta] \right\}$$

is bounded above. The *length* $\ell := \mu(c)$ of c is then defined to be the supremum of all such Riemann sums. As we remarked in §4.3 the concepts of curve and rectifiability are independent of which norm is used on X. The length is heavily dependent on the particular norm.

Next we add some further terminology related to curves. Sometimes we shall make use of the fact that a given rectifiable curve c can be parametrized in the canonical way by arc length. To distinguish this parametrization we shall use s for arc length and ψ for the parametrization. Thus $c = \{x \in X : x = \psi(s), 0 \leq s \leq \mu(c)\}$. The point $\psi(0)$ is the *initial* point of c, $\psi(\ell) = \psi(\mu(c))$ is its *end point* and c is described as a curve *from* $\psi(0)$ *to* $\psi(\ell)$. The curve is *closed* if $\psi(0) = \psi(\ell)$ and *simple* if ψ is one–one, with the possible exception of the initial and end points.

If $x_1, x_2 \in X$ then the segment $[x_1, x_2]$ is the curve given by $\varphi(t) = (1-t)x_1 + tx_2$, $0 \leq t \leq 1$, which has length $\|x_2 - x_1\|$. If c_1 and c_2 are two curves defined by functions ψ_1 and ψ_2 on $[0, \ell_1]$ and $[0, \ell_2]$ respectively and if $\psi_1(\ell_1) = \psi_2(0)$ then $c_1 \cup c_2$ is a curve defined by the function ψ, where

$$\psi(s) = \begin{cases} \psi_1(s), & 0 \leq s \leq \ell, \\ \psi_2(s - \ell_1), & \ell_1 < s \leq \ell_1 + \ell_2. \end{cases}$$

A *polygon* is a finite sequence of line segments $[x_1, x_2] \cup [x_2, x_3] \cup \cdots \cup [x_{n-1}, x_n]$ joined in the way just described. The points $x_1, x_2, \ldots, x_n$ are the *vertices* of the polygon and the segments $[x_{i-1}, x_i]$ the *edges*.

Definition 9.3.1 *If A is a subset of the Minkowski space X, then the **inner metric** δ_A on A is defined by*

$$\delta_A(x, y) := \inf\{\mu(c) : c \text{ is a rectifiable curve from } x \text{ to } y \text{ in } A\}.$$

If no such curve exists then we set $\delta_A(x, y) = +\infty$.

In order to avoid the use of infima and attendant ε's, we shall make repeated use of the following key theorem (for a proof see, *e.g.*, Busemann [73], pp. 25–26).

Theorem 9.3.2 *If A is a compact subset of a metric space then, for each x, y in A with $\delta_A(x, y) < \infty$, there exists a curve c in A with $\mu(c) = \delta_A(x, y)$. Moreover, each such curve is simple.*

Remarks

(i) Since a curve c is a compact set one can apply the last part of Theorem 9.3.2 (with $A = c$) to show that if c is a curve from x to y with length ℓ then there is a simple curve from x to y with length $\ell' \leq \ell$. Hence in Definition 9.3.1 one need only consider the infimum over simple curves.

(ii) If c is a curve in A then the length of c with respect to δ_A is the same as the length c with the respect to the original norm.

To see this suppose $c = \{\psi(s) : 0 \leq s \leq \ell\}$ and consider a partition $0 = s_0 < s_1 < \cdots < s_n = \ell$. Then, because line segments are geodesics,

$$\|\psi(s_i) - \psi(s_{i-1})\| \leq \delta_A(\psi(s_i) - \psi(s_{i-1})).$$

On the other hand, since $\{\psi(s) : s_{i-1} \leq s \leq s_i\}$ is a curve from $\psi(s_{i-1})$ to $\psi(s_i)$ of length $s_i - s_{i-1}$, we also have $\delta_A(\psi(s_i), \psi(s_{i-1})) \leq s_i - s_{i-1}$. Hence

$$\sum_{i=1}^{n} \|\psi(s_i) - \psi(s_{i-1})\| \leq \sum_{i=1}^{n} \delta_A(\psi(s_i), \psi(s_{i-1})) \leq \sum_{i=1}^{n} (s_i - s_{i-1}) = \ell.$$

Thus c is rectifiable with respect to δ_A and when we take suprema over all partitions we see that the two lengths are equal. It follows that a repetition of the construction yields the same metric.

We are particularly interested in the case when

$$A = \{x \in X : \|x\| = 1\} = \partial B.$$

In this case, we call δ_A the *inner metric of X* and write it as δ_X or simply as δ.

We conclude this section with two theorems from Schäffer [466] that we will state without proof. The first will be needed in the proof of Theorem 9.4.8 and again in the last section. The statement appears intuitively obvious but the proof ([466], pp. 25–27) is far from being so. The radial projection, which might appear to be the obvious tool to use, is *not* contractive (Theorem 3.4.8). The interior of B is denoted by int B.

Theorem 9.3.3 (Schäffer) *If X is a normed space, if $x, y \in \partial B$ and if c is a curve from x to y in $X \setminus (\text{int } B)$ then there is a curve c' from x to y in ∂B with $\ell(c') \leq \ell(c)$.*

The proof of the second, while quite elementary, is not straightforward ([466], pp. 15,17).

Theorem 9.3.4 *If $x, y \in \partial B$ then $\delta(x, y) \leq 2\|x - y\|$.*

This inequality can be extended with a further trivial one on either side:

$$\|x - y\| \leq \delta(x, y) \leq 2\|x - y\| \leq 4. \tag{9.5}$$

Remark. Thus the inner metric δ on ∂B is uniformly equivalent to the norm restricted to ∂B; hence the topologies and uniformities they generate on ∂B are the same.

9.4 The girth, perimeter, inner radius and inner diameter of X

This section introduces the four parameters m, M, R and D defined by Schäffer [459] and presents some of the inequalities among them. All of this material (with

9.4 The girth, perimeter, inner radius and inner diameter of X

the exception of Theorem 9.4.8) can be found in Schäffer's book [466]. We begin with a definition.

Definition 9.4.1 If $x \in \partial B$ then $\delta^*(x) := \sup\{\delta(x, y) : y \in \partial B\}$.

Remarks

(i) By the remark following Theorem 9.3.4, ∂B is compact with respect to the inner metric and therefore the supremum in Definition 9.4.1 is attained.

(ii) Perhaps surprisingly, it is not in general the case that $\delta^*(x) = \delta(x, -x)$. In other words, for some points x in ∂B there are points y farther away (in the surface ∂B) than $-x$.

Example 9.4.2 Let B be the cube $\mathrm{conv}\{(\pm 1, \pm 1, \pm 1, \pm 1, \pm 1)\}$ in $(\mathbb{R}^5)^*$ (the only reason for using the dual space is to avoid transposing all the vectors). Let

$$x := (-1, -1/2, 0, 1/2, 1) \in \partial B \quad \text{and} \quad y := (0, 0, 0, 0, -1).$$

Then $\delta(x, -x) \leq 5/2$ and $\delta(x, y) = 3$.

For the first of these assertions, consider the polygon

$$\wp := [x_0, x_1] \cup [x_1, x_2] \cup \cdots \cup [x_4, x_5],$$

where $x_0 := x$,

$$x_1 := (-1, -1, -\tfrac{1}{2}, 0, \tfrac{1}{2}), \quad x_2 := (-\tfrac{1}{2}, -1, -1, -\tfrac{1}{2}, 0),$$

$$x_3 := (0, -\tfrac{1}{2}, -1, -1, -\tfrac{1}{2}), \quad x_4 := (\tfrac{1}{2}, 0, -\tfrac{1}{2}, -1, -1)$$

and $x_5 := (1, \tfrac{1}{2}, 0, -\tfrac{1}{2}, -1) = -x$. Since each segment lies in ∂B, $\wp$ is a curve from x to $-x$ in ∂B. Also each segment has length $\tfrac{1}{2}$ and so $\delta(x, -x) \leq \mu(\wp) = \tfrac{5}{2}$.

To verify the second assertion, first observe that if c is a curve from x to y in ∂B, then c must have a point on the boundary of the facet $\{z \in \partial B : z = \{\zeta_1, \zeta_2, \zeta_3, \zeta_4, -1\}\}$. Let z be such a point of c. Then $|\zeta_i| = 1$ for some $i = 1, 2, 3, 4$. Therefore, $\mu(c) \geq \|y - z\| + \|z - x\| = 3$. But if we choose $z_0 := (-1, -\tfrac{1}{2}, 0, \tfrac{1}{2}, -1)$ then $\wp_0 := [x, z_0] \cup [z_0, y] \subseteq \partial B$ and has length 3 and so $\delta(x, y) = 3$.

The d-dimensional generalization of this example will be discussed in the next section. More surprising than this example is the fact shown by Schäffer [464] that in infinite dimensions it is possible to have $\sup\{\delta^*(x) : x \in \partial B\} > \sup\{\delta(x, -x) : x \in \partial B\}$. In Example 9.4.2, $\delta^*(y) = 4 = \delta^*(y, -y)$.

Despite these examples, we do have the following inequality between $\delta(x, y)$ and $\delta(x, -x)$.

Proposition 9.4.3 *For all $x, y \in \partial B$ we have the following inequalities:*

(i) $\delta(x, y) \leq \delta(x, -x)/2 + 2$;
(ii) $\delta^*(x) \leq \delta(x, -x)/2 + 2$.

Proof. Let Y be the two-dimensional subspace containing x and y. Then $Y \cap \partial B$ contains an arc from x to y and another from y to $-x$ whose combined length (by Theorem 4.3.6) is less than 4, *i.e.* $\delta(-x, y) + \delta(y, x) \leq 4$. Hence

$$2\delta(x, y) = \delta(x, y) + \delta(y, x) \leq 4 + \delta(y, x) - \delta(-x, y) \leq 4 + \delta(x, -x),$$

which proves (i). The second inequality is a direct consequence of (i) and Definition 9.4.1. ∎

Definition 9.4.4 (Schäffer) *Let (X, B) be a normed linear space. Let*

$$m(X) := \inf\{\delta(x, -x) : x \in \partial B\},$$
$$M(X) := \sup\{\delta(x, -x) : x \in \partial B\},$$
$$R(X) := \inf\{\delta^*(x) : x \in \partial B\},$$
$$D(X) := \sup\{\delta^*(x) : x \in \partial B\}.$$

*The number $2m(X)$ is called the **girth** of X (or of B); likewise $2M(X)$ is called the **perimeter** of X (or of B); $R(X)$ and $D(X)$ are, respectively, the (inner) **radius** and (inner) **diameter** of ∂B.*

Remark. Since all these terms apply in a general normed space we have kept the definitions in terms of infima and suprema. In the setting of finite dimensional spaces each extremum is attained. For example, there exists $x_0 \in \partial B$ such that $\delta(x_0, -x_0) = m(X)$; not only that but also, by Theorem 9.3.2, there is then a curve c_0 from x_0 to $-x_0$ in ∂B with $\mu(c_0) = \delta(x_0, -x_0) = m(X)$.

Proposition 9.4.5 *Let $(X, \|.\|)$ be a Minkowski space; then*

(i) $2 \leq m(X) \leq M(X) \leq D(X) \leq 4$;
(ii) $m(X) \leq R(X) \leq D(X)$.

Proof. These inequalities follow immediately from Definitions 9.4.4 and 9.4.1 and the inequalities (9.5). ∎

Proposition 9.4.6 *Let $(X, \|.\|)$ be a Minkowski space; then*

$$D(X) \leq M(X)/2 + 2 \quad \text{and} \quad R(X) \leq m(X)/2 + 2.$$

Proof. These inequalities follow from Proposition 9.4.3. ∎

Remark. If dim $X = 2$ then $m(X) = M(X) = R(X) = D(X) = \mu(\partial B)/2$ and this common value lies in $[3, 4]$.

If dim $X = 3$ it is also possible to strengthen Theorem 9.4.3 and, consequently, 9.4.6.

Proposition 9.4.7 *If $(X, \|.\|)$ is a three-dimensional normed space then*

(i) $2\delta(x, y) \leq \delta(x, -x) + \delta(y, -y)$;
(ii) $2\delta^*(x) \leq \delta(x, -x) + M(X))$;
(iii) $2R(X) \leq m(X) + M(X)$;
(iv) $D(X) = M(X)$.

Proof. Statements (ii) and (iii) follow immediately from (i) and Definitions 9.4.4. Statement (iv) then follows from (ii) and 9.4.4. It remains to prove (i).

Let $x, y \in \partial B$ and let c_1 and c_2 be curves from x to $-x$ and y to $-y$ of lengths $\delta(x, -x)$ and $\delta(y, -y)$ respectively. Then $c_1 \cup (-c_1)$ is a closed curve in ∂B of length $2\delta(x, -x)$. By Theorem 9.3.2, it contains a simple, closed curve c whose length is also $2\delta(x, -x)$. By the Jordan curve theorem, $\partial B \setminus c$ consists of two components and the mapping $z \mapsto -z$ interchanges these components and leaves c invariant. Hence either y (and $-y$) $\in c$ or y and $-y$ are in different components. In either case, $c_2 \cap c \neq \emptyset$. Let $p \in c_2 \cap c$ (see Figure 9.1). Then

$$2\delta(x, y) = \delta(x, y) + \delta(-y, -x)$$
$$\leq \delta(x, p) + \delta(p, y) + \delta(-y, p) + \delta(p, -x)$$
$$= \delta(x, -x) + \delta(y, -y). \blacksquare$$

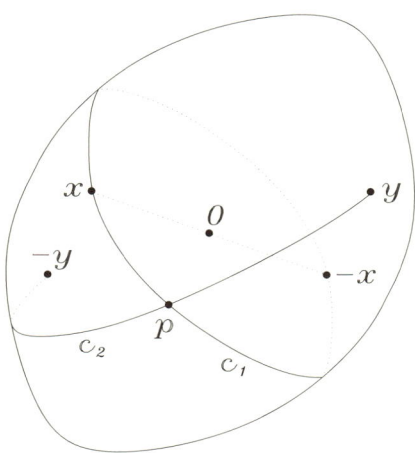

Figure 9.1

We close this section with a more recent result of Witsenhausen [534]. It is an inequality relating the length of an arbitrary closed curve in a Minkowski space to the girth of X. First an item of convenient notation. If c is a simple closed curve of length ℓ parametrized by arc length $c := \{\psi(s) : 0 \leq s \leq \ell\}$ then, for each $x = \psi(s) \in c$, we let $x^\sharp := \psi(s + \ell/2)$, where the addition is modulo ℓ. We also let $\kappa(c) := \min\{\|x - x^\sharp\| : x \in c\}$.

Theorem 9.4.8 (Witsenhausen) *If c is a simple closed rectifiable curve in a Minkowski space $(X, \|.\|)$ then $\mu(c) \geq m(X)\kappa(c)$. Moreover, this inequality is the best possible.*

Proof. Let c be an arbitrary simple, closed, rectifiable curve in X of length ℓ. Suppose $c := \{\psi(s) : 0 \leq s \leq \ell\}$ and let $a_0 := \psi(0) = \psi(\ell)$.

Given $\varepsilon > 0$, choose k so large that $2^{-k}\ell < \varepsilon\kappa(c)$. For each $i = 1, 2, \ldots, 2^k$ let
$$a_i := \psi(i 2^{-k}\ell).$$
Then, with the notation above, $a^\sharp = a_{j(i)}$, where $j(i) = i + 2^{k-1} \pmod{2^k}$. Let
$$x_i := (a_i - a_i^\sharp)/2 \quad \text{and} \quad y_i := (a_i + a_i^\sharp)/2$$
so that
$$a_i = x_i + y_i \quad \text{and} \quad a_i^\sharp = y_i - x_i \quad (i = 0, 1, 2, \ldots, 2^k).$$
Therefore, since $\|a_i - a_i^\sharp\| \geq \kappa(c)$, the points x_i form a finite sequence *outside* the ball $2^{-1}\kappa(c)B$. We also have $x_{j(i)} = -x_i$ and $x_{2^k} = x_0$. Now
$$\begin{aligned}
2\|x_i - x_{i-1}\| &\leq \|x_i - x_{i-1} + (y_i - y_{i-1})\| + \|(y_i - y_{i-1}) - (x_i - x_{i-1})\| \\
&= \|a_i - a_{i-1}\| + \|a_i^\sharp - a_{i-1}^\sharp\| \\
&\leq 2^{-k}\ell + 2^{-k}\ell.
\end{aligned} \tag{9.6}$$
Thus $\|x_i - x_{i-1}\| \leq 2^{-k}\ell < \varepsilon\kappa(c)$. Hence the polygon
$$\wp := [x_0, x_1] \cup [x_1, x_2] \cup \cdots \cup [x_{2^k-1}, x_{2^k}]$$
is a symmetric closed curve outside the ball $(2^{-1} - \varepsilon)\kappa(c)B$. Theorem 9.3.3 now implies that
$$\mu(\wp) \geq (2^{-1} - \varepsilon)\kappa(c) 2m(X) = (1 - 2\varepsilon)\kappa(c)m(X). \tag{9.7}$$
On the other hand,
$$\mu(c) \geq \sum_{i=1}^{2^k} \|a_i - a_{i-1}\| = \sum_{i=1}^{2^{k-1}} \{\|a_i - a_{i-1}\| + \|a_i^\sharp - a_{i-1}^\sharp\|\}$$
$$\geq 2 \sum_{i=1}^{2^{k-1}} \|x_i - x_{i-1}\|, \tag{9.8}$$

the last inequality coming from (9.6). However, since $\wp$ is symmetric

$$2 \sum_{i=1}^{2^{k-1}} \|x_i - x_{i-1}\| = \mu(\wp). \tag{9.9}$$

Since ε is arbitrary, combining (9.7), (9.8) and (9.9) gives the desired inequality.

The last statement of the theorem is a consequence of the remark following Definition 9.4.4 because if c_0 is a curve from x_0 to $-x_0$ with $\mu(c_0) = \delta(x_0, -x_0) = m(X)$ then $c_0 \cup c_0$ is a simple closed curve with

$$\mu(c) = 2m(X) = m(X)\kappa(c).$$

∎

Corollary 9.4.9 *If c is a symmetric, (simple) closed curve in $X \setminus \mathrm{int}\, B$ then $\mu(c) \geq 2m(X)$.*

Proof. Since c is symmetric, $x^\sharp = -x$ and so $\kappa(c) \geq 2$. ∎

9.5 Five examples in $\mathbb{R}^d$

The aim of this section is to determine the four parameters m, M, R and D for five of the unit balls considered as examples in §1.1. These examples are all in the d-dimensional linear space $\mathbb{R}^d$ and we suppose throughout that $d \geq 2$. The first, Example 1.1.12, is the Euclidean norm whose unit ball E_d is self-dual. The next two are C_d and CP_d corresponding to the ℓ_∞ and ℓ_1 norms respectively and are dual to each other (see Example 1.1.13). Finally, there are the two balls Z_d and $S_d - S_d$ from Example 1.1.17 which are also dual to each other and whose norms are given by Equations (1.10) and (1.9) respectively. Although the last four examples are in two dual pairs, in this section we shall regard each of them as lying in $\mathbb{R}^d$ and modify some of the notation from §1.1 accordingly. For example, instead of Equation (1.11), we shall write

$$Z_d := [u] + \sum_{i=1}^{d} [e_i],$$

where $u := \sum_{i=1}^{d} e_i$ and $[x] := [-x/2, x/2]$. The vector u retains this connotation throughout the section.

In addition to the information about Z_d and $S_d - S_d$ given in Example 1.1.17, we shall need the following facts. The ball Z_d is an "affine-facet-regular" body in the sense that if F_1 and F_2 are two facets of Z_d then there is an invertible matrix that maps Z_d to itself and F_1 to F_2; i.e. the isometries of the space $(\mathbb{R}^d, Z_d)$ are transitive on the facets of the ball. The facets of Z_d consist of $(d-1)$-dimensional cubes that are parallel to sums of $d-1$ of the basis vectors and *sloping* facets consisting of facets containing edges parallel to u. Clearly, two cubic facets are affinely equivalent by means of a permutation matrix and likewise two sloping facets are equivalent. We show that a cubic facet is equivalent to a sloping facet.

Consider the $d \times d$ matrix

$$T := \begin{pmatrix} -1 & 1 & 0 & & \ldots & 0 \\ -1 & 0 & 1 & 0 & \ldots & 0 \\ -1 & 0 & 0 & 1 & \ldots & 0 \\ \vdots & & & \ddots & \ddots & 1 \\ -1 & 0 & 0 & 0 & 0 & 0 \end{pmatrix} = (\tau_{ij}),$$

where $\tau_{ij} = -1$ if $j = 1$, $\tau_{ij} = 1$ if $j - i = 1$ and $\tau_{ij} = 0$ otherwise. Then $Te_j = e_{j-1}(j = 2, 3, \ldots, d)$, $Te_1 = -u$ and $T(u) = -e_d$. Thus T permutes the summands of Z_d cyclically and hence maps Z_d onto itself. Moreover, the cubic facet of Z_d that is perpendicular to e_d (and has edges parallel to $e_1, e_2, \ldots, e_{d-1}$) is mapped by T into a sloping face with edges parallel to $u, e_1, e_2, \ldots, e_{d-1}$.

Dually, the isometries of the space $(\mathbb{R}^d, S_d - S_d)$ are transitive on the vertices of $S_d - S_d$. The facets of $S_d - S_d$ and the vertices of Z_d are not uniform in this sense.

Finally we remark that, since Z_d is a zonotope with $(d + 1)$ generators, it can be realized as a projection of the cube C_{d+1}. Hence its dual $S_d - S_d$ can be realized as a d-dimensional cross-section of CP_{d+1}.

After this introduction, we come to the determination of the four parameters m, M, R and D for each of these five unit balls.

Theorem 9.5.1 *If $(X, |.|)$ is a Euclidean space then $m(X) = M(X) = R(X) = D(X) = \pi$.*

Proof. It is well known that if $x, y \in X$ with $|x| = |y| = 1$ then $\delta(x, y) = \arccos(\langle x, y \rangle)$. Hence $\delta^*(x) = \delta(x, -x) = \pi$ for all $x \in \partial B$. ∎

Theorem 9.5.2 (Schäffer) *If $(X, \|.\|_\infty)$ is a d-dimensional Minkowski space whose unit ball is a parallelotope then*

(i) $m(X) = 2d/(d-1)$;
(ii) $M(X) = D(X) = 4$;
(iii) $R(X) = 3$ if $d \geq 3$, $R(X) = 4$ if $d = 2$.

Proof. By choosing as basis vectors (and their negatives) the mid-points of the facets of B we may assume that $X = \mathbb{R}^d$ and that $B = C_d$. For each $i = 1, 2, \ldots, d$, let $F_i^+ := \{x \in \partial B : \xi_i = +1\}$, $F_i^- := -F_i^+$. Then ∂B is made up of two compact, connected sets $\partial B^+ := \bigcup_{i=1}^d F_i^+$ and $\partial B^- := \bigcup_{i=1}^d F_i^-$. With these notational preliminaries we first prove (ii).

Since, by Proposition 9.4.5, $M(X) \leq D(X) \leq 4$, it is sufficient to show that

9.5 Five examples in $\mathbb{R}^d$ 295

$M(X) \geq 4$ and that is achieved by showing that $\delta(u, -u) \geq 4$. Let $c := \{\psi(s) : 0 \leq s \leq \ell\}$ be a curve from u to $-u$ in ∂B. In the inner metric topology, u is an interior point of ∂B^+ and $-u$ is an interior point of ∂B^-. Since ψ is continuous, there exists s_0 with $0 < s_0 < \ell$ and with $\psi(s_0) \in \partial B^+ \cap \partial B^-$ so that $\psi(s_0)$ has at least one coordinate $+1$ and at least one coordinate -1. Hence $\ell \geq \|u - \psi(s_0)\| + \|\psi(s_0) + u\| = 4$. Hence $\delta(u, -u) \geq 4$.

We remark that if y is a point of $\partial B^+ \cap \partial B^-$ then the polygon $[u, y] \cup [y, -u]$ is a curve from u to $-u$ in ∂B of length 4.

Next we prove (i). Let $x_0 = (\xi_1^0, \xi_2^0, \ldots, \xi_d^0)^t$, where $\xi_i^0 := (d+1-2i)/(d-1)$. (The special case with $d=5$ was given in Example 9.4.2.) Let T be the matrix

$$T := \begin{pmatrix} 0 & 0 & 0 & \ldots & 0 & -1 \\ 1 & 0 & 0 & \ldots & 0 & 0 \\ 0 & 1 & 0 & \ldots & 0 & 0 \\ 0 & 0 & 1 & \ldots & 0 & 0 \\ \vdots & \vdots & \vdots & \ddots & \vdots & \vdots \\ 0 & 0 & 0 & \ldots & 1 & 0 \end{pmatrix} = (\tau_{ij}),$$

where $\tau_{1d} = -1$, $\tau_{ij} = 1$ if $i - j = 1$ and $\tau_{ij} = 0$ otherwise. Let $x_k := T^k x_0$, $k = 1, 2, \ldots, d$. It is a matter of routine verification that the polygon $\wp := [x_0, x_1] \cup [x_1, x_2] \cup \cdots \cup [x_{d-1}, x_d]$ is a curve from x_0 to $-x_0$ in ∂B and that the length of each segment is $2/(d-1)$. Hence $m(X) \leq \delta(x_0, -x_0) \leq \mu(\wp) = 2d/(d-1)$.

The reverse inequality takes more ingenuity. Let y_0 be a point of ∂B for which $\delta(y_0, -y_0) = m(X)$. There exists a curve of length $m(X)$ from y_0 to $-y_0$ and such a curve must be a polygon $[y_0, y_1] \cup [y_1, y_2] \cup \cdots \cup [y_{k-1}, y_k]$ with each edge lying in a facet $F_j^\pm$ and each vertex y_i lying in at least one such facet. Let $J := \{j : \exists i$ with $y_i \in F_j^+ \cup F_j^-\}$, let n be the cardinality of J ($1 \leq n \leq d$) and let $y_i(j)$ be the jth coordinate of y_i ($0 \leq i \leq k, 1 \leq j \leq d$). The definition of J means that for each j in J there is at least one $i(j)$ such that $|y_{i(j)}(j)| = 1$. Hence, for j in J,

$$\sum_{i=1}^{k} |y_i(j) - y_{i-1}(j)| = \sum_{i=1}^{i(j)} |y_i(j) - y_{i-1}(j)| + \sum_{i(j)+1}^{k} |y_i(j) - y_{i-1}(j)|$$
$$\geq |y_{i(j)}(j) - y_0(j)| + |y_k(j) - y_{i(j)}(j)|$$
$$\geq |y_{i(j)}(j) - y_0(j) - y_k(j) + y_{i(j)}(j)|$$
$$= 2|y_{i(j)}(j)|$$
$$= 2.$$

The penultimate equality is because $y_k = -y_0$. Thus

$$\sum_{j \in J} \sum_{i=1}^{k} |y_i(j) - y_{i-1}(j)| \geq 2n. \tag{9.10}$$

On the other hand, for each i, $|y_i(j) - y_{i-1}(j)| \leq \|y_i - y_{i-1}\|$. But the edge $[y_{i-1}, y_i]$ lies in a facet of ∂B and hence there is at least one coordinate (from J) that is constant (± 1) on this edge. Hence

$$\sum_{j \in J} |y_i(j) - y_{i-1}(j)| \leq (n-1)\|y_i - y_{i-1}\|$$

and so

$$\sum_{i=1}^{k} \sum_{j \in J} |y_i(j) - y_{i-1}(j)| \leq \sum_{i=1}^{k} (n-1)\|y_i - y_{i-1}\| = (n-1)m(X). \quad (9.11)$$

From (9.10) and (9.11) we get $m(X) \geq 2n/(n-1) \geq 2d/(d-1)$.

Finally we prove (iii). The statement for $d = 2$ follows from Theorem 4.3.6 so we suppose that $d \geq 3$. Let $x \in \partial B$. By means of a suitable reflection, we may assume $x = (\xi_1, \xi_2, \ldots, \xi_d)^t$ with $\xi_i \geq 0$ for all i. Let c be a minimal curve from x to $-u$ in ∂B. As in the proof of (ii), there is a point z of c with $z \in \partial B^+ \cap \partial B^-$. Hence

$$\mu(c) \geq \|x - z\| + \|z + u\| \geq (1 + \min \xi_i) + 2 \geq 3.$$

Thus $\delta^*(x) \geq 3$ for all $x \in \partial B$ and $R(X) \geq 3$.

For the reverse inequality choose $x_0 := (1, 1, 0, \ldots)^t \in \partial B$ (here is where we use $d \geq 3$). If $y = (\eta_1, \eta_2, \ldots, \eta_d)^t$ is in ∂B then there is an isometry of X that leaves x_0 invariant and transforms y so that $\eta_3 \geq 0$. Since $\|y\| = 1$, there exists k such that $|\eta_k| = 1$. By interchanging η_1 and η_2 if necessary (by an isometry that leaves x_0 invariant) we may suppose $k > 1$. Let $x' := (1, 0, 1, 0, \ldots, 0)^t$ and $x'' := (0, \eta_2, 1, \eta_4, \ldots \eta_d)^t$. It is easy to verify that $[x_0, x'] \cup [x', x''] \cup [x'', y]$ is a curve from x_0 to y in ∂B with length $1 + 1 + \max\{|\eta_1|, 1 - \eta_3\} \leq 3$. Hence $\delta^*(x_0) \leq 3$ and $R(X) \leq 3$. ∎

Theorem 9.5.3 (Schäffer) *If $(X, \|.\|_1)$ is the d-dimensional Minkowski space whose unit ball is the convex hull of d basis vectors and their negatives then (i) $M(X) = D(X) = 4$ and (ii) $m(X) = R(X) = 2d/(d-1)$.*

Proof. We assume $X = (\mathbb{R}^d, \|.\|_1)$. For (i), as in the proof of Theorem 9.5.2, it is sufficient (by Proposition 9.4.5) to show that there exists $x_0 \in \partial B$ with $\delta(x_0, -x_0) \geq 4$. For this, let $x_0 := e_1 = (1, 0, 0, \ldots, 0)^t$. A minimal curve from x_0 to $-x_0$ is a polygon $\wp$ which must cross the hyperplane $\{x : \xi_1 = 0\}$. Let $y = (0, \eta_2, \eta_3, \ldots, \eta_d)^t$ be a vertex of $\wp$ on this hyperplane. Then $\sum_{i=2}^{d} |\eta_i| = 1$ and $\delta(x_0, -x_0) = \mu(\wp) \geq \|x_0 - y\| + \|y + x_0\| = 4$.

The proof of (ii) can be found in Schäffer's book [466]. Again, it uses Schäffer's very ingenious choice of polygonal curves and estimation of lengths. The calculations are too extensive to include here. ∎

9.5 Five examples in $\mathbb{R}^d$

Theorem 9.5.4 *If $X = (\mathbb{R}^d, \|.\|_Z)$ is the d-dimensional Minkowski space whose unit ball is the zonotope Z_d then (i) $M(X) = D(X) = 3$ and (ii) $m(X) \leq 3$ with strict inequality if d is sufficiently large.*

Proof. Since $m(X) \leq M(X) \leq D(X)$, to prove (i) and the first part of (ii) it is sufficient to show that $D(X) \leq 3$ and that $\delta(u, -u) \geq 3$.

To prove $D(X) \leq 3$, let $x \in \partial Z_d$. Since each facet of Z_d is affinely equivalent to every other facet, we may assume that $x = (\xi_1, \xi_2, \ldots, \xi_{d-1}, 1)^t$ with $\xi_i \geq 0$ $i = 1, 2, \ldots, d-1$. Let $y = (\eta_1, \eta_2, \ldots, \eta_d,)^t \in \partial Z_d$. There are various cases to consider depending on the values of the η_i. We use the lattice theoretic notation introduced in Example 1.1.17.

(a) Suppose that $y^+ = 0$ and $\eta_d = -1$ so that y lies in the facet opposite to x. We may assume that $|\eta_1| \geq \xi_1$ (otherwise we consider $-y$ and $-x$). Consider the polygon $\wp := [x, x_1] \cup [x_1, z] \cup [z, y_1] \cup [y_1, y]$, where

$$x_1 := (0, \xi_2, \xi_3, \ldots, 1)^t = x - \xi_1 e_1,$$

$$y_1 := y - (1 + \eta_1) e_1 \quad \text{and} \quad z := (-1, \zeta_2, \zeta_3, \ldots, \zeta_{d-1}, 0)^t$$

with

$$\zeta_i := \begin{cases} \eta_i & \text{if } \xi_i - \eta_i < 1, \\ \xi_i - 1 & \text{if } \xi_i - \eta_i \geq 1. \end{cases}$$

It can be verified that each edge of $\wp$ lies in a facet of Z_d so that $\wp$ is a curve from x to y in ∂Z_d. Also, using Equation (1.10) for evaluating the norms, we get $\|x - x_1\| = \xi_i$, $\|x_1 - z\| = 1$, $\|z - y_1\| = 1$ and $\|y - y_1\| = 1 + \eta_1 \leq 1 - \xi_1$. Thus $\mu(\wp) \leq 3$.

(b) If $\eta_i = -1$ for some $i \neq d$ then the polygonal path

$$[x, e_d] \cup [e_d, y - \eta_d e_d] \cup [y - \eta_d e_d, y]$$

lies in ∂Z_d and has each edge of length at most 1.

(c) If $y^- = 0$ then the two segments $[x, u] \cup [u, y]$ have combined length at most 2 and form a curve from x to y in Z_d.

(d) Finally, suppose that $y^+ \neq 0$ and $y^- \neq 0$. Suppose the maximal value of η_i is at i_0. Let $z = (\zeta_1, \zeta_2, \ldots, \zeta_d)^t$, where

$$\zeta_{i_0} := 1,$$
$$\zeta_i := \eta_i \quad \text{if } \eta_i \geq 0 \; i \neq i_0,$$
$$\zeta_i := 0 \quad \text{if } \eta_i < 0 \; i \neq i_0.$$

Therefore, the polygon $[x, u] \cup [u, z] \cup [z, y]$ lies in ∂Z_d and has length ≤ 3.

Next we prove that $\delta(u, -u) \geq 3$. The polytope Z_d has d facets

$$F_i^+ := \{x = (\xi_1, \xi_2, \ldots, \xi_d)^t : \xi_i = +1, x^- = 0\}$$

and d corresponding opposite facets $F_i^- := -F_i^+$. Any polygon $\wp$ from u to $-u$ must have a last vertex x_1 on $\partial Z_d^+ := \bigcup_{i=1}^d F_i^+$ and, subsequent to x_1, a first vertex x_2 on $\partial Z_d^- := -\partial Z_d^+$. Hence $\mu(\wp) \geq \|u - x_1\| + \|x_1 - x_2\| + \|x_2 + u\|$. Since both x_1 and x_2 have at least one zero coordinate, $\|u - x_1\| = \|x_2 + u\| = 1$; since x_1 has at least one coordinate equal to $+1$ and all coordinates of x_2 are ≤ 0, $\|x_2 - x_2\| \geq 1$. Hence $\mu(\wp) \geq 3$ and $\delta(u, -u) \geq 3$.

For the last statement of the theorem, we prove a stronger version – namely, that if d is even and $d \geq 4$ then $m(X) \leq 2d/(d-2)$. For such a d let $n := (d-2)/2$ and let

$$x_0 := (2n)^{-1}(n, n, n-2, n-4, \ldots, n-2k, \ldots, -n+2, -n, -n, \ldots, n-4, n-2)^t,$$

i.e. $x_0 = (\xi_0, \xi_1, \xi_2, \ldots, \xi_{2n+1})^t$, where $\xi_0 := 2^{-1}$, and

$$\xi_i := [n - 2(i-1)]/2n \qquad (i = 1, 2, \ldots n),$$
$$\xi_{n+j} = -\xi_{j-1} \qquad (j = 1, 2, \ldots, n+1).$$

Let $x_1 := (\xi_{2n+1}, \xi_0, \xi_1, \ldots, \xi_{2n})^t$ and continue to permute the coordinates cyclically until, after $n+1$ steps, $x_{n+1} = -x_0$. Then the polygon

$$\wp := [x_0, x_1] \cup [x_1, x_2] \cup \cdots \cup [x_n, x_{n+1}]$$

is a curve from x_0 to $-x_0$ in ∂Z_d. Each edge has length $2/n$ and so

$$m(X) \leq \delta(x_0, -x_0) \leq \mu(\wp) = 2(n+1)/n = 2(1 + n^{-1}) = 2d/(d-2).$$

Hence $m(X) < 3$ if $d \geq 8$ and d is even. Evidently if d is odd one can modify this example by adding a zero coordinate to each vector. Thus if $d \geq 8$, $m(X) < 3$. ∎

Another result of Schäffer's [466], which we state in the next section, is that if d is odd then $m(X) \geq 2d/(d-1)$; i.e. the cube gives not only the largest possible value of M but also the smallest possible value of m. In particular, for $d = 3$ we have $m(X) \geq 3$. Combining this fact with the preceding theorem we get the original motivation for the calculations and perhaps the most interesting aspect of Theorem 9.5.4.

Corollary 9.5.5 *If $\mathbb{R}^3$ is equipped with the rhombic dodecahedron Z_3 as unit ball then $m(X) = M(X) = R(X) = D(X) = 3$ and $\delta(x, -x) = \delta^*(x) = 3$ for all x in ∂Z_3.*

Proof. These statements follow from Theorems 9.5.4 and 9.6.9. ∎

Thus there are finite dimensional spaces other than Euclidean ones for which $\delta(x, -x)$ is constant.

The final theorem in this section shows that the facts given in Theorem 9.5.4 for Z_d are also true for the dual ball.

Theorem 9.5.6 *If $X = (R^d, \|.\|)$ is the d-dimensional Minkowski space whose unit ball is $S_d - S_d$ then (i) $M(X) = D(X) = 3$ and (ii) $m(X) \leq 3$ with strict inequality if d is sufficiently large.*

Proof. The pattern of proof is the same as in Theorem 9.5.4 but we need fewer cases. Let $x \in \partial B$. By a suitable isometry of X we may assume that the non-zero coordinates of x^+ are the first i. Thus $x = (\xi_1, \xi_2, \ldots, \xi_i, -\xi_{i+1}, \ldots, -\xi_d)^t$ with $\xi_k \geq 0$ for all k; we may also assume (otherwise consider $-x$) that $\|x^+\| = 1$, $\|x^-\| \leq 1$. Let $y = (\eta_1, \eta_2, \ldots, \eta_d)^t \in \partial B$ and let $J := \{j : \eta_j \geq 0\}$. Consider the polygon

$$\wp := [x, x'] \cup [x', z] \cup [z, z'] \cup [z', y],$$

where

$$x' := x + \sum_{j=i+1}^{d} \xi_j e_j = (\xi_1, \xi_2, \ldots, \xi_i, 0, \ldots, 0)^t \qquad (x' = x \text{ if } i = d),$$

$$z := (\zeta_1, \zeta_2, \ldots, \zeta_d) \quad \text{with} \quad \zeta_k := \begin{cases} 0 & \text{if } k \notin J, \\ \eta_k (\sum_{j \in J} \eta_j)^{-1} & \text{if } k \in J, \end{cases}$$

and

$$z' := (\zeta_1', \zeta_2', \ldots, \zeta_d')^t, \qquad \text{where } \zeta_k' := \begin{cases} \eta_k & \text{if } k \notin J, \\ \zeta_k & \text{if } k \in J. \end{cases}$$

Since each edge of $\wp$ lies in a facet of $S_d - S_d$ we have

$$\delta(x, y) \leq \mu(\wp) = \|x - x'\| + \|x' - z\| + \|z - z'\| + \|z' - y\|$$

$$= \sum_{j=i+1}^{d} \xi_j + \|x' - z\| + \sum_{k \notin J} |\eta_k| + \sum_{k \in J} \eta_k \left(\left[\sum_{j \in J} \eta_j \right]^{-1} - 1 \right)$$

$$\leq \sum_{j=i+1}^{d} \xi_j + \|x' - z\| + 1 + \left(1 - \sum_{j \in J} \eta_j \right).$$

Here we have used Equation (1.9) to evaluate the norms. Now $\|x' - z\| = 1$ if $J \subseteq \{i+1, \ldots d\}$; otherwise

$$\|x' - z\| \leq \max \left\{ \sum_{j=1}^{i} \xi_j, \sum_{k \in J} \eta_k \left[\sum_{j \in J} \eta_j \right]^{-1} \right\} = 1.$$

Thus $\delta(x, y) \leq 3 + \sum_{j=i+1}^{d} \xi_j - \sum_{j \in J} \eta_j$.

Therefore, if $\sum_{j\in J} \eta_j \geq \sum_{j=i+1}^{d} \xi_j$, i.e. if $\|y^+\| \geq \|x^-\|$, then $\delta(x, y) \leq 3$. In the case when $\|y^+\| < \|x^-\|$, we have $\|y^+\| < 1$ and hence $\|y^-\| = 1$. But then the original argument applies to the pair $(-y, -x)$ and so in both cases $\delta(x, y) \leq 3$. Therefore, $D(X) \leq 3$.

Next, consider the vector $e_d = (0, \ldots, 0, 1)^t$. Any polygon $\wp$ from e_d to $-e_d$ in ∂B must have a vertex $x = (\xi_1, \xi_2, \ldots, \xi_{d-1}, 0)^t$ on the hyperplane $\{x : \xi_d = 0\}$. Let $J := \{i : \xi_i \geq 0\}$. Then

$$\delta(e_d, -e_d) = \mu(\wp) \geq \|e_d - x\| + \|x + e_d\|$$
$$= 1 + \sum_{i \notin J} |\xi_i| + 1 + \sum_{i \in J} \xi_i \geq 3$$

because, since $\|x\| = 1$, either $\sum_{i\in J} \xi_i$ or $\sum_{i \notin J} |\xi_i| = 1$.

Finally, if d is even, $d \geq 4$ and $n := d/2 - 1$, consider x_0, where

$$x_0 := n^{-1}(1, 1, \ldots, 1, 0, -1, -1, -1, \ldots, -1, 0)^t.$$

As in the preceding theorem, let $\wp := [x_0, x_1] \cup [x_1, x_2] \cup \cdots \cup [x_n, x_{n+1}]$ be the polygonal curve from x_0 to $x_{n+1} = -x_0$ in ∂B where each x_i is obtained from x_{i-1} by a cyclical shift of the coordinates. Then we have

$$m(X) \leq \delta(x_0, -x_0) \leq \mu(\wp) = 2(n+1)/n = 2d/(d-2). \qquad \blacksquare$$

Remark. Again using Theorem 9.6.9 in the case when $d = 3$ we have $m(X) = M(X) = R(X) = D(X)$ and $\delta^*(x) = \delta(x, -x) = 3$ for all $x \in \partial B$ when B is the cubo-octahedron in $\mathbb{R}^3$.

9.6 Relationships with the Banach–Mazur distance and extreme values

In this section we first consider isomorphic spaces X and Y. In our finite dimensional setting this means that X and Y have the same dimension. We show that if we have any one of the parameters m, M, R and D for both spaces then we can estimate the Banach–Mazur distance Δ between the spaces. In the final part we list some of the facts about extreme values of the parameters in each dimension (isomorphism class).

If T is an isometry of (X, B_1) onto (X, B_2) and if c is a curve from x_1 to x_2 in ∂B_1 then $T(c)$ is a curve of the same length from Tx_1 to Tx_2 in ∂B_2 and so $\delta_1(x_1, x_2) = \delta_2(Tx_1, Tx_2)$.

Proposition 9.6.1 *If X is a linear space with two norms $\|.\|_1$ and $\|.\|_2$ whose unit balls are B_1 and B_2 respectively, and if $B_1 \subseteq \alpha B_2$, $B_2 \subseteq \beta B_2$ then, for all*

9.6 Relationships with the Banach–Mazur distance and extreme values

$x_1, y_1 \in \partial B_1$, we have

$$(\alpha\beta)^{-1} \leq \frac{2 + \delta_1(x_1, y_1)}{2 + \delta_2(x_2, y_2)} \leq \alpha\beta,$$

where $x_2 := x_1/\|x_1\|_2$ and $y_2 := y_1/\|y_1\|_2$.

Proof. Since $x_2, y_2 \in \partial B_2$, αx_2 and $\alpha y_2 \in \partial(\alpha B_2)$ (see Figure 9.2). If c is a path of minimal length from x_2 to y_2 in ∂B_2, then αc is a path of minimal length from αx_2 to αy_2 in $\partial(\alpha B_2)$. Hence, c' where

$$c' := [x_1, \alpha x_2] \cup (\alpha c) \cup [\alpha y_2, y_1]$$

is a path from x_1 to y_1 in $X \setminus (\text{int } B_1)$. Therefore, by Theorem 9.3.3, we have

$$\delta_1(x_1, y_1) \leq \mu_1(c') = \|x_1 - \alpha x_2\|_1 + \mu_1(\alpha c) + \|\alpha y_2 - y_1\|_1.$$

However, the fact that $B_2 \subseteq \beta B_1$ means that $\|z_1 - z_2\|_1 \leq \beta\|z_1 - z_2\|_2$ for all z_1, z_2 in X. Also $B_1 \subseteq \alpha B_2 \subseteq \alpha\beta B_1$ and so

$$\|\alpha x_2 - x_1\|_1 \leq \|\alpha\beta x_1 - x_1\|_1 = (\alpha\beta - 1),$$

and similarly for $\|\alpha y_2 - y_1\|$. Using these inequalities, we get

$$\delta_1(x_1, y_1) \leq 2(\alpha\beta - 1) + \alpha\mu_1(c)$$
$$\leq 2(\alpha\beta - 1) + \alpha\beta\mu_2(c)$$
$$= 2\alpha\beta - 2 + \alpha\beta\delta_2(x_2, y_2),$$

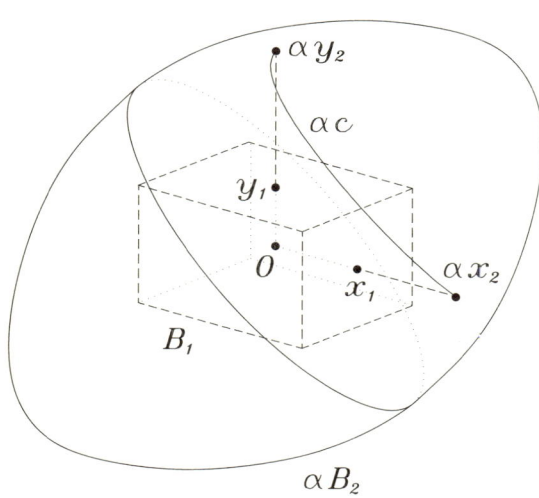

Figure 9.2

which, upon rearrangement, is the right-hand inequality. Similarly, by beginning with a curve of minimal length from x_1 to y_1 in ∂B_1, we get the other inequality. ∎

Theorem 9.6.2 *If (X, B_1) and (Y, B_2) are two normed spaces and if T is an isomorphism of X onto Y then, for all x_1, x_2 in ∂B,*

$$(\|T\| \|T^{-1}\|)^{-1} \leq \frac{2 + \delta_X(x_1, x_2)}{2 + \delta_Y((Tx_1)', (Tx_2)')} \leq \|T\| \|T^{-1}\|,$$

where $(Tx_i)' := Tx_i / \|Tx_i\|_Y \in \partial B_Y$, $i = 1, 2$.

Proof. We first consider Y with the norm induced by the unit ball $T(B_1)$ and call this space Y_1. Thus T is an isometry of X onto Y_1 and so $\delta_X(x_1, x_2) = \delta_{Y_1}(Tx_1, Tx_2)$. From the definition of $\|T\|$ and $\|T^{-1}\|$ we have

$$T(B_1) \subseteq \|T\| B_2 \quad \text{and} \quad B_2 \subseteq \|T^{-1}\| TB_1.$$

Now apply Proposition 9.4.1 to Y with the two balls (TB_1) and B_2. ∎

Corollary 9.6.3 *If X and Y are finite dimensional spaces of the same dimension then*

$$|\log(2 + \Phi(X)) - \log(2 + \Phi(Y))| \leq \Delta(X, Y),$$

where Φ is any one of m, M, R or D.

Proof. The proofs of all four cases are similar. To be specific we let $\Phi := D$. Let T be an isomorphism from X onto Y; then, by Theorem 9.4.2,

$$2 + \delta_X(x_1, x_2) \leq \|T\| \|T^{-1}\|(2 + \delta_Y((Tx_1)', (Tx_2)'))$$
$$\leq \|T\| \|T^{-1}\|(2 + D(Y)).$$

Hence $2 + D(X) \leq \|T\| \|T^{-1}\|(2 + D(Y))$. Rearranging this inequality and taking the logarithm of both sides yields half the required inequality. Interchanging X and Y gives the other half. ∎

For the parameter m, a better inequality than this is obtained from Corollary 9.4.9.

Proposition 9.6.4 *If X and Y are finite dimensional normed spaces of the same dimension then $|\log m(X) - \log m(Y)| \leq \Delta(X, Y)$.*

Proof. Let x be a point in ∂B_X with $\delta_X(x, -x) = m(X)$ and let c be a simple curve from x to $-x$ in ∂B_X of length $m(X)$. Then $c \cup (-c)$ contains a symmetric, simple, closed curve $\tilde{c}$ with $\mu_X(\tilde{c}) = 2m(X)$.

Because $B_Y \subseteq \|T^{-1}\|T(B_X)$, $\|T^{-1}\|T(c)$ is a symmetric, simple, closed curve in $Y \setminus (\text{int } B_Y)$. Hence, Corollary 9.4.9 gives us

$$2m(Y) \le \|T^{-1}\|\mu_Y(T(c)) \le \|T^{-1}\| \|T\|\mu_X(c) = 2m(X)\|T\| \|T^{-1}\|.$$

The argument is now completed in the same way as in Corollary 9.6.3. ∎

Remark. Although the various extrema may not be attained, these proofs hold for general normed spaces X and Y that are isomorphic (see Schäffer [466], p. 44).

Finally, we consider the range of possible values of the four parameters for each dimension d. We consider the metric space $(\mathcal{D}_d, \Delta)$ defined at the end of §2.4. First we show that Corollary 9.6.3 implies that $\Phi(X)$ is continuous on $(\mathcal{D}_d, \Delta)$ for each value of Φ.

Proposition 9.6.5 *The functions $m(X)$, $M(X)$, $R(X)$ and $D(X)$ are each continuous on $(\mathcal{D}_d, \Delta)$.*

Proof. Let Φ be any one of the functions. From Corollary 9.6.3 we have

$$2 + \Phi(X) \le (2 + \Phi(Y))e^{\Delta(X,Y)}$$

and hence

$$\Phi(X) - \Phi(Y) \le (2 + \Phi(Y))[e^{\Delta(X,Y)} - 1] \le 6[e^{\Delta(X,Y)} - 1],$$

where the last inequality comes from Proposition 9.4.5. Since X and Y may be interchanged it follows that Φ is continuous with respect to Δ. ∎

It can be shown that the metric space $(\mathcal{D}_d, \Delta)$ is both compact and connected. With this information the following corollary is immediate.

Corollary 9.6.6 *The range of each of the functions m, M, R and D on $(\mathcal{D}_d, \Delta)$ is a closed, bounded interval.*

Definition 9.6.7 *If Φ is any one of the functions, m, M, R and D let*

$$\Phi_*(d) := \min\{\Phi(X) : X \in \mathcal{D}_d\}; \qquad \Phi^*(d) := \max\{\Phi(X) : X \in \mathcal{D}_d\}.$$

The particular examples of the preceding section give us the following information about the various extreme values.

Theorem 9.6.8 *If $d \ge 2$ then the following inequalities and equations hold:*

(i) $D_*(d) \le 3$, $D^*(d) = 4$;
(ii) $M_*(d) \le 3$, $M^*(d) = 4$;

(iii) $m_*(d) \leq 2d/(d-1)$, $m^*(d) \geq \pi$;
(iv) $R_*(d) \leq 2d/(d-1)$, $R^*(d) \geq \pi$.

Proof. (i) and (ii) follow from Theorems 9.5.4 and 9.5.2 (or Theorems 9.5.6 and 9.5.3) and Proposition 9.4.5; (iii) follows from Theorems 9.5.2 (or 9.5.3) and 9.5.1; likewise, (iv) follows from Theorems 9.5.3 and 9.5.1. ∎

Remark. If $d = 2$ then $\Phi_*(2) = 3$, $\Phi^*(2) = 4$ for $\Phi = m, M, R$, or D from Theorem 4.3.6.

Schäffer [466] gives more information about these parameters. In particular he proves (pp. 93–95) the following theorem.

Theorem 9.6.9 (Schäffer) *For all $d \geq 2$, $m_*(d) = 2 + [d/2]^{-1}$, where [.] is the greatest integer function. For $d = 2$ and for all odd $d \geq 3$, $R_*(d) = 2 + [d/2]^{-1}$.*

9.7 Notes

The seven parameters discussed in this chapter form only a selection of those that might appear here. The selection was made mainly on the grounds of connections with topics in the earlier chapters. An extension of the list could easily be made by consulting, *e.g.*, the survey article of Lindenstrauss and Milman [327]. Kogan, in his review [298] of the article of Kadeč [277], describes the first section "as an attempt at a systematic presentation of the quantitative theory of finite-dimensional normed spaces" and lists many of the parameters that Kadeč considers. A very comprehensive survey of constants measuring either the degree of convexity or the degree of smoothness of the unit ball in infinite dimensional spaces (the so-called *moduli of convexity and smoothness*) is that of Milman [380]. Other constants to measure the size of the unit ball were introduced by Whitley [531] and Kottman [302]. The paper of Rutovitz [452], which we have already used heavily in §§9.1 and 9.2, has a further section on the *basis constant*. Yet another very interesting idea is that explored by Sobczyk [492]. He considers finite subsets of a Minkowski space and asks if these can be isometrically embedded in some (perhaps higher-dimensional) Euclidean space. The sizes of the sets so embeddable provide another measure of the Minkowski geometry. Bohnenblust [44] discusses a parameter he calls *Jung's constant* and gives an upper bound of $2d/d + 1$ for this parameter. Leichtweiss [317] characterizes the spaces for which this maximum is attained. They are the unit balls which are "close" in a certain sense to $S_d - S_d$. Grünbaum [222] established similar results for the *expansion* constant. Chakerian [101] studies an affine-invariant functional which is somewhat related to the functional τ that he and Talley [106] used to give a lower bound for $\mu_B(\partial B)$ (Theorem 7.4.2).

9.7 Notes

The way in which detailed quantitative results for the isometric theory of finite dimensional spaces become much more general qualitative statements in the isomorphic theory of infinite dimensional normed spaces is well illustrated by the behaviour of ϖ_s. First, for infinite dimensional spaces, we must modify the definition of $\varpi(X, Y)$ by limiting the projections to bounded projections onto *closed* subspaces Y of X. Using the uniform boundedness principle (see, e.g., Dunford and Schwartz [130], p. 52), Davis, Dean and Singer [122] showed that if there is a bounded projection onto every closed subspace of a Banach space X then the norms of those projections can be bounded, i.e. $\varpi(X) < \infty$. Using this result Lindenstrauss and Tzafriri [328] showed further that if X has a bounded linear projection onto every closed subspace then X is isomorphic to a Hilbert space. In other words, if X is a Banach space either X is isomorphic to a space H with $\varpi_s(H) = 1$ or $\varpi_s(X) = \infty$ in the sense that there is a closed subspace Y with no bounded linear projection onto Y; i.e. Y has no closed complement in X. Much earlier, Murray [396] had shown that both the L_p and ℓ_p ($p \neq 2$) spaces had such non-complemented closed subspaces and Sobczyk [490] had shown that c_0 (the space of real-valued sequences converging to 0 with the uniform norm) was not complemented in ℓ_∞.

Theorem 9.1.5 is due to Bohnenblust [44]. The theorem that the only Banach spaces X with $\varpi^s(X) = 1$ are spaces of continuous functions on extremally disconnected compact Hausdorff spaces is due independently to Nachbin [397] and Goodner [194]. Goodner used a geometric approach and showed that the extreme points of the unit ball are all separated by a distance of 2. Nachbin used a lattice-theoretic approach. The final piece of the theorem was supplied by Kelley [283]. Zippin [545, 546] proved the following "stable" version of 9.1.11. There is an $\varepsilon > 0$ and a positive function $\varphi(t)$ defined on $(1, \varepsilon)$ with $\lim_{t \to 1} \varphi(t) = 0$ such that for all d if X is a d-dimensional Minkowski space with $\varpi^s(X) < t$ then $\Delta(X, \ell_\infty^d) < \varphi(t)$.

The result of Kadeč and Snobar [278] that $\varpi^s(X) \leq \sqrt{d}$ was improved by König and Lewis [299], who showed that the inequality is strict, and then by König and Tomczak-Jaegermann [300], who proved the conjecture in [299] that there is a positive constant c such that $\varpi^s(X) \leq \sqrt{d} - c/\sqrt{d}$. More details of these and related results can be found in the survey article of Lindenstrauss and Milman [327].

As with the projection constant, which is often denoted by λ, we have a notational problem with Macphail's constant because it is frequently denoted by μ. Mayer [360] introduced a parameter that, although defined differently, is equivalent to Macphail's and therefore it seemed appropriate to revert to Mayer's original notation. The most recent upper bounds for ϖ^s that were mentioned above can be used with Theorem 9.2.3 to give upper bounds for ϱ. One of the many interesting results of Garling and Gordon [175] dealing with connections among various parameters associated with Minkowski spaces is that equality holds in 9.2.3 if there are *enough* isometries of X (meaning that the only operators which commute with all of them are multiples of $\mathbf{1}$). Franchetti and Votruba [162] also give connections among ϱ, ϖ^s and (in two dimensions) the perimeter of B. One of their results is that for two-dimensional spaces $\varrho(X) = 1/m(X)$ and hence, using the result of Garling and Gordon just mentioned, $\varpi(X) = 4/(m(X)$ if X also has enough isometries. These results and Corollary 4.3.9 show that in these cases $\varrho(X) = \varrho(X^*)$ and $\varpi(X) = \varpi(X^*)$.

J. J. Schäffer and his collaborators together with Harrell and Karlovitz have explored the values of m, M, R and D for infinite dimensional spaces with special emphasis on the girth m. Schäffer and Sundaresan [467] showed that if $m(X) > 2$ then X is reflexive but that there are also reflexive spaces for which $m(X) = 2$. Harrell and Karlovitz define a Banach space X to be *flat* if $m(X) = 2$ and there is an $x \in X$ with $\|x\| = 1$ and a curve in the unit sphere from x to $-x$ of length 2. They showed [246] that a flat space is not reflexive but that there are non-reflexive spaces (*e.g.* ℓ_1) that are not flat. A stronger concept than reflexivity is James' notion of *super-reflexivity* [268]. James and Schäffer showed in [269] that $m(X) > 2$ if and only if X is super-reflexive. The geometry of flat Banach spaces is interesting in its pathology and has been investigated by Harrell and Karlovitz [245, 247]. Karlovitz [282] showed that if X is flat then so is X^* and that X^* is not separable. Schäffer showed [463] that if ν is a measure on some space then $L_1(\nu)$ is flat if and only if ν is not purely atomic and, further, that if ν has no atoms then the inner and norm metrics coincide on the surface of the ball. In [400] Nyikos and Schäffer characterized those spaces of continuous functions that are flat. Another surprising fact concerns M and D. In [464] Schäffer showed that if X is the space of continuous functions on $[0, 1]$ that vanish at 0 then $M(X) = 2$ (and $\delta(x, -x) = 2$ for all $x \in \partial B$) but $D(X) = 3$. By Proposition 9.4.6 this discrepancy is maximal.

The particular fact from Theorem 9.6.9 that was used in Corollary 9.5.5 and again in the remark following Theorem 9.5.6 – namely that if $\dim X = 3$ then $m(X) \geq 3$ – was proved in [460]. The results on extreme values in §9.6 are all due to Schäffer and can be found in [466] but were originally proved in [459]–[463].

There are many unresolved problems associated with the parameters discussed in this chapter. The least specific is the need for a more thorough exploration of connections between them (of the type illustrated by Theorem 9.2.3 and Proposition 9.4.5). In [466] and also in his papers Schäffer offers a large number of conjectures and questions. The two that I find most interesting are: (i) Is it true for every normed space X that $m(X) = m(X^*)$? (ii) Is it true for every Minkowski space (or for every reflexive Banach space) X that $M(X) = D(X)$? With regard to the first, Schäffer [466], p. 109, shows that if the conjectured equality holds for all Minkowski spaces then it holds in general. He also points out the relatively easy relationship $m(X) = m(X^{**})$.

10
Fifty problems

Throughout this book a variety of problems have been mentioned. The purpose of this short chapter is to collect them into a convenient location where connections among them can be seen more easily. There are also some additional, related problems. Almost all of these 50 problems have appeared elsewhere; very few originate with me. Many of them are well known and some have a long history of partial and conjectured solutions. In most of these cases some of the history has been given in the notes to the various chapters. It will not be repeated here. As far as I know these problems are still open. Many of them appear to be difficult.

The first two digits of the problem number refer to the section where either the problem was stated or where the material to which it is related can be found. For cross-references among the problems, "Problem" is abbreviated as P.

Problem 2.3.1 If K is a symmetric convex body in $\mathbb{R}^d$ is it true that $\lambda(K)\lambda^*(K^\circ) \geq 4^d/d!$?

Problem 2.4.1 For $d \geq 3$, what is the diameter of the metric space $(\mathcal{D}_d, \Delta)$?

Problem 2.4.2 What is the geometry of $(\mathcal{D}_d, \Delta)$? E.g. for what pairs of spaces is the diameter attained? Are there spaces X in $\mathcal{D}_d$ such that, for all Y in $\mathcal{D}_d$,

$$\Delta(X, Y) \leq \text{diam}(\mathcal{D}_d)/2?$$

Problem 2.4.3 For specific (symmetric) convex bodies B (e.g. the cross-polytope CP_d) what is the distance (Hausdorff, Δ_1 or Banach–Mazur) from B to the class of zonoids?

Problem 3.1.1 If $\|\cdot\|_1$ and $\|\cdot\|_2$ are two norms on $\mathbb{R}^d$ and if T is an isometry of the $\|\cdot\|_1$-unit sphere onto the $\|\cdot\|_2$-unit sphere, is T the restriction of an isometry (necessarily linear) between the two spaces?

Problem 3.1.2 Are there other conditions on the domain and range of an isometry that imply that it is linear?

Problem 3.3.1 (see P2.4.2) For particular d-dimensional Minkowski spaces X (in particular, for the Examples in §1.1) what is $\max\{\Delta(X, Y) : Y \in \mathcal{D}_d\}$?

Problem 3.3.2 (see P2.4.3) John's theorem states that the maximal Δ_1 distance of a unit ball B from the class of ellipsoids is $\frac{1}{2}\log d$; what is $\max\{\Delta_1(B, Z) : Z$ a zonoid $\}$? (Or replace Δ_1 by Δ.)

Problem 4.1.1 Does every d-dimensional Minkowski space admit the construction of a regular d-simplex? I.e. are there unit vectors $x_1, \ldots, x_d$ such that $\|x_i - x_j\| = 1$?

Problem 5.1.1 How large is the set Γ of functions γ that satisfy Requirements 5.1.1?

Problem 5.1.2 Is it possible to describe the set Γ precisely? In particular, what are the extreme points of Γ?

Problem 5.1.3 Requirements 5.1.1(a)–(c) do not imply that (d) is satisfied (Example 5.1.4(v)). Is it possible that a convexity condition on σ (e.g. requiring that flat surfaces minimize area in all dimensions) implies that the corresponding γ is continuous, i.e. satisfies 5.1.1(b)?

Problem 5.4.1 With Γ as in P5.1.1, is it possible to describe the set of isoperimetrices $\{\mathbf{I}_\gamma(B) : \gamma \in \Gamma\}$?

Problem 5.4.2 If the isoperimetrices $\{\mathbf{I}_\gamma(B) : \gamma \in \Gamma\}$ are suitably scaled (e.g., consider the smallest multiple of $\mathbf{I}_\gamma(B)$ that contains B) is $\mathbf{I}(B) = \Pi(B^\circ)$ the largest?

Problem 5.4.3 Are there elements γ_1, γ_2 in Γ such that
$$\mathbf{I}_{\gamma_1}(B) = (\Pi B)^\circ? \qquad \mathbf{I}_{\gamma_2}(B) = I(B^\circ)?$$
If the first is yes, is $(\Pi B)^\circ$ the smallest element in $\{\mathbf{I}_\gamma(B) : \gamma \in \Gamma\}$?

Problem 5.4.4 Figures 5.3 and 5.4 show similarities between $\Pi(B^\circ)$ and $(IB)^\circ$ for various B. Is it possible to describe or quantify this similarity?

Problem 5.4.5 If B is a symmetric convex body in $\mathbb{R}^d$ and if $\tilde{\mathbf{I}}(B)$ is the normalized isoperimetrix arising from either $\mathbf{I}(B) = \Pi(B^\circ)$ or $\mathbf{I}(B) = (IB)^\circ$ or $\mathbf{I}_\gamma(B)$ for other γ, is it true that $\tilde{\mathbf{I}}^n(B)$ converges? (For the first case, see the examples in Figures

5.5, 5.6, 5.7 and 5.8.) If so, does $\tilde{\mathbf{I}}^n(B)$ converge to an ellipsoid? If so, how is that ellipsoid related to B?

Problem 5.4.6 Chapter 6 is concerned with the map $\mathbf{I}(B) = \Pi(B^\circ)$. Chapter 7 is concerned with the map $\mathbf{I}(B) = (IB)^\circ$. Since each of Π, I and $^\circ$ maps sets in X to sets in X^* what can be said about the other compositions: $B \mapsto (\Pi B)^\circ$, $B \mapsto I(B^\circ)$ (already mentioned) and $B \mapsto \Pi(IB)$, $B \mapsto I(\Pi(B))$?

Problem 5.5.1 Are there connections between the ellipsoid arising from the Laplacian differential operator and other ellipsoids associated with B? Does this Laplacian ellipsoid have any bearing on P5.4.5?

Problem 5.5.2 Under what conditions on K or on the unit ball B is the minimal position of K unique? Under what conditions on K or on B is the criterion in Theorem 5.5.13 sufficient to guarantee that the position is minimal?

Problem 6.1.1 For $1 < k < d-1$, is the function σ^k (discussed in §6.1) convex in the sense of Equation (6.10)? *I.e.* if F_1 is a k-dimensional polytope with boundary ∂F_1 and if K is a union $K = \bigcup_{i=2}^{n} F_i$ of k-dimensional polytopes such that $\partial K = \partial F_1$, does it follow that

$$\mu_B^k(F_1) \leq \sum_{i=2}^{n} \mu_B^k(F_i)?$$

Problem 6.2.1 Are either of the mappings $\tilde{\mathbf{I}}$ or $\tilde{\mathbf{I}}^2$ contractive with respect to the Banach-Mazur distance?

Problem 6.2.2 Is the mapping $\mathbf{I}^2$ contractive with respect to Δ_1?

Problem 6.3.3 Is the equation $\mu_B(\partial B) = \mu_{B^\circ}(\partial B^\circ)$ (for all B) (together with Requirements 5.1.1) sufficient to characterize μ_B completely in all dimensions? Or, if not, for $d = 3$?

Problem 6.3.4 If μ_B is such that $\mu_B(\partial B) = \mu_{B^\circ}(\partial B^\circ)$ for all symmetric bodies B in $\mathbb{R}^d$, does it follow that $\gamma(K) = \gamma(K^\circ)$ for symmetric convex bodies K in $\mathbb{R}^{d-1}$.

Problem 6.5.1 What are the precise bounds for $\mu_B(\partial B)$ in all dimensions? For large values of d, is it true that $\mu_B(\partial B)$ is maximal for ellipsoids and minimal for parallelotopes?

Problem 6.5.2 What can be said about the volume product of central cross-sections of symmetric convex bodies B? In particular, are there estimates for

$$\min_{B} \max_{f} \{\lambda(B \cap f^{\perp})\lambda^*(B \cap f^{\perp})^\circ\}?$$

Problem 6.5.3 Is Petty's conjectured projection inequality true?

Problem 6.5.4 For $d \geq 3$ if either $\mathbf{I}^2(B)$ or $\mathbf{I}(B)$ is homothetic to B, is B an ellipsoid?

Problem 6.5.5 For symmetric bodies B (and $\mathbf{I}(B) = \Pi(B^\circ)$), is it true that $\lambda(\mathbf{I}B)\lambda^*(\mathbf{I}B)^\circ) \geq \lambda(B)\lambda^*(B^\circ)$?

Problem 6.5.6 If $i(B)$ denotes the Minkowski isoperimetric ratio $i(B) := \mu_B(\partial B)^d / \mu_B(B)^{d-1}$, is it true that for all B the sequence $(i(\mathbf{I}^n B))$ is eventually decreasing?

Problem 6.5.7 If $\pi(B)$ is defined by

$$\pi(B) := \lambda(B)^{d-1}\lambda(\Pi(B)^\circ),$$

is it true, for B not an ellipsoid, that $\pi(B) < \pi(\Pi(B)^\circ)$?

The problems arising from Chapter 7 are almost identical to those from Chapter 6 but using the Busemann definition and $\mathbf{I}(B) = (IB)^\circ$.

Problem 7.1.1 For $1 < k < d - 1$, is the function σ^k convex in the sense of Equation (6.10)? I.e. if F_1 is a k-dimensional polytope with boundary ∂F_1 and if K is a union $K = \bigcup_{i=2}^{n} F_i$ of k-dimensional polytopes such that $\partial K = \partial F_1$, does it follow that

$$\mu_B^k(F_1) \leq \sum_{i=2}^{n} \mu_B^k(F_i)?$$

Problem 7.2.1 Are either of the mappings $\tilde{\mathbf{I}}$, $\tilde{\mathbf{I}}^2$ contractive with respect to Δ?

Problem 7.2.2 Is the mapping $\mathbf{I}^2$ contractive with respect to Δ_1?

Problem 7.4.1 What is the precise lower bound for $\mu_B(\partial B)$ in all dimensions? For large values of d, is it true that $\mu_B(\partial B)$ is minimal for ellipsoids?

Problem 7.4.2 Does $\tilde{\mathbf{I}}(B) \subseteq B$ imply that B is an ellipsoid?

Problem 7.4.3 Can Busemann's intersection inequality be strengthened to

$$\lambda(K)^{d-1}\lambda(IK)^\circ) \geq (\epsilon_d/\epsilon_{d-1})^d?$$

Problem 7.4.4 For $d \geq 3$, if either $\mathbf{I}^2(B)$ or $\mathbf{I}(B)$ is homothetic to B, is B an ellipsoid?

Problem 7.4.5 For symmetric bodies B is it true that
$$\lambda(\mathbf{I}B)\lambda^*(\mathbf{I}B)^\circ) \geq \lambda(B)\lambda^*(B^\circ)?$$

Problem 7.4.6 If $i(B)$ denotes the Minkowski isoperimetric ratio (P6.5.6), is it true that $i(B) \geq i(\tilde{\mathbf{I}}(B))$?

Problem 7.4.7 If $b(B)$ is defined by $b(B) := \lambda(B)^{d-1}\lambda((IB)^\circ)$, is it true that $b(I(B)^\circ) \leq b(B)$?

Problem 7.5.1 Is there an absolute constant c such that if K_1 and K_2 are symmetric convex bodies with $\lambda^{d-1}(K \cap f^\perp) \leq \lambda^{d-1}(K_2 \cap f^\perp)$ for all linear functionals f then $\lambda(K_1) \leq c\lambda(K_2)$?

Problem 8.4.1 What are "satisfactory" requirements for the measure of angle in Minkowski spaces?

Problem 8.4.2 What is the "most satisfactory" definition of angle measure in Minkowski spaces?

Problem 9.1.1 What are the exact values (or good bounds) for the projection constant ϖ^s for particular spaces?

Problem 9.4.4 Is $m(X) = m(X^*)$ for all Minkowski spaces X?

Problem 9.4.5 Does $M(X) = D(X)$ hold for every Minkowski space X? for every reflexive Banach space X?

Problem 9.4.6 If $\dim X = 3$ is $R(X) = m(X)$?

A final question that does not have a home in a chapter:

Problem 50 One form of Archimedes' theorem (the one reputedly on his gravestone) is that the surface area of a sphere is equal to the surface of an open circumscribing cylinder (and this is true for each slice of the sphere and the cylinder). Is there a generalization of this result to (some) other Minkowski spaces or does this theorem characterize Euclidean space? For the solution in Euclidean space, see Knothe [296], and for that in Riemannian geometry see Vanhecke [517] and Vanhecke and Djorić [518].

REFERENCES

[1] Ader, O.B., An affine invariant of convex regions, *Duke Math. J.* **4** (1938), 291–299.
[2] Aitchison, P.W., Petty, C.M. and Rogers, C.A., A convex body with a false centre is an ellipsoid, *Mathematika* **18** (1971), 50–59.
[3] Aleksandrov, A.D., A theorem on convex polyhedra (Russian), *Trudy Fiz.-Mat. Inst. Steklov.* **4** (1933), 87.
[4] Aleksandrov, A.D., Zur Theorie der gemischten Volumina von konvexen Körpern, III (Russian), *Mat. Sbornik* **3** (1938), 27–46.
[5] Aleksandrov, A.D., Smoothness of the convex surface of bounded Gaussian curvature, *C.R. (Doklady) Acad. Sci. URSS* **36** (1942), 195–199.
[6] Aleksandrov, A.D., On filling of space by polytopes (Russian), *Vestnik Leningrad. Univ. (Ser. Mat. Fiz. Khim.)* **9** (1954), 33–43.
[7] Alexander, R., Zonoid theory and Hilbert's fifth problem, *Geom. Dedicata* **28** (1988), 199–211.
[8] Amir, D., *Characterizations of Inner Product Spaces*, Birkhäuser, Basel (1986).
[9] Asplund, E., Comparison of plane symmetric convex bodies and parallelograms, *Math. Scand.* **8** (1960), 171–180.
[10] Asplund, E. and Grünbaum, B., On the geometry of Minkowski planes, *Enseign. Math.* (2) **6** (1960), 299–306.
[11] Auerbach, H., Sur une propriété caractéristique de l'ellipsoide, *Studia Math.* **9** (1940), 17–22.
[12] Ball, K.M., Cube slicing in $\mathbb{R}^n$, *Proc. Amer. Math. Soc.* **97** (1986), 465–473.
[13] Ball, K.M., Some remarks on the geometry of convex sets, in *Geometrical Aspects of Functional Analysis* (eds. Lindenstrauss, J. and Milman, V.D.), *Lecture Notes in Mathematics* **1317**, Springer, Berlin (1988), 224–231.
[14] Ball, K.M., Logarithmically concave functions and sections of convex sets, *Studia Math.* **88** (1988), 69–84.
[15] Ball, K.M., Volumes of sections of cubes and related problems, in *Geometrical Aspects of Functional Analysis* (eds. Lindenstrauss, J. and Milman, V.D.), *Lecture Notes in Mathematics* **1376**, Springer, Berlin (1989), 251–260.
[16] Ball, K.M., Volume ratios and a reverse isoperimetric inequality, *J. London Math. Soc.* **44** (1991), 351–359.
[17] Ball, K.M., Shadows of convex bodies, *Trans. Amer. Math. Soc.* **327** (1991), 891–901.
[18] Bambah, R.P., Polar reciprocal convex bodies, *Proc. Cambridge Phil. Soc.* **51** (1955), 377–378.
[19] Banach, S., Sur les opérations dans les ensembles abstraits et leur applications aux équations intégrales, *Fund. Math.* **3** (1922), 133–181.
[20] Banach, S., Sur les fonctionelles linéaires I, II, *Studia Math.* **1** (1929), 211–216 and 223–239.
[21] Banach, S., *Théorie des opérations linéaires*, Warsaw (1932).
[22] Barthel, W., Zum Inhaltsbegriff in der Minkowskischen Geometrie, *Math. Z.* **58** (1953), 358–375.

[23] Barthel, W., Zum Busemannschen und Brunn-Minkowskischen Satz, *Math. Z.* **70** (1958–1959), 407–429.
[24] Barthel, W., Zur Minkowski-Geometrie, begründet auf dem Flächeninhaltsbegriff, *Monatsh. Math.* **63** (1959), 317–343.
[25] Barthel, W., Zur isodiametrischen und isoperimetrischen Ungleichung in der Relativgeometrie, *Comment. Math. Helv.* **33** (1959), 241–257.
[26] Barthel, W. and Franz, G., Eine Verallgemeinerung des Busemannschen Satzes vom Brunn-Minkowskischen Typ, *Math. Ann.* **144** (1961), 183–198.
[27] Barthel, W. and Pabel, H., Das isodiametrische Problem in der Minkowski-Geometrie, *Results Math.* **12** (1987), 252–267.
[28] Behrend, F., Über einige Affininvarianten konvexer Bereiche, *Math. Ann.* **113** (1936–1937), 713–747.
[29] Behrend, F., Über die kleinste umbeschriebene und die größte einbeschriebene Ellipse eines konvexen Bereiches, *Math. Ann.* **115** (1938), 379–411.
[30] Benson, R.V., *Euclidean Geometry and Convexity*, McGraw-Hill, New York (1966).
[31] Berger, M., Convexity, *Amer. Math. Monthly* **97** (1990), 650–678.
[32] Bertrand, M., *Journal de Liouville* **7** (1842), 215–216.
[33] Besicovitch, A.S., On the definition and value of the area of a surface, *Quart. J. Math.* **16** (1945), 96–102.
[34] Biberstein, O., *Elements de géométrie différentielle minkowskienne*, Ph.D. Thesis, Univ. of Montréal, Montréal (1957).
[35] Birkhoff, G., Orthogonality in linear metric spaces, *Duke Math. J.* **1** (1935), 169–172.
[36] Blaschke, W., Räumliche Variationsprobleme mit symmetrischer Transversalitätsbedingung, *Ber. Verh. Sächs. Akad. Wiss. Leipzig Math.-Naturw. Kl.* **68** (1916), 50–55.
[37] Blaschke, W., Über affine Geometrie I: Isoperimetrische Eigenschaften von Ellipse und Ellipsoid, *Ber. Verh. Sächs. Akad. Wiss. Leipzig Math.-Phys. Kl.* **68** (1916), 237–239.
[38] Blaschke, W., *Kreis und Kugel*, Veit, Leipzig (1916).
[39] Blaschke, W., Über affine Geometrie VII: Neue Extremeigenschaften von Ellipse und Ellipsoid, *Ber. Verh. Sächs. Akad. Wiss. Leipzig Math.-Phys. Kl.* **69** (1917), 306–318.
[40] Blaschke, W., Affine Geometrie IX, Verschiedene Bemerkungen und Aufgaben, *Ber. Verh. Sächs. Akad. Wiss. Leipzig Math.-Phys. Kl.* **69** (1917), 412–420.
[41] Blaschke, W., *Vorlesungen über Differentialgeometrie II: Affine Differentialgeometrie*, Springer, Berlin (1923).
[42] Blaschke, W., Integralgeometrie 11: zur Variationsrechnung, *Abh. Math. Sem. Univ. Hamburg* **11** (1936), 359–366.
[43] Bliss, G.A., A generalization of the notion of angle, *Trans. Amer. Math. Soc.* **7** (1906), 184–196.
[44] Bohnenblust, P., Convex regions and projections in Minkowski spaces, *Ann. of Math.* (2) **39** (1938), 301–308.
[45] Bohnenblust, P., A characterization of complex Hilbert spaces, *Portugal. Math.* **3** (1942), 103–109.
[46] Boju, V. and Funar, L., Generalized area in Minkowski spaces, *Beiträge Algebra Geom.* **31** (1991), 33–38.
[47] Bolker, E.D., A class of convex bodies, *Trans. Amer. Math. Soc.* **145** (1969), 323–345.
[48] Bollobás, B., *Linear Analysis: An Introductory Course*, Cambridge Univ. Press, Cambridge (1990).
[49] Boltyanski, V.G., Martini, H. and Soltan, P.S., *Excursions into Combinatorial Geometry*, Springer, Berlin (in press).
[50] Bonnesen, T., Über eine Verschärfung der isoperimetrischen Ungleichung des Kreises in der Ebene und auf der Kugeloberfläche nebst einer Anwendung auf eine Minkowskische Ungleichheit für konvexe Körper, *Math. Ann.* **84** (1921), 216–227.
[51] Bonnesen, T. and Fenchel, W., *Theory of Convex Bodies*, BCS Associates (Moscow, U.S.), 1987, transl. by L. Boron, C. Christenson and B. Smith of *Theorie der konvexen Körper*, Springer, Berlin (1934).
[52] Botts, T., Convex sets, *Amer. Math. Monthly* **49** (1942), 527–535.

[53] Bouligand, G. and Choquet, G., Problèmes liés à des métriques variationelles, *C.R. Acad. Sci. Paris* **218** (1944), 696–698.
[54] Bourgain, J. and Lindenstrauss, J., Projection bodies, in *Geometric Aspects of Functional Analysis* (eds. Lindenstrauss, J. and Milman, V.D.), *Lecture Notes in Mathematics* **1317**, Springer, Berlin (1988), 250–270.
[55] Bourgain, J. and Milman, V.D., Sections euclidiennes et volume des corps symétriques convexes dans $\mathbb{R}^n$, *C.R. Acad. Sci. Paris* **300** (1985), 435–438.
[56] Bourgain, J. and Milman, V.D., New volume ratio properties for convex symmetric bodies in $\mathbf{R}^n$, *Invent. Math.* **88** (1987), 319–340.
[57] Bourgain, J. and Szarek, S.J., The Banach–Mazur distance to the cube and the Dvoretzky–Rogers factorization, *Israel J. Math.* **62** (1988), 169–180.
[58] Brass, P., Erdös distance problems in normed spaces (preprint, 1994).
[59] Brothers, J.E. and Morgan, F., The isoperimetric theorem for general integrands, *Michigan Math. J.* (in press).
[60] Brunn, H., *Über Ovale und Eiflächen*, Inaugural dissertation, Munich (1887).
[61] Brunn, H., *Über Kurven ohne Wendepunkte*, Habilitationsschrift, Munich (1889).
[62] Burago, Yu.D. and Zalgaller, V.A., *Geometric Inequalities*, Springer, Berlin (1988) (transl. from Russian edition: Nauka, Leningrad, 1980).
[63] Busemann, H., *Metric Methods in Finsler Spaces and in the Foundations of Geometry*, Ann. of Math. Studies **8**, Princeton Univ. Press, Princeton, NJ (1942).
[64] Busemann, H., Local metric geometry, *Trans. Amer. Math. Soc.* **56** (1944), 200–274.
[65] Busemann, H., Intrinsic area, *Ann. of Math.* (2) **48** (1947), 234–267.
[66] Busemann, H., The isoperimetric problem in the Minkowski plane, *Amer. J. Math.* **69** (1947), 863–871.
[67] Busemann, H., A theorem on convex bodies of Brunn–Minkowski type, *Proc. Nat. Acad. Sci. U.S.A.* **35** (1949), 27–31.
[68] Busemann, H., The isoperimetric problem for Minkowski area, *Amer. J. Math.* **71** (1949), 743–762.
[69] Busemann, H., Angular measure and integral curvature, *Canad. J. Math.* **1** (1949), 279–296.
[70] Busemann, H., The geometry of Finsler spaces, *Bull. Amer. Math. Soc.* **56** (1950), 5–16.
[71] Busemann, H., The foundations of Minkowskian geometry, *Comment. Math. Helv.* **24** (1950), 156–187.
[72] Busemann, H., Volumes in terms of concurrent cross-sections, *Pacific J. Math.* **3** (1953), 1–12.
[73] Busemann, H., *The Geometry of Geodesics*, Academic Press, New York (1955).
[74] Busemann, H., *Convex Surfaces*, Wiley-Interscience, New York (1958).
[75] Busemann, H., Areas in affine spaces III. The integral geometry of affine area, *Rend. Circ. Mat. Palermo* (2) **9** (1960), 226–242.
[76] Busemann, H., Convexity on Grassmann manifolds, *Enseign. Math.* (2) **7** (1961), 139–152.
[77] Busemann, H., Ewald, G. and Shephard, G.C., Convex bodies and convexity on Grassmann cones: I–IV, *Math. Ann.* **151** (1963), 1–41.
[78] Busemann, H., Ewald, G. and Shephard, G.C., Convex bodies and convexity on Grassmann cones: V, *Arch. Math.* **13** (1962), 512–526.
[79] Shephard, G.C., Convex bodies and convexity on Grassmann cones: VI, *J. London Math. Soc.* **39** (1964), 307–319.
[80] Ewald, G., Convex bodies and convexity on Grassmann cones: VII, *Abh. Math. Sem. Univ. Hamburg* **27** (1964), 167–170.
[81] Shephard, G.C., Convex bodies and convexity on Grassmann cones: VIII, *J. London Math. Soc.* **39** (1964), 417–423.
[82] Ewald, G., Convex bodies and convexity on Grassmann cones: IX, *Math. Ann.* **157** (1964), 219–230.
[83] Busemann, H. and Shephard, G.C., Convex bodies and convexity on Grassmann cones: X, *Ann. Mat. Pura Appl.* **70** (1965), 271–294.
[84] Busemann, H., Convex bodies and convexity on Grassmann cones: XI, *Math. Scand.* **24** (1969), 93–101.

[85] Busemann, H. and Petty, C.M., Problems on convex bodies, *Math. Scand.* **4** (1956), 88–94.
[86] Busemann, H. and Shephard, G.C., Convexity on non-convex sets, in *Proc. Colloq. on Convexity Copenhagen 1965*, Copenhagen (1967), 20–33.
[87] Busemann, H. and Straus, E.G. Area and normality, *Pacific J. Math.* **10** (1960), 35–72.
[88] Bynum, W.L., A class of spaces lacking normal structure, *Compositio Math.* **25** (1972), 233–236.
[89] Campi, S., Recovering a centered convex body from the areas of its shadows: a stability estimate, *Ann. Mat. Pura Appl.* (4) **151** (1988), 289–302.
[90] Carathéodory, G., Über den Variabilitätsbereich der Koeffizienten von Potenzreihen die gegebene Werte nicht annehmen, *Math. Ann.* **64** (1907), 95–115.
[91] Carathéodory, G., Über das lineare Mass von Punktmengen: eine Verallgemeinerung des Längenbegriffs, *Nachr. Ges. Wiss. Göttingen* (1914), 404–426.
[92] Cartan, H., Sur la mesure de Haar, *C. R. Acad. Sci. Paris* **211** (1940), 759–762.
[93] Cassels, J.W.S., *An Introduction to the Geometry of Numbers*, Springer, Berlin (1959).
[94] Castaing, C. and Valadier, M., *Convex Analysis and Measurable Multifunctions, Lecture Notes in Mathematics* **580**, Springer, Berlin (1977).
[95] Ceder, J.G., A property of planar convex bodies, *Israel J. Math.* **1** (1963), 248–253.
[96] Chakerian, G.D., The isoperimetric problem in the Minkowski plane, *Amer. Math. Monthly* **67** (1960), 1002–1004.
[97] Chakerian, G.D., Integral geometry in the Minkowski plane, *Duke Math. J.* **29** (1962), 375–382.
[98] Chakerian, G.D., Sets of constant width, *Pacific J. Math.* **19** (1966), 13–21.
[99] Chakerian, G.D., Sets of constant width and constant relative brightness, *Trans. Amer. Math. Soc.* **129** (1967), 26–37.
[100] Chakerian, G.D., The mean volume of boxes and cylinders circumscribed about a convex body, *Israel J. Math.* **12** (1972), 249–256.
[101] Chakerian, G.D., On a certain affine invariant functional for convex bodies, *Stud. Scient. Math. Hungar.* **8** (1973), 91–93.
[102] Chakerian, G.D., Mixed areas and the self-circumferences of a plane convex body, *Arch. Math.* **34** (1980), 81–83.
[103] Chakerian, G.D. and Filliman, P., The measure of the projections of a cube, *Studia Sci. Math. Hungar.* **21** (1986), 103–110.
[104] Chakerian, G.D. and Ghandehari, M.A., The Fermat problem in Minkowski spaces, *Geom. Dedicata* **17** (1985), 227–238.
[105] Chakerian, G.D. and Groemer, H., Convex bodies of constant width, in *Convexity and Its Applications* (eds. Gruber, P.M. and Wills, J.M.), Birkhäuser, Basel (1983), 49–96.
[106] Chakerian, G.D. and Talley, W.K., Some properties of the self-circumference of convex sets, *Arch. Math.* **20** (1969), 431–443.
[107] Charzyński, Z., Sur les transformations isométriques des espaces du type F, *Studia Math.* **13** (1953), 94–121.
[108] Cheng, S.-Y. and Yau, S.-T., On the regularity of the solution of the n-dimensional Minkowski problem, *Commun. Pure Appl. Math.* **29** (1976), 495–516.
[109] Chilakamarri, K.B., Unit distance graphs in Minkowski metric spaces, *Geom. Dedicata* **37** (1991), 345–356.
[110] Chilton, B.L. and Coxeter, H.S.M., Polar zonohedra, *Amer. Math. Monthly* **70** (1963), 946–951.
[111] Cieslik, D., *Steiner Minimal Trees*, Fachrichtungen Mathematik/Informatik, Universität Greifswald, Greifswald (manuscript) (1995), 170 pp.
[112] Clack, R., *Some Minkowski Geometry and the Isoperimetric Problem*, Ph.D. Thesis, Dalhousie Univ., Halifax (1989).
[113] Clack, R., Minkowski surface area under affine transformations, *Mathematika* **37** (1990), 232–238.
[114] Cohn, D.L., *Measure Theory*, Birkhäuser, Basel (1980).
[115] Constantin, R.F., Variational aspects and applications in isoperimetric problems in the Minkowski plane, in *Lucr. Conf. Nat. Geom. Topologie, Timisoara, 1984*, Univ. of Timisoara (1984), 40–45.
[116] Constantin, R.F., On the convex hull of n aleatory points belonging to a convex domain in M_2 plane, in *Lucr. Conf. Nat. Geom. Topologie, Timisoara, 1984*, Univ. of Timisoara (1984), 46–51.

References

[117] Coxeter, H.S.M., *Regular Polytopes*, Pitman, London (1948).
[118] Coxeter, H.S.M., The classification of zonohedra by means of projective diagrams, *J. Math. Pura Appl.* (9) **41** (1962), 137–156.
[119] Croft, H.T., Two problems on convex bodies, *Proc. Cambridge Phil. Soc.* **58** (1962), 1–7.
[120] Danzer, L., Grünbaum, B. and Klee, V., Helly's theorem and its relatives, in *Convexity*, Proc. Symp. Pure Math. VII, Amer. Math. Soc., Providence, RI (1963), 233–270.
[121] Danzer, L., Laugwitz, D. and Lenz, H., Über das Löwnersche Ellipsoid und sein Analogon unter den einem Eikörper einbeschriebenen Ellipsoiden, *Arch. Math.* **8** (1957), 214–219.
[122] Davis, W.J., Dean, D.W. and Singer, I., Complemented subspaces and Λ systems in Banach spaces, *Israel J. Math.* **6** (1968), 303–309.
[123] Day, M.M., Polygons circumscribed about closed convex curves, *Trans. Amer. Math. Soc.* **62** (1947), 315–319.
[124] Day, M.M., Some characterizations of inner product spaces, *Trans. Amer. Math. Soc.* **62** (1947), 320–337.
[125] Day, M.M., Some criteria of Kasahara and Blumenthal for inner product spaces, *Proc. Amer. Math. Soc.* **10** (1959), 92–100.
[126] Day, M.M., *Normed Linear Spaces* (3rd ed.), Springer, Berlin (1973).
[127] de Figueiredo, D.G. and Karlovitz, L.A., On the radial projection in normed linear spaces, *Bull. Amer. Math. Soc.* **73** (1967), 364–368.
[128] Dinghas, A., Über einen geometrischen Satz von Wulff für die Gleichgewichtsform von Kristallen, *Zeitschrift für Kristallographie* **105** (1944), 304–314.
[129] Dinghas, A. and Schmidt, E., Einfacher Beweis der isoperimetrischen Eigenschaft der Kugel im n-dimensionalen euklidischen Raum, *Abh. Preuss. Akad. Wissenschaften Math. Nat. Kl.* **7** (1944), 18 pp.
[130] Dunford, N. and Schwartz, J.T., *Linear Operators Part I*, Wiley-Interscience, New York (1958).
[131] Durier, R. and Michelot, C., Geometrical properties of the Fermat–Weber problem, *European J. Oper. Res.* **20** (1985), 332–343.
[132] Durier, R. and Michelot, C., Sets of efficient points in normed spaces, *J. Math. Anal. Appl.* **117** (1986), 506–528.
[133] Dvoretzky, A., Some results on convex bodies and Banach spaces, in *Proc. Internat. Sympos. Linear Spaces, Jerusalem 1960*, Hebrew Univ., Jerusalem (1961), 123–160.
[134] Dvoretzky, A. and Rogers, C.A., Absolute and unconditional convergence in normed linear spaces, *Proc. Nat. Acad. Sci. U.S.A.* **36** (1950), 192–197.
[135] Eckhoff, J., Helly, Radon, and Carathéodory type theorems, in *Handbook of Convex Geometry* (eds. Gruber, P.M. and Wills, J.M.), North-Holland, Amsterdam (1993), Vol. A, 389–448.
[136] Edelstein, M., On fixed and periodic points under contractive mappings, *J. London Math. Soc.* **37** (1962), 74–79.
[137] Eggleston, H.G., A proof of Blaschke's theorem on the Reuleaux triangle, *Quart. J. Math.* (2) **3** (1952), 296–297.
[138] Eggleston, H.G., *Convexity*, Cambridge Univ. Tracts in Math and Math. Phys. **47**, Cambridge (1958).
[139] Eggleston, H.G., Notes on Minkowski geometry I: Relations between the circumradius, diameter, inradius and minimal width of a convex set, *J. London Math. Soc.* **33** (1958), 76–81.
[140] Eggleston, H.G., Sets of constant width in finite dimensional Banach spaces, *Israel J. Math.* **3** (1965), 163–172.
[141] Eidelheit, M., Zur Theorie der konvexen Mengen in linearen normierten Räumen, *Studia Math.* **6** (1936), 104–111.
[142] El-Ekhtiar, Ali, *Integral Geometry in Minkowski Space*, Ph.D. Thesis, Univ. of California, Davis (1992).
[143] Erdös, P., On sets of distances of n points, *Amer. Math. Monthly* **53** (1946), 248–250.
[144] Erdös, P., Problems and results in combinatorial geometry, in *Discrete Geometry and Convexity* (eds. Goodman, J.E., Lutwak, E., Malkevitch, J. and Pollack, R.), *Ann. New York Acad. Sci.* **440** (1985), 1–11.
[145] Ewald, G., On Busemann's theorem of the Brunn–Minkowski type, in *Proc. Colloq. on Convexity Copenhagen 1965*, Copenhagen (1967), 69–71. See also [80] and [82].

[146] Ewald, G. and Shephard, G.C., Normed vector spaces consisting of classes of convex sets, *Math. Z.* **91** (1966), 1–19.
[147] Falconer, K.J., Applications of a result on spherical integration to the theory of convex sets, *Amer. Math. Monthly* **90** (1983), 690–693.
[148] Fallert, H., Goodey, P.R. and Weil, W., Spherical projections and centrally symmetric sets (preprint, 1994).
[149] Fedorov, E.S., The numerical relationship between the zones and faces of polyhedra, *Mineralogical Magazine* **18** (1919), 99–110.
[150] Fenchel, W., Convexity through the ages, in *Convexity and Its Applications* (eds. Gruber, P.M. and Wills, J.M.), Birkhäuser, Basel (1983), 120–130.
[151] Fenchel, W. and Jessen, B., Mengenfunktionen und konvexe Körper, *Danske Vid. Selskab. Mat.-fys. Medd.* **16** (1938), 3.
[152] Figiel, T., On non-linear isometric embeddings of normed linear spaces, *Bull. Acad. Polon. Sci. Ser. Math. Astron. Phys.* **16** (1968), 185–188.
[153] Figiel, T., Some remarks on Dvoretzky's theorem on almost spherical sections of convex bodies, *Colloq. Math.* **24** (1972) 241–252.
[154] Figiel, T., A short proof of Dvoretzky's theorem, in *Séminaire Maurey-Schwartz, 1974–5, Espaces L^p, applications radonifiantes et géometrie des espaces de Banach* **23**, Centre Math. Ecole Polytech., Paris (1975), 6 pp.
[155] Figiel, T., Local theory of Banach spaces and some operator ideals, *Proc. Internat. Congress Math. 1983*, Vol. 2, Polish Scientific Publishers, North-Holland, Amsterdam (1984), 961–976.
[156] Filliman, P., Extremum problems for zonotopes, *Geom. Dedicata* **27** (1988), 251–262.
[157] Filliman, P., The volumes of duals and sections of polytopes, *Mathematika* **39** (1992), 67–80.
[158] Finsler, P., Über eine Verallgemeinerung des Satzes von Meusnier, *Vierteljschr. Naturforsch. Ges. Zürich* **85** (1940), 155–164.
[159] Firey, W.J., The brightness of convex bodies, Tech. Report 19, Oregon State University, Corvallis (1963).
[160] Firey, W.J., Blaschke sums of convex bodies and mixed bodies, in *Proc. Colloq. on Convexity Copenhagen 1965*, Copenhagen (1967), 94–101.
[161] Firey, W.J., Approximating convex bodies by algebraic ones, *Arch. Math.* **25** (1974), 424–425.
[162] Franchetti, C. and Votruba, G.F., Perimeter, Macphail number and projection constant in Minkowski planes, *Boll. Un. Mat. Ital. B* (5) **13** (1976), 560–573.
[163] Fréchet, M., Sur les ensembles des fonctions et les opérations linéaires, *C.R. Acad. Sci. Paris* **144** (1907), 1414–1416.
[164] Fréchet, M., Sur les opérations linéaires, *Trans. Amer. Math. Soc.* **8** (1907), 433–446.
[165] Funk, P., Über Flächen mit lauter geschlossenen geodätischen Linien, *Math. Ann.* **74** (1913), 278–300.
[166] Furstenberg, H. and Tzkoni, I., Spherical functions and integral geometry, *Israel J. Math.* **10** (1971), 327–338.
[167] Gardner, M., *Mathematical Carnival: From Penny Puzzles, Card Shuffles and Tricks of Lightning Calculators to Roller Coaster Rides into the Fourth Dimension* (1st ed.), Knopf, New York (1975).
[168] Gardner, R.J., Intersection bodies and the Busemann–Petty problem, *Trans. Amer. Math. Soc.* **342** (1994), 435–445.
[169] Gardner, R.J., On the Busemann–Petty problem concerning central sections of centrally symmetric convex bodies, *Bull. Amer. Math. Soc.* **30** (1994), 222–226.
[170] Gardner, R.J., A positive answer to the Busemann–Petty problem in three dimensions, *Ann. of Math.* (2) **140** (1994), 435–447.
[171] Gardner, R.J., Geometric Tomography, *Notices Amer. Math. Soc.* **42** (1995), 422–429.
[172] Gardner, R.J., *Geometric Tomography, Encyclopedia of Mathematics and Its Applications* **54**, Cambridge Univ. Press, New York (1995).
[173] Gardner, R.J. and Volčič, A., Determination of convex bodies by their brightness functions, *Mathematika* **40** (1993), 161–168.

[174] Gardner, R.J. and Volčič, A., Tomography of convex and star bodies, *Adv. Math.* **108** (1994), 367–399.
[175] Garling, D.J.H. and Gordon, Y., Relationships between some constants associated with finite dimensional Banach spaces, *Israel J. Math.* **9** (1971), 346–361.
[176] Ghandehari, M.A., *Geometric Inequalities in the Minkowski Plane*, Ph.D. Thesis, Univ. of California, Davis (1983).
[177] Giannopoulos, A.A., On the Banach–Mazur distance to the cube (preprint, 1993).
[178] Gluck, H., The generalized Minkowski problem in differential geometry in the large, *Ann. Math.* **96** (1972), 245–276.
[179] Gluck, H., Manifolds with preassigned curvature – a survey, *Bull. Amer. Math. Soc.* **81** (1975), 313–329.
[180] Gluskin, E.D., The diameter of the Minkowski compactum is approximately equal to n, *Functional Anal. Appl.* **15** (1981), 57–58 (translation), 72–73 (original Russian).
[181] Golab, S., Quelques problèmes métriques de la géometrie de Minkowski, *Trav. l'Acad. Mines Cracovie* **6** (1932), 1–79.
[182] Golab, S., Sur la longueur de l'indicatrice dans la géometrie plane de Minkowski, *Colloq. Math.* **15** (1966), 141–144.
[183] Golab, S., Sur un problème de la métrique angulaire dans la géometrie de Minkowski, *Aequationes Math.* **6** (1971), 121–129.
[184] Golab, S. and Tamássy, L., Eine Kennzeichnung der euklidischen Ebene unter dem Minkowskischen Ebenen, *Publ. Math. Debrecen* **7** (1960), 187–193.
[185] Golubyatnikov, V.P., On reconstructing the shape of a body from its projections, *Dokl. Akad. Nauk SSSR* **262** (1982), 521–522 (Russian). English transl. *Soviet Math. Dokl.* **25** (1982), 62–63.
[186] Golubyatnikov, V.P., On unique recoverability of convex and visible compacta from their projections, *Mat. Sbornik* **182** (1991) (Russian). English transl. *Math. USSR Sbornik* **73** (1992), 1–10.
[187] Goodey, P.R., Instability of projection bodies, *Geom. Dedicata* **20** (1986), 295–305.
[188] Goodey, P.R., Lutwak, E. and Weil, W., Functional analytic characterization of classes of convex bodies, *Math. Z.* (in press).
[189] Goodey, P.R., Schneider, R. and Weil, W., Projection functions on higher rank Grassmannians (preprint, 1994).
[190] Goodey, P.R. and Weil, W., Centrally symmetric convex bodies and the spherical Radon transform, *J. Differential Geom.* **35** (1992), 675–688.
[191] Goodey, P.R. and Weil, W., Zonoids and generalizations, in *Handbook of Convex Geometry* (eds. Gruber, P.M. and Wills, J.M.), North-Holland, Amsterdam (1993), Vol. B, 1297–1326.
[192] Goodey, P.R. and Weil, W., Intersection bodies and ellipsoids (preprint, 1994).
[193] Goodman, J.E. and Pollack, R., Foundations of a theory of convexity on affine Grassmann manifolds (manuscript, 1993).
[194] Goodner, D.A., Projections in normed linear space, *Trans. Amer. Math. Soc.* **69** (1950), 89–108.
[195] Gordon, Y., On the projection and Macphail constants of ℓ^p spaces, *Israel J. Math.* **6** (1968), 295–302.
[196] Gordon, Y., Meyer, M. and Reisner, S., Zonoids with minimal volume product, a new proof, *Proc. Amer. Math. Soc.* **104** (1988), 273–276.
[197] Goryachev, A.P., Undistorted linear normed spaces, *Vestnik Moskov. Univ. Ser. I Mat. Mekh.* **25** (1970), 21–28 (Russian, English summary).
[198] Graham, R.L., Witsenhausen, H.S. and Zassenhaus, H.J., On tightest packings in the Minkowski plane, *Pacific J. Math.* **41** (1972), 699–715.
[199] Green, J.W., Length and area of a convex curve under affine transformations, *Pacific J. Math.* **3** (1953), 393–402.
[200] Greub, W.H., *Multilinear Algebra*, Springer, Berlin (1967).
[201] Grinberg, E.L., Isoperimetric inequalities for k-dimensional cross-sections and projections of convex bodies, *Math. Ann.* **291** (1991), 75–87.
[202] Grinberg, E.L. and Rivin, I., Infinitesimal aspects of the Busemann–Petty problem, *Bull. London Math. Soc.* **22** (1990), 478–484.

[203] Grinberg, E.L. and Zhang, G., Convolutions, transforms and convex bodies, *Duke Math. J.* (in press).
[204] Groemer, H., On multiple space subdivisions by zonotopes, *Monatsh. Math.* **86** (1978), 185–188.
[205] Groemer, H., Stability theorems for ellipsoids and spheres, *J. London. Math. Soc.* (2) **49** (1994), 357–370.
[206] Groemer, H., *Geometric Applications of Fourier Series and Spherical Harmonics, Encyclopedia of Mathematics and Its Applications*, Cambridge Univ. Press, New York (in press).
[207] Gromov, M., Isoperimetric inequalities in Riemannian manifolds, Appendix I, in *Asymptotic Theory of Finite Dimensional Normed Spaces* by Milman, V.D. and Schechtman, G., *Lecture Notes in Mathematics* **1200**, Springer, Berlin (1986).
[208] Gruber, P.M., Über kennzeichnende Eigenschaften von euklidischen Räumen und Ellipsoiden. I, *J. reine angew Math.* **265** (1974), 61–83.
[209] Gruber, P.M., Über kennzeichnende Eigenschaften von euklidischen Räumen und Ellipsoiden. II, *J. reine angew Math.* **270** (1974), 123–142.
[210] Gruber, P.M., Über kennzeichnende Eigenschaften von euklidischen Räumen und Ellipsoiden. III, *Monatsh. Math.* **78** (1974), 311–340.
[211] Gruber, P.M., Approximation of convex bodies, in *Convexity and Its Applications* (eds. Gruber, P.M. and Wills, J.M.), Birkhäuser, Basel (1983), 131–162.
[212] Gruber, P.M., Minimal ellipsoids and their duals, *Rend. Circ. Mat. Palermo* (2) **37** (1988), 35–64.
[213] Gruber, P.M., History of convexity, in *Handbook of Convex Geometry* (eds. Gruber, P.M. and Wills, J.M.), North-Holland, Amsterdam (1993), Vol. A, 3–15.
[214] Gruber, P.M., The space of convex bodies, in *Handbook of Convex Geometry* (eds. Gruber, P.M. and Wills, J.M.), North-Holland, Amsterdam (1993), Vol. A, 301–318.
[215] Gruber, P.M., Aspects of approximation of convex bodies, in *Handbook of Convex Geometry* (eds. Gruber, P.M. and Wills, J.M.), North-Holland, Amsterdam (1993), Vol. A, 319–346.
[216] Gruber, P.M., Geometry of numbers, in *Handbook of Convex Geometry* (eds. Gruber, P.M. and Wills, J.M.), North-Holland, Amsterdam (1993), Vol. B, 739–764.
[217] Gruber, P.M., Characterizations of ellipsoids in convexity and their relations to other areas of mathematics, in *Proc. Conf. on Convex and Discrete Geometry, Bydgoszcz, 1994*.
[218] Gruber, P.M. and Höbinger, J., Kennzeichnungen von Ellipsoiden mit Anwendungen, in Jahrb. Überblicke Math., Bibliographisches Institut, Mannheim (1976), 9–29.
[219] Gruber, P.M. and Lekkerkerker, C.G., *Geometry of Numbers* (2nd ed.), North-Holland, Amsterdam (1987).
[220] Gruber, P.M. and Lettl, G., Isometries of the space of convex bodies in Euclidean space, *Bull. London Math. Soc.* **12** (1980), 455–462.
[221] Grünbaum, B., Borsuk's partition conjecture in Minkowski planes, *Bull. Res. Council Israel* **F7** (1957/8), 25–30.
[222] Grünbaum, B., On some covering and intersection properties in Minkowski spaces, *Pacific J. Math.* **9** (1959), 487–494.
[223] Grünbaum, B., Projection constants, *Trans. Amer. Math. Soc.* **95** (1960), 451–465.
[224] Grünbaum, B., Self-circumference of convex sets, *Colloq. Math.* **13** (1964), 55–57.
[225] Grünbaum, B., The perimeter of Minkowski unit discs, *Colloq. Math.* **15** (1966), 135–139.
[226] Guggenheimer, H., Pseudo-Minkowski differential geometry, *Ann. Mat. Pura Appl.* **70** (1965), 305–370.
[227] Guggenheimer, H., *The Analytic Geometry of the Minkowski Plane*, Lecture Notes, Univ. of Minnesota, Minneapolis (1967).
[228] Guggenheimer, H., Hill equations with coexisting periodic solutions, *J. Differential Equations* **5** (1969), 159–166.
[229] Guggenheimer, H., A formula of Furstenberg–Tzkoni type, *Israel J. Math.* **14** (1973), 281–282.
[230] Guggenheimer, H., On plane Minkowski geometry, *Geom. Dedicata* **12** (1982), 371–381.
[231] Guggenheimer, H., Elementary geometry of the unsymmetric Minkowski plane, *Rev. Un. Mat. Argentina* **29** (1984), 270–281.
[232] Gurariĭ, N.I. and Sozonov, Y.I., On normed spaces which have no bias of the unit sphere (Russian), *Mat. Zametki* **7** (1970), 307–310. English transl. *Math. Notes* **7** (1970), 187–189.

[233] Gustin, W., An isoperimetric minimax, *Pacific J. Math.* **3** (1953), 403–405.
[234] Haar, A., Der Massbegriff in der Theorie der kontinuierlichen Gruppen, *Ann. of Math.* (2) **34** (1933), 147–169.
[235] Hadwiger, H., Über ausgezeichnete Vektorsterne und reguläre Polytope, *Comment. Math. Helv.* **13** (1940), 90–107.
[236] Hadwiger, H., Mittelpunktspolyeder und translative Zerlegungsgleichheit, *Math. Nachr.* **8** (1952), 53-58.
[237] Hahn, H., Über lineare Gleichungssysteme in linearen Räumen, *J. reine angew. Math.* **157** (1927), 214–219.
[238] Halmos, P.R., The range of a vector measure, *Bull. Amer. Math. Soc.* **54** (1948), 416–421.
[239] Halmos, P.R., *Measure Theory*, Van Nostrand, New York (1950).
[240] Halmos, P.R., *Finite Dimensional Vector Spaces*, Van Nostrand, New York (1958).
[241] Hammer, P.C., Maximal convex sets, *Duke Math. J.* **22** (1955), 103–106.
[242] Hammer, P.C., Convex curves of constant Minkowski breadth, *Proc. Symp. Pure Math.* **7**, Amer. Math. Soc., Providence, RI (1963), 291–304.
[243] Hammer, P.C., Unsolved problems, *Proc. Symp. Pure Math.* **7**, Amer. Math. Soc., Providence, RI (1963), 498–499.
[244] Hardy, G.H., Littlewood, J.E. and Pólya, G., *Inequalities* (2nd ed.), Cambridge Univ. Press, New York (1952).
[245] Harrell, R.E. and Karlovitz, L.A., Girths and flat Banach spaces, *Bull. Amer. Math. Soc.* **76** (1970), 1288–1291.
[246] Harrell, R.E. and Karlovitz, L.A., Non-reflexivity and the girths of spheres, in *Inequalities, Vol. 3, Proc. 3rd. Sympos. Inequalities U.C.L.A. 1969*, Academic Press, New York (1972), 121–127.
[247] Harrell, R.E. and Karlovitz, L.A., The geometry of flat Banach spaces, *Trans. Amer. Math. Soc.* **192** (1974), 209–218.
[248] Hausdorff, F., *Grundzüge der Mengenlehre*, Veit, Leipzig (1914).
[249] Hausdorff, F., Dimension und äusseres Mass, *Math. Ann.* **79** (1919), 157–179.
[250] Hausdorff, F., *Mengenlehre,* de Gruyter, Berlin (1927).
[251] Heil, E., Eigenvalue estimates for Hill's equation, *J. Differential Equations* **18** (1976), 179–187.
[252] Heil, E. and Martini, H., Special convex bodies, in *Handbook of Convex Geometry* (eds. Gruber, P.M. and Wills, J.M.), North-Holland, Amsterdam (1993), Vol. A, 347–385.
[253] Helgason, S., *Groups and Geometric Analysis,* Academic Press, Orlando, FL (1984).
[254] Helly, E., Über Systeme linearer Gleichungen mit unendlich vielen Unbekannten, *Monatsh. Math. Phys.* **31** (1921), 60–91.
[255] Helly, E., Über Mengen konvexer Körper mit gemeinschaftlichen Punkten, *Jber. Deutsch. Math. Verein.* **83** (1923), 175–176.
[256] Hensley, D., Slicing the cube in $\mathbb{R}^n$ and probability, *Proc. Amer. Math. Soc.* **73** (1979), 95–100.
[257] Hensley, D., Slicing convex bodies – bounds for slice area in terms of the body's covariance, *Proc. Amer. Math. Soc.* **79** (1980), 619–625.
[258] Herring, C., Some theorems on the free energy of crystal surfaces, *Phys. Rev.* **82** (1951), 87–93.
[259] Hewitt, E. and Ross, K.A., *Abstract Harmonic Analysis*, Springer, Berlin (1963).
[260] Hölder, E., Über einen Mittelwertsatz, *Nachr. Akad. Wiss. Göttingen Math. Phys.* (1889), 38–47.
[261] Holmes, R.B., *Geometric Functional Analysis*, Springer, Berlin (1975).
[262] Holmes, R.D. and Thompson, A.C., N-dimensional area and content in Minkowski spaces, *Pacific J. Math.* **85** (1979), 77–110.
[263] Holsztyński, W., Linearization of isometric embeddings of Banach spaces, metric envelopes, *Bull. Acad. Polon. Sci. Ser. Math. Astron. Phys.* **16** (1968), 189–193.
[264] Husain, T., *Introduction to Topological Groups*, Saunders, Philadelphia (1966).
[265] James, R.C., Orthogonality in normed linear spaces, *Duke Math. J.* **12** (1945), 291–302.
[266] James, R.C., Orthogonality and linear functionals in normed linear spaces, *Trans. Amer. Math. Soc.* **61** (1947), 265–292.
[267] James, R.C., A non-reflexive Banach space isometric to its second conjugate, *Proc. Nat. Acad. Sci. U.S.A.* **37** (1951), 174–177.
[268] James, R.C., Super-reflexive Banach spaces, *Canad. J. Math.* **24** (1972), 896–904.

[269] James, R.C. and Schäffer, J.J., Super-reflexivity and the girths of spheres, *Israel J. Math.* **11** (1972), 398–401.
[270] John, F., Moments of inertia of convex regions, *Duke Math. J.* **2** (1936), 447–452.
[271] John, F., An inequality for convex bodies, *Univ. Kentucky Res. Club Bull.* **8** (1942), 8–11 (*Math. Rev.* **4** (1943), 252).
[272] John, F., Extremum problems with inequalities as subsidiary conditions, in *Studies and Essays Presented to R. Courant on His 60th Birthday* (eds. Friedrichs, K.O., Neuegebauer, O. and Stoker, J.J.), Interscience, New York (1948), 187–204.
[273] Johnson, K. and Thompson, A.C., On the isoperimetric mapping in Minkowski space, *Intuitive Geometry, Šiofok Colloq. Math. Soc. János Bolyai* **48** (1985), 273–287.
[274] Johnson, R.A., A circle theorem, *Amer. Math. Monthly* **23** (1916), 161–162.
[275] Joichi, J.T., More characterizations of inner product spaces, *Proc. Amer. Math. Soc.* **19** (1968), 1185–1186.
[276] Jordan, P. and von Neumann, J., On inner products in linear metric spaces, *Ann. of Math.* (2) **36** (1935), 719–723.
[277] Kadeč, M.I., Geometry of normed spaces, *Mat. Anal. Itogi Nauk i Tekh. Moscow* **13** (1975), 99–127.
[278] Kadeč, M.I. and Snobar, M.G., Certain functionals on the Minkowski compactum, *Mat. Zametki* **10** (1971), 453–457 (Russian). English transl. *Math. Notes* **10** (1971), 694–696.
[279] Kakutani, S., Ein Beweis des Satzes von M. Edelheit über konvexe Mengen, *Proc. Imp. Acad. Tokyo* **13** (1937), 93–94.
[280] Kakutani, S., On the uniqueness of Haar's measure, *Proc. Imp. Acad. Tokyo* **14** (1938), 27–31.
[281] Kakutani, S., Some characterizations of Euclidean space, *Japan J. Math.* **16** (1939), 93–97.
[282] Karlovitz, L.A., On the duals of flat Banach spaces, *Math. Ann.* **202** (1973), 245–250.
[283] Kelley, J.L., Banach spaces with the extension property, *Trans. Amer. Math. Soc.* **53** (1952), 323–326.
[284] Kelley, J.L., *General Topology*, Van Nostrand, New York (1957).
[285] Kelly, J.B., Hypermetric spaces, in *The Geometry of Metric Linear Spaces* (ed. Kelly, L.M.), *Lecture Notes in Mathematics* **490**, Springer, Berlin (1974), 17–31.
[286] Kelly, L.M. and Moser, W.O.J., On the number of ordinary lines determined by n points, *Canad. J. Math.* **10** (1958), 210–219.
[287] Kelly, P.J., On Minkowski bodies of constant width, *Bull. Amer. Math. Soc.* **55** (1949), 1147–1150.
[288] Kelly, P.J., A property of Minkowski circles, *Amer. Math. Monthly* **57** (1950), 677–678.
[289] Klain, D.A., *Star Measures and Dual Mixed Volumes*, Ph.D. Thesis, MIT, Cambridge, MA (1994).
[290] Klee, V., Maximal separation theorems for convex sets, *Trans. Amer. Math. Soc.* **134** (1968), 133–147.
[291] Klee, V., Separation and support properties of convex sets – a survey, in *Control Theory and the Calculus of Variations* (ed. Balakrishnan, A.), Academic Press, New York (1969), 235–304.
[292] Klee, V., Can a plane convex body have two equichordal points? *Amer. Math. Monthly* **76** (1969), 54–55.
[293] Klein, E. and Thompson, A.C., *Theory of Correspondences, Including Applications to Mathematical Economics* (Canad. Math. Soc. Monograph **2**), Wiley, New York (1984).
[294] Kneser, H. and Süss, W., Die Volumina in linearen Scharen konvexer Körper, *Mat. Tidsskr. B.* (1932), 19–25.
[295] Knothe, H., Contributions to the theory of convex bodies, *Michigan Math. J.* **4** (1957), 39–52.
[296] Knothe, H., Inversion of two theorems of Archimedes, *Michigan Math. J.* **4** (1957), 53–56.
[297] Kolmogoroff, A., Beiträge zur Masstheorie, *Math. Ann.* **107** (1932), 351–356.
[298] Kogan, H., Review of Kadeč [277], *Math. Reviews* **58** (1979), no. 30064.
[299] König, H. and Lewis, D.R., A strict inequality for projection constants, *J. Funct. Anal.* **73** (1987), 328–332.
[300] König, H. and Tomczak-Jaegermann, N., Bounds for projection constants and 1-summing norms, *Trans. Amer. Math. Soc.* **320** (1990), 799–823.

[301] Köthe, G., *Topological Vector Spaces I*, Springer, New York (1969).
[302] Kottman, C.A., Packing and reflexivity in Banach spaces, *Trans. Amer. Math. Soc.* **150** (1970), 565–576.
[303] Kottman, C.A., A characterization of the Euclidean topology among the affine topologies, *Israel J. Math.* **10** (1971), 212–217.
[304] Krause, E.F., *Taxicab Geometry*, Addison-Wesley, Reading, MA (1975).
[305] Krein, M. and Milman, D., On extreme points of regularly convex sets, *Studia Math.* **9** (1940), 133–138.
[306] Kubota, T. and Hemmi, D., Some problems of minima concerning the oval, *J. Math. Soc. Japan* **5** (1953), 372–389.
[307] Kuratowski, K., Les fonctions semi-continues dans l'espace des ensembles fermés, *Fund. Math.* **18** (1932), 148–160.
[308] Kuratowski, K., *Topology* (2 vols.), Academic Press, New York (1966 and 1968).
[309] Larman, D.G. and Rogers, C.A., The existence of a centrally symmetric convex body with central sections that are unexpectedly small, *Mathematika* **22** (1975), 164–175.
[310] Lassak, M., Solution of Hadwiger's covering problem for centrally symmetric convex bodies in E^3, *J. London Math. Soc.* **32** (1985), 501–511.
[311] Lassak, M., Approximation of plane convex bodies by centrally symmetric bodies, *J. London Math. Soc.* **40** (1989), 369–377.
[312] Lassak, M., On five points in a plane convex body pairwise in at least unit relative distances, *Intuitive Geometry, Colloq. Math. Soc. János Bolyai* **63** (1991), 3 pp.
[313] Laugwitz, D., Konvexe Mittelpunktsbereiche und normierte Räume, *Math. Z.* **61** (1954), 235–244.
[314] Laugwitz, D., *Differentialgeometrie in Vektorräumen, unter besonderer Berücksichtigung der unendlichdimensionalen Räume*, Friedr. Vieweg, Braunschweig (1965), 89 pp.
[315] Lawlor, G. and Morgan, F., Paired calibrations applied to soap films, immiscible fluids, and surfaces or networks minimizing other norms, *Pacific J. Math.* **166** (1994), 55–83.
[316] Lebesgue, H., Sur le problème des isopérimètres et sur les domaines de largeur constante, *Bull. Soc. Math. France C.R.* (1914), 72–76.
[317] Leichtweiss, K., Zwei Extremalprobleme der Minkowski-Geometrie, *Math. Z.* **62** (1955), 37–49.
[318] Leichtweiss, K., Über die affine Exzentrizität konvexer Körper, *Arch. Math.* **10** (1959), 187–199.
[319] Leichtweiss, K., Selbstadjungierte Banach-Räume, *Math. Z.* **71** (1959), 335–360.
[320] Lewis, D.R., Finite dimensional subspaces of L_p, *Studia Math.* **63** (1978), 207–212.
[321] Lewis, D.R., Ellipsoids defined by Banach ideal norms, *Mathematika* **26** (1979), 18–29.
[322] Lewy, H., On the existence of a closed surface realizing a given Riemannian metric, *Proc. Nat. Acad. Sci. USA* **24** (1938), 104–106.
[323] Li, C-K., Some aspects of the theory of norms, *Linear Algebra Appl.* **212–213** (1994), 71–100.
[324] Liapunov, A.A., Sur les fonctions-vecteurs complètement additives, *Izvestia Akad. Nauk SSSR* **4** (1940), 465–478 (Russian, French summary).
[325] Liebmann, H., Der Curie-Wulff'sche Satz über Combinationsformen von Krystallen, *Z. Krystallogr. und Mineral.* **53** (1914), 171–177.
[326] Lindenstrauss, J., A short proof of Liapounoff's convexity theorem, *J. Math. Mechanics* **15** (1966), 971–972.
[327] Lindenstrauss, J. and Milman, V.D., The local theory of normed spaces and its application to convexity, in *Handbook of Convex Geometry* (eds. Gruber, P.M. and Wills, J.M.), North-Holland, Amsterdam (1993), Vol. B, 1149–1220.
[328] Lindenstrauss, J. and Tzafriri, L., On the complemented subspaces problem, *Israel J. Math.* **9** (1971), 263–269.
[329] Lippmann, H., Zur Winkeltheorie in zweidimensionalen Minkowski- und Finsler-Räumen, *Nederl. Akad. Wetensch. Proc. Ser. A* **60** = *Indag. Math.* **19** (1957), 162–170.
[330] Lippmann, H., Metrische Eigenschaften verschiedener Winkelmasse im Minkowski- und Finslerraum I, II, *Nederl. Akad. Wetensch. Proc. Ser. A* **61** = *Indag. Math.* **20** (1958), 223–238.

[331] Loomis, L.H., Abstract congruence and the uniqueness of Haar measure, *Ann. Math.* **46** (1945), 348–355.
[332] Lusternik, L. Die Brunn-Minkowskische Ungleichung für beliebige messbare Mengen, *C.R. (Doklady) Acad. Sci. URSS* N.S. **3** (1935), 55–58.
[333] Lutwak, E., Dual mixed volumes, *Pacific J. Math.* **58** (1975), 531–538.
[334] Lutwak, E., Mixed projection inequalities, *Trans. Amer. Math. Soc.* **287** (1985), 91–106.
[335] Lutwak, E., Volumes of mixed bodies, *Trans. Amer. Math. Soc.* **294** (1986), 487–500.
[336] Lutwak, E., On some affine isoperimetric inequalities, *J. Differential Geom.* **23** (1986), 1–13.
[337] Lutwak, E., Intersection bodies and dual mixed volumes, *Adv. Math.* **71** (1988), 232–261.
[338] Lutwak, E., On a conjectured inequality of Petty, *Contemp. Math.* **113** (1990), 171–182.
[339] Lutwak, E., On quermassintegrals of mixed projection bodies, *Geom. Dedicata* **33** (1990), 51–58.
[340] Lutwak, E., Intersection and generalized intersection bodies (preprint, 1991).
[341] Lutwak, E., On some ellipsoid formulas of Busemann, Furstenberg and Tzkoni, Guggenheimer, and Petty, *J. Math. Anal. Appl.* **159** (1991), 18–26.
[342] Lutwak, E., Selected affine isoperimetric inequalities, in *Handbook of Convex Geometry* (eds. Gruber, P.M. and Wills, J.M.), North-Holland, Amsterdam (1993), Vol. A, 151–176.
[343] Macbeath, A.M., A compactness theorem for affine equivalence classes of convex regions, *Canad. J. Math.* **3** (1951), 54–61.
[344] MacKenzie, D., Triquetras and porisms, *College Math. J.* **23** (1992), 118–131.
[345] Macphail, M.S., Absolute and unconditional convergence, *Bull. Amer. Math. Soc.* **53** (1947), 121–123.
[346] Mahler, K., Ein Übertragungsprinzip für konvexe Körper, *Časopis Pěst. Mat. Fyz.* **68** (1939), 93–102.
[347] Makai, E. and Martini, H., A new characterization of convex plates of constant width, *Geom. Dedicata* **34** (1990), 199–209.
[348] Makai, E. and Martini, H., On the number of antipodal or strictly antipodal pairs of points in finite subsets of $\mathbb{R}^d$, in *Applied Geometry and Discrete Mathematics, the "Victor Klee Festschrift"*, DIMACS Ser. Discr. Math. Theor. Comp. Sci. (eds. Gritzmann, P. and Sturmfels, B.), **4** (1991), 457–470.
[349] Makai, E. and Martini, H., On the number of antipodal or strictly antipodal pairs of points in finite subsets of $\mathbb{R}^d$ II, *Period. Math. Hungarica* **27** (1993), 185–198.
[350] Makai, E. and Martini, H., The cross-section body, plane sections of convex bodies, approximation of convex bodies and a new proof of the Petty projection inequality (manuscript, 1993).
[351] Mankiewicz, P., On extensions of isometries in normed linear spaces, *Bull. Acad. Polon. Sci. Ser. Math. Astron. Phys.* **20** (1972), 367–371.
[352] Mann, H., Untersuchungen über Wabenzellen bei allgemeiner Minkowskischer Metrik, *Monatsh. Math. Phys.* **42** (1935), 417–424.
[353] Martini, H., Zur Bestimmung konvexer Polytope durch die Inhalte ihrer Projektionen, *Beiträge Algebra Geom.* **18** (1984), 75–85.
[354] Martini, H., Some results and problems around zonotopes, in *Intuitive Geometry Šiofok, Colloq. Math. Soc. János Bolyai* **48** (1985), 383–418.
[355] Martini, H., Some characterizing properties of the simplex, *Geom. Dedicata* **29** (1989), 1–6.
[356] Martini, H., On inner quermasses of convex bodies, *Arch. Math.* **52** (1989), 402–406.
[357] Martini, H., A new view on some characterizations of simplices, *Arch. Math.* **55** (1990), 389–393.
[358] Martini, H., Convex polytopes whose projection bodies and difference sets are polars, *Discrete Comput. Geom.* **6** (1991), 83–91.
[359] Martini, H., Cross-sectional measures, in *Intuitive Geometry Szeged, Colloq. Math. Soc. János Bolyai* **63** (1991), 269–310.
[360] Mayer, A., Grösste Polygone mit gegebenen Seitenvektoren, *Comment. Math. Helv.* **10** (1938), 288–301.
[361] Mazur, S., Über konvexe Mengen in linearen normierten Räumen, *Studia Math.* **4** (1933), 70–84.
[362] Mazur, S. and Ulam, S., Sur les transformations isométriques d'espaces vectoriels normés, *C. R. Acad. Sci. Paris* **194** (1932), 946–948.

[363] McGuigan, R.A., On the connectedness of isomorphic classes, *Manuscripta Math.* **3** (1970), 1–5.
[364] McKinney, J.R., On maximal simplices inscribed to a central convex set, *Mathematika* **21** (1974), 38–44.
[365] McMullen, P., On zonotopes, *Trans. Amer. Math. Soc.* **159** (1971), 91–109.
[366] McMullen, P., Space tiling zonotopes, *Mathematika* **22** (1975), 202–211.
[367] McMullen, P., Valuations and Euler-type relations on certain classes of convex polytopes, *Proc. London Math. Soc.* **35** (1977), 113–135.
[368] McMullen, P., Convex bodies which tile space by translation, *Mathematika* **27** (1980), 113–121.
[369] McMullen, P., Convex bodies which tile space by translation: Acknowledgement of priority, *Mathematika* **28** (1981), 191.
[370] McMullen, P., Valuations and dissections, in *Handbook of Convex Geometry* (eds. Gruber, P.M. and Wills, J.M.), North-Holland, Amsterdam (1993), Vol. B, 933–988.
[371] McMullen, P. and Schneider, R. Valuations on convex bodies, in *Convexity and Its Applications* (eds. Gruber, P.M. and Wills, J.M.), Birkhäuser, Basel (1983), 170–247.
[372] McMullen, P. and Shephard, G.C., *Convex Polyhedra and the Upper Bound Conjecture*, London Math. Soc. Lecture Notes **3**, Cambridge (1971).
[373] McShane, E.J., *Integration*, Princeton Univ. Press, Princeton, NJ (1944).
[374] Meissner, E., Über Punktmengen konstanter Breite, *Vierteljahresschr. Naturforsch. Ges. Zürich* **56** (1911), 42–50.
[375] Meyer, M., Une caractérisation volumique de certains espaces normés de dimension finie, *Israel J. Math.* **55** (1986), 317–326.
[376] Meyer, M. and Pajor, A., On Santaló's inequality, in *Geometric Aspects of Functional Analysis* (eds. Lindenstrauss, J. and Milman, V.D.), *Lecture Notes in Mathematics* **1376**, Springer, Berlin (1989), 261–263.
[377] Meyer, M. and Pajor, A., On the Blaschke–Santaló inequality, *Arch. Math.* **55** (1990), 82–93.
[378] Meyer, M., Reisner, S. and Schmuckenschläger, M., The volume of the intersection of a convex body with its translates, *Mathematika* **40** (1993), 278–289.
[379] Miles, R.E., A simple derivation of a formula of Furstenberg and Tzkoni, *Israel J. Math.* **14** (1973), 278–280.
[380] Milman, V.D., Geometric theory of Banach spaces II: Geometry of the unit sphere, *Uspekhi Mat. Nauk* **26** (1971), 79–163 (Russian). English transl. *Russian Math. Surveys* **26** (1971), 73–150.
[381] Milman, V.D., A new proof of a theorem of A. Dvoretzky on sections of convex bodies, *Funktsional. Anal. Priložen.* **5** (1971), 28–37 (Russian).
[382] Milman, V.D. and Pajor, A., Isotropic position and inertia ellipsoids and zonoids of the unit ball of a normed n-dimensional space, in *Geometric Aspects of Functional Analysis* (eds. Lindenstrauss, J. and Milman, V.D.), *Lecture Notes in Mathematics* **1376**, Springer, Berlin (1989), 64–104.
[383] Milman, V.D. and Schechtman, G., *Asymptotic Theory of Finite Dimensional Normed Spaces*, Lecture Notes in Mathematics **1200**, Springer, Berlin (1986).
[384] Milman, V.D. and Wolfson, H., Minkowski spaces with extreme distance from the Euclidean space, *Israel J. Math.* **29** (1978), 113–131.
[385] Minkowski, H., Allgemeine Lehrsätze über die konvexen Polyeder, *Nachr. Ges. Wiss. Göttingen, Math.-Phys. Kl.* (1897), 198–219; Gesammelte Abhandlungen, Vol. II, Teubner, Leipzig (1911), 103–121.
[386] Minkowski, H., Über die Begriffe Länge, Oberfläche und Volumen, *Jahresber. Deutsche Math. Ver.* **9** (1901), 115–121. Gesammelte Abhandlungen, Vol. II, Teubner, Leipzig (1911), 122–127.
[387] Minkowski, H., Sur les surfaces convexes fermées, *C. R. Acad. Sci. Paris* **132** (1901), 21–24. Gesammelte Abhandlungen, Vol. II, Teubner, Leipzig (1911), 128–130.
[388] Minkowski, H., Volumen und Oberfläche, *Math. Ann.* **57** (1903), 447–495. Gesammelte Abhandlungen, Vol. II, Teubner, Leipzig (1911), 230–276.
[389] Minkowski, H., *Geometrie der Zahlen*, Teubner, Leipzig (1910).
[390] Minkowski, H., *Theorie der konvexen Körper, insbesondere Begründung ihres Oberflächenbegriffs*, Gesammelte Abhandlungen Vol. II, Teubner, Leipzig (1911), 131–229.

[391] Monna, A.F., *Functional Analysis in Historical Perspective*, Wiley, New York (1973).
[392] Morgan, F., *Geometric Measure Theory, a Beginner's Guide*, Academic Press, San Diego (1988).
[393] Morgan, F., Minimal surfaces, crystals, and norms on $\mathbb{R}^n$, in *Proc. 7th. Annual Symposium on Computational Geometry, North Conway, N.H., 1991*, ACM Press, Baltimore (1991), 204–213.
[394] Morgan, F., Minimal surfaces, crystals, shortest networks and undergraduate research, *Math. Intelligencer* **14** (1992), 37–44.
[395] Mürner, P., Translative Parkettierungspolyeder und Zerlegungsgleichheit, *Elem. Math.* **30** (1975), 25–27.
[396] Murray, F.J., On complementary manifolds and projections in spaces L_p and ℓ_p, *Trans. Amer. Math. Soc.* **41** (1937), 138–152.
[397] Nachbin, L., A theorem of Hahn–Banach type for linear transformations, *Trans. Amer. Math. Soc.* **68** (1950), 28–46.
[398] Naumann, H., Über Vektorsterne und Parallelprojektionen regulärer Polytope, *Math. Z.* **67** (1957), 75–82.
[399] Nirenberg, L., The Weyl and Minkowski problems in differential geometry in the large, *Commun. Pure Appl. Math.* **6** (1953), 337–394.
[400] Nyikos, P. and Schäffer, J.J., Flat spaces of continuous functions, *Studia Math.* **42** (1972), 221–229.
[401] Ohira, K., On some characterizations of abstract Euclidean spaces by properties of orthogonality, *Kumamoto J. Sci.* **1** (1952), 23–26.
[402] Ohmann, D., Extremalprobleme für konvexe Bereiche der euklidischen Ebene, *Math. Z.* **55** (1952), 346–352.
[403] Owens, O.G., The integral geometry definition of arc length for two-dimensional Finsler spaces, *Trans. Amer. Math. Soc.* **73** (1952), 199–210.
[404] Peano, G., *Calcolo geometrico secondo l'Ausdehnungslehre di H. Grassmann preceduto dalle Operazioni della Logica Deduttiva*, Turin (1888).
[405] Peetre, J., Une caractérisation abstraite des opérateurs différentiels, *Math. Scand.* **7** (1959), 211–218; Rectification, *ibid.* **8** (1960), 116–120.
[406] Pelcyński, A., Structural theory of Banach spaces and its interplay with analysis and probability, *Proc. Internat. Congress Math. 1983*, Vol. 1, Polish Scientific Publishers, North-Holland, Amsterdam (1984), 237–269.
[407] Pelcyński, A. and Szarek, S.J., On parallelopipeds of minimal volume containing a symmetric body in $\mathbb{R}^n$, *Proc. Cambridge Philos. Soc.* **109** (1991) 125–148.
[408] Peri, C., Integral geometry in Minkowski plane, *Rend. Sem. Mat. Univ. Politec. Torino* **45** (1987), 107–117.
[409] Peri, C., Buffon's needle problem in the Minkowski plane, *Publ. Inst. Statist. Univ. Paris* **32** (1987), 91–108.
[410] Peri, C., On the minimal convex shell of a convex body, *Canad. Math. Bull.* **36** (1993), 466–472.
[411] Peri, C., Wills, J.M. and Zucco, A., On Blaschke's extension of Bonnesen's inequality, *Geom. Dedicata* **48** (1993), 349–357.
[412] Petty, C.M., *On Minkowski Geometries*, Ph.D. Thesis, Univ. of Southern California, Los Angeles (1952).
[413] Petty, C.M., Geometry of the Minkowski plane, *Riv. Mat. Univ. Parma* (4) **6** (1955), 269–292.
[414] Petty, C.M., Surface area of a convex body under affine transformations, *Proc. Amer. Math. Soc.* **12** (1961), 824–828.
[415] Petty, C.M., Centroid surfaces, *Pacific J. Math.* **11** (1961), 1535–1547.
[416] Petty, C.M., Projection bodies, in *Proc. Colloq. on Convexity Copenhagen 1965*, Copenhagen (1967), 234–241.
[417] Petty, C.M., Equilateral sets in Minkowski space, *Proc. Amer. Math Soc.* **29** (1971), 369–374.
[418] Petty, C.M., Isoperimetric problems, in *Proc. Conf. Convexity and Combinatorial Geometry, Univ. of Oklahoma 1971*, Norman (1972), 26–41.
[419] Petty, C.M., Geominimal surface area, *Geom. Dedicata* **3** (1974), 77–97.
[420] Petty, C.M., Ellipsoids, in *Convexity and Its Applications* (eds. Gruber, P.M. and Wills, J.M.), Birkhäuser, Basel (1983), 264–276.

[421] Petty, C.M., Affine isoperimetric problems, in *Discrete Geometry and Convexity* (eds. Goodman, J.E., Lutwak, E., Malkevitch, J. and Pollack, R.), *Ann. New York Acad. Sci.* **440**, New York (1985), 113–127.
[422] Petty, C.M., and Barry, J.E., A geometrical approach to the second-order linear differential equation, *Canad. J. Math.* **14** (1962), 349–358.
[423] Petty, C.M. and Crotty, J.M., Characterization of spherical neighbourhoods, *Canad. J. Math.* **22** (1970), 431–435.
[424] Petty, C.M. and McKinney, J.R., Convex bodies with circumscribing boxes of constant volume, *Portugal. Math.* **44** (1987), 447–455.
[425] Phadke, B.B., Equidistant loci and the Minkowskian geometries, *Canad. J. Math.* **24** (1972), 312–327.
[426] Phadke, B.B., A triangular world with hexagonal circles, *Geom. Dedicata* **3** (1975), 511–520.
[427] Phelps, R.R., *Lectures on Choquet's Theorem*, Mathematical Studies **7**, Van Nostrand, Princeton, NJ (1966).
[428] Phillips, R.S., A characterization of Euclidean spaces, *Bull. Amer. Math. Soc.* **46** (1940), 930–933.
[429] Pincherle, S., *Le operazioni distributive e le lore applicazioni all'analisi*, Bologna (1901).
[430] Pisier, G., *The Volume of Convex Bodies and Banach Space Geometry*, Cambridge Tracts in Mathematics **94**, Cambridge Univ. Press, Cambridge (1989).
[431] Pogorelov, A.V., On a regular solution of the n-dimensional Minkowski problem, *Dokl. Akad. Nauk SSSR* **199** (1971), 785–788. English transl. *Soviet Math. Doklady* **12** (1971), 1192–1196.
[432] Radon, J., Über eine besondere Art ebener konvexer Kurven, *Ber. Sächs. Akad. Wiss. Leipzig* **68** (1916), 131–134.
[433] Radon, J., Mengen konvexer Körper, die einen gemeinsamen Punkt enthalten, *Math. Ann.* **83** (1921), 113–115.
[434] Radon, J., Annäherung konvexer Körper durch analytisch begrenzte, *Monatsh. Math. Phys.* **43** (1936), 340–344.
[435] Reidemeister, K., Über die singulären Randpunkte eines konvexen Körpers, *Math. Ann.* **83** (1921), 116–118.
[436] Reimann, H., Eine Abschätzung für den Flächeninhalt von Eichbereichen Banach-Minkowskischer Ebenen, *Wiss. Z. Pädagog. Hochsch. Erfurt/Mühlhausen Math.-Natur. Reihe* **23** (1987), 124–132.
[437] Reisner, S., Random polytopes and the volume-product of symmetric convex bodies, *Math. Scand.* **57** (1985), 386–392.
[438] Reisner, S., Zonoids with minimal volume product, *Math. Z.* **192** (1986), 339–346.
[439] Reisner, S., Minimal volume-product in Banach spaces with 1-unconditional basis, *J. London Math. Soc.* (2) **36** (1987), 126–136.
[440] Riesz, F., Sur une espèce de géometrie analytique des systèmes de fonctions sommables, *C.R. Acad. Sci. Paris* **144** (1907), 1409–1411.
[441] Riesz, F., Untersuchungen über Systeme integrierbarer Funktionen, *Math. Ann.* **69** (1910), 449–492.
[442] Riesz, F., Über lineare Funktionalgleichungen, *Acta Math.* **41** (1918), 71–78.
[443] Riesz, F., Zur Theorie des Hilbertschen Raumes, *Acta Sci. Math. Szeged* **7** (1934), 34–38.
[444] Rogers, C.A., Sections and projections of convex bodies, *Portugal. Math.* **24** (1965), 99–103.
[445] Rogers, C.A., *Hausdorff Measures*, Cambridge Univ. Press, Cambridge (1970).
[446] Rogers, C.A. and Shephard, G.C., The difference body of a convex body, *Arch. Math.* **8** (1957), 220–233.
[447] Rogers, C.A. and Shephard, G.C., Convex bodies associated with a given body, *J. London Math. Soc.* **33** (1958), 270–281.
[448] Rolewicz, S., A generalization of the Mazur–Ulam theorem, *Studia Math.* **31** (1968), 501–505.
[449] Ruiz, L.E., Affine rhombic dodecahedric spheres, *Rev. Integr. Temas Mat.* **11** (1993), 39–54.
[450] Rund, H., Zur Begründung der Differentialgeometrie der Minkowskischen Räume, *Arch Math.* **3** (1952), 60–69.
[451] Rund, H., *The Differential Geometry of Finsler Spaces*, Springer, Berlin (1959).

[452] Rutovitz, D., Some parameters associated with finite-dimensional Banach spaces, *J. London Math. Soc.* **40** (1965), 241–255.
[453] Saint Raymond, J., Sur le volume des corps convexes symétriques, in *Séminaire Initiation à l'Analyse*, Univ. Pierre et Marie Curie, Paris, **11** (1987), 1–25.
[454] Sallee, G.T., The maximal set of constant width in a lattice, *Pacific J. Math.* **28** (1969), 669–674.
[455] Sallee, G.T., Sets of constant width, the spherical intersection property and circumscribed balls, *Bull. Austral. Math. Soc.* **33** (1986), 369–371.
[456] Santaló, L.A., An affine invariant for convex bodies in n-dimensional space, *Portugal. Math.* **8** (1949), 155–161.
[457] Santaló, L.A., *Integral Geometry and Geometric Probability, Encyclopedia of Math. and Its Appl.* **1**, Addison-Wesley, Reading, MA (1976).
[458] Schäffer, J.J., Another characterization of Hilbert spaces, *Studia Math.* **25** (1965), 271–276.
[459] Schäffer, J.J., Inner diameter, perimeter, and girth of spheres, *Math. Ann.* **173** (1967), 59–79.
[460] Schäffer, J.J., Symmetric curves, hexagons and the girths of spheres, *Israel J. Math.* **6** (1968), 202–205.
[461] Schäffer, J.J., Minimum girth of spheres, *Math. Ann.* **184** (1969–1970), 169–171.
[462] Schäffer, J.J., Spheres with maximum inner diameter, *Math. Ann.* **190** (1970–1971), 242–247.
[463] Schäffer, J.J., On the geometry of spheres in L-spaces, *Israel J. Math.* **10** (1971), 114–120.
[464] Schäffer, J.J., More distant than the antipodes, *Bull. Amer. Math. Soc.* **77** (1971), 606–609.
[465] Schäffer, J.J., The self-circumference of polar convex bodies, *Arch. Math.* **24** (1973), 87–90.
[466] Schäffer, J.J., *Geometry of Spheres in Normed Spaces, Lecture Notes in Pure and Applied Math.* **20**, Marcel Dekker, New York (1976).
[467] Schäffer, J.J. and Sundaresan, K., Reflexivity and the girth of spheres, *Math. Ann.* **184** (1969–1970), 163–168.
[468] Schäfke, R. and Volkmer, H., Asymptotic analysis of the equichordal problem, *J. Reine Angew. Math.* **425** (1992), 9–60.
[469] Schmuckenschlaeger, M., A simple proof of an approximation theorem of H. Minkowski, *Geom. Dedicata* **48** (1993), 325–335.
[470] Schmuckenschlaeger, M., Petty's projection inequality and Santaló's affine isoperimetric inequality (manuscript, 1993).
[471] Schneider, R., Zu einem Problem von Shephard über die Projektionen konvexer Körper, *Math. Z.* **101** (1967), 71–82.
[472] Schneider, R., Über eine Integralgleichung in der Theorie der konvexen Körper, *Math. Nachr.* **44** (1970), 55–75.
[473] Schneider, R., Zonoids whose polars are zonoids, *Proc. Amer. Math. Soc.* **50** (1975), 365–368.
[474] Schneider, R., Isometrien des Raumes der konvexen Körper, *Colloq. Math.* **33** (1975), 219–224.
[475] Schneider, R., Boundary structure and curvature of convex bodies, in *Contributions to Geometry, Proc. Geom. Symp. Siegen 1978* (eds. Tölke, J. and Wills, J.M.), Birkhäuser, Basel (1979), 13–59.
[476] Schneider, R., Random hyperplanes meeting a convex body, *Z. für Wahrscheinlichkeitsth. verw. Geb.* **61** (1982), 379–387.
[477] Schneider, R., Smooth approximation of convex bodies, *Rend. Circ. Mat. Palermo* (2) **33** (1984), 436–440.
[478] Schneider, R., Geometric inequalities for Poisson processes of convex bodies and cylinders, *Results Math.* **11** (1987), 165–185.
[479] Schneider, R., *Convex Bodies: The Brunn–Minkowski Theory, Encyclopedia of Math. and Its Appl.* **44**, Cambridge Univ. Press, New York (1993).
[480] Schneider, R. and Weil, W., Über die Bestimmung eines konvexen Körpers durch die Inhalte seiner Projektionen, *Math. Z.* **116** (1970), 338–348.
[481] Schneider, R. and Weil, W., Zonoids and related topics, in *Convexity and Its Applications* (eds. Gruber, P.M. and Wills, J.M.), Birkhäuser, Basel (1983), 296–317.
[482] Schneider, R. and Wieacker, J.A., Integral geometry, in *Handbook of Convex Geometry* (eds. Gruber, P.M. and Wills, J.M.), North-Holland, Amsterdam (1993), Vol. B, 1349–1390.
[483] Schneider, R. and Wieacker, J.A., Integral geometry in Minkowski spaces, *Adv. Math.* (in press).
[484] Schoenberg, I.J., A remark on M.M. Day's characterization of inner product spaces and a conjecture of L.M. Blumenthal, *Proc. Amer. Math. Soc.* **3** (1952), 961–964.

[485] Shephard, G.C., Inequalities between mixed volumes of convex sets, *Mathematika* **7** (1962), 125–138.
[486] Shephard, G.C., Shadow systems of convex bodies, *Israel J. Math.* **2** (1964), 229–236. See also [79] and [81].
[487] Shephard, G.C., Space-filling zonotopes, *Mathematika* **21** (1974), 261–269.
[488] Shephard, G.C. and Webster, R.J., Metrics for sets of convex bodies, *Mathematika* **12** (1965), 73–88.
[489] Sholander, M., On certain minimum problems in the theory of convex curves, *Trans. Amer. Math. Soc.* **73** (1952), 139–173.
[490] Sobczyk, A., Projection of the space m on its subspace c_0, *Bull. Amer. Math. Soc.* **47** (1941), 938–947.
[491] Sobczyk, A., Projections in Minkowski and Banach spaces, *Duke Math. J.* **8** (1941), 78–106.
[492] Sobczyk, A., Minkowski planes, *Math. Ann.* **173** (1967), 181–190.
[493] Soltan, V.P., A theorem on full sets, *Dokl. Akad. Nauk SSSR* **234** (1977), 320–322.
[494] Spivak, M., *Calculus on Manifolds,* Benjamin, New York (1965).
[495] Stone, M.H., *Linear Transformations in Hilbert Space*, Amer. Math. Soc. Colloquium Publ. **15**, New York (1932).
[496] Stone, M.H., *Convexity*, Lecture Notes, Univ. of Chicago (1946).
[497] Stromquist, W., The maximum distance between two-dimensional spaces, *Math. Scand.* **48** (1981), 205–225.
[498] Szankowski, A., On Dvoretzky's theorem on almost spherical sections of convex bodies, *Israel J. Math.* **17** (1974), 325–338.
[499] Szarek, S.J., Spaces with large distance to ℓ_∞^n and random matrices, *Amer. J. Math.* **112** (1990), 899–942.
[500] Szarek, S.J. and Talagrand, M., An "isomorphic" version of the Sauer–Shelah lemma and the Banach–Mazur distance to the cube, in *Geometric Aspects of Functional Analysis* (eds. Lindenstrauss, J. and Milman, V.D.), *Lecture Notes in Mathematics* **1376**, Springer, Berlin (1989), 105–112.
[501] Tamássy, L., Ein Problem der zweidimensionalen Minkowskischen Geometrie, *Ann. Pol. Math.* **9** (1960–1961), 39–48.
[502] Taylor, A.E., A geometric theorem and its application to biorthogonal systems, *Bull. Amer. Math. Soc.* **53** (1947), 614–616.
[503] Taylor, J.E., Unique structure of solutions to a class of nonelliptic variational problems, *Proc. Sympos. Pure Math.* **27** (1975), 419–427.
[504] Taylor, J.E., Crystalline variational problems, *Bull. Amer. Math. Soc.* **84** (1978), 568–588.
[505] Taylor, J.E., Zonohedra and generalized zonohedra, *Amer. Math. Monthly* **99** (1992), 108–111.
[506] Thompson, A.A. and Thompson, A.C., The divergence theorem and the Laplacian in Minkowski space, *Geom. Dedicata* (in press).
[507] Thompson, A.C., On certain contraction mappings in a partially ordered vector space, *Proc. Amer. Math. Soc.* **14** (1963), 438–443.
[508] Thompson, A.C., An equiperimetric property of Minkowski circles, *Bull. London Math. Soc.* **7** (1975), 271–272.
[509] Thompson, A.C., Applications of various inequalities to Minkowski geometry, *Geom. Dedicata* **46** (1993), 215–231.
[510] Tingley, D., Isometries of the unit sphere, *Geom. Dedicata* **22** (1987), 371–378.
[511] Tomczak-Jaegermann, N., The Banach–Mazur distance between the trace classes c_p^n, *Proc. Amer. Math. Soc.* **72** (1978), 305–308.
[512] Tomczak-Jaegermann, N., The Banach–Mazur distance between symmetric spaces, *Israel J. Math.* **46** (1983), 40–66.
[513] Tomczak-Jaegermann, N., *The Banach–Mazur Distance and Finite Dimensional Operator Ideals, Monographs and Surveys in Pure and Applied Mathematics* **38**, Longman, Harlow (1989).
[514] Tychonoff, A., Ein Fixpunktsatz, *Math. Ann.* **111** (1935), 767–776.
[515] Vaaler, J.D., A geometric inequality with applications to linear forms, *Pacific J. Math.* **83** (1979), 543–553.
[516] Valentine, F.A., *Convex Sets*, McGraw-Hill, New York (1964).

[517] Vanhecke, L., Archimedes-like theorems in Riemannian geometry, in *Geometry Seminars, Sessions on Topology and Geometry of Manifolds (Ital.) Bologna 1990*, Univ. Stud., Bologna (1992).
[518] Vanhecke, L. and Djorić, M., A theorem of Archimedes about spheres and cylinders and two-point homogeneous spaces, *Bull. Austral. Math. Soc.*, **43** (1991) 283–294.
[519] Venkov, B.A., On a class of Euclidean polytopes (Russian), *Vestnik Leningrad. Univ. (Ser. Mat. Fiz. Khim.)* **9** (1954), 11–31.
[520] Vietoris, L., Bereiche zweiter Ordnung, *Monatsh. Math. Phys.* **31** (1921), 173–204.
[521] Vietoris, L., Stetige Mengen, *Monatsh. Math. Phys.* **32** (1922), 258–280.
[522] Vietoris, L., Kontinua zweiter Ordnung, *Monatsh. Math. Phys.* **33** (1923), 49–62.
[523] von Laue, M., Der Wulffsche Satz für die Gleichgewichtsform von Kristallen, *Z. Kristallogr.* **105** (1943), 124–133.
[524] von Neumann, J., Zum Haarschen Maß in topologischen Gruppen, *Compositio Math.* **1** (1934), 106–114.
[525] von Neumann, J., The uniqueness of Haar's measure, *Mat. Sbornik* N.S. (1) **43** (1936), 721–734.
[526] Wallen, L.J., Kepler, the taxicab metric and beyond: an isoperimetric primer, *College Math. J.* **26** (1995), 178–190.
[527] Weil, A., La mesure invariante dans les espaces de groupes et les espaces homogènes, *Enseign. Math.* **35** (1936), 241.
[528] Weil, W., Über die Projektionkörper konvexer Polytope, *Arch. Math.* **22** (1971), 664–672.
[529] Weil, W., Centrally symmetric convex bodies and distributions, *Israel J. Math.* **24** (1976), 352–367.
[530] Wernicke, B., Triangles and Reuleaux triangles in Banach–Minkowski planes, in *Intuitive Geometry Szeged, Colloq. Math. Soc. János Bolyai* **63** (1991), 505–511.
[531] Whitley, R., The size of the unit sphere, *Canad. J. Math.* **20** (1968), 450–455.
[532] Wieacker, J.A., Translative Poincaré formulae for Hausdorff rectifiable sets, *Geom. Dedicata* **16** (1984), 231–248.
[533] Wieacker, J.A., Geometric measures on finite dimensional normed spaces (preprint, 1990).
[534] Witsenhausen, H.S., On closed curves in Minkowski spaces, *Proc. Amer. Math. Soc.* **35** (1972), 240–241.
[535] Witsenhausen, H.S., Metric inequalities and the zonoid problem, *Proc. Amer. Math. Soc.* **40** (1973), 517–520.
[536] Witsenhausen, H.S., A support characterization of zonotopes, *Mathematika* **25** (1978), 13–16.
[537] Woods, A.C., A characteristic property of ellipsoids, *Duke Math. J.* **36** (1969), 1–6.
[538] Wulff, G., Zur Frage der Geschwindigkeit des Wachsthums und der Auflösung der Krystallflächen, *Z. Krystallogr. Mineral.* **34** (1901), 449–530.
[539] Zaustinsky, E.M., Spaces with non-symmetric distance, *Memoir Amer. Math. Soc.* **34** (1959), 91 pp.
[540] Zhang, G., Restricted chord projection and affine inequalities, *Geom. Dedicata* **39** (1991), 213–222.
[541] Zhang, G., Intersection bodies and the four dimensional Busemann–Petty problem, *Duke Math. J.* **71** (1993), 233–240.
[542] Zhang, G., Centred bodies and dual mixed volumes, *Trans. Amer. Math. Soc.* **345** (1994), 777–801.
[543] Zhang, G., Geometric inequalities and inclusion measures of convex bodies, *Mathematika* **41** (1994), 95–116.
[544] Zhang, G., Intersection bodies and the Busemann–Petty inequalities in $\mathbf{R}^4$, *Ann. of Math.* (2) **140** (1994), 331–346.
[545] Zippin, M., The range of a projection of small norm in ℓ_1^n, *Israel J. Math.* **39** (1981), 349–358.
[546] Zippin, M., The finite dimensional P_λ spaces with small λ, *Israel J. Math.* **39** (1981), 359–365, and **48** (1984), 255–256.

NOTATION INDEX

This index does not list standard notation for set theoretic, algebraic or analytic operations such as $\cup$, $\cap$, $+$, $/$, lim, sup. Nor does it list *ad hoc* notation used in a single proof.

$:=$, equal by definition, xii
1, the identity map, 1
$+$, addition of convex sets, 6
$\times$, Cartesian product, 148
$\oplus$, direct sum, 4
$\otimes$, tensor product, 179
$\cdot$, Blaschke scalar product, 198
$\dotplus$, Blaschke sum, 198
$\widetilde{+}$, radial addition, 234
$\widetilde{+}$, Blaschke radial addition, 237
$*$, suspension operation, 200
$(.)^\circ$, polar reciprocal, 50
$(.)^{\circ\circ}$, $= ((.)^\circ)^\circ$, 50
$\#\{A\}$, cardinality of A, 210
$\vec{\nabla}$, gradient, 179, 219
$\dashv$, is normal to, 78
$\triangleleft$, is transversal to, 125, 172
$\leq$, order relation on ∂B, 266
$\langle .\,,.\rangle$, inner product, 3, 192
$\langle f, x \rangle$, $(=f(x))$, 5
$\triangle abc$, triangle with a, b, c as vertices, 102
$[x, y]$, a line segment in X, 6
$[v]$, $= [-2^{-1}v, 2^{-1}v]$, 7
$\sum[0, x_i]$, the parallelotope spanned by the x_i's, 7
$[x_1, x_2, \ldots, x_n]$, polygon with vertices at x_i's, 112

a, a vector in A, 76
(a, b, c), Euclidean unit linear functional on $\mathbb{R}^3$, 169
A, arbitrary subset of X, 60
$\mathcal{A}$, vector space of alternating k-linear forms, 191

$(b_1, b_2, \ldots, b_d)$, a basis for X, 1
b_i^*, elements of the dual basis, 2
B, C, K, convex sets, 1

$B[0, \rho]$, ball centre 0 radius ρ, 15
$B[x, \alpha]$, 15

c_0, sequences converging to 0, 94
c_i, arbitrary constants, 61
$c\ell(.)$, closure, 61
cm, Minkowski cosine, 252
cone(A), cone subtended by A, 145
conv$(.)$, convex hull, 8
c, (rectifiable) curve, 111, 270, 286
$c_1 \cup c_2$, join of two curves, 112
$c(xy)$, arc of ∂B from x to y, 102
c_0, the oriented unit circle, 127
C, cube, 79
C, $= \partial B$ in two-dimensional space, 104
C, cylinder, 172
$C(K)$, cross-section body of K, 246
$C^\infty(X)$, space of infinitely differentiable real-valued functions on X, 173
$C_e(S^{d-1})$, space of even continuous functions on S^{d-1}, 197
$C_e^+(S^{d-1})$, positive functions in $C_e(S^{d-1})$, 235
$\mathbb{C}$, field of complex numbers, 77
$\mathcal{C}$, non-empty compact, convex sets, 46
$\mathcal{C}_0$, closed convex sets containing 0, 46
$\mathcal{C}_0^*$ $(=\mathcal{C}_0$ in $X^*)$, 51
$\mathcal{C}_b$, convex bodies, 46
$\mathcal{C}_i$, closed convex sets with 0 in the interior, 46

d, the dimension of the space under consideration, 1
det, determinant, 71
diam, the diameter of a set, 76, 238
div, Minkowski divergence, 173
dg_x, differential of g at x, 175
$df/d\theta_y$, derivative of f with respect to angle measure, 266

$df/d\theta'_{y'}$, derivative of f with respect to alternative angle measure, 266
∂K, boundary of K, 58
D, inner diameter, 290
D, Blaschke difference operator, 199
$\check{\mathbf{D}}$, dual Blaschke difference, 237
$\mathcal{D}_d$, d-dimensional isometric equivalence classes, 64

$(e_1, e_2, \ldots, e_d)$, the usual basis for $\mathbb{R}^d$, 1
E, ellipsoid, 54
E_k, k-dimensional ellipsoid, 137

f, g, elements of X^*, 2
$f^\perp$, null space of f, 3
$\hat{f}$, Euclidean unit vector in X^*, 58, 188
$\hat{f}_x$, unit normal to surface at x, 143
$\tilde{f}$, norm 1 functional with respect to $\tilde{\mathbf{I}}_B$, 125
$\hat{f}_H$, parameter for hyperplane, 206
F, finite set, 65
F, facet, 186
F_i, facet of polytope, 141
$\mathcal{F}$, smooth vector field, 173

$g_{|L}$, the function g restricted to the subspace L, 4
G, centroid of triangle, 105

h_K, support function of K, 48
$h_B(K, .)$, Minkowski support functional of K, 106
H, hyperplane, 77
H, hexagon, 108
$H[f, \alpha]^+$, $H[f, \alpha]^-$, closed half-spaces determined by f and α, 8
$H(f, \alpha)^\pm$, corresponding open half-spaces, 8
H^+, H^-, closed half-spaces with $\alpha = 0$, 8
$\mathcal{H}$, set of hyperplanes, 206

i, j, k, m, n, natural numbers used as indices, 67
int B, interior of B, 288
I(.), intersection body, 151
$\mathbf{I}_B$, solution to isoperimetric problem, 120, 138, 144
$\tilde{\mathbf{I}}\,(=\tilde{\mathbf{I}}_B = \tilde{\mathbf{I}}(B))$, the normalized isoperimetrix, 121, 149

J, the natural isomorphism from X to X^{**}, 4
J, finite set of integral indices, 24, 285

K, convex body, 6
$[K]$, equivalence class containing K, 198
K_h, convex set corresponding to functional h, 49
K_α, parallel body, 124
$\mathcal{K}$, non-empty compact sets, 61
$\mathcal{K}_a$, bounded part of $\mathcal{K}$, 68

l, a line, 89
$\ell := \mu(c)$, length of curve, 102, 286

ℓ_p^d, the space $(\mathbb{R}^d, \|.\|_p)$, 277
L, arbitrary subspace, 1
$L^\perp$, annihilator of L, 4
L, line through 0, 136
$\mathcal{L}$, set of lines, 206

m, (half the) girth, 290
m_d, minimal volume product for d-dimensional symmetric convex bodies, 213
mdet, Minkowski determinant, 256
$m \wedge$, Minkowski wedge product, 256
M, arbitrary subspace, 1, 189
M, translate of a subspace, 86
M, (half the) perimeter, 290
M_T, matrix for T, 5
$\mathcal{M}$, surface, 143
$\mathcal{M}_k$, rectifiable surface of dimension k, 211

n, natural number, occasionally a dimension, 5
N, kernel of a linear map, 90
N, nine-point centre of triangle, 106

O, orthocentre of a triangle, 104

p_K, Minkowski functional of K, 47
$\wp$, polygon, 112, 211, 289
P, parallelogram, 125, 138, 254
P, parallelotope, 54, 253
P, polytope, 9
P, projection, 278
Proj, projection, 87, 189
Proj$_1$, projection onto first term of direct sum, 51
Proj$_L$, projection onto L, 89

$\mathbb{Q}$, field of rational numbers, 37

$r(K)$, inner radius of K, 124
$\mathbb{R}$, field of real numbers, 1
R, radial projection, 92
R, Reuleaux triangle, 108
$R\varphi$, Radon transform of φ, 236
R, inner radius, 290
$R(K)$, outer radius of K, 124
$\mathcal{R}$, flat measurable region, 138

s, arc length parameter, 102, 287
sm, Minkowski sine, 253
supp g, support of g, 175
S, symmetry (= isometry), 147
S, T, linear transformations, 63
$S[x_1, x_2, \ldots, x_k] = S[x_i]_{i=1}^k$, simplex with vertices at x_i's, 7
S^{d-1}, Euclidean unit sphere, 59
S_d, "standard" simplex in $\mathbb{R}^d$, 7
S_{xy}, sector between x and y, 266
$\mathcal{S}$, rectifiable surface, 138
$\mathcal{S}$, finite sequence, 283
$\mathcal{S}$, set of symmetric star bodies, 235
S, shadow boundary, 91, 193

Notation index

$(.)^t$, transpose of a matrix (or vector), 2
T, a linear transformation, 5
T^*, the dual transformation, 5
T, isometry, 135
T, triangle, 102
T, regular tetrahedron, 198
T, matrix, 262

u, the vector $\sum_{i=1}^{d} e_i = (1, 1, \ldots, 1)$, 26, 293
$u(x_0)$, Minkowski unit normal to surface at x_0, 173
$\hat{u}$, Euclidean unit vector in X, 58
$\hat{u}_L$, parameter for line L, 206
U, Borel set, 59, 191
U, measurable set, 135, 240

v, w, x, y, z, vectors, elements of X, 1
$V(., ., .,)$, mixed volume, 55, 57
V_B, Minkowski mixed volume, 123, 203
$\tilde{V}(., ., .,)$, dual mixed volume, 235

w, vector, 87
$w(K, g)$, Euclidean width of K in direction g, 106
$w_B(K, f)$, Minkowski width, 106
$W_{B,n}$, projection measure integrals, 204
$W'_{B,n-1}$, projection measure integrals of projections, 204

$x^{\perp}$, annihilator of x, 4
$\hat{x}$, Euclidean unit normal to B°, 188
$x^{\dashv}$, vector normal to x, 261
$(x, x^{\dashv})$, normal basis for two-dimensional space, 261
x_L, parameter for line L, 206
X, (normed) linear space, 1
X^*, dual space of X, 1
X^{**}, second dual of X, 4
X_x, tangent space at x, 173
X^*_x, cotangent space at x, 173
(X, δ), metric space, 76

Y, (normed) linear space, 1

$Z[v_i]_{i=1}^n = Z[v_1, v_2, \ldots, v_n] (= \sum_{i=1}^n [v_i])$, zonotope, 7
Z_d, special zonotope, 26, 293

α, β, arbitrary real numbers, 1
αK, scalar multiple of K, 6
$\alpha(.)$, supremum of Minkowski sine, 256
α_H, parameter for hyperplane, 206
α_i, area of facet F_i, 156, 188

γ, function such that $\gamma(B) = \mu_B(B)$, 136, 187

δ_{ij}, Kronecker delta symbol, 2
$\delta(C, D)$, the Hausdorff metric, 61
$\delta(.)$, general metric, 76, 238
δ_A, inner metric on set A, 287

$\delta = \delta_X$, inner metric on unit sphere of X, 288
$\delta^*(x) := \sup\{\delta(x, y) : y \in \partial B\}$, 289
Δ, Banach–Mazur metric, 63
Δ_1, Δ_2, metrics on balls in X, 62, 137
Δ_B, Minkowski Laplacian, 175

ϵ, ε, arbitrary (small) real numbers, 61, 292
$\epsilon_d := \pi^{2/d}/\Gamma(d/2+1)$, volume of a d-dimensional Euclidean ball, 55

η, arbitrary (small) scalar, 61, 214
$(\eta_1, \eta_2, \ldots, \eta_d)$, coordinates of y, 2

θ, even measure on S^{d-1}, 210
$\theta(x, y)$, angle measure between x and y, 266
ϑ, alternative angle measure, 268
ϑ_x, smooth k-form, 173

$\kappa, = \pm 1$, 188
κ, κ', Minkowski curvatures, 270
κ_B, κ_i, particular constants, 200, 203

λ, specific Haar measure, Lebesgue measure, volume, 41, 53
λ^*, dual Haar measure, 41, 53
λ^k, k-dimensional Lebesgue measure, 136
Λ, map from X to X^* (when X is two-dimensional), 120, 259
Λ^k, dual space of $\mathcal{A}$, 191

$\mu = \mu_B$, Minkowski measure, 135, 187
μ_B^k, Minkowski k-dimensional measure depending on ball B, 136
$\mu^1(c) = \mu_B(c)$, Minkowski length of curve, 111, 141

ν, surface area measure, 188
ν_K, surface area measure of K, 59, 143
$\nu(\mathcal{H})$, measure of set of hyperplanes, 206
$\nu(\mathcal{L})$, measure of set of lines, 206
ν_f, measure on subsets of S^{d-1}, 191
ν_m, m-dimensional Hausdorff outer measure, 238
ν, vector measure, 191

ξ, arbitrary scalar, 189
$(\xi_1, \xi_2, \ldots, \xi_d)$, coordinates of x, 2

$\varpi_s(X), \varpi^s(X)$, projection constants, 277, 279
$\Pi(K)$, projection body of K, 58, 156

ρ, radius of circle, 105
ρ_K, radial function of K, 52
$\rho_B(.)$, Minkowski radial function, 258
ρ, ρ', Minkowski radii of curvature, 270
$\rho^d(B)$, multiple of $\mu_B^d(B)$, 221
$\varrho(X)$, Macphail's constant, 283

$\sigma (= \sigma_B) = \sigma_B^{d-1}$, ratio of μ_B^{d-1} to λ^{d-1}, support function of $\mathbf{I}$, 119, 138

σ_B^k, ratio of μ_B^k to λ^k, 136
σ_B^d ($=\mu_B^d/\lambda_B^d$), a single number, 149
σ_K, support function of projection body of K, 57
$\tilde{\sigma}$, support function of $\tilde{\mathbf{I}}$, 253
ς, symbol (of operator), 177

τ, a topology, 28
τ_{ij}, entries in matrix T, 262

υ, an evaluation, 182

ϕ, φ, real-valued functions, 58, 101

$(\phi_1, \phi_2, \ldots, \phi_d)$, coordinates of f, 2
Φ, element of X^{**}, 34
Φ, one of m, M, R or D, 302
$\Phi^*(d)$, max. value of Φ, 303
$\Phi_*(d)$, min. value of Φ, 303

ψ, real-valued function, 59
$\psi(s)$, curve parametrized by arc length, 102

ω, alternating k-linear form, 191
ω, smooth 1-form, 173
$\Omega^1(X)$, space of smooth 1-forms, 173

AUTHOR INDEX

Ader, O. B., 185
Aitchison, P. W., 97
Aleksandrov, A. D., 196, 222, 224
Alexander, R., 226
Amir, D., 84, 85, 89, 95, 97
Asplund, E., 74, 100, 104, 130
Auerbach, H., 95, 96

Ball, K. M., 95, 96, 185, 225, 229, 249
Bambah, R. P., 72
Banach, S., 32, 33, 42, 43, 44, 73, 283
Barry, J. E., 273
Barthel, W., 132, 139, 182, 230, 238, 246, 250, 272
Behrend, F., 96, 185
Benson, R. V., 118, 132, 140, 156
Berger, M., 71
Bertrand, M., 86, 97
Besicovitch, A. S., 142, 250
Biberstein, O., 265, 273
Birkhoff, G., 77, 95, 130
Blaschke, W., 54, 69, 72, 74, 77, 91, 97
Bliss, G. A., 272
Bohnenblust, P., 89, 131, 278, 304, 305
Boju, V., 182
Bolker, E. D., 187, 190, 222
Bollobás, B., 96
Boltyanski, V. G., 131, 134
Bolzano, B., 9
Bonnesen, T., 10, 45, 53, 60, 65, 71, 123, 124, 133, 202, 230, 272
Botts, T., 32, 46, 71
Bouligand, G., 139, 229, 245
Bourgain, J., 72, 96, 187, 197, 213, 225
Brass, P., 130
Brothers, J. E., 184
Brunn, H., 10, 55, 72, 86, 87, 97
Burago, Yu. D., 73
Busemann, H., 118, 122, 130, 132, 133, 139, 144, 182, 183, 184, 185, 192, 193, 194, 195, 205, 216, 223, 224, 225, 229, 230, 239, 240, 241, 242, 244, 246, 247, 248, 250, 254, 256, 257, 258, 260, 265, 266, 271, 272, 286, 287
Bynum, W. L., 43

Campi, S., 225
Carathéodory, G., 10, 39, 77, 95, 130, 250
Cartan, H., 44
Cassels, J. W. S., 72
Castaing, C., 74
Ceder, J. G., 130
Chakerian, G. D., 95, 108, 109, 131, 132, 133, 205, 222, 243, 244, 304
Charzyński, Z., 95
Cheng, S.-Y., 224
Chilakamarri, K. B., 130
Chilton, B. L., 160, 184
Choquet, G., 139, 197, 229, 245
Cieslik, D., 134
Clack, R., 96, 146, 179, 181, 185, 196
Cohn, D. L., 37, 44
Constantin, R. F., 273
Coxeter, H. S. M., 160, 184, 188, 222
Croft, H. T., 246
Crotty, J. M., 100, 106, 110, 111, 132

Danzer, L., 10, 81
Davis, W. J., 305
Day, M. M., 31, 86, 92, 94, 95, 129, 133
Dean, D. W., 305
de Figueiredo, D. G., 92, 97, 133
Dinghas, A., 184
Djorić, M., 311
Dunford, N., 42, 305
Durier, R., 133
Dvoretzky, A., 72, 73, 95, 97, 284

Eckhoff, J., 11
Edelstein, M., 226
Eggleston, H. G., 45, 53, 71, 124, 130, 131

Eidelheit, M., 71
El-Ekhtiar, A., 205, 207, 208, 210, 225
Erdös, P., 129, 130
Ewald, G., 74, 183, 192, 193, 194, 195, 223, 246

Falconer, K. J., 248
Fallert, H., 247
Fedorov, E. S., 222
Fenchel, W., 10, 45, 53, 60, 65, 71, 202, 224, 230
Figiel, T., 74, 94, 98
Filliman, P., 222, 227, 246
Finsler, P., 271, 273
Firey, W. J., 65, 198
Franchetti, C., 306
Franz, G., 246
Fréchet, M., 10, 42
Fredholm, I, 42
Funar, L., 182
Funk, P., 248
Furstenberg, H., 250

Gardner, R. J., 132, 184, 187, 214, 222, 224, 225, 229, 236, 244, 247, 249, 250
Garling, D. J. H., 305, 306
Ghandehari, M. A., 133, 134
Giannopoulos, A. A., 96
Gluck, H., 224
Gluskin, E. D., 73
Golab, S., 111, 113, 130, 132, 273
Goodey, P. R., 222, 224, 225, 247, 248
Goodman, J. E., 223
Goodner, D. A., 282, 305
Gordon, Y., 72, 284, 286, 305, 306
Goryachev, A. P., 97
Graham, R. L., 134, 273
Grassmann, H., 9, 42
Green, J. W., 185
Greub, W. H., 189, 223
Grinberg, E. L., 74, 86, 87, 184, 246, 247, 250
Groemer, H., 96, 131, 222, 236, 247, 248
Gromov, M., 72, 184
Gruber, P. M., 10, 72, 73, 74, 77, 83, 91, 96, 97
Grünbaum, B., 10, 44, 100, 104, 130, 132, 134, 282, 304
Guggenheimer, H., 43, 120, 123, 130, 132, 133, 250, 260, 263, 265, 272, 273
Gurariĭ, N. I., 97
Gustin, W., 185

Haar, A., 36, 44
Hadamard, J., 42
Hadwiger, H., 222
Hahn, H., 33, 34, 42, 43, 44
Halmos, P. R., 10, 44, 191
Hammer, P. C., 71, 131, 132
Hardy, G. H., 43
Harrell, R. E., 276, 306
Hausdorff, F., 73, 74, 250
Heil, E., 72, 97, 131, 227, 273
Hein, P., 23

Helgason, S., 175
Helly, E., 10, 42, 43
Hemmi, D., 131
Hensley, D., 249
Herring, C., 183
Hewitt, E., 44
Hilbert, D, 42
Höbinger, J., 97
Hölder, E., 21, 43
Holmes, R. B., 32, 46, 71
Holmes, R. D. (Ref. [262]), 122, 132, 133, 138, 140, 151, 156, 160, 183, 184, 187, 201, 215, 220, 221, 227, 272
Holsztiński, W., 94
Husain, T., 44

Ingalls, B., 126, 156, 184

James, R. C., 31, 95, 276, 306
Jessen, B., 224
John, F., 73, 75, 81, 84, 85, 96, 185, 282
Johnson, K. (Ref. [273]), 185, 220
Johnson, R. A., 105
Joichi, J. T., 89
Jordan, P., 86

Kadeč, M. I., 282, 286, 304, 305
Kakutani, S., 44, 71, 89
Karlovitz, L. A., 92, 97, 133, 276, 306,
Kelley, J. L., 61, 305
Kelly, J. B., 206
Kelly, L. M., 222
Kelly, P. J., 129, 130
Klain, D. A., 246
Klee, V., 10, 71, 132
Klein, E. (Ref. [293]), 60, 61, 67, 73, 74
Kneser, H., 72
Knothe, H., 72, 311
Kogan, H., 304
Kolmogoroff, A., 183, 195, 233
König, H., 305
Köthe, G., 10, 43
Kottman, C. A., 97, 304
Krause, E. F., 134
Krein, M., 11
Kubota, T., 73, 131
Kuratowski, K., 73

Laguerre, E. N., 9, 42
Larman, D. G., 249
Lassak, M., 73, 96, 130
Laugwitz, D., 81, 132, 273
Lawlor, G., 129
Lebesgue, H., 41, 131, 249
Leichtweiss, K., 44, 96, 131, 304
Lekkerkerker, C. G., 72
Lenz, H., 81
Lettl, G., 73
Lewis, D. R., 96, 178, 305
Lewy, H., 224

Li, C.-K., 43
Liapunov, A. A., 191
Liebmann, H., 184
Lindenstrauss, J., 74, 98, 187, 191, 197, 225, 304, 305
Lippmann, H., 272, 273
Littlewood, J. E., 43
Loomis, L. H., 44
Löwner, K., 75, 80, 81
Lusternik, L., 183
Lutwak, E., 140, 151, 187, 202, 204, 214, 217, 222, 223, 225, 226, 229, 233, 234, 235, 237, 244, 246, 247, 248, 249, 250

Macbeath, A. M., 73, 74
MacKenzie, D., 104, 105, 130
Macphail, M. S., 283
Mahler, K., 54, 72, 212
Makai, E., 130, 131, 226
Mankiewicz, P., 94
Mann, H., 97
Martini, H., 97, 130, 131, 134, 222, 224, 226, 227, 246, 247, 249
Mayer, A., 286, 305
Mazur, S., 42, 43, 44, 71, 75, 76
McGuigan, R. A., 73
McKinney, J. R., 95, 97
McMullen, P., 72, 182, 222
McShane, E. J., 240
Meissner, E., 130
Meyer, M., 72, 97
Michelacci, G., 132
Michelot, C., 133
Miles, R. E., 250
Milman, D., 11
Milman, V. D., 72, 74, 96, 98, 178, 213, 229, 230, 249, 304, 305
Minkowski, H., 10, 11, 22, 42, 55, 57, 72, 73, 74, 182, 224, 267
Monna, A. F., 9, 42, 44
Morgan, F., 129, 184, 237, 238
Moser, W. O. J., 222
Mürner, P., 222
Murray, F. J., 305

Nachbin, L., 305
Naumann, H., 222
Nirenberg, L., 224
Nyikos, P., 306

Ohira, K., 95
Ohmann, D., 107, 109, 131
Overington, J., 151, 184
Owens, O. G., 226

Pabel, H., 250
Pajor, A., 72, 178, 229, 230, 249
Peano, G., 9, 42
Peetre, J., 175
Pelcyński, A., 74, 95

Peri, C., 133, 205
Petty, C. M., 72, 95, 96, 97, 99, 100, 101, 103, 106, 110, 111, 123, 124, 129, 130, 132, 133, 178, 181, 185, 187, 196, 212, 216, 225, 226, 242, 246, 248, 250, 260, 265, 271, 272, 273
Phadke, B. B., 133
Phelps, R. R., 11
Phillips, R. S., 89
Pickel, P. F., 184
Pincherle, S., 42
Pisier, G., 74
Pogorelov, A. V., 224
Pollack, R., 223
Pólya, G., 43

Radon, J., 10, 11, 128, 133
Reidemeister, K., 59, 143
Reimann, H., 131
Reisner, S., 54, 72, 97
Riesz, F., 10, 30, 42, 43
Rivin, I, 249
Rogers, C. A., 72, 95, 97, 183, 225, 227, 237, 238, 239, 249, 250, 284
Rogers, J. C., 132
Rolewicz, S., 95
Ross, K. A., 44
Ruiz, L. E., 43
Rund, H., 246, 273
Rutovitz, D., 282, 284, 286, 304
Rychlik, M., 132

Saint Raymond, J., 41, 44, 72
Sallee, G. T., 131
Santaló, L. A., 54, 72, 225
Schäffer, J. J., 73, 97, 101, 112, 114, 118, 132, 201, 276, 288, 289, 290, 294, 296, 298, 303, 304, 306
Schäfke, R., 106, 132
Schechtman, G., 74
Schmidt, E., 42, 183
Schmuckenschlaeger, M., 74, 97, 226
Schneider, R., 6, 10, 45, 51, 53, 59, 60, 71, 72, 73, 74, 97, 143, 183, 187, 188, 196, 205, 206, 207, 210, 222, 224, 225, 226, 248, 250
Schoenberg, I. J., 97
Schwartz, J. T., 42, 305
Shephard, G. C., 72, 73, 74, 183, 192, 193, 194, 195, 222, 223, 225, 227, 246, 248
Sholander, M., 131
Singer, I., 305
Snobar, M. G., 282, 286, 305
Sobczyk, A., 277, 278, 304, 305
Soltan, P. S., 131, 134
Soltan, V., 130, 131
Sozonov, Y. I., 97
Spivak, M., 142, 172, 223
Steiner, J., 86
Stone, M. H., 10, 32, 71
Straus, E. G., 183, 185, 223, 250, 258
Stromquist, W., 73

Sundaresan, K., 276, 306
Süss, W., 72
Szankowski, A., 98
Szarek, S. J., 85, 95, 96

Talagrand, M., 85, 96
Talley, W. K., 243, 244, 304
Tamássy, L., 130
Taylor, A. E., 77
Taylor, B., 156, 184
Taylor, J. E., 183, 184, 222
Thompson, A. A., 172, 185
Thompson, A. C. (Refs. [506]–[509]), 73, 117, 132, 140, 151, 187, 201, 214, 227; see also Holmes, R. D., Johnson, K. and Klein, E.
Thompson, H., 78
Tingley, D., 94
Tomczak-Jaegermann, N., 74, 305
Tychonoff, A., 38, 44
Tzafriri, L., 305
Tzkoni, I., 250

Ulam, S., 43, 75, 76, 130

Vaaler, J. D., 249
Valadier, M., 74
Valentine, F. A., 32, 46
Vanhecke, L., 311
Venkov, B. A., 222
Vietoris, L., 73
Vitali, G., 240

Volčič, A., 132, 224
Volkmer, H., 106, 132
Volterra, V., 42
von Laue, M., 184
von Neumann, J., 44, 86
Votruba, G. F., 306

Wallen, L. J., 132, 133, 273, 274
Webster, R. J., 74
Weil, A., 44
Weil, W., 187, 188, 196, 222, 224, 225, 226, 247, 248, 249
Wernicke, B., 131
Whitley, R., 304
Wieacker, J. A., 183, 205, 211, 212, 225, 226, 241, 250
Wills, J. M., 133
Witsenhausen, H. S., 134, 226, 273, 276, 292
Wolfson, H., 96
Woods, A. C., 97
Wulff, G., 184

Yau, S.-T., 224

Zalgaller, V. A., 73
Zassenhaus, H. J., 134, 273
Zaustinsky, E. M., 43, 272
Zhang, G., 74, 226, 229, 234, 247, 248, 249
Zippin, M., 280, 281, 305
Zucco, A., 133

SUBJECT INDEX

affine invariant, 54, 95, 96, 122, 137, 139, 140
affine isometry, 94
affine transformation, 178, 185
Aleksandrov's theorem, 196
algebra, multilinear, 191
alternating k-linear form, 173, 191, 223
angle, 251
angle measure, 122, 251, 265, 268, 269, 311
    axioms for, 272
annihilator, 4
antipodal point, 130
arc length, 226, 287
    Euclidean, 119
Archimedes' theorem, 311
area, 73, 141
    continuity of, 64, 67
    equal in equal time, 133, 274
    intrinsic, 182
    maximal, 118
    minimal, 138, 144
    minimized by flat surfaces, 141, 223
    normalized, 122
    problem of, 145
    projected, 73
    translation-invariant, 118
area of convex body, 58, 66
area function, characterization of, 183
area of sector, 265
arithmetic–geometric mean inequality, 82, 232
associativity, 6
asymptotic properties, 74
atom, 191
Auerbach system, 95
axiom of choice, 44

Banach space
    flat, 276, 306
    local theory of, 74, 98
    reflexive, 311; *see also* reflexive
    separable, 280

Banach–Mazur distance, 63, 73–5, 85, 96, 138–9, 196, 234, 276, 277, 282, 300, 309, 310
    to cube, 96
    to Euclidean space, 75, 98
Banach–Mazur metric, *see* Banach–Mazur distance
basis, 1
    dual, 2, 10, 127, 178
    mutually normal, 79
    normal, 261, 262
    orthonormal, 80
    usual, 1
basis constant, 304
basis of eigenvectors, 176
Bertrand's theorem, 86
Blaschke addition, 198, 223, 225, 237
    harmonic, 199, 201
    radial, 237
Blaschke difference body, 199
Blaschke scalar product, 198
Blaschke selection theorem, 46, 67, 69, 74, 81, 181
Blaschke sum, *see* Blaschke addition
Blaschke–Santaló inequality, 54, 72, 122, 212, 217, 259
body
    convex, *see* convex body
    intersection, *see* intersection body
    projection, *see* projection body
Bonnesen's inequality, 124, 133, 272
Borel measure, 40
    regular, 40, 237; translation-invariant, 36
Borel set, 29, 39, 40, 53, 191, 198
Borel σ-algebra, 53, 135
bounded, 14, 29
    totally, 29, 67, 68
B–P property, *see* Busemann–Petty problem
Brunn–Minkowski inequality, 55, 57, 72, 184, 199, 225, 230, 232, 238

Brunn–Minkowski theorem, *see* Brunn–Minkowski inequality
Brunn–Minkowski theory, 144
  generalizations of, 183
Brunn's theorem, 87, 97
Busemann definition, 139, 150, 212, 229
Busemann intersection inequality, 214, 244–5, 250
  strengthening of, 310
Busemann–Petty problem, 246–9

Carathéodory condition, 39, 237
Carathéodory's theorem, 8
cardinality, 210
Cartesian product, 220, 221
Cauchy sequence, 14, 29, 67
Cauchy's area formula, 60, 73, 190, 201, 202, 204
Cauchy's theorem, *see* Cauchy's area formula
Cauchy–Schwarz inequality, 20
centre
  false, 97
  metric, 76
  nine-point, 100
  pseudo, 97
centroid, 105
chords, mid-points of, 86, 87
circle
  of Apollonius, 126
  nine-point, 106, 130
  six-point, 106, 130
circumcentre, 100, 106
circumcircle, 104, 105
circumference, 100
circumradius, 96, 104–5, 131
combinatorial problems, 129
commutativity, 6
compact, 14, 67
compact Hausdorff space
  continuous functions on, 280
  extremally disconnected, 280, 305
compact set, 29, 61
  space of, 73
compactness, sequential, 46, 67, 69, 74
compactness of $(\mathcal{D}_d, \Delta)$, 73
compactness of $(\mathcal{K}_a, \delta)$, 68
compactness of unit ball, 30, 31
completeness, 29, 67
completeness of $(\mathcal{K}, \delta)$, 74
component, 193
cone, 192
  double, 129, 170; dual of, 159
  double, over hexagon, 167–9
  double, intersection body of, 154; dual of, 154
  double, projection body of dual, 159
  double, over regular n-gon, 184
  subtended, 145
connectedness of $(\mathcal{D}_d, \Delta)$, 73
continuous even functions, 197

contractive mapping, 196, 224, 226, 234, 288, 309, 310
convergence,
  absolute, 283
  unconditional, 283, 284
convergence of sequence of sets, 73
convex, 6
  strictly, 9, 87
convex body, 46; *see also* convex set
  boundary of, 142
  sequence of, 70
  smooth, 144
  symmetric, 55
convex function, 193
convex hull, 7, 194
convex set, 1
  with analytic boundary, 64, 74; density of, 65
  approximation of, 46, 74
  closed, 46; with 0 an interior point, 46
  compact, 11; non-empty, 46
  dimension of, 8
convex sets, space of, 74
  non-empty compact, 69; uniformly bounded, 69
convexity, degree of, 304
convexity requirement, 139
convexity of σ, 138, 139, 172
convexity of σ^k, 183
convexity of unit ball, 17
core, 32
cosine curve, solid of revolution of, 160, 184
cosine function, 252, 256
cosine transform, 247
cotangent space, 173
cross-polytope, 21, 55, 72, 85, 213, 214, 222, 227, 282
  inscribed, 80, 140; maximal, 95
cross-product, 120
cross-section, 138, 147, 148, 171, 187, 214, 229, 243, 244, 248, 250, 280
  area of, 151
  congruent, 250
  dual of, 151, 188
  homothetic, 250
  n-dimensional, 98
  rhombic, 148
  two-dimensional, 86, 97
cross-section body, 246–7
cross-section function, 246
cross-section of intersection body, 247
cross-section of polytopes, 246
cross-section and projection, duality between, 72
crystallography, 183–4, 222
cube, 9, 55, 96, 138, 141, 148, 151, 170, 171, 176, 198, 214, 219, 222, 227, 247, 249, 289, 293, 298
  circumscribed, 96; unit, 80
  cross-sections of, 249
  dual of, 158
  high dimensional, 222

Subject index 341

intersection body of, 153; dual of, 153
projection of, 294
projection body of dual, 158
unit, 20, 27, 70, 170
cubo-octahedron, 148, 170, 215, 227, 300
affine regular, 24
regular inscribed, 104
curvature, 111, 117, 270
Gaussian, 10, 224, 249
radius of, 270
curve, 102, 112, 276, 286, 288, 289, 306
closed, 287
convex, 112
end point of, 287
initial point of, 287
length of, 97, 102, 111, 112, 141, 287; minimizing, 185
rectifiable, 211, 270, 286, 287
simple, 287; closed, 112, 291; rectifiable, 292
curve of constant width, 131
cylinder, 138, 147, 148, 172, 184, 189
dual of, 160
intersection body of, 155; dual of, 155
projection body of dual, 160

derived geometry, first, 263
determinant, 173
diameter, 85, 96, 131, 237, 238
diameter of $(\mathcal{D}_d, \Delta)$, 73
diametrically maximal, 130
differentiability, 29
differential, 175
differential equation, 271
differential forms, 173
differential formulas, 268
differential geometry of Minkowski spaces, 273
differential operator, 175, 178
translation-invariant, 175
dimension, 1
dimension of convex set, 8
distance to Euclidean ball, maximal, 96
distance, Hausdorff, *see* Hausdorff metric
distance, shortest, 125
distinct distances, minumum number of, 130
distributive law, 6
divergence, 173, 178, 185
divergence theorem, 72, 177, 184, 185
dodecahedron, 166, 167
rhombic, *see* rhombic dodecahedron
double cone, *see* cone, double
dual ball, 19, 21
representative of, 149, 184
dual basis, *see* basis, dual
dual body, projection of, 151, 186
dual mixed volume, 234, 235, 237, 247, 248
dual norm, 13, 23, 26, 43
dual space, 1, 10, 13, 31, 43
algebraic, 14, 31
topological, 14, 31
dual transformation, 5, 41

duality, 123, 140
Dvoretzky's theorem, 97

Edelstein's contractive mapping theorem, 226
edges of $\mathbf{I}_B$, 187
edges of polygon, 112, 287
ellipse, 86, 133, 138, 259, 274
ellipsoid, 54, 70, 72, 86, 91, 137, 138, 178, 197, 214, 216, 227, 246, 250, 258, 259, 309, 310, 311
associated, 178
average, 82
Blaschke sum of, 247
close to, 98
intrinsic, 176
Löwner, *see* Löwner ellipsoid
Minkowski sum of, 247
radial sum of, 247
unique circumscribed of minimal volume, 75, 81
unique inscribed of maximal volume, 75, 81
volume of, 250
elliptic operator, 176–7
energy density, 183
equichordal point, 100, 106, 110–11
set with two, 106, 110, 132
equichordal problem, in Euclidean space, 106, 132
equichordal set, 100, 106
equivalence class, 198
affine, 74; of convex sets, 74
homothetic, 63
isometric, 64
Euclidean ball, 70, 225, 245, 249, 285
unit, volume of, 55
Euclidean geometry, 3
Euclidean measure, plane, 119; *see also* Lebesgue measure
Euclidean norm, 20, 119, 190, 293
Euclidean structure, 3, 53
Euler line, 100, 105
evaluation map, 192
expansion constant, 304
extension of linear operators, 279, 282–3
extreme point, 9, 20, 21, 114, 139, 281, 282, 305
convex hull of, 11
extreme subset, 8, 15

face, 9
facet, 9, 20, 21, 24, 141, 151, 156, 177, 188, 200, 226
facets of $\mathbf{I}_B$, 189
facets of Z_d, 293
facial structure, 24, 188
Fermat's problem, 133, 134
finite dimensional space, characterizations of, 10, 36
Finsler geometry, 95, 226, 246
Finsler space, 183, 246
flat, 7, 135

flat regions, 138
Fubini's theorem, 40, 202
fundamental theorem of Minkowski geometry, 35, 41, 50
Funk–Hecke theorem, 248

Gauss divergence theorem, 172
Gaussian curvature, *see* curvature, Gaussian
girth, 276, 288, 290, 292, 306
Golab's theorem, 111, 212
gradient, 179
Gram–Schmidt orthogonalization, 9
Grassmann cone, 192
Grassmann manifold, 136, 138, 223
group, 14
    commutative, 14
    locally compact, 14, 36; abelian, 37
    symmetry, 147
    topological, 14
group of isometries, *see* isometry, group of
group of translations, 36

Haar measure, 14, 36, 37, 41, 53, 118, 123, 135, 187, 188, 229, 238, 239
    normalized, 123, 136, 140
    outer, 238
Hadwiger conjecture, 96
Hadwiger's covering problem, 222
Hahn–Banach theorem, 14, 30, 32, 33, 44-7, 71, 79, 117, 252, 257, 278, 279, 280
half-space, closed, 8, 48
half-space, open, 8
    intersection of, 71
Hausdorff dimension, 239
Hausdorff limit, 247
Hausdorff linear topology, 13, 28, 44, 45, 53
Hausdorff measure, 183, 229, 237, 238, 241, 250
    m-dimensional outer, 238
Hausdorff metric, 45, 61, 62, 73, 138, 139, 189, 196, 224
Helly's theorem, 10, 11, 278
hemisphere, 188
hexagon, 117, 131
    B-, 108; based on H, 108
    circumscribed, 109; symmetric, 109
    inscribed, 114; regular, 100
    regular, 73, 100, 108, 194, 212; affine, 24, 27, 100, 114, 216, 245, 260
hexagonal curve, 194
Hilbert's fifth problem, 226
Hill's equation, 273
hodograph, 273
Hölder's inequality, 21, 23, 43, 235
Holmes–Thompson definition, 140, 151, 187, 272
hull, convex, *see* convex hull
hypermetric, 206, 226
hyperplane, 4, 7, 34, 141
hyperspace, 45
hypersurface, 142

icosahedron, 164, 165
illumination problem, 222
injectivity of **I**, 196, 234, 236
inner diameter, 276, 288, 290
inner metric, 276, 286–9
inner product, 2, 3, 9, 20, 95, 192, 236
inner product space, 80, 86, 87, 88, 89, 92, 94, 95, 96
    complete, 10
inner radius, 124, 276, 288, 290
inradius, 96, 131
integral geometry, 205, 225
integro-geometric formulas, 183
interior, relative, 8
intersection body, 120, 151, 184, 222, 223, 229, 234, 246–9
    cross-section of, 247
    dual of, 151, 229
involution, 196, 277, 234
isepiphane problem, 10
isodiametrical inequality, 237, 238, 250
isometric, 137
isometric classification, 13
isometric embedding, 304
isometric isomorphism, 14, 280
isometric sub-spaces, 147
isometric theory, 305
isometry, 15, 16, 20, 35, 43, 73, 94, 135, 196, 281, 300, 307
    affine, 94
    group of, 75; finite, 83
    left inverse of, 94
    linear, 82
    linearity of, 76, 308
    nonlinear, 94
    onto, 76
    reflectional, 136
    rotational, 136
isomorphic invariant, 13
isomorphic theory, 13, 305
isomorphism, 283
    natural, 4, 10, 31, 35, 44, 50–1
isoperimetric inequality, 123
isoperimetric problem, 10, 100, 118, 132, 183, 187, 217, 229
    solution of, 144, 151
isoperimetric ratio, 217, 310, 311
isoperimetric theorem, 121
isoperimetrix, 99, 121, 122, 123, 149, 150, 178, 183, 215, 247, 251, 254, 258, 263, 273, 274
    normalized, 133, 216
iterate, 156, 226
iterations, 161–9

John's theorem, 84, 96, 308
Jordan curve theorem, 291
Jung's constant, 304

Kakutani's theorem, 89, 91
kernel, 3, 30

Kneser–Süss inequality, 199, 225
Kolmogoroff's monotonicity condition, 183, 195, 225, 233
Kubota's formula, 73, 190, 202, 204–5

Laplacian, 175, 178, 185, 309
lattice, 24
lattice structure, 73
lattice theoretic notation, 305
Lebesgue measure, 53, 119, 136, 141, 188, 192
length, 141, 286
   Minkowski, 119
length of curve, *see* curve, length of
length of polygon, 112
Liapunov's theorem, 191
limits, two fundamental, 264
line segment, 6, 7, 104, 156, 188, 190
   sum of, 7
linear closure, 32
linear functional, 1
   bounded, space of, 19; *see also* dual space
   continuous, 10
linear spaces, partially ordered, 73
linear structure, 13
linear transformation, 1, 18, 145
   bounded, 18, 29
   continuous, 13, 18, 29
   of determinant 1, 179
   non-singular, 54
   rank 1, 179
   sequence of, 70
Lipschitz constant, 195, 196, 233, 234
Lipschitz mapping, 29, 195, 211, 233, 242
local compactness, 14
locally compact topological space, 30
Löwner ellipse, 87
Löwner ellipsoid, 80, 82, 84, 96, 140

Macphail's constant, 276, 283, 305
Mahler's conjecture, 54, 72, 212, 242, 245
Mahler–Reisner inequality, 54, 122, 213, 215, 217
manifold, smooth, 142
MAPLE, 126
Mathematica, 156
matrix, positive definite symmetric, 197
Mazur–Ulam theorem, 17, 76, 94, 137
measure
   angle, *see* angle measure
   atomic, 224
   Borel, *see* Borel measure
   even, 247
   Hausdorff, *see* Hausdorff measure
   invariant, 37; under isometry, 135
   Lebesgue, *see* Lebesgue measure
   Minkowski, *see* Minkowski measure
   outer, 39
   product, 41, 42
   surface area, *see* surface area measure
   symmetric, on sphere, 196
   translation-invariant, 36–7, 53, 118, 205, 239
   vector, 191
measure of set of hyperplanes, 205–6, 209
measure of set of lines, 205–7
metric centre, *see* centre, metric, 76
metric structure, 13
metric
   Banach–Mazur, *see* Banach–Mazur distance
   Hausdorff, *see* Hausdorff metric
   inner, *see* inner metric
   from norm, 15
   topologically equivalent, 65
   uniformly equivalent, 61
minimal position, *see* position, minimal
minimal surfaces with prescribed boundary, 184
Minkowski addition, 198, 247
Minkowski area, 121, 123
Minkowski content, minimized by flat regions, 141, 193, 223
Minkowski cosine, 252, 260, 261, 271, 272, 273
Minkowski curvature, 270
Minkowski determinant, 256
Minkowski divergence, 175
Minkowski functional, 17, 52
Minkowski inequality, 22, 23
Minkowski inequality for mixed volumes, 57, 60, 121, 123, 144, 215, 245
   dual of, 235
Minkowski isoperimetric ratio, 226
Minkowski Laplacian, 175
Minkowski length, 119
Minkowski measure, 141
Minkowski mixed volume, 123, 149
Minkowski problem, 224
   generalized, 224,
Minkowski radial function, 258
Minkowski sine, 172, 253, 261, 263, 271, 273
Minkowski surface area, 143, 144, 204
Minkowski trigonometric functions, 273
Minkowski volume, 145, 204
Minkowski width, 106
   bodies of constant, 100, 106, 109
Minkowski space, 15
Minkowski speed, constant, 274
Minkowski's theorem, 196, 198, 204, 206, 207, 210, 223, 237
mixed area measure, 204
mixed volume, 10, 45, 55, 73, 120, 144, 146, 181, 182, 185, 190, 192, 202, 204, 220, 237, 247, 286
   continuity of, 182
   dual, *see* dual mixed volume
   linearity of, 146
   Minkowski, *see* Minkowski mixed volume, 123
   monotonicity of, 121
moduli of convexity and smoothness, 304
moment problem, 44
monotonic function, 101
monotonic valuation, 182

monotonicity requirement, 182
multiplicity, 195, 223

natural map, *see* isomorphism, natural
nearest point, 89
non-expansive mapping, 92
non-reflexive, 306
non-symmetric bodies, 198
norm, 13, 14, 19
  dual, *see* dual norm
  equivalent, 14, 31, 44; uniformly, 61
  Euclidean, *see* Euclidean norm
  ℓ_1, 21
  ℓ_∞, 21
  ℓ_p, 21
  ℓ_q, 23
  non-symmetric, 118
  supremum, 20
  uniform, 20, 21, 62, 248
norm topology, 15, 27
normal, 78, 251, 253, 257, 258, 261
  mutually, 127
  unit, 141, 143, 173, 178, 179, 188, 193, 201, 220; outward, 119, 173, 203, 224
normal basis, *see* basis, normal
normality, 75, 77, 87, 100, 178, 251
  symmetric, 92
  symmetry of, 129, 215, 258
normalization, suitable, 118
normalization of **I**, 145, 149
normalizing factor, 122
normed space, infinite dimensional, 98
normed spaces, isomorphic, 63
null space, 3

octahedron, 138, 148, 169, 170–1, 199
  dual of, 157; projection body of, 157
  intersection body of, 152; dual of, 152
  iterations of the map **I**, 161–3
  maximal inscribed, 96
orbit, 83
  dense, 83
order reversing, 18, 195, 233
orientation, 102, 118, 174, 193, 252–3
orthocentre, 100, 104–6, 130
  construction of, 104
  metric definition of, 104
orthogonal, 3, 277
orthogonal complement, 261
orthogonal projection, 156, 277
orthogonality, 105, 125, 126
  concepts of, 95
  isosceles, 95
  pythagorean, 95
  symmetry of, 99
outer radius, 124

parallel body, 124
parallelogram, 57, 104, 114, 115, 116, 125, 138, 148, 212, 213, 254, 262, 267
  circumscribed, 114
parallelogram law, 86
parallelotope, 7, 26, 41, 54, 70, 72, 77, 79, 130, 174, 189, 192, 212–14, 242, 251, 253, 255, 294, 309
  circumscribed, 75, 77, 80, 140; minimal, 95; with constant volume, 95
  volume of, 182
parameters, 275–6
  asymptotic behaviour of, 275
  extreme values of, 300
partition, 112
Petty projection inequality, 212, 214, 215, 216, 226, 244, 245
  complementary, 226
  conjectured, 217, 226, 310
perimeter, 111, 276, 288, 290
  minimal, 118
perimeter of unit ball, 100, 111
perpendicular, 3, 252
perpendicular bisector, 126
perpendicular height, 125, 171
perpendicularity, 75, 100, 256
piecewise smooth hypersurface, 142
plane measure, Euclidean, 119
polar reciprocal, *see* dual; reciprocal polar
polygonal arc, 211
polygon, 112, 287, 289, 292
  circumscribed, 113
  inscribed, 113
  length of, 112
polyhedral, 284
polynomial, homogeneous, 55
polytope, 8, 46, 64, 156, 188, 200, 218, 221, 280, 281
  approximation by, 188, 189
  density of, 64, 66
position, 71, 96, 178, 181
  area minimizing, 96
  minimal, 178–80, 309; existence of, 181; uniqeness of, 181
  sequence of, 71
positive definite, 9
positive definite matrix, 176, 178
projection, 151, 222, 223, 278–80, 283
  bounded linear, 305
  orthogonal, 156, 277
  radial, *see* radial projection
projection, area of, 224, 225
projection body, 58, 120, 184, 187, 190, 209, 222, 223, 225, 234, 247, 249
projection constant, 275, 305, 311
projection and cross-section, duality between, 72
projection along f, 151
projection function, 246
projection measure integral, 190, 204
projection of norm 1, 89, 91–2, 99, 215, 275, 276, 281
projection onto subspace, 275
property, two-dimensional, 86

pseudo-centre, 97
Ptolemaic inequality, 97
Pythagorean theorem, 95

radial addition, 234, 247
radial function, 52, 119, 120, 151, 234, 236, 247, 248
　analytic, 247
radial limit, 247
radial projection, 92, 97, 288
radial sum, 235
Radon curve, 77, 100, 127–30, 133, 215, 227, 259
Radon transform, 236, 247
　dense range of, 236
　self-adjointness of, 236
　spherical, 247
Radon's theorem, 10
range of **I**, 196
rearrangement, 283
reciprocal, polar, 45, 48, 50, 120; *see also* dual
rectifiable, 112, 288
rectifiable curve, *see* curve, rectifiable
rectifiable set of dimension k, 211
rectifiable surface, 225
reflexive, 31, 35, 118, 306
reflexive Banach space, 311
regular Borel measure, 40, 237
regularity of measure, 40
requirements, geometrical, 137
Reuleaux polygon, 131
Reuleaux triangle, 107, 108, 131
　minimal property of, 109
rhombic dodecahedron, 148, 170, 198, 215, 227, 245, 298
　affine regular, 27
Riesz' lemma, 30
Rogers and Shephard's inequality, 227
rotation, 120

sector, 265
self-circumference, 111, 171
　of ball and dual, 132
self-surface-area, 214–15, 242
　of ball and dual, 202
semi-group structure, 73
semi-linear structure, 198
separable Banach space, 280
separation of sets, 46–7
separation theorem, 32, 47, 49, 71; *see also* Hahn–Banach theorem
series, absolutely convergent, 283
series, unconditionally convergent, 283
set, equilateral, 129
shadow boundary, 91, 151, 189, 193, 194, 210
　relative to x, 151
　sharp, 193; oriented, 193
simple curve, *see* curve, simple
simple elements, 192
simple harmonic motion, 271, 273
simplex, 7, 21, 24, 55, 142

　characterization of, 227
　dimension of, 8
　equilateral 4-, 129
　inscribed, 77; maximal, 78, 95
　regular, 104, 129, 185, 308
　2-, 100,
sine formula, 251, 261
sine function, 252, 256
　second, 263
six-point circle, 106
smooth, 36, 89, 107, 172, 174, 175, 219–20, 224, 251, 257, 258, 260
smooth ball, 264
smooth body, 46
smooth hypersurface, 142
smooth k-form, 173
smooth manifold, 142
smooth 1-form, 173
　set of, 173
smooth radial function, 249
smooth surface, 172, 174
smooth vector field, 173
smoothness, 252
　degree of, 304
solid of revolution, 147
space-filling polytope, 222
spherical harmonics, 248
star body, *see* star-shaped body, 235
star-shaped, 234
star-shaped body, 235, 247, 248
　set of, 246
star-shaped set, 130, 234, 250
Steiner problem, 134
Steiner symmetrization, 72, 86, 95
Stokes' theorem, 172, 174
Stone's lemma, 32, 71
Stone–Weierstrass theorem, 197
strictly convex, 9, 87, 104, 172, 174, 179, 219, 220, 257, 258
strictly monotonic, 226
structure
　additive, 62
　lattice, 73
　linear, 13
　metric, 13
　multiplicative, 62
　semi-group, 73
　semi-linear, 198
subadditive, 192
sublinear functional, 32, 33
　continuity of, 72
subspace, 1, 7
　closed, 305
　complementary, 4
　complemented, 305
　non-complemented, 305
super-circles, 23
superreflexive, 276, 305
support function, 10, 34, 45, 48, 62, 106, 120, 145
　Minkowski, 106

support hyperplane, 34, 71
support theorem, 47, 71; *see also* Hahn–Banach theorem
supporting functional, 179
surface, 142, 145
surface area, 66, 182; *see also* area
surface area measure, 63, 143, 188, 190, 196, 197, 198, 201, 224
surface tension function, 183
suspension, 221
symbol, 177
symmetric, 7
symmetric bilinear functional, 9
symmetric matrix, 176
   positive definite, 176
   of rank 1, 179
symmetry, 147
symmetry group, 147
symmetry of metric, 15
symmetry of unit ball, 17

tangent hyperplane, unique, 142
tangent space, 173
taxicab metric, 134
Taylor's theorem, 77
tensor product, 179
tetrahedron, 247
   regular, 198, 199
three circles theorem, 130
topology
   Euclidean, 13
   Hausdorff linear, 13, 28, 44, 45, 53
   independent of norm, 61
   linear, 27
   norm, 15, 27
   quotient, 74
   Vietoris, 73
total energy, 183
trace, 180, 285
transitive, 83, 294
translation, 16, 118, 135
transpose, 2
transversal, 125, 172, 257–8
transversalität, 95
transversality, 100, 125, 133, 172, 178, 185, 251, 258
triangle, 105
   altitude of, 130
   equilateral, 100, 129, 131, 185, 194
   geometry of, 104, 130
triangle inequality, 15, 111, 138
trigonometric addition formulas, 260
trigonometric function, 251, 260
   calculus of, 264
trigonometric identities, 251
Tychonoff's theorem, 38

uniform boundedness principle, 305
uniform norm, *see* norm, uniform
uniformity, 29, 61

uniformly dense, 197
uniformly equivalent, 61, 288
unit ball, 13, 15
   non-symmetric, 132
   perimeter of, 100, 111
unit circle, oriented, 127
unit cube, *see* cube, unit
unit distances, maximum number of, 130
unit normal, *see* normal, unit
unit sphere, 15
   isometry on, 94
Urysohn's lemma, 40

valuation, 182, 183
   monotonic, 182
vector
   column, 2
   row, 2
vector area, 193
vector measure, 190, 191
vector operations on convex sets, 6
vector space, finite dimensional, 1
vertex, 9, 26, 148, 151
vertices of I_B, 188
vertices of polygon, 112, 287
Vitali's covering theorem, 240
volume, 53, 73, 141
   continuity of, 64, 67
   dual, 53
   maximal, 144
   mixed, *see* mixed volume
   smallest, 200
volume product, 54, 72, 122, 139, 203, 217, 220, 226, 242, 244, 273, 310, 311
   normalized, 195

weakly dense, 196
wedge product, 191, 256
width, 96
   mean, 96
   Minkowski, *see* Minkowski width
   sets of constant, 106, 130, 131;
     characterization of, 131
Wronskian, 273

zone, 189, 218, 219
zone of facets, 189
zonohedron, 184, 222
zonoid, 54, 55, 72, 97, 156, 189, 195, 196, 197, 209, 210, 212, 213, 214, 215, 222, 223, 225, 226, 244, 247, 307, 308
   dual of, 206, 212, 213, 214, 215, 226
   polar of, 97
zonotope, 7, 26, 54, 156, 184, 188, 189, 195, 196, 214, 222, 245, 249, 276, 285, 294, 297
   dual of, 214, 276
   limit of, 189
   maximum problems for, 227
Zorn's lemma, 34